BUILDING SERVICES HANDBOOK

The tenth edition of Hall and Greeno's leading textbook has been reviewed and updated in relation to the latest building, wiring and water regulations, new technology and new legislation. For this edition, new updates include electric vehicle charging, overheating in buildings, waste water treatment, energy performance certification, mechanical ventilation, food waste recycling, electro-magnetic lifts, biogas processing and much more.

Building Services Handbook summarises the application of all common elements of building services practice, technique and procedure to provide an essential information resource for students as well as practitioners working in building services, building management and the facilities administration and maintenance sectors of the construction industry. Information is presented in a highly illustrated and accessible style.

The comprehensive reference for all construction and building services students, *Building Services Handbook* is ideal for a wide range of courses including T Level, NVQ and BTEC National through Higher National Certificate and Diploma to Foundation and three-year Degree level. The clear illustrations and complementary references to industry Standards combine essential guidance with a resource base for further reading and development of specific topics.

Fred Hall had a varied career in the building services industry before lecturing on craft, technician and professional examination courses at Guildford College. His numerous books on the subject have assisted countless students to successfully pursue their qualification objectives.

Roger Greeno is a well-known author of construction texts. He has extensive practical and consultancy experience in the industry, in addition to lecturing at several colleges of further and higher education, and the University of Portsmouth. He has also examined for City & Guilds, Edexcel, the Chartered Institute of Building, the University of Reading and the University College of Estate Management.

BUILDING SERVICES HANDBOOK

10th edition

Fred Hall

and

Roger Greeno

LONDON AND NEW YORK

Designed cover image: © Shutterstock

Tenth edition published 2024
by Routledge
4 Park Square, Milton Park, Abingdon, Oxon, OX14 4RN

and by Routledge
605 Third Avenue, New York, NY 10158

Routledge is an imprint of the Taylor & Francis Group, an informa business

First edition published by Butterworth-Heinemann 2001
Ninth edition published by Routledge 2017

British Library Cataloguing-in-Publication Data
A catalogue record for this book is available from the British Library

Library of Congress Cataloging-in-Publication Data
Names: Greeno, Roger, author.
Title: Building services handbook / Roger Greeno.
Description: 10th edition. | New York, NY : Routledge, 2023. | Previously published: Routledge, 2017. | Includes bibliographical references and index.
Identifiers: LCCN 2023015163 (print) | LCCN 2023015164 (ebook) | ISBN 9781032563046 (hardback) | ISBN 9781032548371 (paperback) | ISBN 9781003434894 (ebook)
Subjects: LCSH: Building--Handbooks, manuals, etc. | Buildings--Environmental engineering. | Buildings--Mechanical equipment--Installation--Handbooks, manuals, etc.
Classification: LCC TH6010 .H249 2023 (print) | LCC TH6010 (ebook) | DDC 690/.24--dc23/eng/20230520
LC record available at https://lccn.loc.gov/2023015163
LC ebook record available at https://lccn.loc.gov/2023015164

ISBN: 978-1-032-56304-6 (hbk)
ISBN: 978-1-032-54837-1 (pbk)
ISBN: 978-1-003-43489-4 (ebk)

DOI: 10.1201/9781003434894

Typeset in Chudley
by KnowledgeWorks Global Ltd.

CONTENTS

Contents

Contents

Contents

Contents

Contents

Contents

Contents

PREFACE

The *Building Services Handbook* originated as Fred Hall's *Essential Building Services and Equipment* some 35 years ago. Since then, under its new title, the content has been regularly expanded, updated and revised retaining the original presentation of simple illustrations, easily accessible text, tables and charts, calculations and references for further study. In combination with the *Building Construction Handbook*, this book is an essential reference for the building industry and for all students pursuing building services and construction-related courses.

Building services encompass a range of professions and specialised practices, the extent of which is impossible to contain in one comprehensive volume. This book is a learning resource that presents aspects of the services most commonly encountered in existing and new buildings. It is not intended to be prescriptive, neither is it extensive. A library of texts and reference material is needed to develop this subject in full, and many excellent specialised texts exist for this purpose. As a handbook, it provides the reader with an understanding and appreciation of the importance of building services to the environment in which we live, work and play.

Building services are the dynamics in a static structure. They provide facilities for light, comfort, movement, communications and convenience. The impact of services in a modern building can be measured against the amount of space they occupy and the cost of installation, notwithstanding the maintenance bill thereafter. In spatial terms, the equivalent of one floor in six (i.e. over 15% of a building's volume) may be attributed to accommodating cables, ducts, pipes, etc. As a proportion of the capital cost of constructing a highly serviced structure, such as a hospital or a sports centre, in excess of 75% is not unusual. A typical modern office block can require about 50% of the construction budget for its services.

Historically, building services have been little more than a few cables, pipes and ducts. In the past half-century, the role of mechanical and electrical (M&E) engineers, as they were known, has transformed to architectural design team consultants and construction site coordinators. The complexity and impact of building services has not always been appreciated, as indicated by the architect Louis Kahn when in 1964 he wrote disparagingly in *World Architecture*: 'I do not like ducts, I do not like pipes. I hate them so thoroughly, I feel that they have to be given their place. If I just hated them and took no care, I think they would invade the building and completely destroy it'. Designers now accept services as an integral part of a building and that they can be accommodated within raised flooring and suspended ceilings. Some architects have even featured the pipes and ducts on the outside of their designs, namely, the Pompidou Centre in Paris (Renzo Piano and Richard Rogers) and the Lloyds Building in London (Rogers).

Preface

Today, sustainable design has political, economic and social importance to modern buildings. Conservation and the control of diminishing fossil fuels, atmospheric pollution and other 'green' issues are at the forefront of research and development. This new edition contains examples of contemporary practice designed to attain these objectives, including guidance on a variety of alternative and renewable energy concepts and initiatives.

Roger Greeno

1 INTRODUCTION

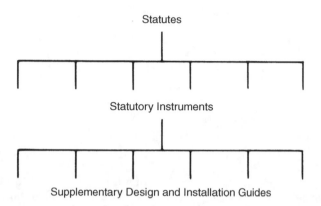

Statutes

Statutory Instruments

Supplementary Design and Installation Guides

THE INDUSTRY

BUILDING SERVICES IN THE CONSTRUCTION PROCESS

LEGISLATIVE AND SUPPORT DOCUMENTS

HEALTH AND SAFETY AT WORK, ETC., ACT

THE BUILDING ACT

THE WATER INDUSTRY ACT

BRITISH, EUROPEAN AND INTERNATIONAL STANDARDS

BUILDING RESEARCH ESTABLISHMENT

LOSS PREVENTION CERTIFICATION BOARD

DESIGN AND INSTALLATION STANDARDS

ENERGY PERFORMANCE CERTIFICATE

DOI: 10.1201/9781003434894-1

1

The Industry

The building services industry is based on engineering principles that are applied to the construction of buildings and the built environment. In many respects, building services are responsible for the artificial environment in which we live and work, and associated with that, the environmental condition of our planet.

Its origins as a science and technology are well documented, not least the use of Archimedes' 'spiral for movement of water' and the concept of underfloor heating in Roman palaces. More recently, it has evolved in response to the demands of population growth and the expectation of comfortable shelter, convenience and a healthy home and workplace environment. As an industry, it is vast in terms of the diversity of professions and trades that it encompasses.

Fossil fuels became readily and abundantly available during the mid- to latter part of the twentieth century, and building design responded with mechanical and electrical systems that in many instances consumed fuel excessively. Large single-glazed areas caused overheating, glare and solar discomfort in the summer, combined with cold draughts and high heat losses in the winter. Thermostatic control was often rudimentary and compensated by opening and closing windows accordingly. The industry has responded, and buildings are now designed to be sustainable, energy conscious and environmentally friendly. Inevitably this has changed the image of the industry professionals from those that run pipes or cables from one place to another, to that of high-profile consultants on building design with responsibilities for environmental issues, fuel conservation and energy performance.

Progress has been effected through government legislation formulated from consultation with the industry professional bodies and research organisations. Advances have also been made by product manufacturers' research and development in response to market competition.

The industry is generally divided between design and installation. Design is undertaken by specialist consultancies and installation undertaken by specialist contractors. The latter subcontracted on-site under the overall administration of a construction management main contractor. Some design and installation practices exist, which simplify contractual and communication relationships.

2

Building Services in the Construction Process

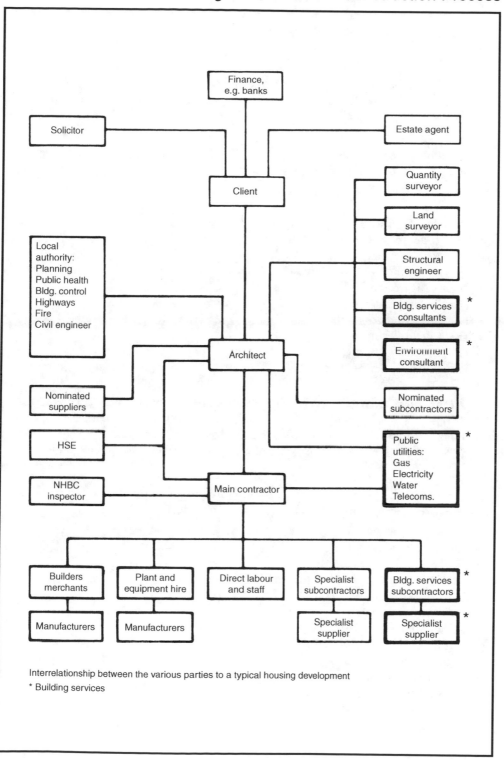

Interrelationship between the various parties to a typical housing development
* Building services

Legislative and Support Documents

Statute – an Act of Parliament that establishes a standard of law. Primary legislation.

Statutory Instrument – a regulation made under a statute to provide guidance that satisfies a particular standard of law. Secondary legislation.

The number of statutes and associated secondary legislation that influence the building services industry is extensive. Some of the most significant include

- The Health and Safety at Work, etc., Act.
- The Building Act.
- The Water Industry Act.
- The Consumer Protection Act.
- The Housing Act.
- The Clean Air Act.
- The Environment Act.
- The Energy Act.
- The Climate Change Act.
- The Building Safety Act.

There is also a category of legislation known as by-laws. These are authorised by a state charter that allows a municipal or corporate administration to effect a standard of law through its own regulation (e.g. Greater London Authority and by-laws relating to Inner London).

Practice guidance documents – often quoted in support of legislation. These too are extensive, some of which include

- British Standards (BS).
- European Standards (BS EN).
- International Standards (BS EN ISO).
- Building Research Establishment:
  Digests
  Good Building Guides
  Good Repair Guides
  Information Papers.
- Loss Prevention Certification Board (Loss Prevention Standards).
- CIBSE Guides.
- British Board of Agrément – Certificates.
- The Institution of Engineering and Technology Wiring Regulations (BS 7671).

This statute is fundamental in application to all persons engaged in the workplace and to any others that could be affected by the work activity. Administration is through the Health and Safety Executive (HSE) from national, regional and locally based offices. The HSE inspectorate is empowered to access building sites and other workplaces to review procedures and safety records. Where appropriate, improvement and prohibition notices can be issued and if necessary, prosecutions.

The Health and Safety at Work, etc., Act incorporates numerous supplementary Statutory Instruments. Some of the more relevant include

- Construction (Design and Management) Regulations.
- Construction (Health, Safety and Welfare) Regulations.
- Workplace (Health, Safety and Welfare) Regulations.
- Management of Health and Safety at Work Regulations.
- Control of Substances Hazardous to Health Regulations (COSHH).
- Control of Asbestos Regulations.
- Manual Handling Operations Regulations.
- Health and Safety (Miscellaneous Amendments) Regulations.
- Work at Height Regulations.
- Health and Safety (Safety Signs and Signals) Regulations.
- Control of Major Accident Hazards Regulations (COMAH).
- Lifting Operations and Lifting Equipment Regulations (LOLER).
- Personal Protective Equipment at Work Regulations.
- Electricity at Work Regulations.
- Gas Safety (Management) Regulations.
- Gas Safety (Installation and Use) Regulations.

Other related statutes – regulation of environmental standards with particular regard to health and safety in the workplace is also effected through the long-established Factories Act and the Offices, Shops and Railway Premises Act. Both are now largely superseded by the H&S at W Act. In addition to many other commitments under these statutes, employers are obliged to provide acceptable levels of lighting, temperature and atmospheric conditions.

Health and Safety at Work etc. Act – Secondary Legislation (1)

- Construction (Design and Management) Regulations.

These are applicable to all building and engineering construction work involving more than 500 employee days or lasting more than 30 days. They create an integrated and planned approach to health and safety, with responsibility apportioned to every person involved in the project workplace. The project client has overall responsibility and must appoint a principal designer (usually the architect), provide them with information on health and safety matters and advise them of perceived hazards and commitments to care of third parties. The principal designer cooperates with the principal contractor (the builder) in preparing a construction phase plan. This must contain specific reference to the identification and assessment of all risks (i.e. health and safety), and information conveying the plan's content to all specialist building services subcontractors and others engaged in the work. See also the *Building Construction Handbook* (R. Chudley and R. Greeno 2020 Routledge).

- Construction (Health, Safety and Welfare) Regulations.

[Now withdrawn as an independent set of regulations, but the requirements are retained and incorporated within the CDM Regulations.] These establish objectives for the well-being of all persons involved in a construction site-related project. The main requirements apply to groundwork, ventilation of workplaces, accessibility of workplaces (ladders), safety at the workplace (scaffold), welfare (first-aid, etc.), accommodation (shelter, rest room, sanitary facilities) and protective clothing. See also Parts 1 and 3 of the *Building Construction Handbook*, 11th edition.

- Workplace (Health, Safety and Welfare) Regulations.

These cover a wide range of health, safety and welfare issues mentioned above, but have particular application to most workplaces other than construction work on building sites. This can include schools, hospitals, offices, factories, hotels, places of entertainment, etc. Responsibility is placed on employers to satisfy certain minimum standards for their employees and also for others on their premises, possibly those attending for purposes of plant maintenance, repair or alterations.

- Management of Health and Safety at Work Regulations.

These regulations provide guidance on the general duties and obligations that employers have to their employees and third parties. They also contain guidance on the responsibilities that employees have to themselves and their colleagues.

- Control of Substances Hazardous to Health Regulations (COSHH).

These require product manufacturers to declare on their product packaging any possible health risk that could be associated with the contents (e.g. plastic pipe solvent jointing adhesive is labelled 'Irritant' and 'Do Not Breath Vapour' among other guidance). Manufacturers of and employers using products having a potential health risk to personnel are required to determine what safety measures and other controls are needed. Where toxins, irritants, solvents, dust, etc., are apparent, users are required to wear appropriate personal protective clothing. Where applications create fumes or other air contaminants, employers are required to monitor exposure, retain records and document procedures.

- Control of Asbestos Regulations.

Since the 1970s, it has become apparent that some people with respiratory health problems have related this to previous employment that exposed them to asbestos insulation on industrial pipes and boilers. Until this time, insulation was commonly produced from a slurry of asbestos fibres mixed with water. The slurry was applied by hand to a wire-reinforced surface. Many people who worked with asbestos, notably pipe-fitters and laggers, have since suffered lung disorders (asbestosis) leading to claims against former employers and manufacturers. There is now a legacy of asbestos in industrial plant rooms, process plant and hospital services. In these workplace situations, an employer is obliged to undertake a risk assessment by survey and analysis. The Regulations provide strict guidelines on handling asbestos and its removal, a procedure normally undertaken by specialist licensed contractors.

- Manual Handling Operations Regulations.

These require employers to provide operatives with training in the correct procedures for handling and carrying equipment without incurring injury, to recognise their limitations and to identify appropriate use of mechanical handling facilities. This includes planning for efficient use of storage and loading areas to avoid unnecessary double handling, and informing suppliers of suitable quantities and package sizes relative to available resources.

- Health and Safety (Miscellaneous Amendments) Regulations.

Reporting of procedures relating to injuries, diseases and dangerous occurrences.

- Work at Height Regulations.

These regulations place emphasis on employers to ensure that suitable and sufficient safe access facilities are provided. For building service applications, this will include work below ground, work at ground level and above. Support to trench excavations, barriers to prevent falls, scaffolding and ladders for use above ground and in all situations suitable means for safe access and egress, must be constructed by experienced and qualified persons. Equipment inspections and safety reports must be undertaken within seven days of use, following adverse weather and after alterations. Reports are mandatory and must be retained on file. See also Part 2 (scaffold) and Part 4 (trench support) of the *Building Construction Handbook*.

- The Health and Safety (Safety Signs and Signals) Regulations.

These require employers to provide health and safety signs, in order to draw attention to potential hazards. They are commonly seen at the entrances to building sites and are colour-coded according to significance:

Prohibition – red and circular, showing what must not be done.
Mandatory – blue and circular, showing what must be done.
Warning – yellow and triangular, showing a risk, hazard or danger.
Safe – green square or oblong, showing escape routes, first-aid, etc.

- Control of Major Accidents Hazards Regulations (COMAH).

COMAH regulations apply specifically to the practical use of dangerous substances or equipment. Every operative must take all necessary measures to prevent accidents with equipment in their charge and to limit their consequences. An example is an awareness of precautionary procedures with the fire and explosive potential of welding equipment, particularly portable units.

- Lifting Operations and Lifting Equipment Regulations (LOLER).

These regulations relate to the increased use of mechanical plants for conveying people (hoists) and materials (hoists and cranes) around building sites. The regulations are principally about the safe and correct use of this equipment with regard to the assessment of risks by trained operatives.

- Personal Protective Equipment at Work Regulations.

Employers are required to provide suitable personal protective equipment (PPE) to employees who may be exposed to any identifiable risk to their health and safety. Self-employed subcontract personnel are required to provide themselves with PPE. Examples of protective clothing include earmuffs, safety helmets, safety footwear, eye shields, gloves and overalls.

- Electricity at Work Regulations.

These regulations ensure that the electrical installation and equipment in places of employment are of a satisfactory standard, suitably detached and insulated to prevent accidental damage. Provision must be made for isolation of individual circuits and overall power cutoff. Responsibility for this is not solely with the employer, as employees and the self-employed also have responsibility for the safe use of electricity in their work situation. The regulations apply to all places of work and are effected under four main areas:

Installation systems.
Connected equipment.
Conductors.
Competence of people using or near electrical equipment.

- Gas Safety (Management) Regulations and the Gas Safety (Installation and Use) Regulations.

The 'management' regulations apply to the conveyance of natural gas to its point of use in domestic and other premises. Four main areas are covered:

Conveyance management through the network of pipes to end users.
Procedures to be adopted during an emergency.
Procedures for dealing with incidents such as a gas escape.
Composition of the gas conveyed.

Both regulations include duties of care by providers, installers and landlords to their customers. Emphasis is on safe installation and maintenance of gas appliances with work only undertaken by qualified persons. For this purpose, the HSE recognises engineers on the 'Gas Safe Register'.

The 'installation and use' regulations specifically require landlords and property managing agents to have installations and appliances in their care checked at least once annually. Gas safety certificates are to be kept for at least 2 years, and tenants are issued with a safety check record within 28 days of check completion.

The Building Act

The Building Act of 1984 consolidates previous by-laws and enables the Secretary of State (Ministry of Housing, Communities and Local Government) and Dept. for Leveling Up, Housing & Communities to make regulations for the design and construction of buildings. This includes the provision of building services facilities and equipment.

Building Regulations – Statutory Instruments made under the Building Act to include approval of building inspectors and inspection fees. The main aspect of the Building Regulations is the establishment of minimum performance standards applicable to the construction and environmental performance of buildings. These standards are supported by practical guidance Approved Documents that are regarded as an acceptable means for compliance.

Approved Documents – England (other UK regions may vary)

Part	Title
	Basements for dwellings
A	Structure
B*	Fire safety: Vol. 1 Dwellings
	Vol. 2 Buildings other than dwellings
C	Site preparation and resistance to contaminants and moisture
D	Toxic substances
E	Resistance to sound
F*	Ventilation: Vol. 1 Dwellings
	Vol. 2 Buildings other than dwellings
G*	Sanitation, hot water safety and water efficiency
H*	Drainage and waste disposal
J*	Combustion appliances and fuel storage systems
K	Protection from falling, collision and impact
L*	Conservation of fuel and power: Vol. 1 Dwellings
	Vol. 2 Buildings other than dwellings
M	Access to and use of buildings: Vol. 1 Dwellings
	Vol. 2 Buildings other than dwellings
O*	Overheating
P*	Electrical safety
Q	Security – Dwellings
R*	Physical infrastructure for high-speed electronic communications networks
S*	Infrastructure for charging electric vehicles

A.D. to Regulation 7* of the Building Act – Materials and workmanship.

*Indicates those particularly relevant to the building services industry.

Until 1999 this statute enabled the various water supply authorities in the UK to create its own by-laws. Since then these diverse by-laws have been consolidated and replaced by the following Statutory Instruments:

- Water Supply (Water Quality) Regulations [England and Wales]
- Water Supply (Water Fittings) Regulations [England and Wales]
- Water Supply (Water Fittings) (Scotland) Byelaws

These apply to all main water supply systems from where the service pipe enters the property boundary to include fittings and appliances connected thereafter. The principal objective of these regulations and by-laws is to prevent water wastage, misuse, excessive consumption and contamination. Building owners, occupiers and installers have a responsibility to ensure that the installation satisfies the regulations. Architects and developers must also satisfy the regulations in advance of future owners. Where an approved contractor (see WRAS below) is engaged a signed certificate will be issued on completion. Notice containing a description of the work must be given to the water undertaker for new installations. Approval must also be sought for any significant changes, alterations or additions. Some examples include the installation of a swimming pool or pond exceeding 10000 litres and automatic garden-watering systems.

The Department for Regional Development in Northern Ireland also has water regulations that are similar to the standards applied elsewhere in the UK. They are made under the Statutory Rules of Northern Ireland, conferred by the Water and Sewerage Services Order as the Water Supply (Water Fittings) Regulations (Northern Ireland) and the Water Supply (Water Quality) Regulations (Northern Ireland).

Water Regulations Advisory Scheme (WRAS) – an advisory body based in Newport, South Wales. Its purpose is to support water supply legislation through communications and publications, particularly the *Water Regulations Guide*. Other activities include consultation with local and national governments, professional and trade organisations and product manufacturers. In addition, development of test criteria for materials and fittings, publication of a directory of approved products, approval and listing of installers in a directory and representing the industry on the development of national and international standards.

Consumer Protection Act – should ensure that products and components are of a quality standard without defects. In the context of building services, it applies to fittings and appliances such as gas and electric cookers as well as central components such as boilers and refrigeration units. A consumer provided with defective or unsafe goods can pursue legal claims for damage to property and other losses caused by the item. Suppliers, manufacturers and importers are all liable.

Housing Act – this contains a number of measures relating to landlords maintaining their properties in a safe and healthy manner to safeguard the interests of tenants and visitors. A housing fitness standard in the form of a hazard and risk assessment plan must be provided in accordance with the Housing Health and Safety Rating System.

Closely associated is the Energy Performance of Buildings (Certificates and Inspections) (England and Wales) Regulations. This statutory instrument has significance for the building services industry by requiring an Energy Performance Certificate to be provided with:

- Newly constructed buildings.
- Existing buildings that are extended.
- Buildings that are altered, e.g. houses converted to flats.
- Refurbished buildings >1000 m^2 floor area where the work includes provision of fuel/energy-consuming equipment.
- Marketing particulars for buildings for sale or rent.

Certificates rate a property on a scale ranging from A at the upper end down to G. Its purpose is to encourage householders to update and refurbish central heating systems, particularly with installation of high-efficiency condensing boilers and thermostatic controls. The overall objective is to reduce fuel bills and the carbon emission impact on the environment. Page 18 has more EPC information.

Clean Air Act – passed in response to the atmospheric pollution/smog of the early 1950s. The causes were, to a large extent, discharge from industrial furnaces and boiler plant, coal-burning electricity power generators and on a smaller scale but more predominant, domestic flues. Local authorities are empowered to prohibit dark smoke, grit, dust and fumes from these sources of pollution by encouraging the use of smokeless fuels to reduce sulphur levels, and construction of tall chimneys.

Environment Act – sets out a strategy for the protection of the environment. Factors and criteria relevant to the building services industry include air quality with regard to flue gases, drainage and pollution control and water resource management.

Energy Act – principle provisions:

- Feed-in tariff – a financial incentive and support for small-scale low-carbon electricity-generating projects up to five megawatts (5MW) capacity. See page 732.

- Renewable heat incentive – allows for the establishment of a financial support programme for renewable heat generated anywhere from households to large industrial sites. Examples include biogas or biomass processing, particularly where the biogas is used to supplement the national gas network. See page 733.

- Smart metres – designed to end estimated gas and electricity bills and metre readings. Smart metres transmit and receive data for the energy supplier and the consumer. They display the amount of fuel energy being used, with an estimate of cost at any time. They also compute carbon dioxide emission information and compare daily, weekly and monthly fuel use. A facility to allow micro-generated energy to be sold back to the grid is included. The objective was for every home to have a smart metre by the end of 2020. The roll-out continues.

- Ofgem – the gas and electricity markets authority. Required to reinforce its contribution to sustainable development. To have amended powers to run offshore transmission licensing more effectively. Transfer of various regulatory functions such as gas and electricity meter testing to the Dept. for Business, Energy & Industrial Strategy.

- Carbon capture and storage – creation of regulations to enable private sector investment in reducing carbon emissions from fossil fuel power-stations.

- Offshore oil and gas licensing – improvement of the licensing process to respond to changes in the commercial environment.

- Nuclear waste – decommissioning costs to be met by power-station operators.

Climate Change Act – the principal objective of this statute is to control greenhouse gas emissions. A target of at least an 80% reduction was established by 2050 (now 100%) relative to 1990 figures with a 34% reduction in carbon emissions by 2020. Greenhouse gases are mainly carbon dioxide (85%), methane (6%), water vapour, nitrous oxide, ozone and halocarbons. The Earth's atmosphere comprises nitrogen (78%) and oxygen (21%); the remaining 1% is greenhouse gases. Although relatively small by percentage, greenhouse gases can have a big impact on climate change.

Building Safety Act – appointment of Building Safety Regulator by HSE to maintain a register of building control approvers and inspectors.

British, European and International Standards

The British Standards Institution was established in 1901 as an independent and impartial body to provide a quality standard for products and practices. Its principles are much the same today, with its well-known Kitemark appearing on items appropriate to all industries and businesses. In building services, the range of BSs is extensive, affecting materials, components, design and installation procedures. BSs are in effect information solutions for guidance, but may be quoted as deemed to satisfy provisions in the Building Regulations and other Statutory Instruments. Reference documents are prefixed BS followed by an allocated number, e.g. BS 1566-1: Copper indirect cylinders for domestic purposes.

Other documents published by the BSI:
- Codes of practice – these are guides for good site practice. They may also have BS status, e.g. BS 5250: Management of moisture in buildings. Code of Practice.
- Publicly available specifications – function as a developmental standard for the creation of management systems, product quality standards or codes of practice. In due course, they are reviewed and either become an established BS or are withdrawn. Prefixed PAS.
- Drafts for development – these are BSs or Codes of practice in the process of completion, where some data is still to be formulated. Prefixed DD or DC (Draft for Completion).
- Published documents – papers not conveniently placed in any of the preceding categories. Prefixed PD.

European standards are gradually replacing standards for one country, as products become harmonised across the European Community. These standards are administered by the Comité Européen de Normalisation (CEN) which incorporates the BSI. Compliant BSs are prefixed BS EN, e.g. BS EN 274-1: Waste fittings for sanitary appliances. Requirements.

The International Organization for Standardization (known as ISO) is a worldwide federation incorporating about 100 national standards bodies to promote international exchange of goods and services by establishing a uniformly acceptable quality standard. ISO documents are compatible with BSs when they are prefixed BS ISO or BS EN ISO, e.g. BS EN ISO 8434-1: Metallic tube connections for fluid power and general use.

The Building Research Station was created in 1921 as a civil service department to develop initiatives for improving housing standards. Over the years, it incorporated several other research bodies including the Fire Research Station. During the 1970s, it changed its name and in 1997 became a private organisation owned by the charity BRE Trust. The Trust has representatives from industry, universities, building owners, managers and users. The BRE has some government funding in addition to income from commercial programmes, consultancy, a bookshop, research contracts, and testing and product certification. The latter includes LPCB fire products and security services (see below), and Building Regulations support documents.

BRE publications are extensive; some of the better-known include
- Digests – up-to-date topics relating to all aspects of construction design and technology. Presented with illustrations and photographs in easy-to-read format, e.g. DG 532: Renewable energy sources: Parts 1 & 2.
- Good Building Guides – highly illustrated practice guidance providing technical advice and solutions, e.g. GG 40: Protecting pipes from freezing.
- Good Repair Guides – illustrated applications of remedial procedures for rectifying common defects, e.g. GR9: Repairing and replacing rainwater goods.
- Information Papers – summary findings of recent BRE research into practical advice and solutions, e.g. IP12/05: Small-scale building integrated wind power systems.

Loss Prevention Certification Board (LPCB) – this organisation originated in the late 1800s when building fire insurers formed a subdivision known as the Fire Offices Committee. This later became known as the Loss Prevention Council (LPC) until changing its name to the LPCB in the 1980s. Long before the Building Regulations came into being, the LPC produced technical standards and specifications for fire prevention and control. These standards are now updated and published as Loss Prevention Standards by BRE Certification Ltd.
See also page 654.

Further Design and Installation Standards (1)

CIBSE – The Chartered Institution of Building Services Engineers incorporates an extensive range of professional practice interests. This includes a structured membership symbolised by individual qualifications and experience, publication of hands-on practical guides, scientific research papers and technical applications relating to design theory. The Institution is also representative of many national and international research and standardisation bodies. CIBSE publications are often provided as support to the Building Regulations and other national standards. They are produced in the following categories:

- Guides
- Technical Memoranda
- Application Manuals
- Knowledge Series
- Commissioning Codes

BBA – The British Board of Agrément is a representative member of EOTA.* The Board's purpose is to provide a facility for accrediting manufacturers' new products, services and innovative use of materials that are not otherwise covered by a BS or European Standard, or other conforming documentation. Agrément Certificates may also be provided as a supplement to show national or international conformity. BBA Certificates are proof of rigorous testing and assessment, ensuring compliance with Building Regulations and European Technical Approvals (ETA*) enabling products to achieve CE* marking. Testing procedures are monitored by UKAS,* e.g. Cert. No. 02/H069: HDPE Twin Wall Drainage System.
*See pages 754 and 755.

IET – The Institution of Engineering and Technology publishes Wiring Regulations as a guide to cost-effective and safe installation practice. The British Standards Institution has adopted these regulations into the national standard BS 7671: Requirements for electrical installations. IET Wiring Regulations. The most recent editions are to a large extent harmonised with the requirements of the IEC (International Electrotechnical Commission) and CENELEC (European Committee for Electrotechnical Standardisation).

CIPHE – The Chartered Institute of Plumbing and Heating Engineering is the representative body for practitioners of plumbing and related employment. Its membership is structured to the various factions within the industry, with a registration scheme for qualified plumbers and plumbing companies. The Institute has many publications that contribute to design and installation practice. These include

- Plumbing Engineering Services Design Guide
- Books and Technical Publications
- Videos
- Member Exclusive Merchandise

GAS SAFE REGISTER – Administers a mandatory 'register' of competent* gas installation and appliance maintenance technicians. Members are required to have suitable qualifications such as a NVQ in Gas Services. Thereafter, competence assessment is through the Nationally Accredited Certification Scheme (ACS) with renewal every 5 years. A technical library of information is available for their members. They also provide gas safety 'Fact Sheets' primarily to promote consumer awareness of gas and carbon monoxide leakage, and as guidance documents for landlords with regard to customer safety certificates and Building Regulations compliance certificates. Originally known as the Council for Registered Gas Installers (CORGI).

OFTEC – The Oil Firing Technical Association has a register of technicians suitably qualified as competent* for oil-fired equipment installation and maintenance. Their publications include Easy Guides, Information Sheets and a range of Standards.

HETAS – The Heating Equipment Testing and Approval Scheme provides a quality standard accreditation service for domestic solid fuel appliances. Appliances are assessed to BS or European Standard for 'safety and fitness for purpose' and were approved, branded with a three-tick logo and listed in the organisation's register. There is also a register of competent* installers.

*Note: The term 'competent' is used in the Building Regulations as recognition of gas safe, OFTEC and HETAS registered personnel for installation of heat-producing appliances.

Energy Performance Certificate – EPC

An EPC is required to be in place prior to selling or letting a property. Undertaken by an approved assessor, qualified through the government's accreditation scheme. The data include many aspects of building services, some listed below. Included will be recommendations and potential savings that could improve the rating. Examples may be installation of a condensing boiler and double glazing.

EPC rating is graded from A (most efficient) down to G (least efficient). A minimum of C being required for rented properties. EPCs are valid for 10 years.

Energy Efficiency %	Current	Potential
92 + A		
81–91 B		
69–80 C		< 70
55–68 D		
39–54 E	< 50	
21–38 F		
1–20 G		

EPC content may include the following:
- Assessment date, reference and address.
- Type of accommodation, e.g. flat or semi-detached house.
- Alterations/extensions/loft conversions if applicable.
- Lighting – fitment types and bulbs.
- Building dimensions and number of floors.
- Glazing type, e.g. single, double, triple, low emissivity.
- Material construction, e.g. timber frame, masonry.
- Type and amount of insulation in roof and walls.
- Roof type – flat or pitched.
- Chimneys and open flues – number.
- Heating and hot water system – fuel type, boiler type, alternative energy.

2 COLD WATER AND SUPPLY SYSTEMS

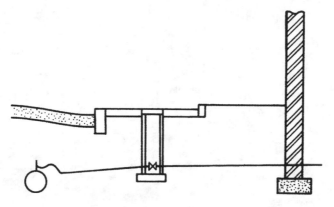

RAIN CYCLE – SOURCES OF WATER SUPPLY
ACIDITY AND ALKALINITY IN WATER
FILTRATION OF WATER
STERILISATION
STORAGE AND DISTRIBUTION OF WATER
DISINFECTION OF WATER
WATER MAINS
VALVES AND TAPS
JOINTS ON WATER PIPES
PIPE JOINTING MATERIALS
DIRECT SYSTEM OF COLD WATER SUPPLY
INDIRECT SYSTEM OF COLD WATER SUPPLY
HARD AND SOFT WATER
WATER SOFTENING
WATER CONDITIONING AND TREATMENT
BACKFLOW PROTECTION
SECONDARY BACKFLOW PROTECTION
COLD WATER STORAGE CISTERNS
COLD WATER STORAGE CALCULATIONS
BOOSTED COLD WATER SYSTEMS
DELAYED ACTION FLOAT VALVE
PIPE SIZING BY FORMULA
HYDRAULICS AND FLUID FLOW

DOI: 10.1201/9781003434894-2

Rain Cycle – Sources of Water Supply

Water is the essence of life.

Surface sources – lakes, streams, rivers, reservoirs, run-off from roofs and paved areas.

Underground sources – shallow wells, deep wells, artesian wells, artesian springs, land springs.

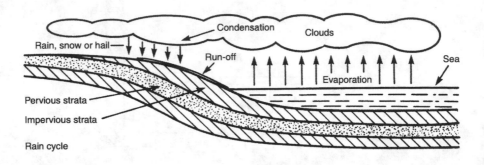

Rain cycle

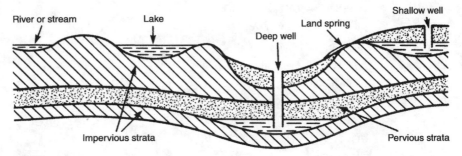

Surface and normal underground supplies

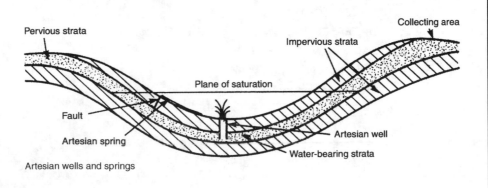

Artesian wells and springs

Acid – a substance containing hydrogen which can be replaced by other elements. Litmus paper in the presence of acidic water turns red.

Alkali – a substance which will neutralise acid by accepting its hydrogen ions (H^+). Litmus paper in the presence of alkaline water turns blue.

More accurate definitions can be obtained by using hydrochemical electric meters. These measure the amount of hydrogen ions (H^+) in a relative proportion of water. This measure of acidity or alkalinity in solution is referred to numerically from 0 to 14 as the pH value.

- pH < 7 indicates acidity
- pH > 7 indicates alkalinity
- pH = 7 chemically pure

The quality of processed water is unlikely to be pure due to contamination at source.

Rainwater – contaminated by suspended impurities as it falls through the air. These impurities are principally carbon dioxide, sulphur and nitrous oxides originating from domestic flue gases and industrial manufacturing processes. The mixture of these impurities and rainfall produces 'acid rain', an occurrence frequently blamed for the destruction of plant life.

Surface and substrata water sources – contaminated by dissolved inorganic materials such as calcium, magnesium and sodium. These are responsible for water hardness as described on pages 44 46. Organic matter from decaying vegetation, animals and untreated waste water can also contaminate ground water supplies. These are normally associated with ammonia compounds in the water or bacteria. Certain types of bacteria present in water can be responsible for outbreaks of typhoid, cholera and dysentery. Chlorination, as described on page 23, is applied to filtered water to destroy any remaining bacterial microbes before general distribution through service reservoirs and mains.

The following table shows the quantity of pollutant microbes present during the stages of water processing, as described on pages 22-24:

Source/process	Typical pollutant microbe count per litre
River	41000
Impounding reservoir	1500
Primary filter	500
Secondary filter	50
Chlorination	0
Service reservoir	0
Distribution main	0

Filtration of Water

Pressure filter – rate of filtration 4 to $12m^3$ per m^2 per hour. To backwash, valve A is closed and valves B and C opened. Compressed air clears the sand of dirt. Diameter = $2\cdot4m$.

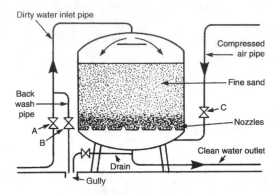

Slow sand filter bed – rate of filtration $0\cdot2$ to $1\cdot15m^3$ per m^2 per hour. Filter beds can occupy large areas and the top layer of sand will require removal and cleaning at periodic intervals.

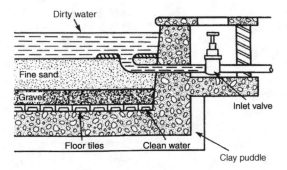

Small domestic filter – the unglazed porcelain cylinder will arrest very fine particles of dirt and even micro-organisms. The cylinder can be removed and sterilised in boiling water for 10 minutes.

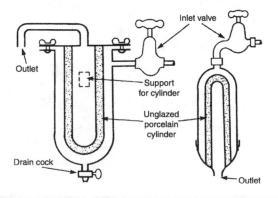

Sterilisation by chlorine injection – water used for drinking must be sterilised to make it completely free of living micro-organisms. Chlorine is generally used for this purpose. A minute quantity of gaseous chlorine (Cl) or sodium hypochlorite (NaClO) in solution, commonly known as bleach (0.1 to 0.3ppm), is added after filtration in absorption towers or small covered reservoirs known as contact tanks. The process takes about two hours before the treated water is released into the water mains or pumped to service reservoirs.

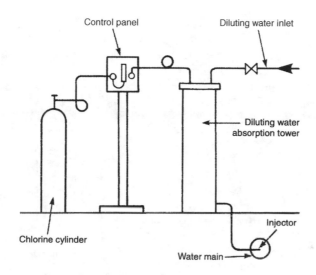

Fluoridation – an additive introduced to drinking water by some supply authorities. Unlike chlorine, it is not added to make supplies safe. The objective is to reduce tooth decay in young children, as fluoride is known to make the enamel covering of their teeth tougher. When added, the amount is between 0.5 and 1mg per litre of water (0.5 to 1.0ppm). This amount is not enough to affect the appearance, taste or smell of water. Arguments against using fluoride are the availability of fluoride toothpastes and that it is wasteful to treat water supplies when over 99% of water is used for other purposes than cleaning teeth. Others include the suggestion that too much exposure can cause staining and mottling of the teeth, even bone disorders and other health issues.

Storage and Distribution of Water

Gravitational distribution – the water from upland gathering grounds is impounded in a reservoir. From this point, the water is filtered and chlorinated before serving an inhabited area at lower level. There are no pumping costs.

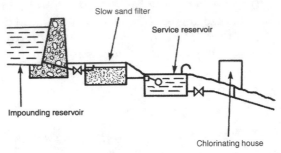

Pumped distribution – water extracted from a river is pumped into a settlement tank, subsequently filtered and chlorinated. Pump maintenance and running costs make this process more expensive than gravity systems.

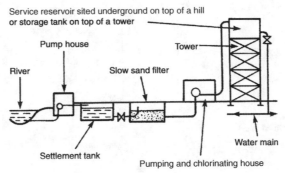

Ring main distribution – water mains supplying a town or village may be in the form of a grid. This is preferable to radial distribution as sections can be isolated with minimal disruption to the remaining system, and there is no more opportunity for water to maintain a flow.

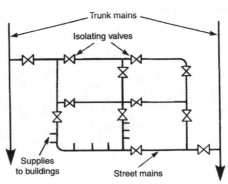

Disinfection – the process of inactivating bacterial or viral cells, either by destruction or by eliminating their pathogenic properties.

- Off-line, system not in use –
  1. Oxidising disinfectants including sodium hypochlorite (chlorination) and chlorine dioxide. Bromine and ozone can be used to treat water in industrial processing plants. Water systems containing these and chlorine disinfectants at concentrations greater than that approved in drinking water (see note) should be fitted with a backflow prevention device (see page 55).
  2. Thermal disinfection, also known as pasteurisation, requires stored water to be maintained between 60°C and 70°C. This may be acceptable for industrial processes, but for domestic use and washing facilities in general, it is impractical due to the possibility of scalding at hot water outlets.

- On-line, system active –
  Either continuous dosing with chlorine additives (see note) or an electrolytic treatment as described on page 51.

- Ultraviolet – exposing water to a UV light with a dedicated output of 254 nanometer wavelength. An irradiating process that damages the DNA of bacterial and viral cells, inactivating them and preventing their reproduction. Of limited dispersal, therefore most suited to small circulation installations such as garden ponds.

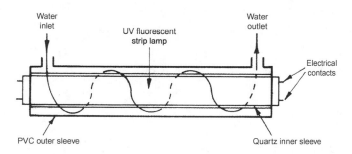

Note: Water that contains disinfectants with concentrations in excess of those acceptable for drinking purposes is known as Category 3 quality, as defined under Schedule 1 of the Water Supply (Water Fittings) Regulations. See page 27.

Ref. BS EN 806-4: Specifications for installations inside buildings conveying water for human consumption. Installation. (Part 6.3 Disinfection).

Cold Water Supply

Water for drinking, washing or for food preparation must be of a wholesome quality. Reclaimed greywater, captured or harvested rainwater and water abstracted directly from wells, springs, boreholes and other accessible water courses is regarded as less than wholesome, but it can be used for other purposes. Pipework and equipment used to convey less than wholesome water must be appropriately marked and identified as such. All supplies are to be reliable and of sufficient pressure and flow rate to operate end-use appliances efficiently.

Definitions:
- Wholesome – water complying with regulations made under Section 67 (Standards of wholesomeness) of the Water Industry Act. The term wholesome is often used instead of potable, i.e. fit to drink. Category 1.[*]

- Greywater – water from showers, baths, taps and washing machines, collected, treated, stored and recycled as an alternative to using wholesome water for sanitary appliances (WCs) and for outdoor uses (gardening). Category 5.[*]

- Captured or harvested rainwater – rainwater collected and stored from roofs and other external surfaces. An old technology that has evolved to become integral with contemporary building design. Used for flushing WCs, washing machines and garden watering (see pages 391 to 393). Category 5.[*]

*Water Supply (Water Fittings) Regulations – see next page.

Refs. Building Regulations Part G, Approved Document G1: Cold Water Supply.

BS 8525-1: Greywater systems. Code of practice.

Water Regulations Advisory Scheme (WRAS) Guidance Note 9-02-05.

In recent years, water consumption in the UK has amounted to about 150 litres per person per day. Each household uses about 100000 litres ($100 \, m^3$) per year.

Total UK annual consumption is about 16.5 billion m^3 with some 13.5 billion m^3 attributed to non-domestic users.

Data source: Office for National Statistics.

Schedule 1 of the Water Supply (Water Fittings) Regulations, categories of fluids:

- Category 1 – wholesome water supplied by an approved water undertaker that complies with standards of wholesomeness defined in Section 67 of the Water Industry Act. Suitable for domestic consumption and for food preparation purposes. Obtained directly from the water company's main.

- Category 2 – water that is not considered to be a health hazard, although it is not suitable for drinking. Water originating from a category 1 source that has changed in temperature, taste, smell or appearance. Some examples are water that has been subjected to a rise in temperature in a hot water system, mixed cold and hot water and domestic water softened by salt regeneration.

- Category 3 – water that is possibly a health hazard, therefore unsuitable for drinking as it may contain low concentrations of toxic additives. These include ethylene glycol (antifreeze) that may be used in solar systems of hot water supply and sodium hypochlorite disinfectants. Also applies to the water in primary hot water and heating circuits (with or without additives) and commercial water softening by salt regeneration.

- Category 4 – water that is a distinct health hazard due to concentrations of toxic substances or bacterial or viral micro-organisms, e.g. Legionnaires' disease. Unsuitable for drinking or for any domestic uses. Includes water in non-domestic hot water and heating circuits, treated water from processes other than salt regeneration, water from commercial dishwashers and washing machines, water containing herbicides, pesticides and other high concentrations of chemicals and carcinogenic substances.

- Category 5 – the highest level of fluid toxicity and contamination. A serious health hazard from concentrations of pathogenic (disease-carrying) organisms, including bacteria and viruses such as salmonella and cholera. Water containing radioactive and very toxic substances. Many situations may apply, including poorly or unmaintained food-processing machinery, sanitary facilities and medical equipment. Recycled greywater (waste water from basins, baths, shower trays, dishwashers and washing machines) is in this category.

Water Mains

Water mains have been manufactured from a variety of materials. The material selected must be compatible with the water constituents, otherwise corrosion and decomposition of the pipes may occur. Contemporary materials which suit most waters are ductile cast iron and uPVC. The water undertaking or authority must be consulted prior to laying mains to determine suitable materials, laying techniques and pipe diameter. Firefighting and hydrant requirements will prioritise the criteria with a minimum pressure of 30m head (300kPa) from a 75mm diameter pipe supplied from both ends, or 100mm diameter from one end only. Bedding of mains is usually a surround of shingle to accommodate any movement. uPVC pipes are pigmented blue for easy identification in future excavations and cast iron has a blue plastic tape attached for the same reason.

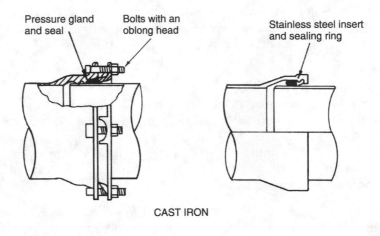

CAST IRON

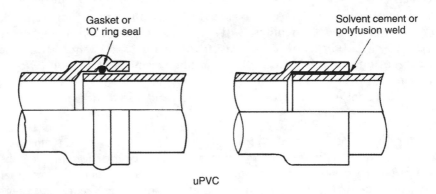

uPVC

Refs. BS EN 545: Ductile iron pipes, fittings, accessories and their joints for water pipelines.

BS EN ISO 1452-2: Plastics piping systems for water supply and for buried and above-ground drainage and sewerage under pressure.

The water authority requires at least seven days' written notice for connection to their supply main. The main is drilled and tapped live with special equipment, which leaves a plug valve ready for connection to the communication pipe. A goose neck or sweeping bend is formed at the connection to relieve stresses on the pipe and valve. At or close to the property boundary, a stop valve is located with an access compartment and cover at ground level. A meter may also be located at this point. The communication and supply pipe should be snaked to allow for settlement in the ground. During warm weather, plastic pipes in particular should be snaked to accommodate contraction after backfilling.

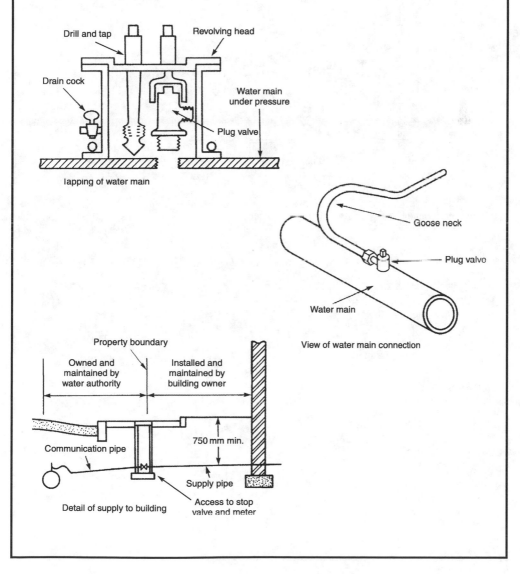

Drill and tap

Revolving head

Drain cock

Water main under pressure

Plug valve

Tapping of water main

Goose neck

Plug valve

Water main

View of water main connection

Property boundary

Owned and maintained by water authority

Installed and maintained by building owner

750 mm min.

Communication pipe

Supply pipe

Detail of supply to building

Access to stop valve and meter

Water Meters

Water meters are installed at the discretion of the local water authority. Most require meters on all new-build and conversion properties, plus existing buildings which have been substantially altered. In time, in common with other utilities, all buildings will have metered water supply. Meters are either installed in the communication pipe, or by direct annular connection to the stopvalve. If underground location is impractical, the water authority may agree internal attachment to the rising main.

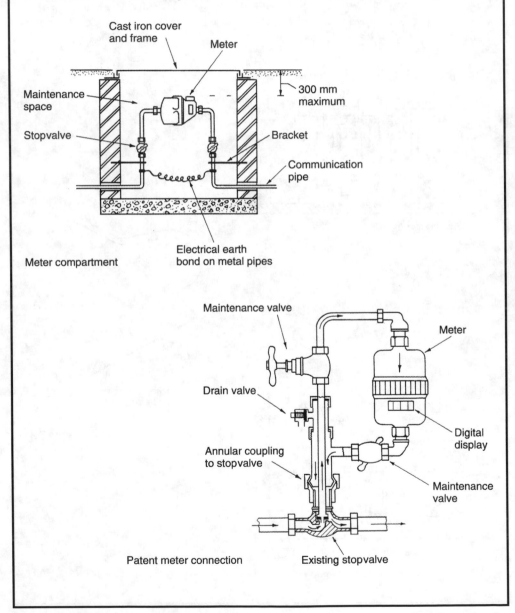

Meter compartment

Patent meter connection

Water is a finite resource with provision determined very much by the fickle nature of the weather. In the UK, demand from an increasing population is imposing considerably on this limited and unreliable resource. In addition, rising levels of affluence and higher standards of living create expectations for luxury goods such as whirlpool baths, power showers, hot tubs and possibly swimming pools in some high specification modern homes. Promotion of water use efficiency is therefore paramount to management of demand.

Fixed sanitary appliances must be designed to prevent undue consumption of water. This places an emphasis on sanitaryware and equipment manufacturers to produce end-use fittings that function efficiently and economically (see page 43).

Building Regulation 17K and associated Approved Document G2 set a target for consumption of wholesome water not exceeding 125 litres per person per day to include a fixed factor of 5 litres per person per day for outdoor use. Tables and charts are used to calculate consumption based on fitment manufacturers' flow rate data. The use of greywater and rainwater is encouraged by offsetting this against consumption of wholesome water.

Water meters – average home consumption is some 15% less than in homes without a meter. Meter use is not a legal requirement, although it is generally standard with all newly built dwellings. Over half of UK homes now have a meter. The proportion is increasing annually.

Refs. WRAS Approved Products and Materials.

Building Regulation 17K (Water efficiency of new dwellings).

Building Regulations Part G, Approved Document G2: Water Efficiency.

National House Building Council (NHBC): Water efficiency in new homes (NF 20).

Control and Drain Valves

The globe-type stopvalve is used to control the flow of water at high pressure. To close the flow of water, the crutch head handle is rotated slowly in a clockwise direction gradually reducing the flow, thus preventing sudden impact and the possibility of vibration and water hammer.

The gate or sluice valve is used to control the flow of water on low-pressure installations. The wheel head is rotated clockwise to control the flow of water, but this valve will offer far less resistance to flow than a globe valve. With use, the metallic gate will wear and, on high-pressure installations, would vibrate.

The drain valve has several applications and is found at the lowest point in pipe systems, boilers and storage vessels.

For temperatures up to 100°C, valves are usually made from brass. For higher temperatures, gunmetal (a type of bronze) is used. Brass contains 50% zinc and 50% copper. Gunmetal contains 85% copper, 5% zinc and 10% tin.

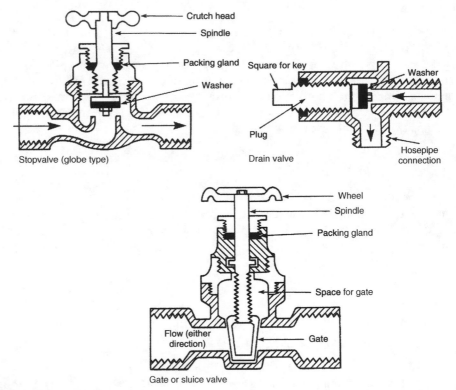

Stopvalve (globe type)

Drain valve

Gate or sluice valve

Refs. BS 5154: Specification for copper alloy globe, globe stop and check, check and gate valves.

BS EN 12288: Industrial valves. Copper alloy gate valves.

Float valves are automatic flow control devices fitted to cisterns to maintain an appropriate volume of water. Various types are in use. The diaphragm type is the least noisy as there is less friction between moving parts. The Portsmouth and Croydon-type valves have a piston moving horizontally or vertically, respectively, although the latter is obsolete and only likely to be found in very old installations. Water outlets must be well above the highest water level (see page 53) to prevent back siphonage of cistern water into the main supply. Nozzle diameters reduce as the pressure increases. High-, medium- and low-pressure valves must be capable of closing against pressures of 1380, 690 and 275kPa, respectively. Produced from copper alloy or ABS plastic (acrylonitrile butadiene styrene) depending on application.

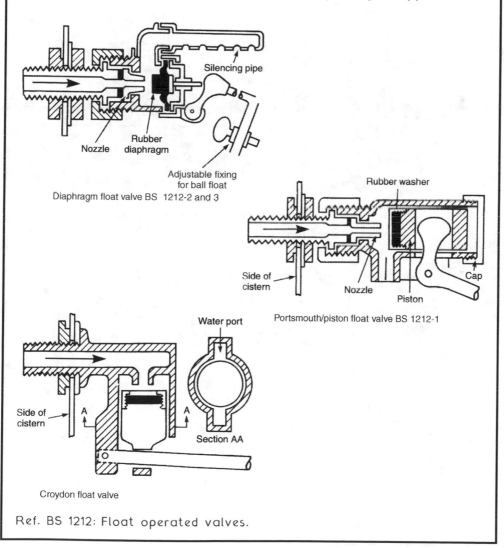

Nozzle

Rubber diaphragm

Adjustable fixing for ball float

Silencing pipe

Diaphragm float valve BS 1212-2 and 3

Rubber washer

Side of cistern

Nozzle

Piston

Cap

Portsmouth/piston float valve BS 1212-1

Water port

Side of cistern

A   A

Section AA

Croydon float valve

Ref. BS 1212: Float operated valves.

Taps

The pillar tap is used to supply water to basins, baths, bidets and sinks. Combined hot and cold pillar taps are available with fixed or swivel outlet. The outlet of these taps must be bi-flow, i.e. separate waterways for hot and cold water to prevent crossflow of water within the pipework.

The bib tap is for wall fixing, normally about 150mm above a sanitary appliance. The 'Supatap' bib tap permits a change of washer without shutting off the water supply. It is also available in pillar format. Quarter-turn taps are easy to operate by hand or elbow, and therefore are suitable for use by the disabled and medical practitioners.

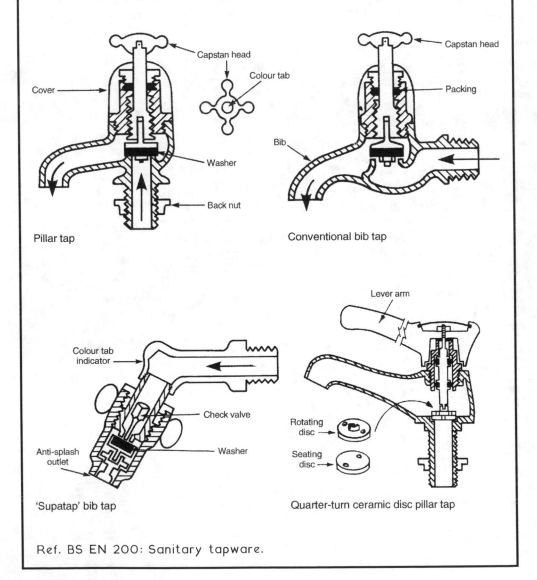

Pillar tap

Conventional bib tap

'Supatap' bib tap

Quarter-turn ceramic disc pillar tap

Ref. BS EN 200: Sanitary tapware.

The blending of two supplies of water at different temperatures using a combination tap with a common outlet can be effected within the body of the tap, or through separate waterways to the spout as shown below.

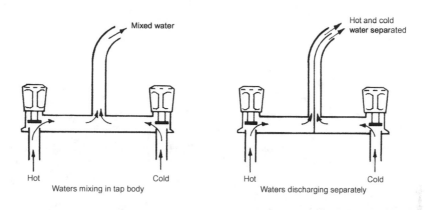

Waters mixing in tap body   Waters discharging separately

Where there is an imbalance of pressures, typical of blending high-pressure mains-supplied cold water with lower pressure cistern-supplied hot water in the body of a combination tap, a check valve should be provided on each supply. These are required to prevent cross-contamination by the higher pressure cold flowing into the hot water supply system and, under fault or vacuum conditions, the hot flowing into the cold water supply.

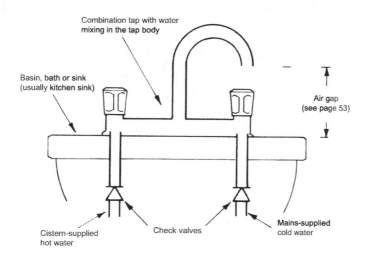

Air gap
(see page 53)

Ball Valve

Operation – a quarter-turn (90°) valve with a lever control handle. The lever is attached by spindle to a ball with a central hole that aligns with the adjacent pipe bore. When fully open, water flows unopposed by internal components and directional changes. Pressure and flow losses are minimal.

Application – an on/off isolating service valve used for system and appliance maintenance. 'Every inlet to a storage cistern, combined feed and expansion cistern, WC flushing cistern or urinal flushing cistern shall be fitted with a servicing valve on the inlet pipe adjacent to the cistern'. Extract from the Water Supply (Water Fittings) Regulations.

Size and function – generally relatively small, up to 75mm nominal bore. Functionally simple, durable and rarely requiring attention. Produced with a chromium-plated brass or ceramic ball that rotates against a seating of PTFE to achieve an effective seal. Valve body is typically a chromium-finished copper alloy known as DZR or brass.

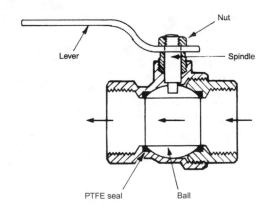

Operation – a quarter-turn (90°) valve used with piped supplies of over 50mm nominal bore, although diameters considerably in excess of this are available. Direct rotation of a centrally positioned disc is by lever handle or indirectly by wheel through a reduction gearbox. The latter is essential with larger diameter valves. Both types can also be operated by motorised actuators.

Application – because of their size, non-domestic situations, particularly those associated with process plant as an end-of-line drainage valve. May also be used with water waste treatment, chemicals, food processing and as flow control for gas-, air- and gravity-fed powders.

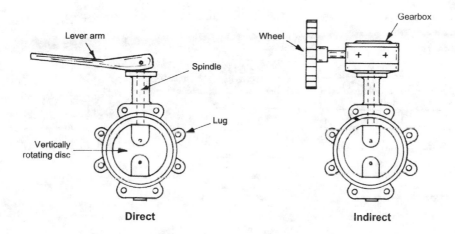

Direct **Indirect**

Produced from copper and aluminium alloys, ductile and cast iron. An EPDM (ethylene propylene diene monomer) synthetic rubber, PTFE (polytetrafluorethylene) or nitrile rubber lining provides for an effective seal.

Joints on Water Pipes

Copper pipes may be jointed by bronze welding. Non-manipulative compression joints are used on pipework above ground and manipulative compression joints are used on underground pipework. The latter are specifically designed to prevent pipes from pulling out of the joint. Push-fit joints are made from polybutylene. These provide simplicity of use and savings in time. Capillary joints have an integral ring of soft solder. After cleaning the pipe and fitting with wire wool and fluxing, heat application enables the solder to flow and form a joint. Solder alloy for drinking water supplies must be lead-free, i.e. copper and tin.

The Talbot joint is a push-fit joint for polythene pipes. A brass ferrule or support sleeve in the end of the pipe retains the pipe shape.

Threaded joints on steel pipes are sealed by non-toxic jointing paste and hemp or polytetrafluorethylene (PTFE) tape. A taper thread on the pipe will help to ensure a watertight joint. Union joints permit slight deflection without leakage.

Lead pipes are no longer acceptable due to the risk of poisoning.

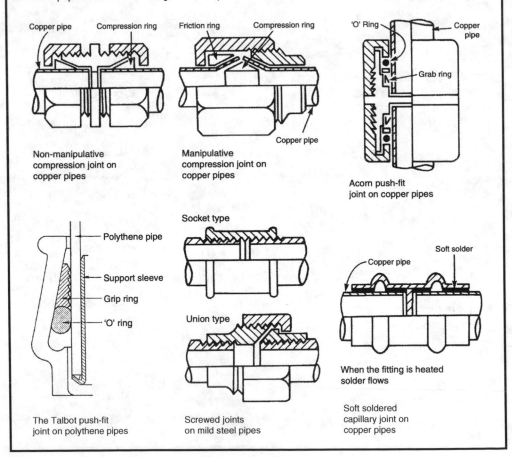

Non-manipulative compression joint on copper pipes

Manipulative compression joint on copper pipes

Acorn push-fit joint on copper pipes

The Talbot push-fit joint on polythene pipes

Screwed joints on mild steel pipes

Soft soldered capillary joint on copper pipes

When the fitting is heated solder flows

Categories of polythene (also referred to as alkathene) and fittings suitable for cold water supplies:

Blue polythene – a medium density polythethylene (MDPE) for use with underground drinking water supplies. Pigmented blue for ease of identification in excavations. Produced in outside diameters from 20mm up to 63mm (nominally 15 to 52mm internal diameter). Not suitable for above-ground applications unless protected, as it can be affected by direct sunlight.

Black polythene – an MDPE for use with above-ground drinking water supplies. Pigmented black to differentiate from blue underground pipe. Produced to the same diameter specifications as for blue MDPE. Resistant to ultraviolet direct sunlight.

MDPE coiled standard lengths – 25, 50 and 100 metres.

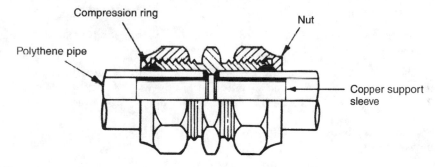

Standard gunmetal straight compression coupling for use with MDPE pipe

Compression ring

Nut

Polythene pipe

Copper support sleeve

Note: Copper alloy pipe fittings/couplings to be made from gunmetal, an alloy of copper (85%), tin (10%) and zinc (5%).

Refs. BS EN 12201 series: Plastic piping systems for water supply.
Polyethylene (PE).
BS EN 1254-3: Copper and copper alloys. Plumbing fittings.
Fittings with compression ends for use with plastic pipes.

Pipe Jointing Materials

Linseed oil 'white' jointing paste – a blend of linseed oil and clay which surface hardens to form a strong, dense joint. Used mainly on threaded steel pipework with fibrous hemp reinforcement between the threads. Microbial action can break down the linseed component and the hemp can degrade, therefore not recommended for use on drinking water supplies. Synthetic reinforcement fibres are more durable. Unreinforced paste is suitable for gas and steam pipe lines. Graphite is sometimes added to the paste for use on steam, as this eases joint breakage when undertaking maintenance and alterations.

Silicone oil jointing paste (acetosilane). Combined with synthetic reinforcement fibres, this compound may be used on drinking water supplies. Also suitable for jointing hot water and gas pipes. Non-setting, non-cracking and flexible, so easily broken for maintenance and alterations.

BS 6956-5: Jointing materials and compounds.

Resin-based compounds – these are specified for chemical and oil pipe joints where the liquid conveyed may contain solvents which could weaken oil-based sealants. Resin and fillers are mixed with a catalyst and, after application to pipe threads, tightened joints will require time to set.

PTFE tape – wound into threads prior to joint tightening. Chemical and temperature resistant with an element of flexibility. Suitable for water and gas pipe joints. Also available as a liquid, but relatively expensive.

BS 7786: Specification for unsintered PTFE tapes for general use.

BS EN 751-3: Sealing materials for metallic threaded joints———.

Solders and fluxes – the established method for economically jointing copper pipe and fittings. Solder types:

- 29% tin + 71% lead. Traditionally used for all joints but now prohibited on drinking water supplies because of the lead content. Melting point = 210°C.
- 63% tin + 37% lead. Bit solder for electronic applications. Melting point = 185°C.
- 99% tin + 1% copper. Lead-free for drinking water supplies. Melting point = 235°C.

BS 6920: Suitability of non-metallic products for use in contact with water———. BS EN ISO 9453: Soft solder alloys. Chemical compositions and forms.

Fluxes are classified as passive or self-cleaning. Available in liquid or paste format and function by preventing cleaned surfaces from tarnishing under heat. Passive fluxes do not contain any free acid and will require heat application to effect cleaning. These are water-soluble organic fluxes and are preferred by gas companies due to the flux's non-corrosive properties. Water-soluble fluxes are preferred for use with lead-free solders and are easily cleaned from finished joints. Self-cleaning fluxes contain an acid to clean tarnished copper as soon as applied. Heat application accelerates the process. Any flux residue must be cleaned from the pipe surface to prevent corrosion. Deposits internally are removed by flushing the system.

For efficient operation, a high-pressure water supply is essential particularly at periods of peak demand. Pipework is minimal and the storage cistern supplying the hot water cylinder need only have 115 litres capacity. The cistern may be located within the airing cupboard or be combined with the hot water cylinder. Drinking water is available at every draw-off point and maintenance valves should be fitted to isolate each section of pipework. With every outlet supplied from the main, the possibility of back siphonage must be considered.

Back siphonage may occur when there is a high demand on the main. Negative pressure can then draw water back into the main from a submerged inlet, e.g. a rubber tube attached to a tap or a shower fitting without a check valve facility left lying in dirty bath water.

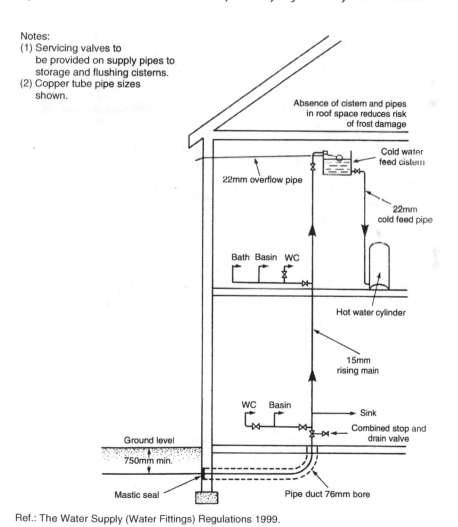

Notes:
(1) Servicing valves to be provided on supply pipes to storage and flushing cisterns.
(2) Copper tube pipe sizes shown.

Absence of cistern and pipes in roof space reduces risk of frost damage

Cold water feed cistern

22mm overflow pipe

22mm cold feed pipe

Bath Basin WC

Hot water cylinder

15mm rising main

WC Basin

Sink

Combined stop and drain valve

Ground level

750mm min.

Mastic seal

Pipe duct 76mm bore

Ref.: The Water Supply (Water Fittings) Regulations 1999.

Indirect System of Cold Water Supply

The indirect system of cold water supply has only one drinking water outlet, at the sink. The cold water storage cistern has a minimum capacity of 230 litres, for location in the roof space. In addition to its normal supply function, it provides an adequate emergency storage in the event of water main failure. The system requires more pipework than the direct system and is therefore more expensive to install, but uniform pressure occurs at all cistern-supplied outlets. The water authorities prefer this system as it imposes less demand on the main. In addition, with fewer fittings attached to the main, there is less chance of back siphonage. Other advantages of lower pressure include less noise and wear on fittings, and the opportunity to install a balanced pressure shower from the cistern.

Notes:
(1) Servicing valves to be provided on supply pipes to storage and flushing cisterns.
(2) Copper tube pipe sizes shown.

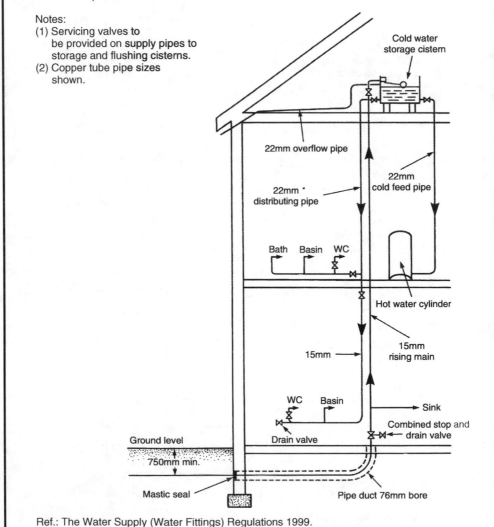

Ref.: The Water Supply (Water Fittings) Regulations 1999.

42

In 2006 the UK government set water consumption objectives across six progressive levels to become mandatory by 2016:

Levels	Litres/person/day
1 and 2	120
3 and 4	105
5 and 6 (2016 objective)	80
Building Regulation compliance (2010)	125

Standard sanitary fittings and appliances are shown earlier in this Part and in Part 9. To conserve water use, some variations include:

Aeration fitting – end-use fitting (tap) that combines entrained air with water to bulk up the discharge. Unsuited to low pressures.

Click tap or water brake – lever-operated tap that has resistance to full opening. Resistance can be overcome if a full flow is required.

Proximity sensor tap – used in public conveniences where an electronic sensor detects a person close to a wash basin to discharge a limited volume of water from a motorised valve through an open tap.

Low-flow tap – a flow restrictor fitted inside the outlet spout or into the tap stem.

Flow regulator – valve with a synthetic rubber 'O' ring that deforms in response to water pressure variation to maintain a consistent flow.

Low-volume WC cistern – dual flush facility of 4 or 6 litres max., the latter determined by the Water Supply (Water Fittings) Regulations.

Reduced-volume bath – a lower than standard height overflow or a reduced base width. Unoccupied capacity limited to 150 litres.

Low-flow shower – shower rose with small holes to encourage aeration and water droplets instead of a continuous spray.

Low water-use washing machine – limited to 60 litres/wash. Efficiency measured in litres/kg load.

Low water-use dishwasher – efficiency measured in litres/place setting.

Hard and Soft Water Characteristics − 1

Hardness in water occurs when calcium or magnesium salts are present. This is most common where water extraction is from boreholes into chalky strata or chalky aquifers.

Measurement

• Parts per million (ppm), i.e. milligrams per litre (mg/l), e.g.

Location	Typical ppm
Bristol	300
Cardiff	100
Hartlepool	460
London	285
Manchester	< 60
Newcastle	160
Scotland	< 50

For a general guide to England and Wales, see map on page 46.

• Clarke's scale − a numerical classification, sometimes referred to as degrees Clarke.

Classification

Type of water	Clarkes	Approx. ppm (see next page)
Soft	< 3·5	< 50
Moderately soft	3·5–7·0	50–100
Slightly hard	7·0–10·5	100–150
Moderately hard	10·5–14·0	150–200
Hard	14·0–21·0	200–300
Very hard	> 21·0	> 300

1 degree Clarke is about 1 part per 70000.

When hard water is heated, the dissolved salts change to solids and deposit on the linings of pipework, boilers and other ancillaries. Kettle scale is an obvious example, but far more significant is its efficiency reduction of hot water and central heating plant. If enough scale is deposited, pipework systems can become completely blocked or 'furred up'. This can have explosive consequences, as safety valves will also be affected. Chalk build-up normally takes years, but in very hard water areas it may be just a few months, depending on the frequency of plant use. Hence the limitations of direct hot water systems (see page 84) where fresh water is continually introduced. Direct systems are only applicable where water hardness is less than 150ppm and water temperatures do not exceed 65°C. The water temperature in modern hot water and heating systems exceeds 80°C, therefore direct systems are effectively obsolete in favour of indirect installations. (see page 85). Indirect systems have the same water circulating throughout the primary and heating pipework and it is only drained off during maintenance and repair.

Temporary hardness – due to the presence of calcium bicarbonate in water. Heating the water to temperatures above 65°C releases the insoluble carbonates and these deposit on the surface of the heating vessel, typical of the scaling found in kettles.

Permanent hardness – due to calcium and magnesium sulphates in water. The water quality is not affected by heating.

Expressions of water hardness – on the previous page a comparison is made between degrees Clarke and approximate parts per million (ppm). Ppm in this context is in milligrams per litre (mg/l) as a calcium carbonate equivalent, often referred to by the initials CCE or as an expression of total hardness. Hardness of water may also be expressed in mg/l as calcium. A comparison is shown below:

Type of water	Hardness as calcium*	Hardness as calcium carbonate equivalent*
Soft	< 20	< 50
Moderately soft	20–40	50–100
Slightly hard	40–60	100–150
Moderately hard	60–80	150–200
Hard	80–120	200–300
Very hard	> 120	> 300

*Mg/l = 1 part per million.

An arithmetical relationship exists between expressions of water hardness in degrees Clarke, calcium and CCE:

CCE × 0·4 = mg/l as calcium

CCE × 0·07 = degrees Clarke

Degrees Clarke × 5·714 = mg/l as calcium

E.g. Water with a CCE of 250 ppm:

Mg/l as calcium = 250 × 0.4 = 100
Degrees Clarke = 250 × 0.07 = 17.5

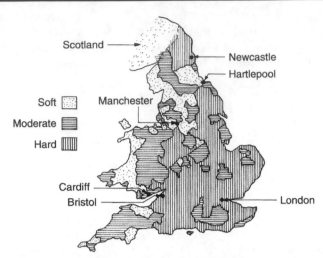

Guide to water hardness in England and Wales

Soft water – in the areas indicated, notably parts of the West Country, Northern England and Wales where the ground conditions are of dense rock or granite, rainfall penetrating the surface is unable to dissolve these rocks and it remains soft.

Hard water – caused by a chemical change as rainfall percolates into chalky ground. The reaction causes the chalk or calcium carbonate to dissolve and change to calcium bicarbonate to give the water extract hardness characteristics.

Characteristics of hard water are:

- difficult to create a lather with normal soap
- scum and tide marks in baths and basins
- chalk staining or streaking on washed glassware.

In hard water areas, these problems can be overcome with the installation of a water softener. These devices are relatively compact and will fit conveniently within the housing under a domestic sink. This location is ideal, as the housing will normally accommodate the rising water main and stop valve. It also provides simple access for replacement of salt granules or blocks. The unit contains a resin bed saturated with sodium chloride or common salt. The salt exchanges calcium and magnesium ions for non-scale-forming sodium ions. Regeneration can be by electric timer, but most domestic softeners have an integral water volume metering device.

Typical Installation of a domestic water softener –

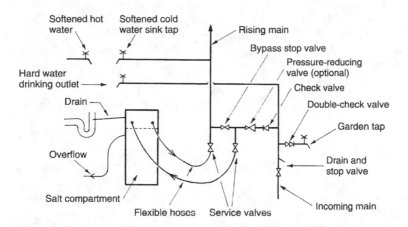

Components of a water softener – resin, sodium chloride (common salt) and a metering device (measures time or volume).

Function – granulated resin is located within a sealed compartment. It is unlikely that this will ever require changing or replenishing during the extensive life of a softener. Resin effectively filters the incoming water by retaining its hardness. Accumulated hardness is automatically washed off with a salt solution and discharged to the drain, leaving the resin recharged with salt. Regeneration is controlled by the metering device and a regulating valve.

Calcium bicarbonate $Ca(HCO_3)_2$ becomes sodium bicarbonate $Na(HCO_3)$.

Sodium bicarbonate remains soluble in water and unlike calcium, does not deposit as scale when heated.

Maintenance – depending on water consumption, the sodium chloride in the form of salt blocks or granules is replenished by the user. This is the only attention that the unit requires.

Water quality – because of the balanced chemical exchange, over-softening cannot occur. The UK limit for sodium is 200mg per litre of drinking water. The amount of sodium added to water through a water softener is unlikely to exceed this. Nevertheless, a separate hard water drinking outlet is recommended.

Domestic Water Conditioners

Hard water is difficult to lather and the combination of stearates in soap with calcium in water will produce a residual scum on sanitary fitments. Where mains-fed water heaters are to be installed, the supply should be tested. If it has a hardness factor greater than 200 ppm or 200 mg/l CCE, the water is unsuitable for use with directly fed water heaters. This includes electric showers and combination boilers. In the UK this affects approximately 65% of households.

Inspecting the inside of a kettle will provide an indication as to whether water hardness is an issue. For more reliable assessment a number of simple tests can be applied. As a guide, these include a dip pad test with colour indicator and a colour-change tablet test. Accurate definition in ppm can be achieved by using a handheld TDS (total dissolved solids) meter or by sample analysis in a laboratory.

As indicated on the preceding page, a water softener can be used to remove water hardness and associated limescale deposits. An alternative is a water conditioner and these are available in three different types:

- Electronic
- Magnetic
- Electrolytic

Compared to a water softener, conditioners have the following characteristics:

1. Scale-forming particles are suspended in water instead of precipitating onto surfaces.
2. Limescale is not eliminated but controlled.
3. Generally of relatively low purchase and installation cost when compared with a water softener.
4. No maintenance, but of limited life.
5. Water quality unaffected as chemicals are not added.

This type of conditioner is attached to the incoming rising main to provide whole-house treatment. It requires no alterations to existing plumbing and no special provision with new installations. The operating and energy requirement is from the electrical mains supply through a standard three-pin power socket. Connected to this is a reduced-voltage (12 volt DC) transformer with a coil of wire or antenna attached to or around the pipe.

Installation –

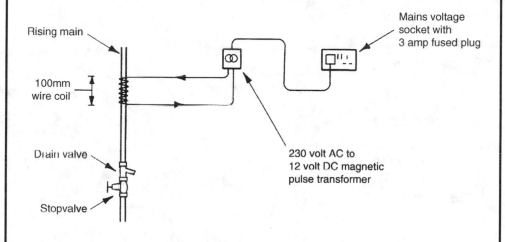

Rising main

100mm wire coil

Drain valve

Stopvalve

Mains voltage socket with 3 amp fused plug

230 volt AC to 12 volt DC magnetic pulse transformer

The coil of wire emits a range of electromagnetic signals through the pipe wall and into the water supply. These audio or radio signals have a sonic frequency modulation between 0·5 and 5kHz. The effect is to energise any suspended or diluted material, preventing it from forming as scale on pipe or appliance surfaces. The water remains chemically unchanged, retaining its minerals and taste. Unlike water softener installations, there is no need for a separate drinking water outlet.

Domestic Water Conditioner – Magnetic

A magnetic type of water conditioner is most effective when applied directly as dedicated water treatment to individual water heating appliances such as an electric shower, a combination boiler or an independent water heater. This type of conditioner is unsuitable for whole-house installations where water is stored.

The unit has a very strong magnetic core of ceramic construction. Water supplying a hot water appliance passes around the core and receives a small electrical induction charge. This is sometimes referred to as the magnetohydronamic process. As the water is heated the charged salts or crystals remain suspended in solution, allowing them to be flushed through, thereby preventing their formation as scale deposits on pipe and heating chamber walls.

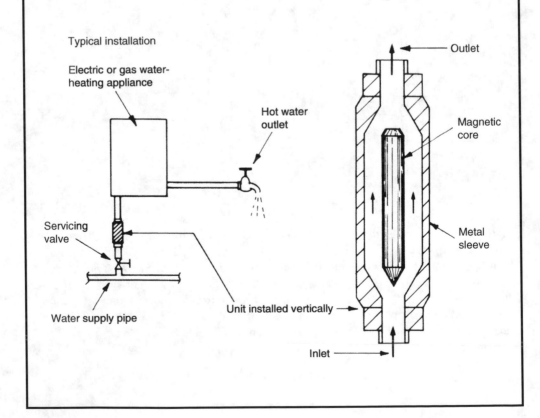

An electrolytic water conditioner provides whole-house treatment. It is installed on the rising main just after the stopvalve and before the first draw-off. An earth bonding cable should be provided to bypass the unit to ensure earth continuity. No other electrical connection is required.

Within the unit is a galvanic cell consisting of a copper cathode and zinc anode (see page 137). Water passing through acts as an electrolyte and gains a small charge. Like the magnetic conditioner, calcium deposits remain in suspension. In addition, zinc ions are produced which attract calcium and magnesium particles to produce suspended crystals of the more plumbing-friendly aragonite. Life expectancy of these units is about 10 years.

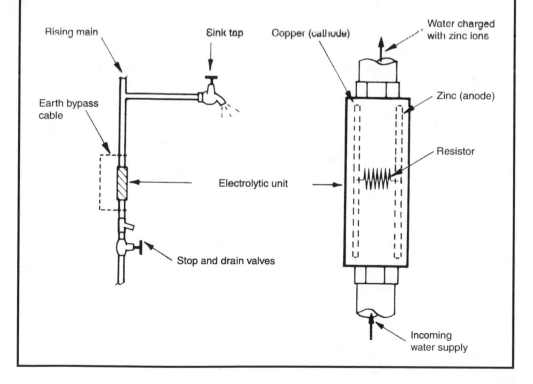

Rising main

Sink tap

Copper (cathode)

Water charged with zinc ions

Earth bypass cable

Zinc (anode)

Resistor

Electrolytic unit

Stop and drain valves

Incoming water supply

Water Treatment – Lime and Soda

The lime and soda process involves relatively large dosing tanks that require regular maintenance and checking. Therefore it is unsuited to individual domestic situations, but it does provide a cost-viable means for reducing the amount of calcium and magnesium in the water supply to industrial and municipal installations.

Lime – used as a reagent to remove temporary water hardness by breaking up the soluble bicarbonates into insoluble carbonates:

Impurity		Reagent		Precipitate
Calcium bicarbonate	+	Hydrated lime	=	Calcium carbonate
$Ca(HCO_3)_2$	+	$Ca(OH)_2$	=	$2CaCO_3 + 2H_2O$

Soda or soda ash – used as a reagent to remove permanent water hardness by exchanging the carbonate from the sodium (soda ash) with the sulphates, chlorides and nitrates of the calcium impurities.

Impurity		Reagent		By-product		Precipitate
Calcium sulphates, chlorides and nitrates	+	Soda ash	=	Sodium sulphates, chlorides and nitrates	+	Calcium carbonate
$CaSO_4$ $CaCl_2$ $Ca(NO_3)_2$	+	$NaCO_3$	=	Na_2SO_4 $2NaCl$ $2NaNO_3$	+	$CaCO_3$

In both processes the precipitate is removed by filtration.

Domestic sanitary appliances – all potable (drinkable) water supplies must be protected against pollution by backflow or back siphonage from water that could be contaminated. Protection is effected by leaving sufficient space or air gap between the lowest point of the control device or tap discharge and the appliance spill-over level.

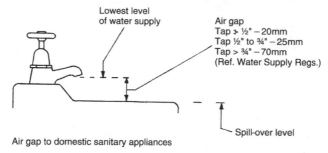

Lowest level of water supply

Air gap
Tap ≯ ½" – 20mm
Tap ½" to ¾" – 25mm
Tap > ¾" – 70mm
(Ref. Water Supply Regs.)

Spill-over level

Air gap to domestic sanitary appliances

British Standard determination of air gap to domestic sanitary appliances:

- Single feed pipe, i.e. one tap, air gap ≥ 20mm or 2 × internal diameter of tap orifice (take greater value).
- Multiple feed pipe, i.e. hot and cold taps, air gap ≥20mm or 2 × sum of orifice diameters (take greater value).

For example, a bath with two taps of 20mm internal diameter inlet orifice:

20mm or 2 × (20 + 20mm) = 80mm. Air gap = 80mm minimum.

Water cisterns or storage vessels – pipework supplying potable water must discharge into an unobstructed air gap between the regulating device water inlet to the vessel and the overflow or warning pipe.

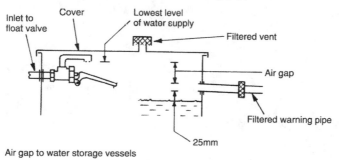

Inlet to float valve

Cover

Lowest level of water supply

Filtered vent

Air gap

Filtered warning pipe

25mm

Air gap to water storage vessels

In this situation the air gap should be ≥ 20mm or 2 × internal diameter of the regulating valve inlet orifice (take greater value).

For example, a 20mm internal diameter orifice:

20mm or 2 × 20mm = 40mm. Air gap = 40mm minimum.

Refs. Water Supply (Water Fittings) Regulations.

BS EN 1717: Protection against pollution of potable water in water installations and general requirements of devices to prevent pollution by backflow.

Secondary Backflow Protection

Secondary backflow or back siphonage protection is an alternative or supplement to the provision of air gaps. It is achieved by using mechanical devices such as double-check valves or a vacuum breaker in the pipeline. Special arrangements of pipework with branches located above the spill level of appliances are also acceptable.

Typical applications – primary heating circuits, washing machines and garden taps.

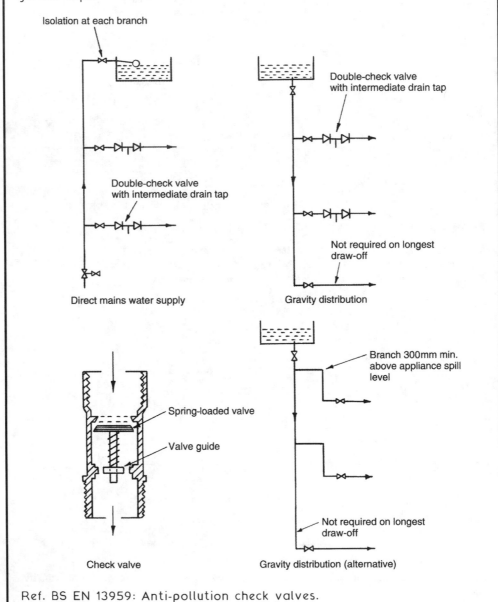

Ref. BS EN 13959: Anti-pollution check valves.

Mains water supply to commercial and industrial premises must be protected against the possibility of contamination by backflow. Where toxic processes exist (e.g. dyeing, chemical manufacture, insecticide preparation, car washing, irrigation systems, etc.), it is imperative that the effects of a pressure reduction on drinking water supplies be contained.

Contamination of domestic water supply situations is prevented by installing double-check valves to appliances or systems which could be a risk. In the interests of public health, the water authorities require greater security measures on mains supplies to industrial processes. Hitherto, a device containing two check valves with an intermediate pressure-relief valve discharging visibly to a tundish has been considered adequate. Current requirements include a modification to verify or check through test points that the fitting is functioning correctly. This modified device is known as a 'Verifiable backflow preventer with reduced pressure zone'. It contains three pressure zones separated by differential obturators (two positively loaded check valves). Each pressure zone has a test point to verify that the valve is functioning correctly.

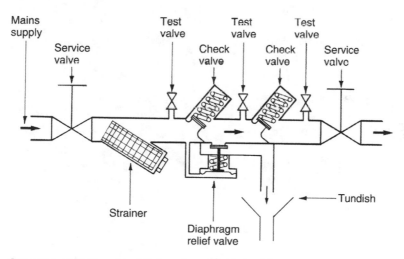

Operating principle and installation of a verifiable backflow prevention device with reduced pressure zone

Refs. The Water Supply (Water Fittings) Regulations.

BS EN 1717: Protection against pollution of potable water in water installations and general requirements of devices to prevent pollution by backflow.

Protecting Cold Water Pipes against Freezing

Cold water installations within a building's thermal envelope should not need to be insulated. This does not include pipes and cisterns in the roof space unless this area is part of the accommodation. Also excluded from the external envelope is space below suspended ground floors, unheated outhouses and garages.

The purpose of insulating pipes is to retain thermal energy in the pipe and in the water it conveys. Supply pipes are buried at least 750mm below ground for this reason, as well as protection against ground movement.

Common pipe-insulating materials:

Material	Thermal conductivity (W/mK)λ
Rigid phenolic foam	0.020–0.025
Polisocyanurate foam	.. ..
Rigid polyurethane foam	.. ..
PVC foam	0.025–0.030
Expanded polystyrene	0.030–0.035
Extruded polystyrene	.. ..
Cross-linked polyethylene foam	.. ..
Expanded nitrile rubber	.. ..
Standard polyethylene foam	0.035–0.040
Expanded synthetic rubber	.. ..
Cellular glass fibre	.. ..

Note: Thermal conductivity (λ) is a measure of the rate that heat energy is conducted through a material under specific conditions in units of W/mK. The lower the value the better the insulating effect.

Comparative guide to minimum provision of pipe insulation:

Pipe dia. (mm)   Thermal conductivity and min. thickness
copper

	Phenolic foam 0.020	Exp. nitrile rubber 0.035	Cellular glass 0.040
15	15	13	25
22	15	19	25
28	20	25	32
35 and over	20	32	32

See also Water Regulations UK insulation calculator spreadsheet at www.waterregsuk.co.uk based on data from BS 5422: Method for specifying thermal insulation materials for pipes, etc., etc.

Cisterns can be manufactured from galvanised mild steel (large non-domestic capacities), polypropylene or glass-reinforced plastics. They must be well insulated and supported on adequate bearers to spread the concentrated load. Plastic cisterns will require uniform support on boarding over bearers. A dustproof cover is essential to prevent contamination.

For large buildings, cisterns are accommodated in a purpose-made plant room at roof level or within the roof structure. This room must be well insulated and ventilated, and be provided with thermostatic control of a heating facility.

Where storage demand exceeds 4500 litres, cisterns must be duplicated and interconnected. In the interests of load distribution this should be provided at much lower capacities. For maintenance and repairs each cistern must be capable of isolation and independent operation.

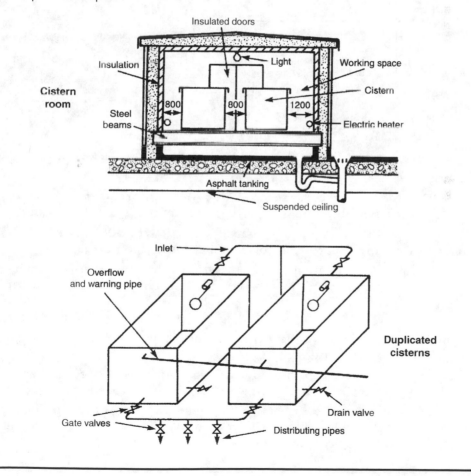

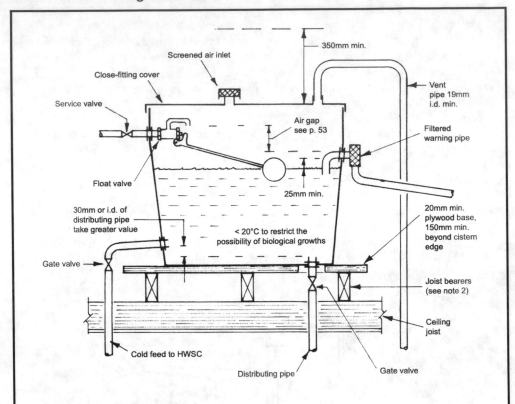

Refs. BS 417-2: Specification for galvanised low carbon steel cisterns, cistern lids, tanks and cylinders.

BS 4213: Cisterns for domestic use. Cold water storage and combined feed and expansion (thermoplastic) cisterns up to 500 litres. Specification.

BS 7181: Specification for storage cisterns up to 500 litres actual capacity for water supply for domestic purposes (see note 3).

BS EN 13280: Specification for glass fibre-reinforced cisterns of one-piece and sectional construction, for the storage, above ground, of cold water.

Note 1: Where installed in an unheated area such as a roof space, an insulating jacket is to be fitted to the cistern and all associated pipework to be fully insulated.

Note 2: Bearers at 350mm c/c max. for galvanised steel cisterns. Plastic and glass fibre cisterns on sheet plywood over bearers.

Note 3: Actual capacity refers to the quantity of water contained when the float valve and overflow/warning pipe are fitted. Nominal capacity refers to the cistern capacity if filled to the brim.

Inlets and outlets:

Water supply pipe/rising main – provided with a servicing valve fitted as close as possible to the float valve to isolate the supply during repair and maintenance.

Inlet valve – water flow control is by a float-operated valve (see page 33) or a motorised valve that responds to the cistern water level. A float valve is the most common and is fitted as high as practicable to maximise storage capacity. An air gap is required between float valve outlet and warning pipe (see page 53) to prevent the possibility of water contamination by backflow. Cistern water level is at least 25mm below the warning pipe overflowing level, effected by adjustment of the ball float.

Outlets – preferably from the cistern base to prevent sediment retention that could contain nutrients for bacteria. Positioned opposite the inlet to encourage cross-flow. This is particularly important with large-capacity cisterns to reduce the possibility of stagnation.

Where hot and cold water feed pipes originate from the same cistern, the hot water connection should be higher than the cold connection in case the float valve seizes allowing the cistern to run dry. Where combined hot and cold water taps and showers are fitted, this will stop hot water flowing before cold water, preventing the possibility of scalding.

Cold feed and distributing outlet pipes are to be provided with a servicing gate valve to prevent water wastage that would otherwise occur from emptying the cistern. An exception is the cold feed from a hot water and/or central heating feed and expansion cistern. In this situation the cold feed to the primary circuit and boiler is not fitted with a valve as this could be inadvertently closed, possibly causing boiling and loss of boiler feed water if the system temperature control were to fail.

Ref. The Water Supply (Water Fittings) Regulations.

Warning and overflow pipes:

Warning pipe – this combines the purpose of providing a conspicuous outfall of water if the float valve malfunctions and safely discharging the surplus water to a suitable place that will not be damaging. A minimum 19mm nominal bore and at least one pipe size above the inlet pipe diameter.

Overflow pipe – for cisterns less than 1000 litres actual capacity, an adequately sized warning pipe is regarded as a suitable overflow. Otherwise, positioned at least 25mm (min. overflowing or invert levels) above the warning pipe with regard to sufficient air gap (see previous page) to discharge potentially damaging and disruptive surplus water to a suitable location. This could be into a rainwater pipe or gutter system with the warning pipe still maintaining a conspicuous discharge.

Cisterns exceeding 5000 litres actual capacity may be fitted with a float switch actuated alarm instead of a warning pipe. This should operate when the cistern water level is within 25mm of the overflow. In this situation the overflow is positioned with its invert level not more than 50mm above normal water level.

Warning and overflow pipes – installed to fall away from the cistern to its point of discharge, preferably to the outside of the building to which it is fitted. For convenience and for visual reasons, discharge may be inside the building over a tundish with an air gap as shown for cisterns on page 53. Alternatively, at least 150mm above a WC rim or other sanitary fitting. A combined bath/overflow manifold may also be used, as shown below:

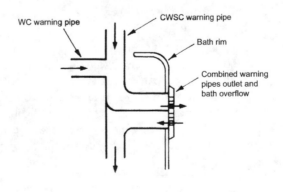

WC warning pipe

CWSC warning pipe

Bath rim

Combined warning pipes outlet and bath overflow

Cold water storage data is provided to allow for up to 24-hour interruption of mains water supply.

Building purpose	Storage/person/24 hrs	
Boarding school	90 litres	
Day school	30	
Department store with canteen	45	(3)
Department store without canteen	40	(3)
Dwellings	90	(1)
Factory with canteen	45	
Factory without canteen	40	
Hostel	90	
Hotel	135	(2) (3)
Medical accommodation	115	
Office with canteen	45	
Office without canteen	40	
Public toilets	15	
Restaurant	7 per meal	

Notes:
(1) 115 or 230 litres min. (see pages 41 and 42).
(2) Variable depending on classification.
(3) Allow for additional storage for public toilets and restaurants.

At the design stage the occupancy of a building may be unknown. Therefore the following may be used as a guide:

Building purpose	Occupancy
Department store	1 person per 30m² net floor area
Factory	30 persons per WC
Office	1 person per 10m² net floor area
School	40 persons per classroom
Shop	1 person per 10m² net floor area

E.g. A 1000m² (net floor area) office occupied only during the day therefore allow 10 hours' emergency supply.

1000/10 = 100 persons × 40 litres = 4000 litres (24 hrs)
= 1667 litres (10 hrs)

Boosted Cold Water System – 1

For medium- and high-rise buildings, there is often insufficient mains pressure to supply water directly to the upper floors. Boosting by pump from a break tank is therefore usually necessary and several more of these tanks may be required as the building rises, depending on the pump capacity. A break-pressure cistern is also required on the down service to limit the head or pressure on the lower fittings to a maximum of 30m (approx. 300kPa). The drinking water header pipe or storage vessel supplies drinking water to the upper floors. As this empties and the water reaches a predetermined low level, the pipeline switch engages the duty pump. A float switch in the break tank protects the pumps from dry running if there is an interruption to mains supply. The various pipe sections are fitted with isolating valves to facilitate maintenance and repairs.

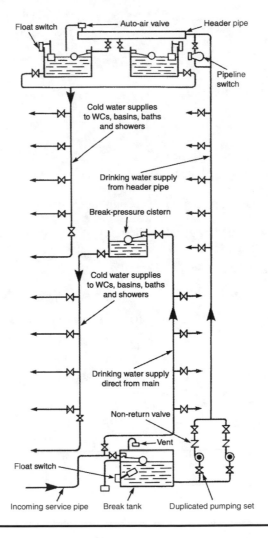

As an alternative to the drinking water header pipe, an auto-pneumatic cylinder may be used. Compressed air in the cylinder forces water up to the float valves and drinking water outlets on the upper floors. As the cylinder empties a low-pressure switch engages the duty pump. When the pump has replenished the cylinder, a high-pressure switch disengages the pump. In time, some air is absorbed by the water. As this occurs, a float switch detects the high water level in the cylinder and activates an air compressor to regulate the correct volume of air. Break-pressure cisterns may be supplied either from the storage cisterns at roof level or from the rising main. A pressure-reducing valve is sometimes used instead of a break-pressure cistern.

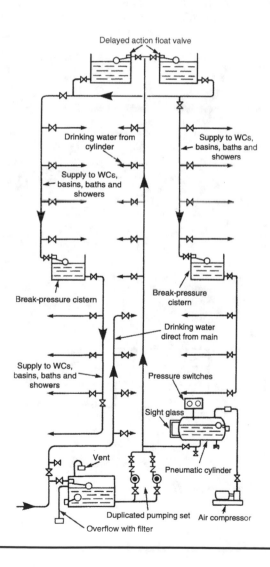

Boosted Cold Water System – 3

In modest-rise buildings of several storeys where water is in fairly constant demand, water can be boosted from a break tank by a continuously running pump. The installation is much simpler and less costly than the previous two systems as there is less need for specialised items of equipment. Sizing of the pump and its delivery rating are critical, otherwise it could persistently overrun, or at the other extreme be inadequate. Modern pumps have variable settings allowing considerable scope around the design criteria. The pump is normally scheduled to run on a timed programme, e.g. in an office block it may commence an hour before normal occupancy and run on for a couple of hours after. Water delivery should be just enough to meet demand. When demand is low a pressure-regulated motorised bleed valve opens to recirculate water back to the break tank.

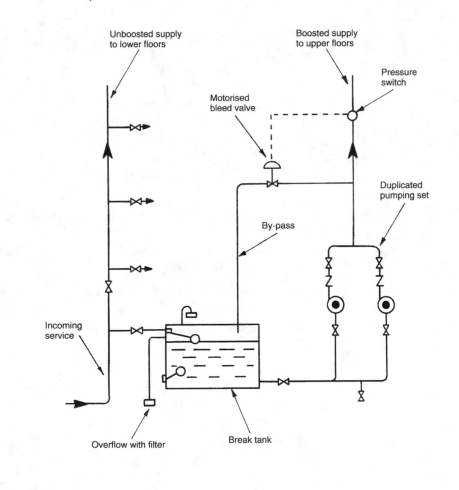

If normal float valves are used to regulate cistern water supply from an auto-pneumatic cylinder (page 63), then cylinder and pump activity will be frequent and uneconomic. Therefore to regulate activity and deliveries to the cistern, a delayed action float valve mechanism is fitted to the storage cistern.

Stage 1. Water filling the cistern lifts hemispherical float and closes the canister valve.

Stage 2. Water overflows into the canister and raises the ball float to close off water supply.

Stage 3. As the cistern empties, the ball float remains closed until low water level releases the hemispherical float. As this float valve drops, water is released from the canister to open the ball float valve to replenish the cistern from the pneumatic supply.

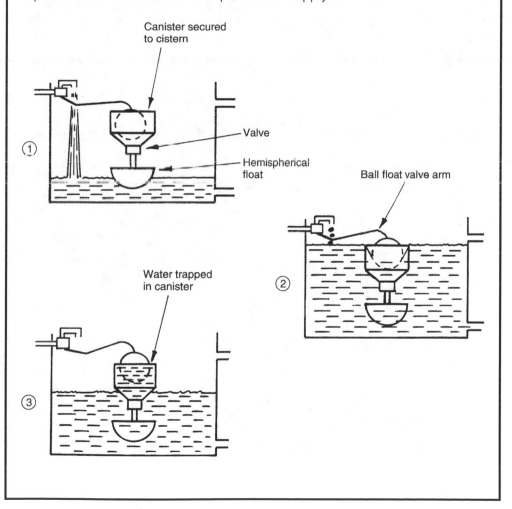

Canister secured
to cistern

Valve

Hemispherical
float

Ball float valve arm

Water trapped
in canister

①

②

③

Non-Return Valve

The high-rise cold water supply systems illustrated on pages 62–64 have a non-return valve fitted to the outlet of each booster pump. This essential fitting will prevent reversal of the flow by gravitation when the pump is inactive. Water flow reversal into the break tank would be wasteful, potentially damaging to the plant room and, with a significant head of water, the pressure could burst pump seals, gaskets and other joints.

When the pump stops its delivery, the head of water above it will attempt to reverse and gravitate. A swing pattern non-return valve of the type shown on page 170 will not function instantly and a small amount of backflowing water will allow the water column to drop. As the disc closes, the column will be halted abruptly and this may produce vibrations or water hammer through the installation.

A rapid functioning spring-assisted type of non-return valve is preferred particularly where the potential head will exceed 50 m. This type of non-return valve is similar in principle to the horizontal lift pattern shown on page 170 but is produced to suit larger diameter pipes. In this format it usually has flanged connections and is known as a recoil valve.

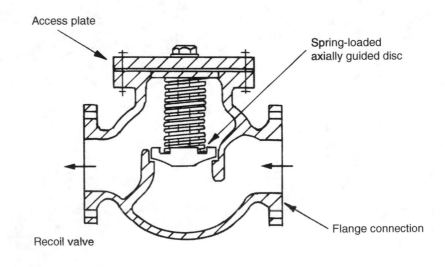

Access plate

Spring-loaded
axially guided disc

Flange connection

Recoil valve

Pump power calculations are based on the physics of work done relative to time. Work done is applied force through distance moved.

Unit of measurement is the joule, the work done when a 1 newton force acts through 1 metre distance, i.e. 1 joule = 1N × 1m.

Time is expressed in seconds. By combining work done over a period of time:

Power = work done ÷ time
      = (force × distance) ÷ seconds
      = (newtons × metres) ÷ seconds [J/s]   where 1 J/s = 1 watt

Force in newtons = kg mass × acceleration due to gravity [9·81m/s^2]
Power expressed in watts = (mass × 9·81 × distance) ÷ time
For example

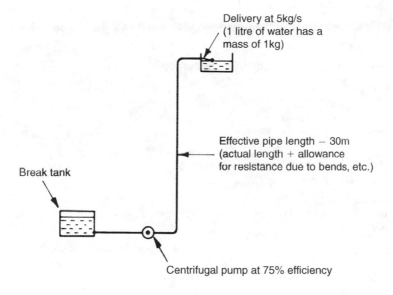

Delivery at 5kg/s
(1 litre of water has a mass of 1kg)

Effective pipe length — 30m
(actual length + allowance for resistance due to bends, etc.)

Break tank

Centrifugal pump at 75% efficiency

Power = (mass × 9·81 × distance) ÷ time
      = (5 × 9·81 × 30) ÷ 1
      = 1471·5 watts

Allowing for the pump efficiency: 1471·5 × (100 ÷ 75) = 1962 watts
Pump rating: 2kW at 5l/s (1962 watts rounded up to nearest kW)

Pump Laws – 1

In normal application with the exception of maintenance and repair, the components of a water pump will remain unchanged during use. If a pump proves unsuitable for purpose, the complete unit is usually replaced with a pump of better specification. A pump with an impellor of constant diameter will have the following characteristics:

- Water quantity (Q) or volume delivered varies directly with the rotational speed (N) or angular velocity (rpm) of the impellor.

$$(Q_2 \div Q_1) = (N_2 \div N_1)$$

- Pressure (P) produced varies as the square of rotational speed (N).

$$(P_2 \div P_1) = (N_2)^2 \div (N_1)^2$$

- Power (W) required varies as the cube of rotational speed (N).

$$(W_2 \div W_1) = (N_2)^3 \div (N_1)^3$$

where: Q_1 and Q_2 = discharge of water delivered (l/s)

N_1 and N_2 = impellor rotational speed (rpm or rps)

P_1 and P_2 = pressure produced (kPa or kN/m^2)

W_1 and W_2 = power absorbed/required (Watts)

E.g. A 2kW pump discharges 5kg/s when the pump impellor speed is 1000rpm. Increasing the impellor speed to 1200rpm will provide the following characteristics:

$$(Q_2 \div Q_1) = (N_2 \div N_1)$$

Transposing: $Q_2 = (N_2 \times Q_1) \div N_1$

$Q_2 = (1200 \times 5) \div 1000 = 6\text{kg/s or } 6\text{l/s}$

$$(W_2 \div W_1) = (N_2)^3 \div (N_1)^3$$

Transposing: $W_2 = (N_2)^3 \times W_1 \div (N_1)^3$

$= (1200)^3 \times 2000 \div (1000)^3$

$= 3456 \text{ Watts or } 3.5\text{kW}$

$$(P_2 \div P_1) = (N_2)^2 \div (N_1)^2$$

Transposing: $P_2 = (N_2)^2 \times P_1 \div (N_1)^2$

At 40kPa pressure at 1000rpm increasing to 1200rpm will produce:

$P_2 = (1200)^2 \times 40 \div (1000)^2$

$P_2 = 57.6 \text{ kPa}$

If a water pump has adaptability to component change and the impellor can be replaced with compatible units of different diameters, the following apply:

- At constant rotational speed (N) the water quantity (Q) delivered varies as the cube of the impellor diameter (D).

$$(Q_2 \div Q_1) = (D_2)^3 \div (D_1)^3$$

- Pressure (P) produced varies as the square of impellor diameter (D).

$$(P_2 \div P_1) = (D_2)^2 \div (D_1)^2$$

- Power (W) required varies as the fifth power of impellor diameter (D).

$$(W_2 \div W_1) = (D_2)^5 \div (D_1)^5$$

Single two-speed pump characteristics:

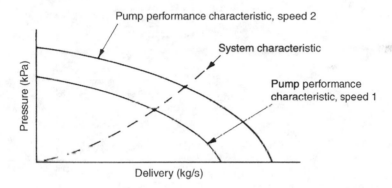

Duplicate pumps of equal characteristics working together in parallel:

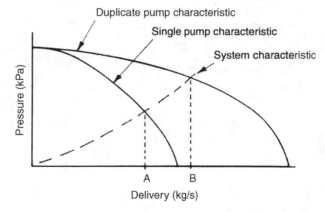

Delivery is theoretically twice that of a single pump, but realistically the pressure or resistance to flow in the system will determine the flow, i.e. flow at B is not twice that at A.

69

Pipe Sizing by Formula

Thomas Box formula:

$$d = \sqrt[5]{\frac{q^2 \times 25 \times L \times 10^5}{H}}$$

where: d = diameter (bore) of pipe (mm)
      q = flow rate (l/s)
      H = head or pressure (m)
      L = length (effective) of pipe (m)
          (actual length + allowance for bends, tees, etc.)

e.g.

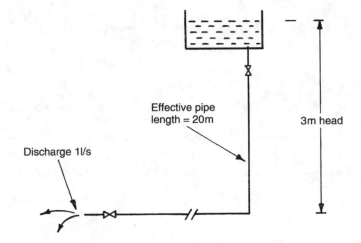

Effective pipe length = 20m

3m head

Discharge 1l/s

$$d = \sqrt[5]{\frac{(1)^2 \times 25 \times 20 \times 10^5}{3}}$$

$$d = \sqrt[5]{16\,666\,667} = 27.83\text{mm}$$

The nearest commercial size above this is 32mm bore steel or 35mm outside diameter copper.

Note: Head in metres can be converted to pressure in kPa by multiplying by gravity, e.g. 3m × 9·81 = 29·43kPa (approx. 30kPa).

Steel pipe (inside dia.)		Copper tube (mm)		Polythene (mm)	
Imperial (")	Metric (mm)	Outside dia.	Bore	Outside dia.	Bore
$\frac{1}{2}$	15	15	13.5	20	15
$\frac{3}{4}$	20	22	20	27	22
1	25	28	26	34	28
$1\frac{1}{4}$	32	35	32	42	35
$1\frac{1}{2}$	40	42	40		
2	50	54	51.5		
$2\frac{1}{2}$	65	67	64.5		
3	80	76	73.5		

Approximate equivalent pipe lengths of some fittings (m).

Pipe bore (mm)	Elbow	Tee	Stop valve	Reduction 4:1	2:1	4:3
15	0·6	0·7	4·5	0.30	0.25	0.10
20	0·8	1·0	7	0.50	0.45	0.15
25	1·0	1·5	10	0.70	0.60	0.20
32	1·4	2·0	13	0.90	0.75	0.25
40	1·7	2·5	16	1.20	0.90	0.30
50	2·3	3·5	22	1.50	1.10	0.35

Notes: Figure given for a tee is the change of direction; straight through has no significant effect. These figures are only intended as a guide; they will vary between materials and design of fittings.

Recommended flow rates for various sanitary appliances (litres/sec)

WC cistern	0·11
Hand basin	0·15
Hand basin (spray tap)	0·03
Bath (19mm tap)	0·30
Bath (25mm tap)	0·60
Shower	0·11
Sink (13mm tap)	0·19
Sink (19mm tap)	0·30
Sink (25mm tap)	0·40

Pipe Sizing – Loading Units

Loading units are factors which can be applied to a variety of appliances. They have been established by considering the frequency of use of individual appliances and the desired water flow rate.

Appliance	Loading units
Hand basin	1·5 to 3 (depends on application)
WC cistern	2
Washing machine	3
Dishwasher	3
Shower	3
Sink (13mm tap)	3
Sink (19mm tap)	5
Bath (19mm tap)	10
Bath (25mm tap)	22

By determining the number of appliances on a pipework system and summating the loading units, an equivalent flow in litres per second can be established from the following conversion graph:

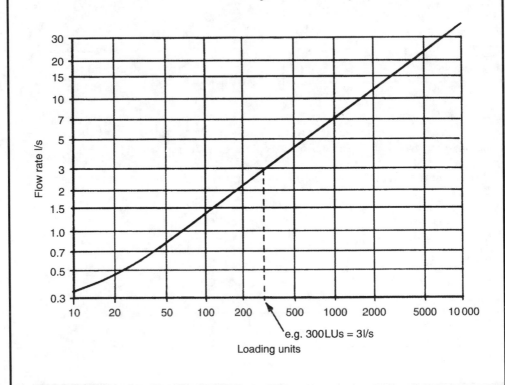

e.g. 300LUs = 3l/s

Loading units

Pressure or head loss in pipework systems can be expressed as the relationship between available pressure (kPa) or head (m) and the effective length (m) of pipework. The formula calculation on page 70 can serve as an example:

Head = 3m. Effective pipe length = 20m. So, 3/20 = 0.15m/m

By establishing the flow rate from loading units or predetermined criteria (1l/s in our example), a nomogram may be used to obtain the pipe diameter. The chart below is for illustration and general use. For greater accuracy, pipe manufacturers' design data should be consulted for different materials and variations in water temperatures.

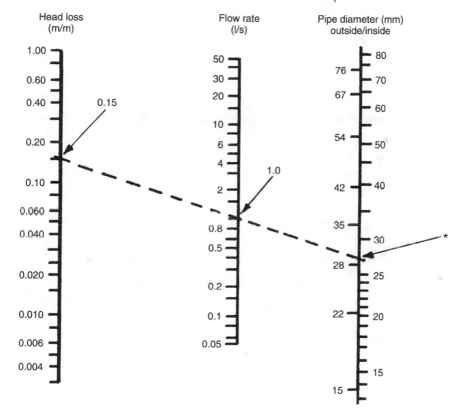

Head loss
(m/m)

Flow rate
(l/s)

Pipe diameter (mm)
outside/inside

* Inside diameter = 27.83 mm
(see page 70)

See also:

BS 8558: Guide to the design, installation, testing and maintenance of services supplying water for domestic use.

BS EN 806-3: Specifications for installations inside buldings conveying water for human consumption. Pipe sizing. Simplified method.

Pipe Sizing – Head Loss and Flow Rate Application

On the preceding page a pipe sizing chart or nomogram is used to verify the size of pipe calculated by the formula on page 70. The nomogram may also be used to determine the size of water mains and supply pipes as shown in the following example:

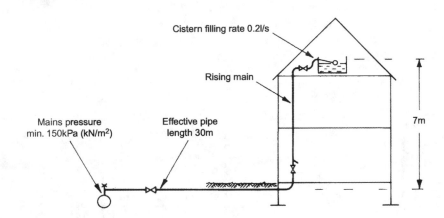

Pressure or residual head at the cistern will be the minimum mains pressure less the height the water has to rise to the cistern. To convert 150kPa to metres head, divide by gravity (9.81), i.e.:

$$150 \div 9.81 = 15.30m$$

Pressure or residual head at the cistern inlet is therefore:

$$15.30m - 7.00m = 8.30m$$

$$Head \div Effective\ pipe\ length = Head\ or\ pressure\ loss\ per\ metre$$

Extending a straight line through the nomogram coordinates of 0.176m/m and the selected cistern filling rate of 0.2l/s indicates that a copper pipe of at least 15mm outside diameter or equivalent is adequate.

74

Hydraulics is the experimental science concerning the study of energy in fluid flow. That is, the force of pressure required to overcome the resistance to fluid flowing through pipes, caused by the friction between the pipe and liquid movement.

The total energy of the liquid flowing in a pipe declines as the pipe length increases, due mainly to friction between the fluid and the pipe wall. The amount of energy or pressure loss will depend on:

- Smoothness/roughness of the internal pipe wall.
- Diameter of pipe or circumference of internal pipe wall.
- Length of pipe.
- Velocity of fluid flow.
- Amount of turbulence in the flow.
- Viscosity and temperature of fluid.

Theories relating to pressure loss by fluids flowing in pipes are diverse, but an established relationship is that the pressure losses (h) caused by friction are proportional to the square of the velocity of flow (v):

$$h \propto v^2$$

From this, for a pipe of constant size it may be seen that by developing the proportional relationship, a doubling (or more) of pressure will increase the velocity accordingly.

h (m)	v (m/s)
4	1·5
8	2·12 (1·5 × $\sqrt{2}$)
12	2·60 (1·5 × $\sqrt{3}$)
16	3·00 (1·5 × $\sqrt{4}$) or (2·12 × $\sqrt{2}$)
24	3·66 (1·5 × $\sqrt{6}$) or (2·60 × $\sqrt{2}$)
32	4·24 (1·5 × $\sqrt{8}$) or (3·00 × $\sqrt{2}$) etc., etc.

It can also be shown that if the condition (temperature and viscosity) of a fluid in a pipe remains constant, the discharge through that pipe is directly proportional to the square root of the fifth power of its diameter:

$$\sqrt{d^5}$$

This relationship can be identified in the Thomas Box pipe sizing formula shown on page 70.

Fluid Flow Formulae – 1

Bernoulli's theorem (see also pages 296 and 297) – the theoretical basis for fluid flow, established with the assumption that there is no fluid flow energy loss due to friction. It therefore applies to the steady motion where a fluid moves in streamlines as depicted in the diagram below. Theoretically, the fixed path of fluid movement passes through given points of known small cross-sectional area (a_1, a_2), pressure (h_1, h_2) and velocity (v_1, v_2).

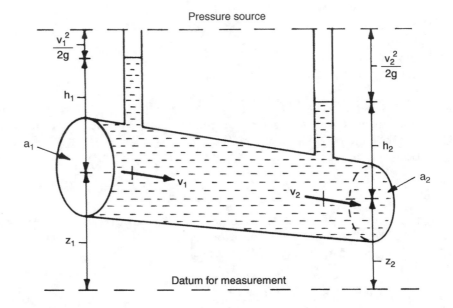

The total energy of unit weight of a fluid in flow can be expressed by the following summation:

Potential energy (z) + Pressure energy (h) + Kinetic energy ($v^2/2g$) = Constant, i.e.: If there is a loss of energy in any category there must be gain in the others for the balance to remain constant.

By formula –

$$z_1 + h_1 + v_1^2/2g = z_2 + h_2 + v_2^2/2g$$

Note: g represents gravitational acceleration of $9.81 m/s^2$.

Bernoulli's theory is approximately true for liquid movement in a short length of straight pipe, but with pipework installations the pressure head decreases over distance due to frictional resistance between the fluid conveyed and the pipe wall. Nevertheless, Bernoulli's principles of pressure differentials have become an established basis for the development of numerous other liquid flow calculations.

Venturimeter – a device developed from Bernoulli's principles of fluid flow for measuring the quantity or discharge of a liquid through a pipe (typically a water main), by comparing pressure differences through a constriction in the pipe. In the direction of flow, the instrument combines a fairly rapidly tapering pipe to reduce the cross-sectional area at the throat. Thereafter, is a relatively long taper to enlarge the cross-section back to the original diameter of the pipe.

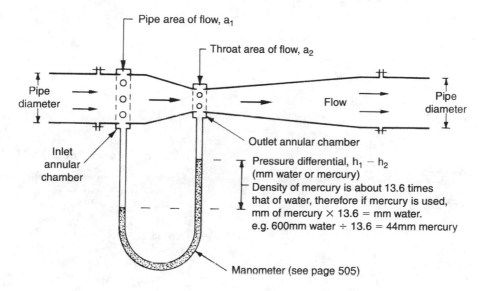

The discharge formula may be expressed as:

$$Q = C \times a_1 \times \sqrt{[2g\,(h_1 - h_2)] \div [(a_1 \div a_2)^2 - 1]}$$

Where:
- Q = Quantity or discharge (m³/s)
- C = Coefficient of discharge velocity, (0.96 to 0.99, 0.98 is usually used for water)
- a_1 and a_2 = area of pipe (m²)
- g = gravitational acceleration (9.81 m/s²)
- h_1 and h_2 = pressure head

E.g. a 100mm diameter pipe (area, a_1 = 0·00785 m²) and an instrument throat diameter of 50mm (area, a_2 = 0·00196 m²). $h_1 - h_2$ = 600mm (0·6 m). C = 0·98.

$$Q = 0.98 \times 0.00785 \times \sqrt{[2 \times 9.81 \times 0.6] \div [(0.00785 \div 0.00196)^2 - 1]}$$

$$Q = 0.007693 \times \sqrt{11.772 \div 15.040}$$

$$Q = 0.0068 \text{ m}^3/\text{s or } 6.8 \text{ l/s}$$

Fluid Flow Formulae – 3

Reynolds number – a coefficient of friction based on the criteria for similarity of motion for all fluids. Relevant factors are related by formula:

$$\frac{density \times velocity \times linear\ parameter\ (diameter)}{viscosity}$$

This is more conveniently expressed as $\quad R = \frac{\rho v d}{\mu}$

Where: R = Reynolds number

ρ = fluid density (kg/m^3)

v = velocity (m/s)

d = diameter of pipe (m)

μ = viscosity of the fluid (Pa s) or (Ns/m^2)

Whatever the fluid type or temperature, an R value of less than 2000 is considered streamline or laminar. A value greater than 2000 indicates that the fluid movement is turbulent.

E.g. 1. A 12mm diameter pipe conveying fluid of density $1000kg/m^3$ and viscosity of 0.013Pa s at 2m/s flow velocity has a Reynolds number of:

$$\frac{1000 \times 2 \times 0\cdot012}{0\cdot013} = 1846\ (streamline\ flow)$$

D'Arcy formula – used for calculating the pressure head loss of a fluid flowing full bore in a pipe, due to friction between fluid and pipe surface.

$$h = \frac{4\ fL\ v^2}{2\ g\ d}$$

Where: h = head loss due to friction (m)

f = coefficient of friction

L = length of pipe (m)

v = average velocity of flow (m/s)

g = gravitational acceleration $(9.81m/s^2)$

d = internal diameter of pipe (m)

Note: 'f', the D'Arcy coefficient, ranges from about 0.005 (smooth pipe surfaces and streamline flow) to 0.010 (rough pipe surfaces and turbulent flow). Tables may be consulted, although a mid-value of 0.0075 is appropriate for most problem solving.

E.g. 2. A 12mm diameter pipe, 10m long, conveying a fluid at a velocity flow of 2m/s:

$$Head\ loss = \frac{4 \times 0\cdot0075 \times 10 \times 2^2}{2 \times 9\cdot81 \times 0\cdot012} = 5\cdot09m$$

Depending on the data available, it is possible to transpose the D'Arcy formula for other purposes. For example, it may be used to calculate pipe diameter in this format:

$$d = \frac{4 f L v^2}{2 g h}$$

Flow rate (Q) – the discharge rate or flow rate of a fluid in a pipe is expressed as the volume in cubic metres (V) flowing per second (s). Q (m^3/s) is dependent on the pipe cross-sectional area dimensions (m^2) and the velocity of fluid flow (m/s). Q may also be expressed in litres per second, where $1 m^3$/s = 1000l/s.

A liquid flowing at an average velocity (v) in a pipe of constant area (A) discharging a length (L) of liquid every second (s) has the following relationship:

Q = V ÷ s   where V = L × A   and   v = L ÷ s
So, Q = L × A ÷ s   where v = L ÷ s,   ∴   Q = v × A
    Q = flow rate (m^3/s), v = velocity of flow (m/s) and
    A = cross-sectional area of pipe (m^2)

E.g. 1. The quantity of water flowing through a 12mm diameter pipe at 2m/s will be:

    Q = v × A, where A = πr^2
    Q = 2 × 0·000113 = 0·000226 m^3/s or 0·226 l/s

Relative discharge of pipes – this formula may be used to estimate the number of smaller branch pipes that can be successfully supplied by one main pipe:

$$N = \sqrt{(D \div d)^5}$$

    where N = number of short branch pipes
          D = diameter of main pipe (mm)
          d = diameter of short branch pipes (mm)

E.g. 2. The number of 32mm short branch pipes that can be served from one 150mm main will be:

$$N = \sqrt{(150 \div 32)^5} = 47$$

E.g. 3. The size of water main required to supply 15 20mm short branch pipes will be by formula transposition:

    D = d $\sqrt[5]{N^2}$
    D = 20 $\sqrt[5]{15^2}$ = 59 (65mm nearest standard)

As shown in the preceding pages, flow calculations for water conveyance through pipes can incorporate several variables, e.g. temperature, density, etc., resulting in quite complex arithmetic. For indicative purposes, there is a simpler mathematical interrelationship between pipe diameter (m), design criteria for flow velocity (m/s) and quantity of flow (m^3/s):

$$Q = v \times A$$
$$\text{where} \quad Q = m^3/s$$
$$v = m/s$$
$$A = m^2$$

Criteria for calculation:

- Pipe internal diameter (m)
- Pipe internal cross-sectional area (m^2)
- Velocity of water flow (m/s)
- Quantity discharge of water (m^3/s)

Formulae:

- Area of internal diameter of pipe (A) = πr^2
- Quantity of water (Q) = Velocity (v) × Area (A)

Referring to example 1 on the previous page, the calculation below is reverted to make the pipe diameter the subject. Q is 0.226 l/s or 0.000226 m^3/s with v = 2 m/s.

$Q = v \times A$ or $A = Q \div v$

So, A = 0.000226 ÷ 2

       = 0.000113 m^2

   A = πr^2

So, $r^2 = A \div \pi$

       = 0.000113 ÷ 3.142 = 0.000036

r = sq. root of 0.000036 = 0.006 m or 6 mm
Therefore, pipe diameter is 12 mm.

Another example with Q = 0.1l/s or 0.0001 m^3/s flowing through a 50 mm diameter pipe, i.e. 25 mm or 0.025 m radius.

$Q = v \times A.$ or $v = Q \div A$

$A = \pi r^2 = 3.142 \times (0.025)^2 = 0.002 \ m^2$

$v = 0.0001 \div 0.002 = 0.05$ m/s

If the same pipe is required to deliver a greater velocity of 1m/s, the quantity (Q) will be:

$Q = v \times A = 1 \times 0.002 = 0.002 \ m^3/s$ or 2 l/s.

3 HOT WATER SUPPLY SYSTEMS

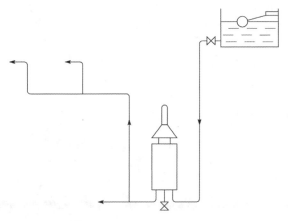

DIRECT SYSTEM OF HOT WATER SUPPLY
INDIRECT SYSTEM OF HOT WATER SUPPLY
UNVENTED HOT WATER STORAGE SYSTEM
EXPANSION AND TEMPERATURE-RELIEF VALVES
HOT WATER STORAGE CYLINDERS
PRIMATIC HOT WATER STORAGE CYLINDERS
INDIRECT HOT WATER SYSTEM FOR A THREE-STOREY
BUILDING
INDIRECT SUPPLEMENTARY HOT WATER SYSTEM
SEALED INDIRECT HOT WATER SYSTEM FOR A HIGH-RISE
BUILDING
PRIMARY THERMAL STORES
TYPES OF BOILER
SECONDARY CIRCULATION
DUPLICATION OF PLANT
ELECTRIC AND GAS WATER HEATERS
SOLAR HEATING OF WATER
HOT WATER STORAGE CAPACITY
BOILER RATING
PIPE SIZING
PRESSURISED SYSTEMS
CIRCULATION PUMP RATING
LEGIONNAIRES' DISEASE IN HOT WATER SYSTEMS
SEDBUK
ENERGY EFFICIENCY LABELING
GREENHOUSE GASES
GALVANIC OR ELECTROLYTIC ACTION
WATER TREATMENT

DOI: 10.1201/9781003434894-3

Expansion of Water

Water expands with changes in temperature. At 4°C water is at its most dense. At temperatures below 4°C down to zero or freezing, water expands about 9% (approximately 1/10) by volume. This is why underground supplies require adequate ground cover and externally exposed water pipes require insulation to prevent damage. At temperatures between 4°C and 100°C or boiling, water expands by about 4% (approximately 1/25) by volume and is significantly less dense – see table below. This degree of expansion and reduction in density is the principle of convective water circulation in elementary hot water systems.

Temperature (°C)	Density (kg/m^3)
0	999·80
4	1000·00
10	999·70
20	998·20
30	995·00
40	992·20
50	987·50
60	983·20
70	977·50
80	971·80
90	965·60
100	958·00

The following formula can be used to calculate the amount that water expands in a hot water system:

$$E = C \times (\rho_1 - \rho_2) \div \rho_2$$

Where: E = expansion (m^3)

C = capacity or volume of water in system (m^3)

ρ_1 = density of water before heating (kg/m^3)

ρ_2 = density of water after heating (kg/m^3)

Example: A hot water system containing 15m^3 of water, initially at 10°C to be heated to 80°C.

$$E = 15 \times (999·70 - 971·80) \div 971·80$$
$$E = 0·430\text{m}^3$$

Hot water and heating systems must incorporate a means for accommodating expansion. A fail-safe mechanism must also be provided should the initial provision malfunction.

Heated wholesome water (defined on page 26) must be available at:

- Sanitary appliances for washing, i.e. bath, basin, bidet and shower.
- Any sink used for the preparation of food.

Systems for generating hot water must be provided with controls to regulate water temperature in normal use. In the event of controls malfunctioning, temperature and pressure safety devices are required. Installations must be able to withstand the effects of higher than normal operating temperature and pressure.

- Vented system – operates at atmospheric pressure with an open vent and hot water expansion pipe above the feed cistern. This pipe contains the excess of hot water on heating. A water temperature control thermostat is fitted to the hot water storage cylinder (set at 60–65°C). The boiler is fitted with a thermostatic control (manually set at about 80°C) to prevent the water from boiling. The boiler pipework has a pressure-relief valve (see page 85 and accompanying note).

- Unvented systems – these are sealed systems that have gained in popularity since the 1980s. Hot water expansion is accommodated by a cushion of air in a spherical vessel (see page 160). In addition to a water temperature control thermostat, the hot water storage cylinder must have a temperature- and pressure-relief valve. These may combine as one valve that satisfies both functions and it should discharge through a tundish and safely into the atmosphere. Further safety features are the boiler control thermostat and a non self-resetting over-temperature energy cut-out to disconnect the supply of heat to the storage vessel.

Note 1: Hot water cylinders specifically for unvented systems are available factory supplied with all relevant safety accessories, whereas traditional vented systems rely on the competence of the installer for the correct fitting of safety devices.

Note 2: Hot water is produced and stored at relatively high temperatures to destroy bacteria. To avoid scalding at draw off taps, the water temperature must be regulated by thermostatic blending with cold water. More detail is shown on pages 440 to 445.

Direct System of Hot Water Supply

The hot water from the boiler mixes directly with the water in the cylinder. If used in a 'soft' water area the boiler must be rust-proofed. This system is not suited to 'hard' waters, typical of those extracted from boreholes into chalk or limestone strata. When heated the calcium precipitates to line the boiler and primary pipework, eventually 'furring up' the system to render it ineffective and dangerous. The storage cylinder and associated pipework should be well insulated to reduce energy losses. If a towel rail is fitted, this may be supplied from the primary flow and return pipes.

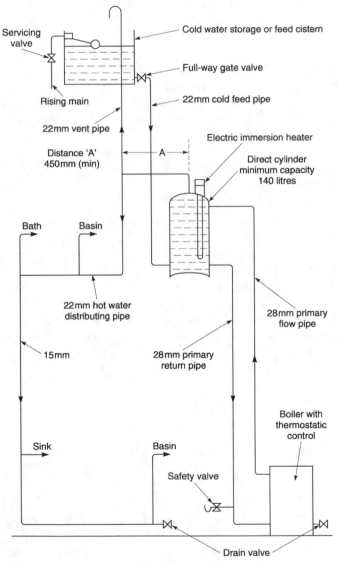

Servicing valve

Cold water storage or feed cistern

Full-way gate valve

Rising main

22mm cold feed pipe

22mm vent pipe

Electric immersion heater

Distance 'A'
450mm (min)

A

Direct cylinder
minimum capacity
140 litres

Bath

Basin

22mm hot water
distributing pipe

28mm primary
flow pipe

15mm

28mm primary
return pipe

Boiler with
thermostatic
control

Sink

Basin

Safety valve

Drain valve

Note: All pipe sizes shown are for copper outside diameter.

This system is used in 'hard' water areas to prevent scaling or 'furring' of the boiler and primary pipework. Unlike the direct system, water in the boiler and primary circuit is not drawn off through the taps. The same water circulates continuously throughout the boiler, primary circuit and heat exchange coil inside the storage cylinder. Fresh water cannot gain access to the higher temperature areas where precipitation of calcium would occur. The system is also used in combination with central heating, with flow and return pipes to radiators connected to the boiler. Boiler water temperature may be set by thermostat at about 80°C.

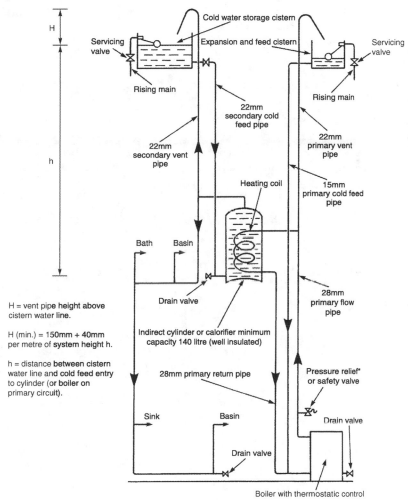

H = vent pipe height above cistern water line.

H (min.) = 150mm + 40mm per metre of system height h.

h = distance between cistern water line and cold feed entry to cylinder (or boiler on primary circuit).

*In the unlikely occurrence of the primary flow and vent becoming obstructed, water expansion could also be accommodated up the cold feed pipe.

Unvented Hot Water Storage System

The Building Regulations, Approved Document G3, permit the installation of packaged unit unvented hot water storage systems which have been accredited by the British Board of Agrément (BBA) or other European Organisation for Technical Approvals (EOTA) member bodies. Components should satisfy BS EN 12897: Water supply. Specification for indirectly heated unvented (closed) storage water heaters. A system of individual approved components is also acceptable. Safety features must include:

1. Stored hot water temperature control between 60 and 65°C.
2. Non-self-resetting over-temperature cut-out to close off the fuel supply if the boiler thermostat fails.
3. Expansion and temperature-relief valves to operate at 95°C.
4. Check valves on water main connections.

Note: A supplementary 95°C boiler limit thermostatic control may be fitted by the manufacturer.

The system is less space consuming than conventional systems and saves installation costs as there are no cold water storage and expansion cisterns. In addition to satisfying the Building Regulations, the local water authority should be consulted for approval and to ensure that there is adequate mains pressure.

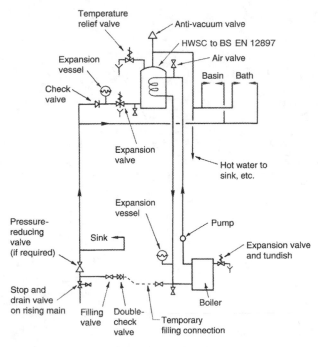

Unvented system with hot water storage capacity in excess of 15 litres, with a sealed primary circuit

Installation – by suitably qualified person in possession of a registered operative identity card/certificate, issued by a recognised assessment body such as the Chartered Institute of Plumbing and Heating Engineering or the Construction Industry Training Board.

Notice of installation – given to the local authority Building Control Department. Building Regulation G3 – Hot Water Supply and Systems, requires a competent installer, precautions to prevent water temperature from exceeding 100°C and any hot water discharge from safety devices to be conveyed safely and visibly.

Water supply – direct feed from water main, therefore no atmospheric vent pipe and no cold water storage cistern.

Water expansion – accommodated by suitably sized expansion vessel. Some units operate with an internal air gap (see next page).

Systems – direct heated by immersion heater, or indirect from a central heating boiler.

Storage cylinder materials – stainless steel, glass/vitreous enamel-coated steel or heavy-gauge copper.

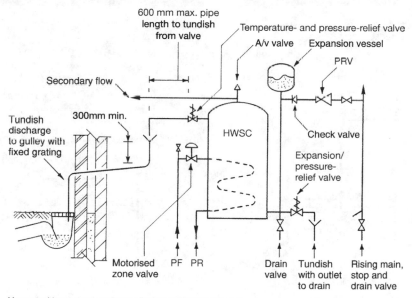

Unvented hot water storage cylinder and tundish detail

Controls –

- Temperature and pressure-relief valve.
- Expansion/pressure-relief valve.
- Cylinder temperature regulating thermostat manually set to operate the zone valve at 60–65°C.
- Over-temperature cut-out thermostat, pre-set to operate the zone valve at 85°C.

UHWSS – Internal Air Gap

For all hot water systems, especially those exceeding 15 litres storage capacity, a purpose-made hot water storage cylinder designed with provision for an `air gap' or `bubble top' is an effective alternative to installing a separate expansion vessel.

Typical installation –

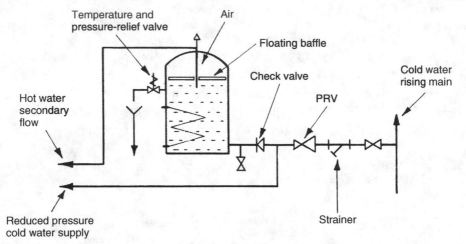

Unvented HWSC incorporating an air gap

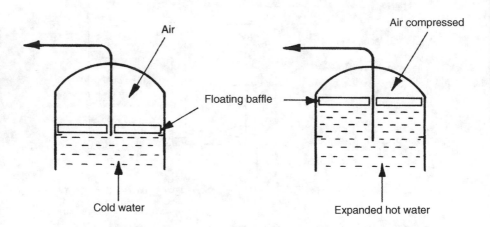

Function of the internal air gap

As the water expands on heating, the volume of trapped air is compressed to provide adequate delivery pressure and flow. After some time, the air may become depleted due to turbulence by water movement through the hot water storage cylinder. This will be noticed by the pressure-relief valve discharging. The `air gap' is recharged by draining the system and refilling. Some manufacturers fit a floating baffle between the water and the air, to reduce the effect of turbulence.

Expansion devices in hot water systems are designed as a safe means for discharging water when system operating parameters are exceeded, i.e. in conditions of excess pressure and/or temperature.

Expansion valve – Care should be taken when selecting expansion or pressure-relief valves. They should be capable of withstanding 1·5 times the maximum pressure to which they are subjected, with due regard for water mains pressure increasing overnight as demand decreases.

Temperature-relief valve – These should be fitted to all unvented hot water storage vessels exceeding 15 litres capacity. They are normally manufactured as a combined temperature and pressure-relief valve. In addition to the facility for excess pressure to unseat the valve, a temperature-sensing element is immersed in the water to respond at a pre-set temperature of 95°C.

Discharge from these devices should be safely controlled and visible, preferably over a tundish, as shown on page 160.

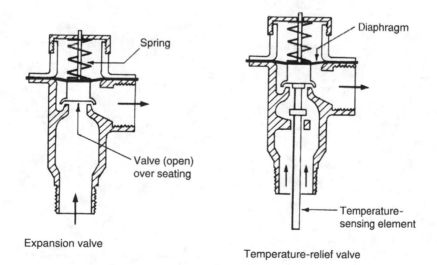

Expansion valve

Temperature-relief valve

Ref. BS 6283-2: Safety and control devices for use in hot water systems. Specifications for temperature-relief valves for pressures from 1 bar to 10 bar.

Pressure-Reducing Valves

Pressure-reducing valves are otherwise known as pressure regulators. PRVs can be applied to many different piped services including gas, compressed air, water and steam. These applications may range from relatively simple installations such as mains water-supplied domestic unvented hot water storage systems, to larger scale industrial steam and district heating schemes.

High pressure is needed to overcome the resistances of long lengths of pipe distribution, changes in direction, valves, etc. For local distribution, the pressure must be reduced to:

- Prevent undue wear and damage to the lighter gauge fittings and fixtures at the end use.
- Provide a maximum safe working pressure to prevent injury to end users.
- Regulate supplies at a constant value or desirable secondary pressure, irrespective of inlet pressure variations and changes in demand.

Function and installation

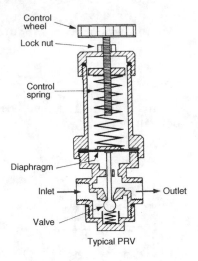

Typical PRV

- Outlet reduced pressure acts on the underside of the diaphragm.

- Control spring opposes the reduced pressure.

- Reduced pressure and control spring setting affect the position of the valve and flow condition.

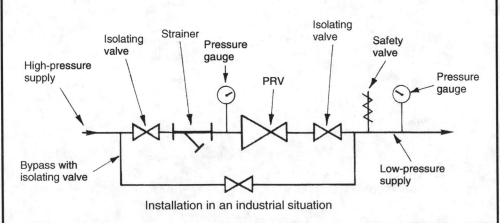

Installation in an industrial situation

A strainer is used to filter out and trap fluid-suspended debris, pipe scale and carbonate deposits from hard water. This facility is essential to prevent component wear by erosion and abrasion, and interference with the efficient operation of pipe system controls. Strainers are a standard installation on processing plant and other industrial applications. There has been little need for strainers in domestic systems, until the use of items such as thermostatic mixing valves, shower mixers, check valves and pressure-reducing valves have become standard. To protect the sensitivity of these units, most manufacturers integrate a means of filtering within the casting. Otherwise, an independent pipeline strainer of the type shown can be installed upstream of the unit.

Typical pipeline strainers

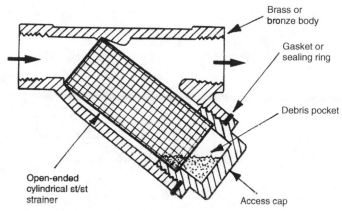

Threaded for domestic and light industrial services

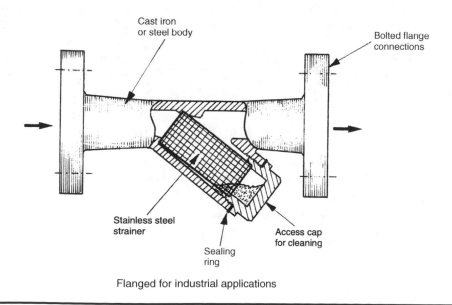

Flanged for industrial applications

Hot Water Storage Cylinders

BS 1566-1: Copper indirect cylinders for domestic purposes. Open-vented copper cylinders. Requirements and test methods.

BS 1566-2: Copper indirect cylinders for domestic purposes. Specification for single-feed indirect cylinders.

BS 417-2: Specification for galvanised low carbon steel cisterns, cistern lids, tanks and cylinders.

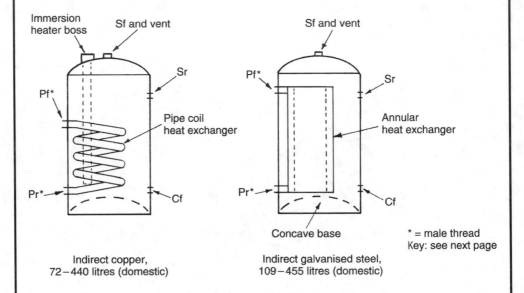

Indirect copper,
72–440 litres (domestic)

Indirect galvanised steel,
109–455 litres (domestic)

* = male thread
Key: see next page

Direct cylinders have no coil or annular heat exchangers. They can be identified with female pipe threads for the primary flow and return connections. For domestic use: copper – 74 to 450 litres capacity, galvanised steel – 73 to 441 litres capacity. Direct and indirect cylinders for industrial and commercial applications are manufactured in copper and galvanised steel in capacities up to 4500 litres.

Notes:

(1) Copper and galvanised (zinc-plated) steel pipes and components should not be used in the same installation. In addition to electrolytic action between the dissimilar metals, pitting corrosion caused by tiny particles of dissolved copper settling on the galvanising will produce local cells which dissolve the zinc and expose the steel to rusting.

(2) Copper and galvanised steel cylinders normally incorporate an aluminium and a magnesium sacrificial anode, respectively. These are designed to deteriorate over sufficient time to allow a protective coating of limescale to build up on the exposed surfaces.

BS 1566-2: Specification for single feed indirect cylinders.

An indirect hot water system may be installed using a 'primatic' or single-feed indirect cylinder. Conventional expansion and feed cistern, primary cold feed and primary vent pipes are not required, therefore by comparison, installation costs are much reduced. Only one feed cistern is required to supply water to be heated indirectly, by water circulating in an integral primary heater. Feed water to the primary circuit and boiler is obtained from within the cylinder, through the primary heater. The heat exchanger inside the cylinder has three airlocks which prevent mixing of the primary and secondary waters. No corrosion inhibitors or system additives should be used where these cylinders are installed.

Key:
Sf = Secondary flow pipe
Pf = Primary flow pipe
Pr = Primary return pipe
He = Heat exchanger
Cf = Cold feed pipe

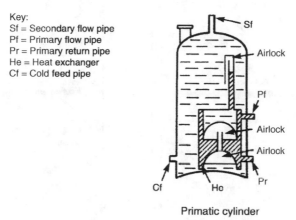

Primatic cylinder

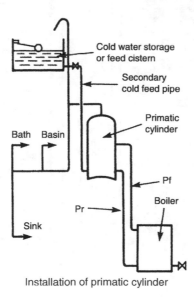

Installation of primatic cylinder

Indirect Hot Water System for a Three-Storey Building

For larger buildings a secondary circuit will be required to reduce 'dead-legs' and to maintain an effective supply of hot water at all outlets. Convection or thermo-siphonage may provide circulation, but for a more efficient service a circulatory pump will be necessary. In buildings which are occupied for only part of the day (e.g. schools, offices, etc.), a time control or programmer can be used to regulate use of the pump. In addition, one of the valves near the pump should be motorised and automatically shut off with the pump and boiler when hot water is not required. All secondary circuits should be well insulated to reduce heat loss through the pipework. A heating installation can operate in conjunction with this system, but may require duplication of boilers or separate boilers for each function.

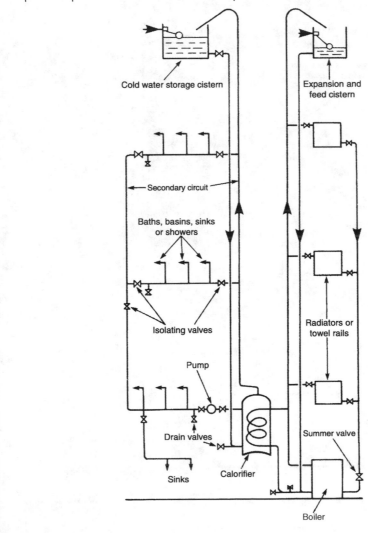

Hot water provision in moderately large buildings such as spacious houses, small hotels, hostels and other situations where demand is periodically high, can be from a large storage cylinder or cylinders installed in duplicate. Alternatively or additionally, depending on requirements, a supplementary storage vessel may be strategically located at high level. This vessel is relatively small, containing no more than 20% of the total design capacity.

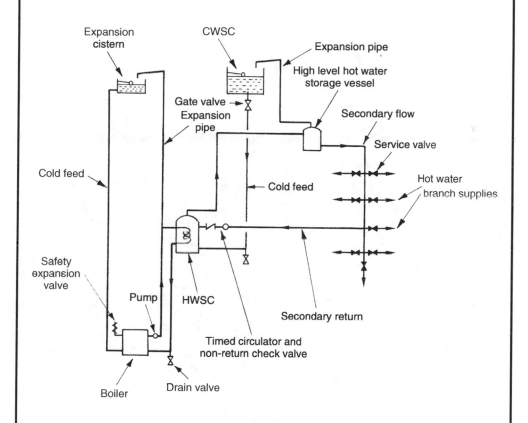

Advantages over a single storage facility:

• Smaller secondary flow and return distribution pipes.
• Less concentrated dead load on the structure.

Sealed Indirect Hot Water System for a High-Rise Building

For convenience and to reduce wear on fittings, the maximum head of water above taps and other outlets is 30m. This is achieved by using intermediate or break-pressure cisterns for each sub-circuit. Head tanks are provided to ensure sufficient volume of stored hot water and adequate delivery to the upper floors. Compared with conventional installations a considerable amount of pipework and fitting time can be saved by using an expansion vessel to absorb expansion of water in the primary circuit. However, the boiler and calorifiers must be specified to a high-quality standard to withstand the water pressure. All pipework and equipment must be well insulated.

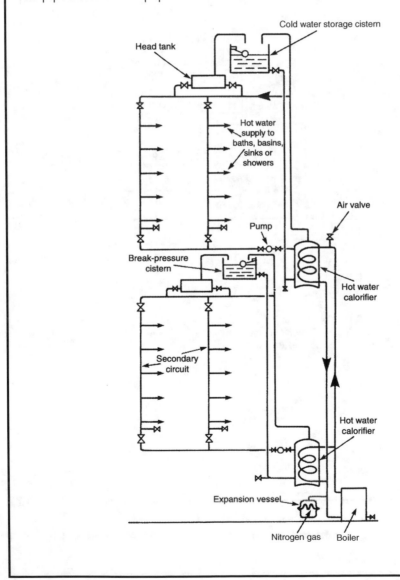

Principle – cold water passes through a heat exchanger (pipe coil) heated by hot water surrounding the coil.

Types:

Natural convection –

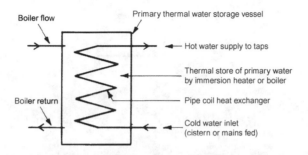

Pumped circulation (internal) –

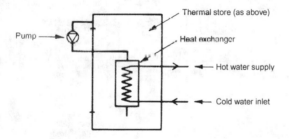

Pumped circulation (external) (see also page 246) –

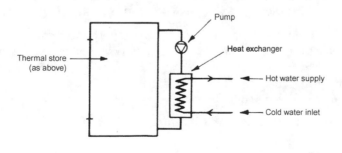

Primary Thermal Store Water Heaters – 2

Principle for providing hot water only –

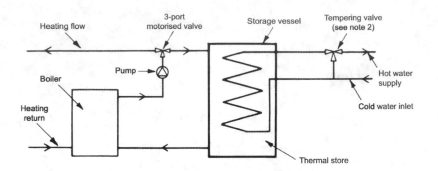

Principle for hot water supply and central heating functions –

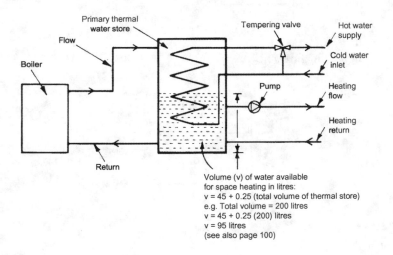

Volume (v) of water available
for space heating in litres:
v = 45 + 0.25 (total volume of thermal store)
e.g. Total volume = 200 litres
v = 45 + 0.25 (200) litres
v = 95 litres
(see also page 100)

Note 1: Examples of boiler circuit filling, expansion and safety facilities omitted, but are shown elsewhere in Part 2 and Part 3.

Note 2: Some thermal store systems, notably those connected to solar heat collectors and to solid fuel boilers may produce stored hot water above 80°C. Therefore, as control of the heat source may not be that reliable, the hot water outlet should be fitted with a tempering valve (see page 444) to maintain the hot water supply at less than 60°C.

Ref. Building Regulations Part G, Approved Document G3.

Thermal store – sometimes referred to as an inertia or a buffer vessel. A hot water storage facility located between primary energy/heat source(s) and secondary heat load, i.e. hot water supply to taps and other fittings and possibly for space heating circuit.

Energy source – conventional fossil-fuelled boiler as shown on the preceding two pages. These installations are more energy efficient where supplemented with a low or zero carbon alternative energy source, such as a heat pump or solar panels. Alternative energy can be the sole source if it is sufficiently effective.

Installation principle combining a conventional fossil-fuelled boiler with a solar energy source –

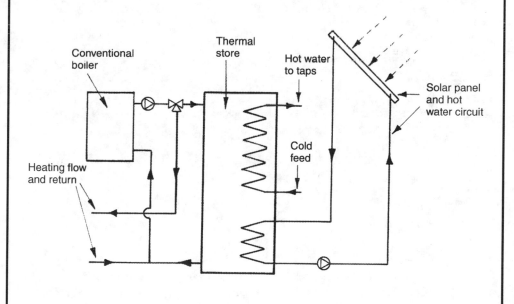

See notes 1 and 2 on previous page.

Primary Thermal Storage for Hot Water and Heating

Schematic principle of combining high- and low-temperature heat sources of energy for hot water thermal storage. Provides a domestic hot water supply to taps and other fittings through a heat exchanger within the thermal store. A separate circuit connected directly to the thermal store is for space heating –

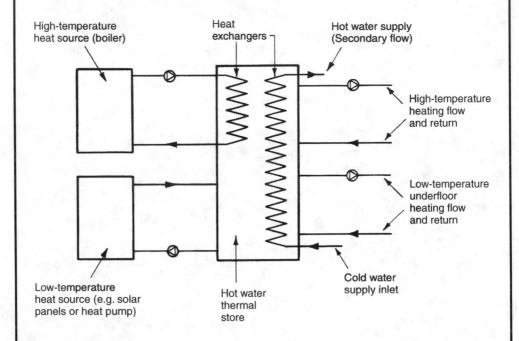

Typical buffer storage capacity of hot water for heating (litres) –

Dwelling heat loss (kW)	5K ΔT	10K ΔT	15K ΔT	20K ΔT
5	143	72	48	36
10	287	143	96	72
15	430	215	143	107
20	573	287	191	143
25	717	358	239	179

Note: ΔT in this context refers to the temperature difference between heating flow and return. With pumped circuit temperatures typically 80°C and 70°C respectively, ΔT = 10K.

Cast iron sectional – made up of a series of hollow sections, joined together with left- and right-hand threaded nipples to provide the heat capacity required. When installed, the hollow sections contain water which is heated by energy transfer through the cast iron from the combusted fuel. Applications: domestic to large industrial boilers.

Steel shell, fire or flame tube – hot combusted fuel and gases discharge through multiple steel tubes to the extract flue. Heat energy from the burnt fuel transfers through the tube walls into cylindrical waterways. Tubes may be of annular construction with water surrounding a fire tube core. Uses: commercial and industrial buildings.

Copper or steel water tube – these reverse the principle of fire tubes. Water circulates in a series of finned tubes while the combusted fuel effects an external heat transfer. These are typical of the heat exchangers in domestic boilers.

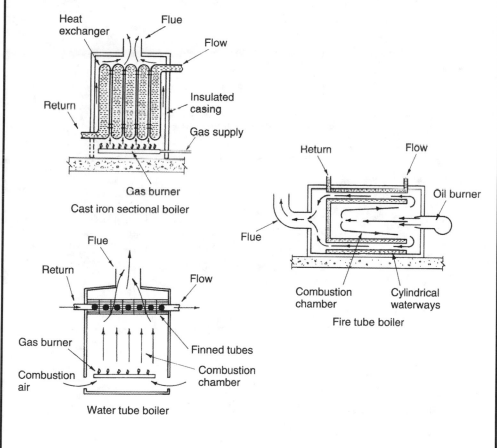

Cast iron sectional boiler

Fire tube boiler

Water tube boiler

All these boiler types may be fired by solid fuel, gas or oil.

Condensing Gas Boilers – Principles

Condensing boilers have a greater area of heat transfer surface than conventional boilers. In addition to direct transfer of heat energy from the burning fuel, heat from the flue gases is used as secondary heating to the water jacket. Instead of the high-temperature (200–250°C) flue gases and water vapour discharging to the atmosphere, they are recirculated around the water jacket by a fan. This fan must be fitted with a sensor to prevent the boiler from firing in the event of failure. Condensation of vapour in the flue gases is drained to a suitable outlet. The overall efficiency is about 90%, which compares well with the 75% expected of conventional boilers. However, purchase costs are higher, but fuel savings should justify this within a few years.

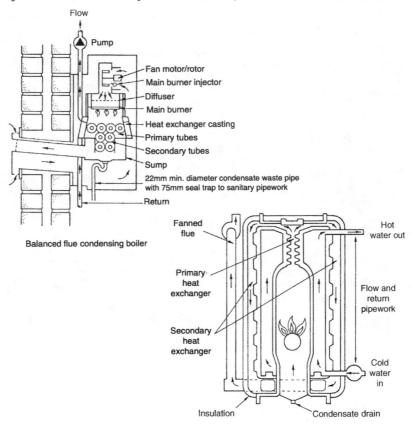

Balanced flue condensing boiler

Conventional flue condensing boiler

Refs. BS 6798: Specification for selection, installation, inspection, commissioning, servicing and maintenance of gas-fired boilers of rated input not exceeding 70kW net.

Building Regulations. Approved Document H1: Foul Water Drainage, Section 1 – Sanitary pipework.

The concept of operating principles for a condensing boiler is illustrated in the schematic diagram below. As indicated, there are two heat exchangers. These combine to optimise nearly all the energy from combustion of fuel by transferring it into the primary and heating water circulation.

Heat exchangers –

* Hot flue gases to water pre-heat exchanger.
* Combusted fuel to water heat exchanger.

Schematic operation of a condensing boiler

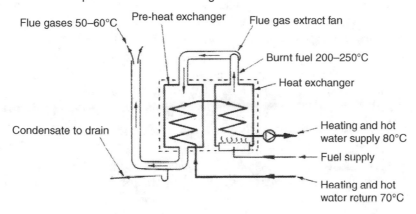

Approximate hot water return temperature to boiler efficiency

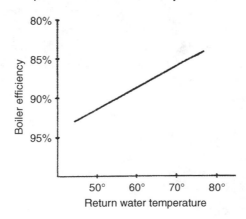

Condensing boilers are manufactured in standard and combination (combi) format as floor standing or wall hung.

Condensing Gas Boilers – Characteristics (1)

Otherwise known as high-efficiency boilers.

Originally developed in the 1930s. Lack of technological advances and less concern about the effect of consuming fuel limited interest until the fuel crises of the 1970s.

Introduced to the domestic market in the early 1980s. Slow to establish due to relatively higher purchase cost. From 2005, it was virtually compulsory for new installations, to satisfy energy efficiency bands A and B. From 2010, band A only has been acceptable.

Extracts heat from flue gases to gain from the secondary heating effect.

Heat exchanger must be corrosion resistant, i.e. stainless steel or aluminium to resist the acidity of condensate. Cast iron and copper are only suitable in non-condensing boilers with high flue gas temperatures which are unaffected by condensation.

Non-corrosive plastic condensate waste pipe required. Waste usually connected to a siphon which discharges condensate in one go from a 150ml sump. This reduces the possibility of a drip discharge freezing.

Least efficient condensing boiler has about the same efficiency as the most efficient non-condensing boiler.

Condensing boilers are most efficient with low return water temperatures as this effects most condensation (see graph on previous page). Therefore, they are best used with modulating controls as described on page 191.

About 80% energy exchange occurs as combusted gas at temperatures above 200°C effects the primary heat exchange. The secondary heat exchange adds about another 5% as the fanned flue gases reduce to about 55°C as they pre-warm the returning system cool water. With this temperature reduction the flue gases condense and dew point occurs (steam turns to water), adding about another 5% in latent energy transfer.

The gas burner has to impart less energy to raise the temperature at the primary heat exchange, hence fuel savings and less CO_2 and NO_x emissions from the flue.

Controls –

Non-condensing boilers are efficiently controlled with thermostatic valves, thermostats and an interlock facility. The boiler is switched on and off relative to internal air temperature. High-temperature water is delivered to emitters.

Condensing boilers are at their most efficient when enabled to run for sustained periods with a moderate flow water temperature and low return water temperature. They are ideally suited to modulating, weather compensated control systems.

Flue discharge has a distinct plume or cloud of moisture droplets. May be a problem with neighbouring properties.

Flue slopes back slightly towards the boiler to discharge any condensation from the flue duct into the condensate drain.

Typical energy efficiency factors:

Modern condensing boiler 90%
Modern non-condensing boiler 75% } see page 132
Older boiler 58%

A non-condensing boiler loses at least 20% of heat energy produced into the flue. Therefore these boilers are 80% efficient at best. Approximately half the heat energy that would be otherwise lost in the flue is recovered by a condensing boiler. Therefore these boilers are approximately 90% efficient.

Approximate number of households in UK with a gas boiler = 14 million.

Typical annual household production of CO_2 with a non-condensing boiler = 5 tonnes.

Total potential CO_2 emissions = 70 million tonnes.

Typical annual household production of CO_2 with a condensing boiler = 3 tonnes.

Total potential CO_2 emissions = 42 million tonnes.

Therefore, in addition to fuel savings, condensing boilers represent a potential for an annual reduction in polluting or greenhouse gases of 28 million tonnes.

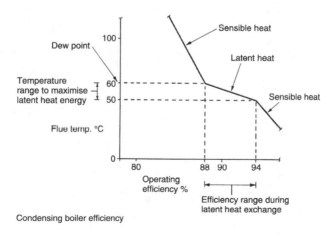

Condensing boiler efficiency

Note: Oil-fired condensing boilers are also marketed with specifications to satisfy current energy use requirements.

105

Combination Boiler

This system saves considerably in installation time and space, as there is no need for cisterns in the roof space, no hot water storage cylinder and associated pipework. The 'combi' gas boiler functions as an instantaneous water heater only heating water as required, thereby effecting fuel savings by not maintaining water at a controlled temperature in a cylinder. Water supply is from the mains, providing a balanced pressure at both hot and cold water outlets. This is ideal for shower installations. Boiler location may be in the airing cupboard, leaving more space in the kitchen. The system is sealed and has an expansion vessel which is normally included in the manufacturer's pre-plumbed, pre-wired package for simple installation. Further control details are shown on page 183.

Note: Where mains supply water has a hardness characteristic greater than 200 ppm (see pages 44–51), there should be provision for its treatment to control limescale deposit.

Ref. Building Regulations Part L, Domestic Building Services Compliance Guide.

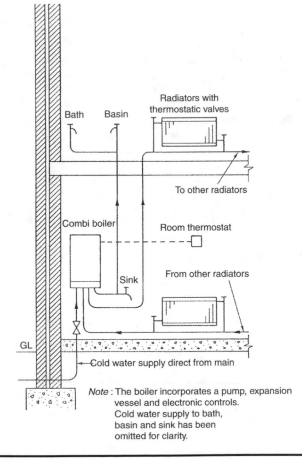

Radiators with thermostatic valves

Bath   Basin

To other radiators

Combi boiler   Room thermostat

Sink   From other radiators

GL

Cold water supply direct from main

Note : The boiler incorporates a pump, expansion vessel and electronic controls. Cold water supply to bath, basin and sink has been omitted for clarity.

To prevent user inconvenience waiting for the cold water 'dead-leg' to run off and to prevent water wastage, long lengths of hot water distribution pipework must be avoided. Where cylinder to tap distances are excessive, a pumped secondary flow and return circuit may be installed with minimal 'dead-legs' branching to each tap. The pipework must be fully insulated and the circulation pump timed to run throughout the working day, e.g. an office system could be programmed with the boiler controls, typically 8.00 a.m. to 6.00 p.m., five days a week. A non-return valve prevents reverse circulation when the pump is not in use.

Nominal inside pipe dia. (mm)	Equivalent copper tube outside dia. (mm)	Max. length of secondary flow without a return (m)
10	12	20
> 10 to 19	> 12 to 22	12
> 19 to 25	> 22 to 28	8
> 25	> 28	3

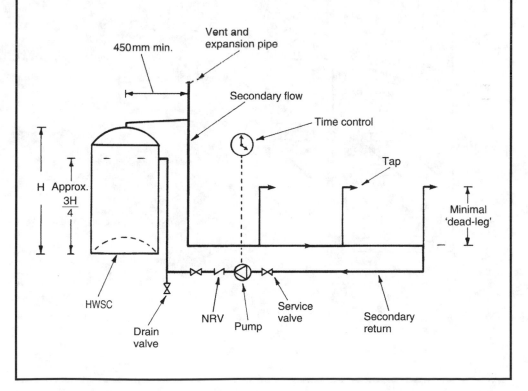

Duplication of Plant

Dual installations or duplication of plant and equipment is required in buildings where operating efficiency is of paramount concern. With this provision, the supply of hot water in hotels, commercial buildings, offices, etc. is ensured at all times, as it is most unlikely that all items of plant will malfunction simultaneously. It may also be necessary to divide the design capacity of plant to reduce the concentration of structural loads. Each boiler and calorifier may be isolated for repair or renewal without disturbing the function of the others. Therefore when designing the system it is usual to oversize plant by up to one-third, to ensure the remaining plant has reasonable capacity to cope with demand. There is also the facility to economise by purposely isolating one boiler and calorifier during periods when a building is only part occupied.

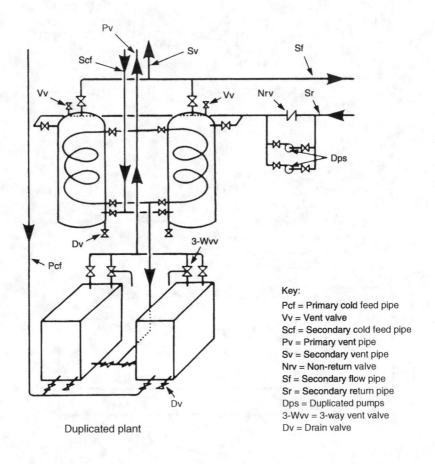

Duplicated plant

Key:

Pcf = Primary cold feed pipe
Vv = Vent valve
Scf = Secondary cold feed pipe
Pv = Primary vent pipe
Sv = Secondary vent pipe
Nrv = Non-return valve
Sf = Secondary flow pipe
Sr = Secondary return pipe
Dps = Duplicated pumps
3-Wvv = 3-way vent valve
Dv = Drain valve

An electric immersion heater may be used within a conventional hot water storage cylinder. Alternatively, individual or self-contained open outlet heaters may be located over basins, baths or sinks. Combined cistern-type heaters can be used to supply hot water to several sanitary appliances. Energy conservation is achieved with an integral thermostat set between 60 and 65°C. This temperature is also sufficient to kill any bacteria. The immersion heater must have a circuit protective conductor to earth and the cable supplying the heating element must be adequate for the power load. A cable specification of 2·5mm² is normally adequate with a 20 amp double pole control switch supplied direct from the consumer unit or fuse box. Overload protection at the consumer unit is a 16 amp fuse or circuit breaker for a 3 kW element and 20 amp for a 4 kW element.

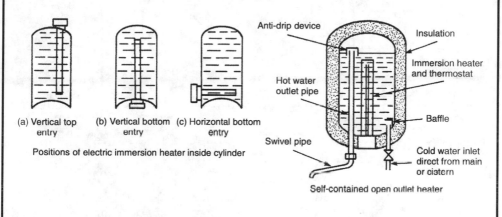

(a) Vertical top entry   (b) Vertical bottom entry   (c) Horizontal bottom entry

Positions of electric immersion heater inside cylinder

Self-contained open outlet heater

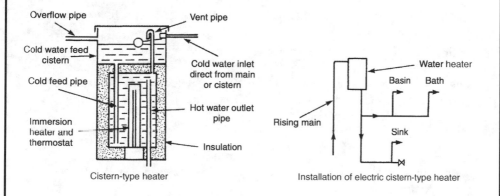

Cistern-type heater

Installation of electric cistern-type heater

Ref. BS 3198: Specification for copper hot water storage combination units for domestic purposes.

The cistern-type heater should be located with the water level at least 1·5m above the water draw-off taps. If there is insufficient space to accommodate this combination unit, a smaller pressure-type water heater may be fitted. These are small enough to locate under the sink or elsewhere in the kitchen. They have two immersion heaters. The upper element of 500 watts rating is for general use supplying hot water to the basin, sink and other small appliances. The lower element of 2500 watts may be on a timed control to provide sufficient hot water for baths. The pressure heater is supplied with cold water from a high-level cistern.

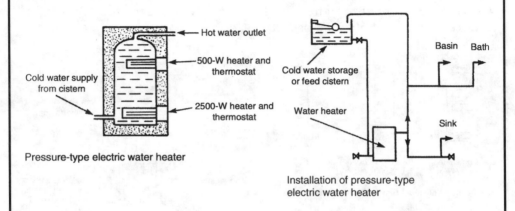

Pressure-type electric water heater

Installation of pressure-type electric water heater

Immersion heaters – safety cut-out. Since 2004, immersion heater manufacturers are required to incorporate an additional integral safety device, independent of the main thermostat. This brings immersion heaters for vented water heating into line with the requirements for unvented water heaters.

Function – if the main thermostat fails, water will boil, with considerable damage potential to personnel, the installation and premises. The manufacturer's pre-set safety cut-out is designed to prevent water in a hot water storage vessel from exceeding 98°C. It must not reset automatically.

Methods – either:

* A 'one-shot' thermal cut-out or thermostat. This is principally a fusible link which melts or ruptures at a predetermined temperature, or
* A manually resettable cut-out or thermostat which responds to critical temperature change to break electrical contact.

Ref. BS EN 60335-2-73: Household and similar electrical appliances. Safety. Particular requirements for fixed immersion heaters.

Instantaneous water heaters are relatively compact non-storage units suitable for use with individual sinks, basins and showers. For user safety they are fitted with a pressure switch to disconnect the electricity if the water supply is interrupted and a thermal cut-out to prevent the water from overheating. Mains pressure to these units should be maintained below 400kPa (4 bar). In some high-pressure supply areas this will require a pressure-reducing valve to be installed on the service pipe. Some expansion of hot water will occur while the unit is in use. This can be contained if there are at least 3 metres of pipework before the unit and the closest cold water draw-off. If this is impractical, an expansion vessel may be used. For more details of electric shower installations see pages 431 and 432.

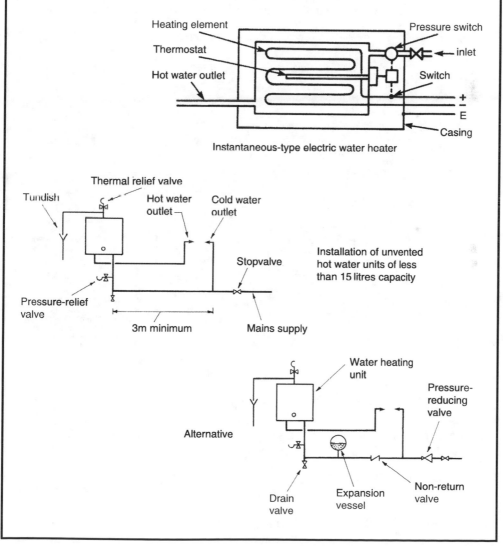

Instantaneous-type electric water heater

Installation of unvented hot water units of less than 15 litres capacity

Industrial, commercial and domestic demand for electricity is considerably reduced overnight. Therefore during this time, the electricity supply companies can market their spare capacity as off-peak electricity by selling it at a reduced rate – approximately half the cost of the standard daytime tariff. Supplies are adapted to operate through a programmer or time control which diverts the electricity to a special off-peak or white meter, usually from midnight to 7 a.m. In order to maximise the benefit, slightly larger than standard capacity hot water storage cylinders of 162 or 190 litres are recommended. To conserve energy, these cylinders must be thoroughly insulated and the immersion heaters fitted with integral thermostatic control. If supplementary hot water is required during the day, this can be provided by a secondary immersion heater at the standard supply tariff.

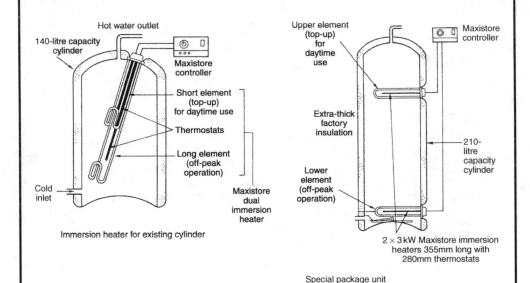

Immersion heater for existing cylinder

Special package unit

Economy 10 – some suppliers provide this as a discounted tariff variation to Economy 7. This operates for three hours in the afternoon, two hours in the evening and five hours overnight. See also page 584.

The secondary immersion heater or boost heater is close to the top of the cylinder to ensure that only a limited quantity of water is heated at standard tariff. To maximise economy, the off-peak thermostat is set at 65°C and the boost thermostat at 60°C.

When the hot water outlet is opened, cold water flows through a venturi fitting. The venturi contains a diaphragm which responds to the flow differential pressure and this opens the gas valve. A pilot flame ignites gas flowing through the burner which heats the water as it passes through the heat exchanger. Installation can be direct from the water main or from a cold water storage cistern. A multi-point system has the hot water outlet suppling several appliances.

A gas circulator can be used to heat water in a storage cylinder. They are usually fitted with an economy or three-way valve. This gives optional use of water circulation through a high or low return pipe for variable hot water storage volume. Domestic installations may be in the kitchen, with vertical flow and return pipes to a storage cylinder in the airing cupboard.

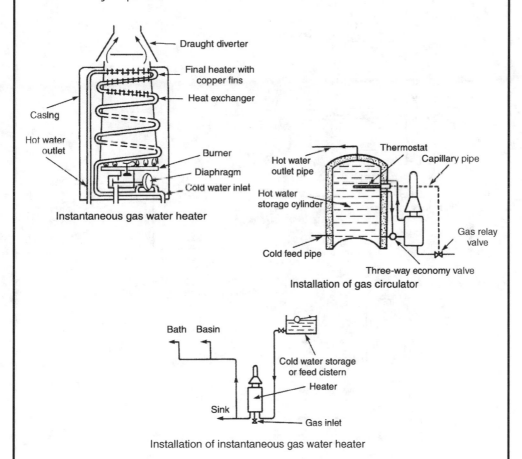

Instantaneous gas water heater

Installation of gas circulator

Installation of instantaneous gas water heater

Ref. BS EN 26: Gas-fired instantaneous water heaters for the production of domestic hot water.

Gas Water Heaters – 2

The storage type of gas water heater is a self-contained unit and is therefore simpler and quicker to install than a gas circulator. Capacities range from 75 to 285 litres. The smaller units are single-point heaters for supplying hot water to an individual sink or basin. Larger, higher rated storage heaters can be used to supply hot water to a bath, basin, sink and shower. These are called multi-point heaters. They may also be installed in flats up to three storeys, with cold water supplied from one cistern. A vent pipe on the cold feed will prevent siphonage. To prevent hot water from the heaters on the upper floors from flowing down to the heater on the ground floor, the branch connection on the cold feed pipe must be above the heaters.

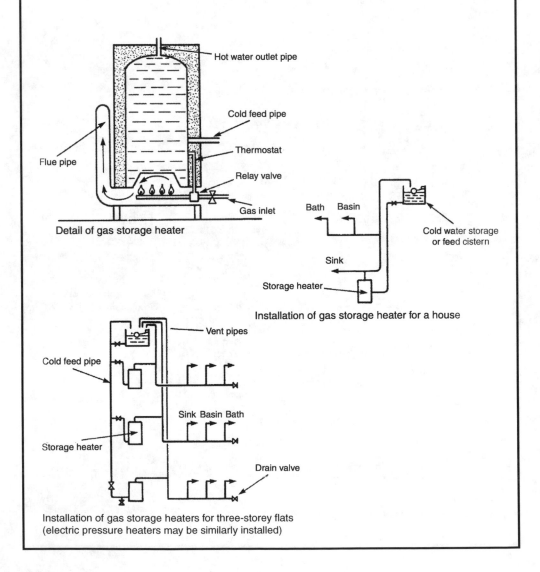

Detail of gas storage heater

Installation of gas storage heater for a house

Installation of gas storage heaters for three-storey flats
(electric pressure heaters may be similarly installed)

Condensing water heater – a variation on the multi-point-type heater. The condensing heater is a hot water storage vessel, capable of very rapid heat recovery.

Application – typical examples include small hotels, schools, residential homes, student halls of residence, camp sites and sports centres.

Function – a fanned gas burner discharges into a stainless steel combustion chamber within a cylindrical water storage vessel. From the combustion chamber the burnt gases descend into a stainless steel spiral to exit at low level through a flue. Condensate from the flue is trapped and discharged to a drain.

Controls –

Automatic electric ignition in response to a water temperature thermostat.

Limit thermostat.

Overheat safety thermostat and warning light.

Fan failure device and warning light.

Manual on/off switch.

Water supply – either:

- Cistern, gravity feed pipe and atmospheric vent and expansion pipe, or
- Direct connection to an unvented mains supply. Unvented supplies require backflow prevention (check valve), an expansion vessel and an expansion valve. A pressure- and temperature-relief valve must also be fitted to the hot water outlet to discharge safely into a tundish.

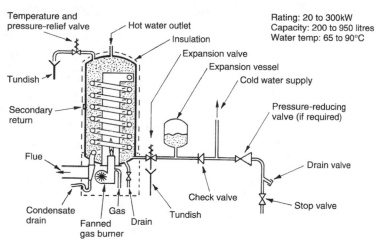

Mains-connected condensing water heater

Continuous Flow Gas Water Heater

Gas and electric instantaneous water heaters are well established as a means for providing *point of use or multi-point* installations. Continuous flow water heaters have developed from these by combining established technology with effective and efficient heat energy transfer from combustion of gas. Close control systems modulate the required hot water flow rate with volume of gas consumption.

Condensing hot water heater operating principles –

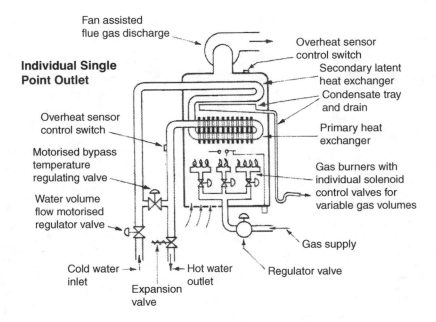

The heat exchange process functions similarly to a condensing boiler. Primary heat transfer is from fuel combustion and secondary heat transfer from the energy in the flue gases. In this application it provides an economical fast response direct supply of drinkable hot water.

Combination of units arranged as a bank or manifold of water heaters serving numerous hot water draw points –

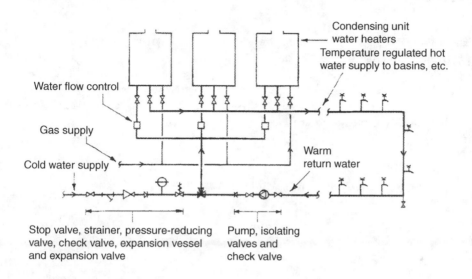

Supplementary solar energy sourced pre-heat thermal store –

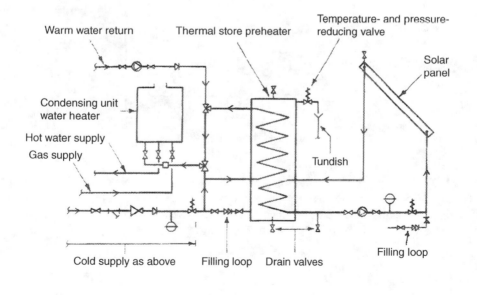

Solar Energy Systems – 1

Solar energy systems reduce the environmental impact of buildings by providing a viable energy alternative to reliance on diminishing fossil fuel reserves. They also provide a fuel cost saving for the building user.

Energy from the sun can be used for heating swimming pools, preheating ventilation air and for space heating. The most common application in the UK is for heating water for domestic use. Many new homes feature solar systems as contributory to their energy performance criteria. However, a conventional back-up heat source from a boiler is still required to compensate when solar energy capture is limited.

Operating and installation principle –

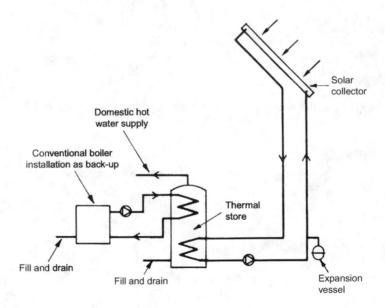

Note: Safety and other over-temperature controls and filling accessories omitted to emphasise the main components. Further details are shown on pages 86, 87, 97, 98 and 120.

Solar collectors – there are many variations, most based on the following types:

- Flat plate – a relatively thin rectangular box with a glass or transluscent cover over a series of small diameter pipes positioned above a black painted absorber plate, insulated on the underside. Installed mainly on south-facing pitched roofs (see next page).

- Batch or bread box – a black-painted storage vessel incorporating an insulated tank lined with glass. Located in any position exposed to the sun. Cold water entry is at the base and hot water is drawn from the top. In effect, the bread box functions as a collector, absorber and store of energy in hot water.

- Transpired – a south-facing exposed external masonry wall used as an active thermal store. Its performance is enhanced by overlaying the wall with a dark-coloured sheet metal plate collector, perforated to draw in outdoor heated air. This air or the heat energy in the wall can be used to directly pre-heat air conditioning/ventilation air or indirectly across the evaporator of a heat pump (see pages 332–338).

- Evacuated glass tube (detailed on page 121).

- Overheating – A correctly sized expansion vessel will be adequate for all but extreme situations. However, the effect of the sun is variable and very high water temperatures and pressures can occur in solar collectors. This could result in excessively high domestic stored hot water and possibly system fractures. Some control can be achieved with automatic blinds, but more reliable self-venting controls are required. Building Regulations Part G, Approved Document G3 specifically mentions solar systems and defines safety back-up requirements. See pages 86 and 87 for safety facilities applied to sealed systems and page 98 for thermal stores.

Approximate sizing of solar collector area (see also pages 122 and 123):
Domestic hot water – 1.0 to 2.0m^2 per person served.
Swimming pool – 0.05 to 0.10m^2 per 1.0m^2 pool surface.
Space heating – 15 to 20% of the heated floor area.

Solar Energy – Flat Plate Collector

Solar energy can contribute significantly to hot water requirements. In some countries it is the sole source of energy for hot water. In the UK its efficiency varies with the fickle nature of the weather, but fuel savings of about 40% are possible. For domestic application, the collector should be 4 to $6m^2$ in area, secured at an angle of 40° to the horizontal and facing south. The solar cylinder capacity of about 200 litres is heated to 60°C. The cylinder and associated pipework must be very well insulated and the solar part of the system should contain a blend of water and non-toxic antifreeze. The pump is switched on when the temperature of water at point X exceeds that at point Y by 2 to 3°C. The solar cylinder and the conventional cylinder may be fitted on the same level, or to save space a combined solar/conventional cylinder can be obtained from specialist suppliers.

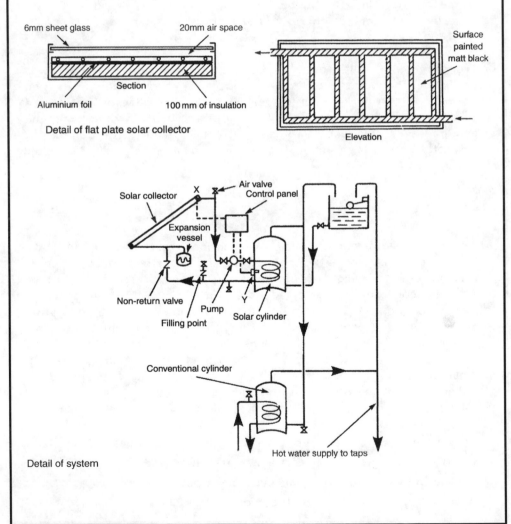

6mm sheet glass

20mm air space

Section

Aluminium foil

100 mm of insulation

Detail of flat plate solar collector

Surface painted matt black

Elevation

Solar collector

X

Air valve

Control panel

Expansion vessel

Non-return valve

Filling point

Pump

Y

Solar cylinder

Conventional cylinder

Hot water supply to taps

Detail of system

Although having the general appearance of a panel, the evacuated glass tube collector functions differently from a flat plate collector. The panel is made up of a series of refrigerant charged copper tube elements as heat exchangers or heat pipes contained concentrically within individual vacuum-sealed glass tubes. The advantage is that a refrigerant is more responsive than water, with better performance in low light conditions. The outer glass tubes provide for greater efficiency at high temperatures.

The refrigerant within the inner heat pipes evaporates in response to solar gain. This generates a convection cycle as the hot vapour gives off its heat energy into water circulating through a header pipe compartment or manifold. The cooling vapour condenses into a fluid, returning to the lower part of the heat pipe to continue the cycle.

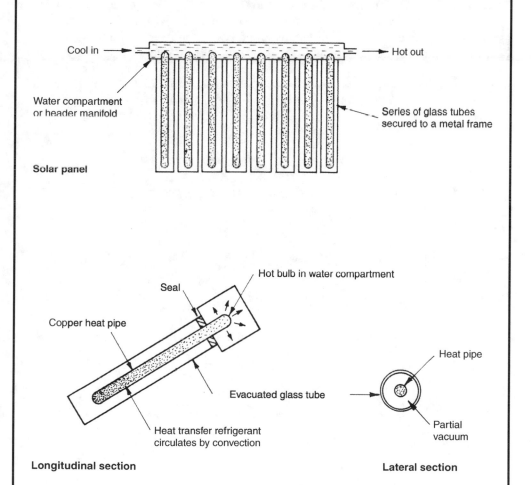

Cool in → Hot out

Water compartment or header manifold

Series of glass tubes secured to a metal frame

Solar panel

Hot bulb in water compartment

Seal

Copper heat pipe

Heat pipe

Evacuated glass tube

Partial vacuum

Heat transfer refrigerant circulates by convection

Longitudinal section

Lateral section

The area of a solar collecting panel should not exceed the potential of the system it serves. A relatively large collector area does not necessarily provide greater effect or efficiency. Oversizing can cause an excess of hot water and possibly overheating (heat stagnation). Heating system components are not designed to withstand persistent and excessively high temperatures and pressures. If they are subjected to this, they may malfunction and create safety issues for the end user, particularly with regard to risk of scalding. See note 2 relating to thermal stores and solar panels on page 98 and the reference to Building Regulation G3.

The basis for sizing a solar system is determined by the daily hot water demand. This can be calculated from data in the DBEIS publication, *Measurement of domestic hot water consumption in dwellings*. This provides guidance on volumes of water by appliances, number of occupants and patterns of use. Alternatively, figures can be used from floor area data based on the DBEIS publication, *Standard Assessment Procedure*. SAP is the methodology used to compare energy performance of dwellings. Developed by BRE, based on a Domestic Energy Model for calculating energy use – known by the initials BREDEM. Guidance is shown on pages 125 and 133.

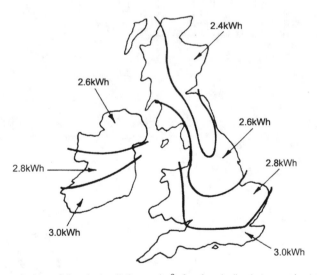

2.4kWh

2.6kWh

2.6kWh

2.8kWh

2.8kWh

3.0kWh

3.0kWh

Average daily solar irradiation on 1m^2 of surface inclined at approximately 30°

Ref. UK government department.
DBEIS – Department for Business, Energy and Industrial Strategy.

Guidance for sizing a solar panel for domestic hot water –

1. Estimate the quantity of energy required (Q) in kWh/day.

Q = Daily requirement for hot water × Shc of water × Temp. diff.

Where: C = Daily requirement or capacity in litres.

Shc = 4180J/kgK

Converting to Wh where 1kWh = 3.6MJ (million joules)
4180 ÷ 3.6 = 1.16Wh/kgK.

Temp. diff. = Difference between desired water temperature (60°C) and incoming water temperature (10°C), i.e. 50K.

E.g. For a hot water storage facility of 200 litres daily use:

Q = 200 × 1.16 × 50 = 11600Wh or 11.6kWh

2. Estimate the size of solar collector panel area (A) in m².

$$A = \frac{\text{No. of days} \times Q \times \text{Solar fraction}}{\text{Annual solar irradiation} \times \text{Average system efficiency}}$$

Where: No. of days is 365 if the panel is in use all year.
Q is taken from the calculation above.
Solar fraction is the amount of energy provided and effectiveness of the collector relative to the total amount of energy that the installation requires. Zero is where there is no solar facility and 1 or 100% is for all energy from a solar source. 55% is a typical figure for the UK.
Annual solar irradiation – see map on previous page or take an annual estimate from the map on page 744.
Average system efficiency – related to that at the equator, up to 60% in the UK.

E.g. Using a figure of 2.6kWh/day (map on previous page):

2.6 × 365 = 949kWh/m²

Applied to the solar collector panel sizing formula:

$$A = \frac{365 \times 11.6 \times 55}{949 \times 60} = \frac{232870}{56940} = 4.09m^2 \text{ ie. } 4.1m^2$$

Properties of Heat – Hot Water

The heat energy properties of water are fundamental for determining pipe sizes and component dimensions in hot water and heating systems.

HEAT is a form of energy, otherwise known as thermal energy.

The standard unit of energy is the joule (J).

1 joule = amount of energy supplied by 1 watt (W) in 1 second (s).

Other units of energy found in older textbooks and product references include:

1 British thermal unit (1Btu) = 1·055kJ
1 calorie (1 cal) = 4·187J
1 kilowatt hour (1kWh) = 3·6MJ
1 therm (1 therm) = 105·5MJ

POWER is a measure of work rate.

Power (W) = heat energy (J) ÷ time in seconds (s)

Thus, 1W = 1 joule/second

TEMPERATURE is measured on a scale between two fixed points. These points are chosen at normal atmospheric pressure to represent water at the melting point of ice as zero, and the boiling point at 100, hence the term centigrade. A point on this scale is known as degrees Celsius (°C). The thermodynamic or absolute scale of temperature is represented in degrees kelvin (K). Temperature intervals are the same as Celsius, but kelvin originates at −273·15°C, the point at which no more internal energy can be extracted from a body. Temperature change intervals of 1°C and 1K are the same, except that

thermodynamic temperature (K) = temperature in °C + 273·15:

e.g. 1: water at 30°C = 303·15K

e.g. 2: a hot water system with primary flow and return temperatures of 80°C and 70°C respectively, has a temperature differential of 10K.

SPECIFIC HEAT CAPACITY (Shc) is the amount of heat energy required to raise 1 kilogram (kg) of a substance by 1K.

Some approximate values of Shc (will vary slightly with temperature and pressure):

Water	4180 J/kg K	Aluminium	910 J/kg K
Ice	2100	Cast iron	500
Nylon	1700	Copper/zinc	385
Air	1010	Lead	126

From the above, it can be seen that it would require over four times as much heat energy to raise 1kg of water 1K, than 1kg of air (4180 ÷ 1010 = 4·14). Conversely, as the Shc of water is relatively high, it is a good medium for storing heat. This is also a reason why hot water plant occupies less space than warm air systems, i.e. pipes are much smaller than air ducts conveying the same amount of energy.

The capacity of hot water storage vessels must be adequate for the building's purpose. Exact requirements are difficult to determine, but reasonable estimates are possible. These should include provision for rate of energy consumption (see table below) and the time taken to reheat the water to the required storage temperature (see boiler rating calculation – next page). Many buildings have variable uses and inconsistent demands. This often creates an overdesign situation, unless care is taken to establish peak use periods and the system calculations adjusted accordingly. With these building types, non-storage instantaneous fittings may be preferred.

For most buildings the following table can be used as guidance:

Building's purpose	Storage capacity (litres/person)	Energy consumption (kW/person)
Dwellings:		
single bath	30	0·75
multi-bath	45	1·00
Factory/office	5	0·10
Hotels	35*	1·00
Hostels	30	0·70
Hospitals	35*	1·00
Schools/colleges:		
day	5	0·10
boarding	25	0·70
Sports pavilions	35	1·00

*Average figures

E.g. A student hall of residence (hostel) to accommodate 50 persons.
Capacity: $50 \times 30 = 1500$ litres

Energy consumption: $50 \times 0.70 = 35$ kW

The nearest capacity storage vessel can be found in manufacturers' catalogues or by reference to BS 1566. For convenience, two or three cylinders of equivalent capacity may be selected.

Boiler Rating

Boilers are rated in kilowatts, where 1 watt equates to 1 joule of energy per second, i.e. W = J/s. Many manufacturers still use the imperial measure of British thermal units per hour for their boilers. For comparison purposes 1kW equates to 3412Btu/h.

Rating can be expressed in terms of gross or net heat input into the appliance. Values can be calculated by multiplying the fuel flow rate (m^3/s) by its calorific value (kJ/m^3 or kJ/kg). Input may be gross if the latent heat due to condensation of water is included in the heat transfer from the fuel. Where both values are provided in the appliance manufacturer's information, an approximate figure for boiler-operating efficiency can be obtained, e.g. if a gas boiler has gross and net input values of 30 and 24kW respectively, the efficiency is 24/30 × 100/1 = 80%.

Oil and solid fuel appliances are normally rated by the maximum declared energy output (kW), whereas gas appliances are rated by net heat input rate (kW[net]).

Calculation of boiler power:

$$kW = \frac{kg \text{ of water} \times Shc \times temp. \text{ rise}}{Time \text{ in seconds}}$$

where: 1 litre of water weighs 1kg

Shc = specific heat capacity of water, 4·2kJ/kgK

K = degrees kelvin temperature interval

Temp. rise = rise in temperature that the boiler will need to increase the existing mixed water temperature (say 30°C) to the required storage temperature (say 60°C).

Time in seconds = time the boiler takes to achieve the temperature rise: 1 to 2 hours is typical, use 1·5 hours in this example.

From the example on the previous page, storage capacity is 1500 litres, i.e. 1500kg of water. Therefore:

$$Boiler \text{ power} = \frac{1500 \times 4.2 \times (60 - 30)}{1.5 \times 3600} = 35kW \text{ net}$$

Given that the boiler has an efficiency of 80%, it will be gross input rated:

$$35 \times 100/80 = 43.75kW$$

Note: The boiler-operating efficiency is the relationship between a unit of fuel energy consumed to produce a unit of heat energy in the appliance hot water. It is not to be compared with the seasonal energy efficiency of a boiler, see page 132.

126

The water in primary flow and return pipework may circulate by convection. This produces a relatively slow rate of movement of about 0·2m/s, depending on pipe length and location of boiler and cylinder. Modern systems are more efficient, incorporating a circulation pump to create a water velocity of between 0·50 and 3·0m/s. This permits smaller pipe sizes and will provide a faster thermal response.

Inside diameter of pipe	Velocity min.	Velocity max. (copper)	Velocity max. (steel)
< 50mm*	0.50m/s	1·0m/s	1·5m/s
> 50mm	1.25m/s	1·5m/s	3·0m/s

Exceeding these recommendations may lead to excessive system noise and possible pipe erosion.

E.g. using the Copper Development Association design chart shown on the next page, with the boiler rating from the previous example of 43.75kW gross heat input and 35kW net heat input.

$$\text{Mass flow rate (kg/s)} = \frac{\text{Boiler net heat input}}{\text{Shc} \times \text{Temp. diff. (pf} - \text{pr)}}$$

Temperature difference between primary flow (pf) and primary return (pr) in pumped water circuits is usually about 10K, i.e. 80°C – 70°C. With convected circulation the return temperature will be about 60°C.

$$\text{Mass flow rate} = \frac{35}{4.2 \times 10} = 0.83\text{kg/s}$$

On the design chart, coordinating 0.83kg/s with a pumped flow rate of 1m/s indicates a 42mm inside diameter copper tube (35mm is just too small).

By comparison, using convected circulation of, say, 0.15m/s and a mass flow rate with a 20K temperature difference of 0·42kg/s, the pipe size would be 76mm.

*See also page 210.

Water Flow Resistance through Copper Tube

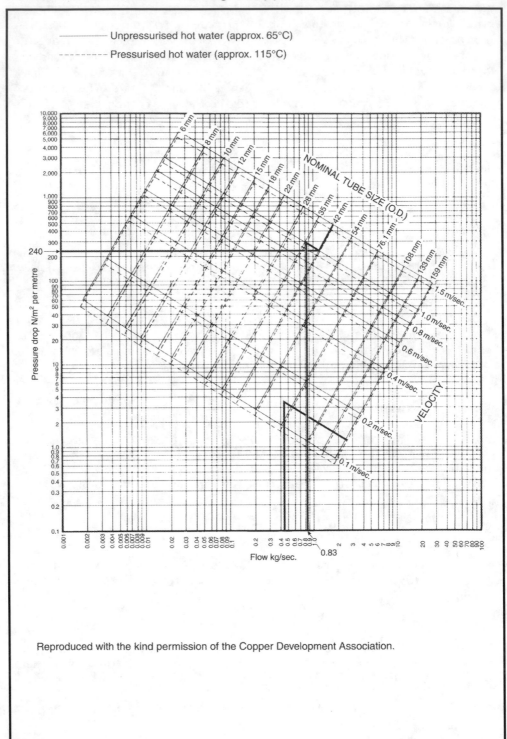

Reproduced with the kind permission of the Copper Development Association.

Where gravity or convection circulation of hot water between boiler and emitter is used, guidance on the circulating pressure can be determined by applying standard gravity of 9.80665m/s^2 (generally taken as 9.81) to the water density differential between boiler flow and return pipes. Reference to page 82 shows water density values between 0°C and boiling point.

Formula:

CP = 9.81 × water density differential between flow and return
CP = Circulating pressure per metre of circulation height

E.g.

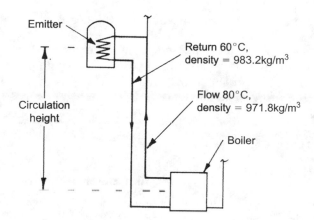

Emitter

Return 60°C,
density = 983.2kg/m³

Flow 80°C,
density = 971.8kg/m³

Circulation
height

Boiler

Water density differential = 983.2 - 971.8 = 11.4kg/m^3

CP = 9.81m/s^2 × 11.4kg/m^3 = 111.8, i.e. 112N/m^2 per m

If, for purposes of this example, the system output is rated at 8.4kW, the mass flow rate will be:

$$\frac{8.4}{4.2 \times 20} = 0.1 \text{kg/s} \qquad \text{(see page 127)}$$

With coordinates of 112N/m^2 per metre and 0.1kg/s, the chart on the previous page indicates that a 22mm outside diameter copper tube could be used for the flow and return pipes. However, this does not allow for the slow circulation velocity, frictional resistance due to fittings and the need for a reasonable heat response time. A more reliable guide compares circulation velocity typically about 0.15m/s with the calculated 0.1kg/s. On the chart this indicates that a 35mm pipe would be more appropriate.

A less arbitrary determination of fluid flow criteria can be obtained from the reference data in Guide C produced by the CIBSE.

Circulation Pump Rating

Circulatory pumps produce minimal pressure in the primary flow and return, but the flow rate is considerably enhanced. The pressure can be ascertained from design charts as a pressure drop in N/m^2 per metre or pascals per metre. $1N/m^2$ equates to 1 pascal (Pa).

From the design chart, circulation in a 42mm copper tube at 1m/s produces a pressure drop of 240Pa per metre. An estimate of the primary flow and return effective pipe length (see page 71) is required to establish the total resistance that the pump must overcome. For example, if the effective pipe length is 20m:

$240 \times 20 = 4800$Pa or 4·8kPa.

Therefore the pump specification would be 0·83kg/s at 4·8kPa.

Manufacturers' catalogues can be consulted to select a suitable pump. To provide for flexibility in installation, a degree of variable performance is incorporated into each model of pump. This range of characteristics can be applied by several different control settings, as shown in the following graphic. See also pages 207 and 208.

Pump performance chart:

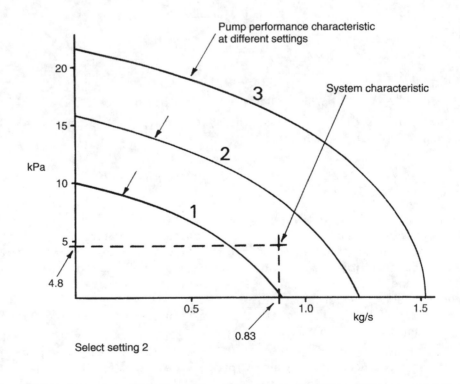

Legionnaires' Disease in Hot Water Systems

Bacterial growths which cause Legionnaires' disease develop in warm, moist, natural conditions such as swamps. They have adapted to living in the built environment in the artificial atmosphere of air-conditioning and hot water systems. A large number of outbreaks of the disease have occurred, with some people suffering a prolonged illness similar to pneumonia. The elderly are particularly vulnerable and many have died, hence the name of the illness which was attributed to a group of retired legionnaires who were infected while attending a reunion in Philadelphia, USA in 1976. Numerous other outbreaks and subsequent deaths have led to strict maintenance and installation controls of services installations. This has been effected by the Health and Safety Executive under the Health and Safety at Work, etc. Act and the Workplace (Health, Safety and Welfare) Regulations. The following measures are recommended for use with hot water systems:

1. Stored hot water temperature 60 to 65°C throughout the storage vessel.

2. Routine maintenance involving heating the water to 70°C as a precaution.

3. Changing the design of cylinders and calorifiers with concave bases. These are suspect, as the lower recesses could provide areas of reduced water temperature with little or no movement.

4. Connections to storage vessels should encourage through movement of water.

5. Pipework 'dead-legs' to be minimal.

6. All pipework to be insulated to reduce water temperature losses.

7. Where secondary circulation is required, supplementary trace element heating tape should be applied to maintain a minimum water temperature of 50°C.

8. Showers with recessed/concave outlet roses to be avoided. Other designs to have a self-draining facility to avoid inhalation of contaminated moisture droplets.

9. Spray taps – similar provision to 8.

Note: Cold water should be kept below 20°C.

Appliance Efficiency

SEDBUK – Seasonal Efficiency of Domestic Boilers in the United Kingdom. This developed under the government's Energy Efficiency Best Practice Programme to provide a manufacturer's data base which represents the efficiency of gas- and oil-fired domestic boilers marketed in the UK. SEDBUK is the average annual in-use efficiency achieved in typical domestic conditions. As an emphasis in time, should not be confused with operating efficiencies that compare gross and net heat energy input values – see page 126.

SSHEE – Seasonal Space Heating Energy Efficiency. This supersedes SEDBUK (Seasonal Efficiency of Domestic Boilers in the UK) as a means for categorising heat-producing appliances in accordance with an EU Energy related Products Directive (ErP) appropriate for EU member countries. Similar to SEDBUK, the ErP emphasises seasonal energy use and efficiency rather than the possibility of a less accurate indication based on a single point in time. SSHEE is defined as 'The ratio between the space heating demand for a designated heating season, supplied by a heater and the annual energy consumption required to meet this demand, expressed in percentage.'

Both measures provide similar percentage efficiency ratings with some slight variations attributed to different manufacturers and their boiler specifications. The ErP superseded SEDBUK in 2015, but many UK appliance producers continue to favour the latter.

Typical appliance efficiencies as certified to UK and/or European standard laboratory tests by an independent notified body:

Co-generation (CHP) space heaters	100
Heat pump space heaters and heat pump combi heaters	110
Low-temperature inverter heat pumps (Note 2)	125
Gas and oil condensing boilers (<70 kW)	98
Gas and oil non-condensing boilers (<10 kW)	≤ 75

Note 1: The objective in the short term is to eliminate all non-condensing boilers from the new and replacement market. Existing boilers remain in the lower category bands G up to B, with new condensing boilers in Band A.

Note 2: Inverter heat pumps are a variant on single speed compressor motors, designed with a motor to speed up and slow down relative to energy demand. The modulating effect saves in energy use by avoiding temperature swings by maintaining a more consistent and constant temperature in a building.

Ref. Energy related Products Directive (ErP) 2009/125/EC.

Energy standards and efficiencies provide a basis for specifiers, installers and consumers with a reference facility to compare different technologies and fuel types. Energy efficiency figures and band guidance labels provide an easy to understand method and simple representation for assessing and comparing different products. In addition to band related energy efficiency, labelling may also show:

- Noise emissions for heat pumps.
- Nitrous oxide emissions from fuel burning appliances.
- Stored hot water volume.
- Stored hot water energy loss.
- QR code – provides data on specific model.

In addition to easy to read labels, supplementary product information must also be available. This is usually in the form of the manufacturer's technical papers or a manual containing guidance on general use, application, maintenance and an account of how the band classification is achieved.

Since 2018, Boiler Plus legislation requires all new gas boiler installations in England to have a minimum efficiency of 92% by ErP standards.

Gas combination (combi) boilers require a specific measure for energy efficiency. At least one of:

- Flue gas heat energy recovery device.
- Weather compensation control.
- Load compensation control.
- Smart controls, i.e. optimisation and automation.

Appliance efficiency and SAP – Efficiency of an appliance is an essential component in calculating the Standard Assessment Procedure for Energy Rating of Dwellings (SAP). Additional factors are ventilation, heat losses through the fabric (U – values) and solar gains. To comply with the Building Regulations, designers and/or building developers are required to submit energy rating calculations to the relevant building control authority. This data is also available to prospective house buyers and tenants as a useful means for comparing and assessing annual fuel costs for hot water and heating. SAP values range from 1 to 100 with figures close to 100 an expectation for new dwellings. Above 100 is possible where the surplus energy produced, e.g. from solar power, provides a net export facility, i.e. surplus energy can be sold back to the grid.

Appliance Efficiency Labelling

Energy labelling – originally introduced for 'white goods' such as washing machines, refrigerators, ovens and tumble dryers. The retailer has responsibility for displaying the energy use labels. This is to provide customers with informed purchasing data when comparing the performance and energy consumption of different products.

Space heaters – includes boilers and other heat-producing appliances such as heat pumps. Consumers are unlikely to purchase these directly from a retailer, therefore the installer is responsible for ensuring that the appliance efficiency label is displayed. Installers are also required to provide a package energy efficiency label where there are combinations of equipment, e.g. boiler and room thermostats.

% efficiency	Band
90 – 100	A
86 – 90	B
82 – 86	C
78 – 82	D
74 – 78	E
70 – 74	F
<70	G

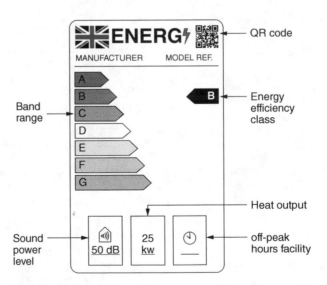

The above is a return to the simple scale of A – G which previously included A ratings extending to A+, A++ and A+++. The revised band ratings for boilers are similar to that shown.

Ref. Energy labelling directive 2017/1369/EU.

Greenhouse gas emissions in the UK (2022) are around 425 million tonnes CO_2 equivalent ($MtCO_2e$). A progressive decline since the established 1990 statistical base for environmental data. CO_2 equivalent is a collective unit for measuring the global warming potential of other greenhouse gases.

Greenhouse gases – trap heat in the atmosphere, leading to global warming.

Atmospheric composition:

Oxygen 21%
Nitrogen 78%

The remaining 1% comprising trace gases, primarily:

Carbon dioxide (CO_2)
Water vapour (H_2O)
Methane (CH_4)
Nitrous oxide (N_2O)
Hydrofluorocarbons (HFC)
Perfluorocarbons (PFC)
Sulphur hexafluoride (SF_6)

Fluoride gases – "F" gases. The latter 3 above can originate from various domestic, commercial and industrial sources. They contain very few greenhouse gases but cannot be ignored as they are potent and can significantly contribute to ozone depletion and heat trapping.

Carbon dioxide and water vapour – the prime contributors to global warming, even though CO_2, has only 0.04% atmospheric content. Water in vapour form can also be minimal in cold climates, less so in humid environments. H_2O vapour can be harmful by displacing some of the other gases. Both identify as a cap or layer that absorbs infrared radiation from Earth, retaining heat in the low atmosphere. Water may have a greater atmospheric presence, but it retains less energy due to a filtering effect. Also, it occurs at relatively low levels whereas CO_2 is effective at higher elevations. Water vapour content cannot be controlled, but CO_2 can be, e.g. by less use of fossil fuels and reducing deforestation.

UK Carbon Change Act 2008 – sets "carbon budgets" representing a limit on greenhouse gas emissions over the period of 5 years. Commenced 2008-2012 with an objective to reduce gas emissions by 80% by 2050 relative to 1990 levels. Now revised to 100%, i.e. a net zero target.

UK Low Carbon Economy

Progressive amendments to the Building Regulations, notably since the 1990s, have introduced considerable improvements to standards of energy efficiency. Several initiatives have been applied to the installation and use of fuel-consuming appliances and attention to details of construction. Buildings have been specifically identified as the a major source of all atmospheric carbon emissions, half of which is attributed to emissions from domestic hot water and heating equipment.

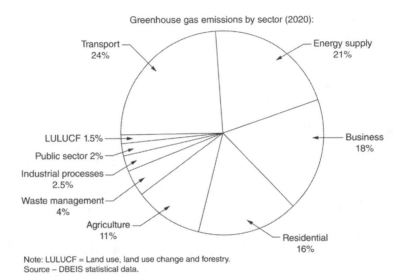

Greenhouse gas emissions by sector (2020):

- Transport 24%
- Energy supply 21%
- Business 18%
- Residential 16%
- Agriculture 11%
- Waste management 4%
- Industrial processes 2.5%
- Public sector 2%
- LULUCF 1.5%

Note: LULUCF = Land use, land use change and forestry.
Source – DBEIS statistical data.

Effective measures:

- Domestic boilers – new and replacement appliances with an energy efficiency rating of at least A, i.e. high-efficiency condensing boilers.
- Insulation standards for new and refurbished buildings improved, e.g. replacement double and triple glazed windows and reduced 'U' values.
- Regular inspection and maintenance of air-conditioning systems.
- Measures to prevent overheating by solar gain.
- Installation of energy recovery systems, e.g. MVHR and heat pumps.
- Restricted use of inefficient appliances, e.g. gas decorative-effect fires.
- Insulation of hot and chilled water pipework and sealing of ductwork joints to prevent air leakage.
- Use of high-efficacy electric lamps and power rating limitations on external and internal lighting.
- Calculation of carbon emission limits from dwellings, re. SAP ratings. For other buildings measures required to show improvements, such as renewable energy use, solar systems and CHP.
- Reduced air leakage through the building envelope, max. 10m³/hour/m² at 50Pa.

Refs. Relevant Building Regulations – See page 10.

DBEIS: Standard Assessment Procedure.

Electrolysis – the corrosion or decomposition of different metals in the presence of water. Three criteria exist which will encourage corrosion:

- Neutral or acidic water, pH value ≤ 7
- Warm or hot water
- Metals widely apart on the electrochemical or galvanic series.

Electrochemical series for metals used in plumbing and hot water services:

Protected end (cathode)	Stainless steel
	Copper
	Gunmetal and bronze
	Tin
	Lead
	Steel
	Cast iron
	Aluminium
	Zinc (galvanising)
Corroded end (anode)	Magnesium

Water functions as an electrolyte, i.e. a solution which conducts an electric current between a cathode and anode of dissimilar metals. Therefore, in water services systems materials must be compatible, otherwise decomposition of pipework and equipment will occur. For example, galvanised steel and copper pipes should never be used together, particularly in hot water installations.

Plumbo-solvency – term used to describe the breakdown of lead pipes conveying water with 'soft' characteristics. This should not be a problem, as for health reasons lead is no longer acceptable as a water services material. However, exposed lead flashings could be affected in areas of 'soft' rainwater.

Cupro-solvency – term used to describe the breakdown of copper pipes where soft water contains dissolved carbon dioxide. This type of water is generally associated with private wells and springs.

Dezincification – this affects brass pipe fittings and valves. Brass is an alloy of copper and zinc (50:50). Electrolytic reaction between the two metals, particularly in high chloride waters, causes corrosion of the zinc. This leaves the fitting unchanged in appearance, but with no strength and possibly porous. Installations in areas known to be prone to this problem should be specified with gunmetal fittings, an alloy of copper, tin and zinc (85:10:5).

Anodic protection – before the introduction of plastic storage cisterns it was common practice to fit a sacrificial anode of magnesium into galvanised cold water storage cisterns if copper pipes were used. As magnesium is below zinc in the electrochemical series, the magnesium dissolves away instead of the galvanising. Sacrificial anodes are fitted as a precautionary measure to the inside of copper hot water storage cylinders.

Water Treatment – System Flushing

As part of the commissioning and testing process (see page 213), new water services to include every length of pipe, cistern, hot water storage cylinder and all connected components should be flushed through with wholesome water. This process is not to be regarded as a substitute for care and cleanliness during installation.

Cisterns in particular should receive special attention. Any debris or deleterious matter must be removed before a cistern and associated system are filled. Failure to undertake this simple check may be the cause of system blockages, pipework corrosion, contamination of the supply, reduced system efficiency and reduced effectiveness of any water treatments. For installations larger than that required for single-family private dwellings, cisterns should be filled with chlorinated water at a dosage of 50 parts chlorine to 1 million parts water (50 mg/litre). Terminal valves and taps are opened to ensure the presence of chlorine by smell, then closed and the system allowed to stand for at least one hour. After this time the chemical smell should again be present at opened terminals (at least 30 ppm by measure). If not, the procedure is repeated. Thereafter, the system is flushed with wholesome water to remove any remaining chemical.

Fluxes used with soldered capillary joints on copper tube will in general dissolve in water, but large deposits can become water repellent and may attract a build-up of surface deposits. In practice there is no need for an excess of flux to be applied.

Filling and draining from the lowest point of an installation is insufficient to ensure complete cleansing. All terminal connections, particularly those at the end of long horizontal runs and 'dead-legs', should be opened and flushed through. Where work has been completed on a building or it is left unoccupied, pipe systems should not be charged with water that could become stagnant. To reduce the possibility of pipework corrosion and water quality issues, unused systems should be flushed regularly, i.e. at least twice during a week.

Ref. Water Supply (Water Fittings) Regulations, Schedule 2, Paragraph 13.

Disinfection – the process of sanitising water by deactivating any living bacteria and micro-organisms in hot or cold water systems. Adding approved chemicals to the system water is the most common method. After testing and flushing, all new installations should be disinfected. An exception is small works such as private dwellings occupied by only one single family. Disinfection also applies to underground supply for human consumption (see page 25 and BS EN 806-4: Specifications for installations inside buildings conveying water for human consumption. Installation. Part 6.3 Disinfection).

Procedures:

- Off-line (chemical) – the use of either sodium hypochlorite or stabilised chlorine dioxide as oxidising disinfectants to produce free residual chlorine. Application as described on the previous page. Bromine and ozone oxidising disinfectants are alternative additives. Precautions during use include system backflow prevention, personal protective equipment and terminals/outlets to be marked DISINFECTION IN PROGRESS – DO NOT USE. Disposal facilities to be agreed with the water authority and the Environment Agency.

- Off-line (thermal) – this is supplementary to disinfecting supply cisterns as described on the previous page. The process is otherwise known as pasteurisation and it requires raising the whole system water temperature to between 60 and 70°C and maintaining this for at least one hour.

- On-line (chemical) – a routine or continuous dosing process (manual or automatic) using chlorine or chlorine dioxide. Where used with a drinking water supply, will require specific approval from the water authority.

- On-line (electrical) – use of an electric water conditioner that releases copper and silver ions through electrodes in the supply pipe.

- On-line (thermal) – see pages 131 and 440.

Water Treatment – System Corrosion Inhibitors

Bacteria – the most common bacteria in water systems is Pseudomonas bacteria. It occurs where there is lack of water circulation or stagnation in discontinuous lengths of pipes and storage vessels. The latter is typical of expansion and feed cisterns in indirect hot water and central heating systems. High ambient temperatures between 20 and 40°C and poorly ventilated roof spaces or compartments are ideal for its development. First indications are usually its highly putrid odour. Inspection usually reveals a brown, slimy film lining the water surface and storage cistern. Eradication is by flushing and disinfection with biocides in solution.

Corrosion inhibitors – see also page 213. Boiler and associated equipment will only operate effectively and efficiently if water in the system is maintained clean and free of impurities. The minimal build-up of scale or magnetite sludge will significantly reduce boiler efficiency and increase its contribution to carbon emissions.

New systems should be flushed to remove debris such as metal filings, flux and loose solder deposits. Filling is with clean water and the manufacturer's recommended dose of corrosion inhibitor, as shown in the illustrations. Following maintenance, repair or modification, existing systems should be treated similarly.

Proprietary corrosion inhibitors may be compounds of sodium silicate, benzoate, nitrite and chromate. Sodium pentachlorophenate is a bactericide or biocide which can be used to prevent the accumulation of hydrogen gas in radiators.

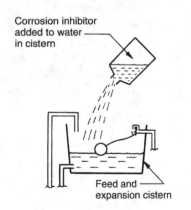

Dosing an open vent system

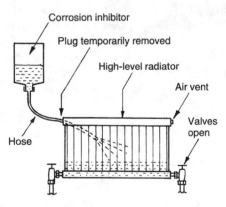

Dosing a sealed system

Refs. BS 7593: Code of practice for the preparation, commissioning and maintenance of domestic central heating and cooling water systems.

Building Regulations, AD L, Vols. 1 and 2: Conservation of fuel and power.

Filters and dirt separators – it is usual practice to install a filtration and dirt separation facility to domestic hot water and heating systems. This has the effect of concentrating and containing sludge, magnetite, filings, suspended jointing paste and contaminants in one easily maintained and accessible place.

Filtration unit – comprises a stainless steel mesh and a drain valve. When the valve is opened for a few seconds, collected debris is removed instantly with no disruption to the system. Sealed (unvented) systems will require a nominal re-charge to replace lost water. The filter separates and removes larger suspended particles and a magnet attracts metallic micro-particles. For domestic installations, the magnet can be a flexible sleeve fitted over the filter body. Filter units for larger installations usually contain an integral accessible magnetic core. Location is generally in the lower part of the primary circuit, close to the boiler.

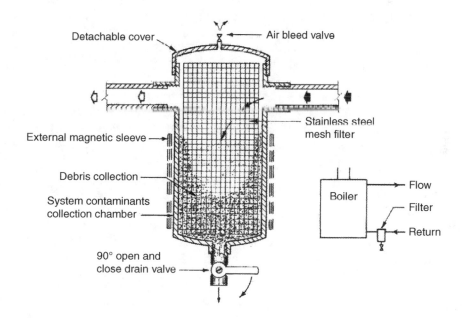

Ref. Building Regulations Part L – Domestic Building Services Compliance Guide: 'A filter can also be fitted to the (*hot water and*) central heating circuit to help maintain the efficiency and reliability of the system.'

Water Treatment – Maintenance

Hot water circulatory systems comprise several interdependent components, each requiring regular maintenance to ensure the system functions effectively. Neglecting system care may in time manifest in any one of, or a combination of, the following:

- Boiler shutting down too frequently.
- Boiler overheating at the top.
- Boiler noises, i.e. gurgling/banging, sometimes known as "kettling".
- Pump failure.
- Cold spots on circulatory pipework.
- Water leakage, usually minor but notably stained.
- Water slow to heat up.
- Higher than expected fuel bills – uneconomic.

As indicated on pages 138-141, a number of installation and aftercare provisions are available to enhance and extend working efficiency. This may include draining, flushing and refilling with a diluted corrosion inhibitor. Around 5-yearly intervals or whenever repairs or alterations occur is appropriate. Accumulation of sludge and debris from corrosion and electrolytic activity in the presence of dissimilar metals cannot be entirely avoided. Inhibitors are effective as a control medium but cannot provide complete protection as they have a limited life, hence the need to drain and replenish periodically.

Refilling with chemical additives and draining after a few hours can be an effective way to dissolve deposits of sludge. Over time, a build-up can occur on boiler and pipe linings, particularly at sharp bends and junctions. Inhibitors can be equally effective, but dissolved sludge remains in the system until drained. Where a filter is fitted, the opportunity for an annual drain can be scheduled for the same time as a yearly boiler maintenance.

Boiler maintenance – recommended annually to ensure maximum efficiency of the heat exchange unit. Combustion of fuel produces small accumulative deposits in the appliance flue and on the heat exchanger surface with some collecting on the burner tray. Light vacuum cleaning will ensure clean surfaces and effective fuel combustion. Appliance component condition can be checked and an analysis recorded of burnt gas composition – see page 530. Only to be undertaken by a qualified "competent" person – see page 17.

4 HEATING SYSTEMS

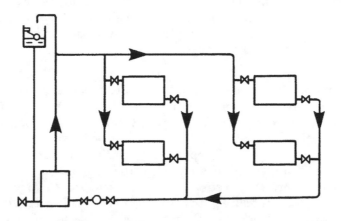

HEAT EMITTERS
LOW-TEMPERATURE, HOT WATER HEATING SYSTEMS
PANEL AND UNDERFLOOR HEATING
EXPANSION FACILITIES IN HEATING SYSTEMS
EXPANSION VESSELS
SOLAR SPACE HEATING
HIGH-TEMPERATURE, PRESSURISED HOT WATER
HEATING SYSTEMS
STEAM HEATING SYSTEMS
DISTRICT HEATING
COMBINED HEAT AND POWER
PIPEWORK EXPANSION
THERMOSTATIC CONTROL OF HEATING SYSTEMS
TIMED CONTROL OF HEATING SYSTEMS
ZONED CONTROLS
PROGRAMMABLE CONTROL
ENERGY MANAGEMENT SYSTEMS
WARM AIR HEATING SYSTEM
HEATING DESIGN
DOMESTIC HEATING CIRCULATOR/PUMP
MODULATING PUMP
OVERHEATING

DOI: 10.1201/9781003434894-4

Radiators and convectors are the principal means of heat emission in most buildings. Less popular alternatives include exposed pipes and radiant panels for use in warehousing, workshops and factories, where appearance is not important. Embedded panels of pipework in the floor screed can also be used to create 'invisible' heating, but these have a slow thermal response as heat energy is absorbed by the floor structure.

Despite the name, radiator, no more than 40% of the heat transferred is by radiation. The remainder is convected, with a small amount conducted through the radiator brackets into the wall. Originally, radiators were made from cast iron in three forms: hospital, column and panel. Hospital radiators were so called because of their smooth, easy-to-clean surface, an important specification in a hygienic environment. Column radiators vary in the number of columns. The greater the number, the greater the heat-emitting surface. Cast iron radiators are still produced to special order, but replicas in cast aluminium can be obtained. Cast iron panels have been superseded by pressed profiled steel-welded panels. These are much slimmer and easier to accommodate than cast iron in the modern house. In addition to the corrugated profile, finned backing will also increase the heating surface and contribute to a higher convected output. Pressed steel radiators are made in single, double and triple panels.

Convectors have a steel casing containing a finned heat exchanger. About 90% of the heat emission is convected, and this may be enhanced if a thermostatically controlled fan is also located in the casing. They are more effective than radiators for heating large rooms, and in this situation, their extra bulk can be accommodated.

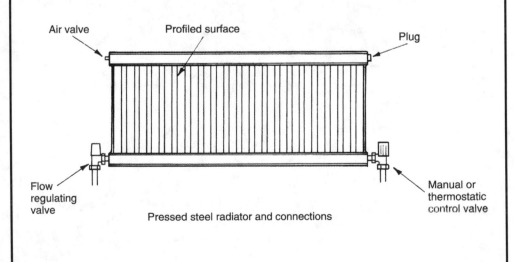

Pressed steel radiator and connections

In temperate and cold climates where there is insufficient warmth from the sun during parts of the year, heat losses from the human body must be balanced. These amount to the following approximate proportions: radiation 45%, convection 30% and evaporation 25%. Internal heat gains from machinery, lighting and people can contribute significantly, but heat emitters will provide the main contribution in most buildings.

Enhancement of radiator performance can be achieved by placing a sheet of reflective foil on the wall between the fixing brackets. Emitter location is traditionally below window openings, as in older buildings, the draughts were warmed as they infiltrated the ill-fitting sashes. With quality double-glazed units, this is no longer so important and in the absence of a window, locating a shelf above the radiator will prevent pattern staining of the wall due to convective currents. Radiant panels and strips are suspended from the ceiling in industrial premises and other situations where wall space is unavailable.

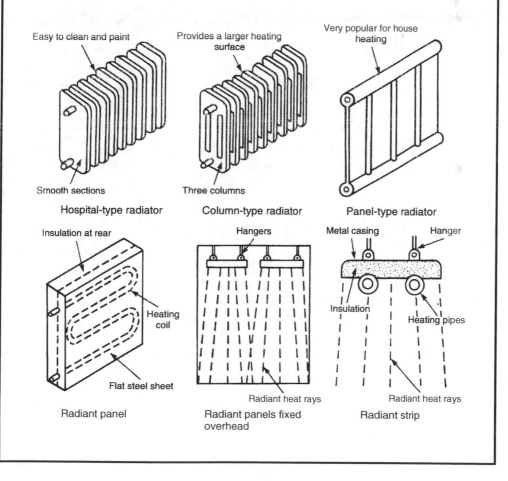

Easy to clean and paint

Smooth sections

Hospital-type radiator

Provides a larger heating surface

Three columns

Column-type radiator

Very popular for house heating

Panel-type radiator

Insulation at rear

Heating coil

Flat steel sheet

Radiant panel

Hangers

Radiant heat rays

Radiant panels fixed overhead

Metal casing   Hanger

Insulation

Heating pipes

Radiant heat rays

Radiant strip

Radiant and convector skirting heaters are unobtrusive at skirting level and provide uniform heat distribution throughout a room. Natural convectors have a heating element at a low level within the casing. This ensures that a contained column of warm air gains velocity before discharging to displace the cooler air in the room. Fan convectors may have the heater at high level with a variable speed fan located below. In summer, the fan may also be used to create air circulation. Overhead unit heaters are used in workshops to free the wall space for benches, machinery, etc. A variation may be used as a warm air curtain across doorways and shop entrances. Individual unit heaters may have a thermostatically controlled inlet valve, or a bank of several units may be controlled with zoning and diverter valves to regulate output in variable occupancy situations.

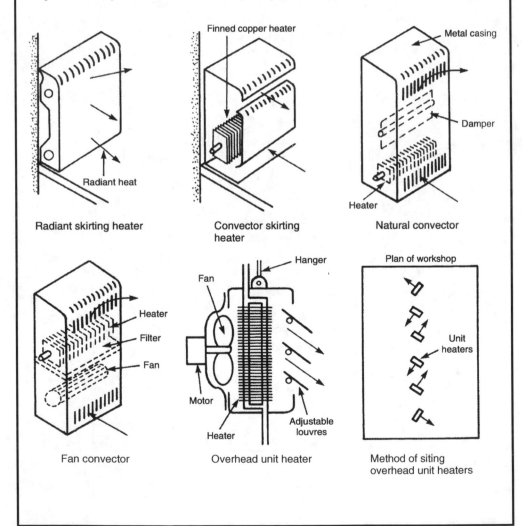

Radiant skirting heater

Convector skirting heater

Natural convector

Fan convector

Overhead unit heater

Method of siting overhead unit heaters

In low-temperature, hot water heating systems, the boiler water temperature is thermostatically controlled to about 80°C. Systems may be 'open' with a small feed and expansion cistern or mains fed 'sealed' with an expansion vessel.

The type of system and pipe layout will depend on the building purpose and space available for pipework. A ring or loop circuit is used for single-storey buildings. Drop and ladder systems are used for buildings of several storeys. The drop system has the advantage of being self-venting, and the radiators will not become airlocked. Traditional solid fuelled systems operate by convection or gravity circulation (otherwise known as thermo-siphonage). Contemporary practice is to install a pump for faster circulation and a more rapid and effective thermal response. This will also complement modern fuel controls on the boiler and allow for smaller pipe sizes. The additional running costs are minimal.

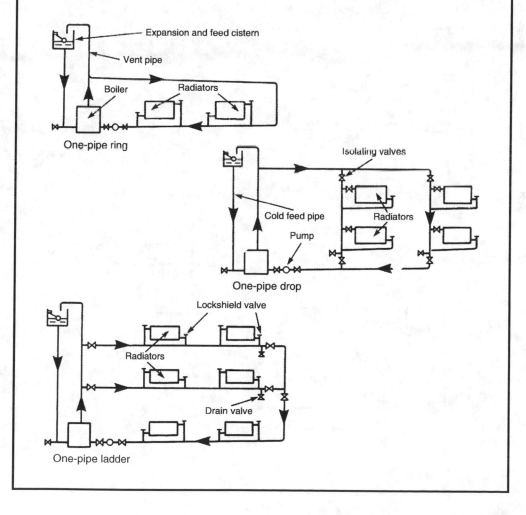

The one- and two-pipe parallel systems are useful where pipework can be accommodated within a floor structure, a raised floor or a suspended ceiling. The disadvantage with all one-pipe systems is the difficulty of supplying hot water to the radiators furthest from the boiler. As the heat is emitted from each radiator, cooling water returns to mix with the hot water supplying subsequent radiators, gradually lowering the temperature around the circuit. Eventually the last or 'index' radiator receives lukewarm water at best, necessitating a very large radiator to provide any effect. Pumped circulation may help, but it will require a relatively large-diameter pipe to retain sufficient hot water to reach the 'index' radiators. Two-pipe systems are less affected, as the cool water from each radiator returns directly to the boiler for reheating. However, radiators will need flow balancing or regulating to ensure an even distribution of hot water. The reverse-return or equal travel system requires the least regulating, as the length of pipework to and from each radiator at each floor level is equal. In all systems, the circulating pump is normally fitted as close to the boiler as possible, either on the heating flow or return. Most pump manufacturers recommend location on the higher temperature flow.

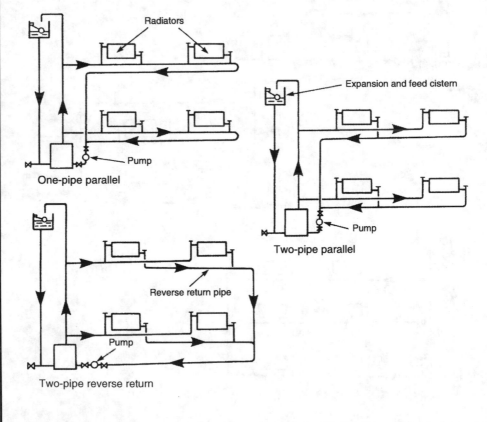

One-pipe parallel

Two-pipe parallel

Two-pipe reverse return

The two-pipe upfeed system is used when it is impractical to locate pipes horizontally at high level. The main heating distribution pipes can be placed in a floor duct or within a raised floor. The two-pipe drop is used where a high-level horizontal flow pipe can be positioned in a roof space or in a suspended ceiling, and a low-level return within a ground floor or basement ceiling. This system has the advantage of self-venting. The two-pipe high-level return system is particularly appropriate for installation in refurbishments to existing buildings with solid ground floors. In this situation, it is usually too time-consuming, impractical and possibly structurally damaging to cut a trough or duct in the concrete.

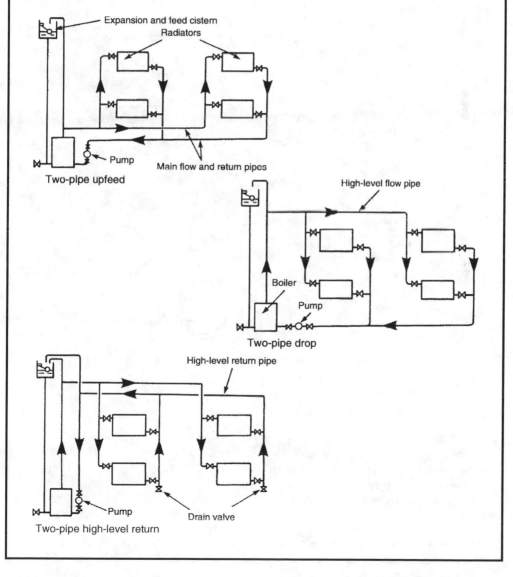

Low-Temperature, Small Bore Hot Water Heating System

Pumped small bore heating systems have 28 or 22mm outside diameter copper tube for the main heating flow and return pipework, with 15mm o.d. branches to each radiator. This compares favourably with the old gravity/convection circulation systems which sometimes required pipes of over 50mm diameter to effect circulation. If cylinder and boiler are separated vertically by floor levels, there will be sufficient pressure for hot water to circulate by convection through the primary flow and return pipes. However, most modern systems combine a pumped primary and heating flow with circulation regulated by thermostats and motorised valves. Variations in one- and two-pipe systems are shown on pages 147–149. Two-pipe systems are always preferred for more effective hot water distribution.

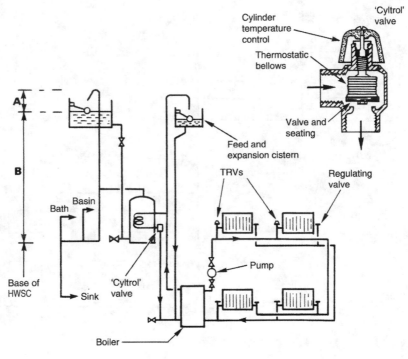

Small bore heating system

Notes:
1. 'Cyltrol' valve to be as close as possible to HWSC, to sense hot water return temperature and maintain stored water at 60°C minimum. Where used with a solid fuel boiler, an unvalved radiator or towel rail is connected across the primary pipes to dissipate excess heat when the 'cyltrol' closes.
2. Min. height of expansion pipe above cistern water level (A) = (B) in metres × 40mm + 150mm. E.g. if (B), cistern water level to base of HWSC is 2·5m, then (A) is 2·5 × 40mm + 150mm = 250mm.

Low-Temperature Microbore Hot Water Heating System

The microbore system also has pumped circulation through 28 or 22mm o.d. copper tube main flow and return pipes to radiators. The diameter depends on the number and rating of emitters connected. The difference between this system and conventional small bore is the application of a centrally located manifold between boiler and emitters. Manifolds are produced with standard tube connections for the flow and return and several branches of 6, 8, 10 or 12mm outside diameter. A combined manifold is also available. This is more compact, having a blank in the middle to separate flow from return. Manifolds are generally allocated at one per floor. Systems may be open vented or fitted with an expansion vessel. The advantage of microbore is ease and speed of installation, as long lengths of small-diameter soft copper tubing are produced in coils. It is also unobtrusive where exposed, very easily concealed and is less damaging to the structure when holes are required. Water circulation noise may be noticeable as velocity is greater than in small bore systems. Pumped circulation is essential due to the high resistance to water flow in the small diameter pipes.

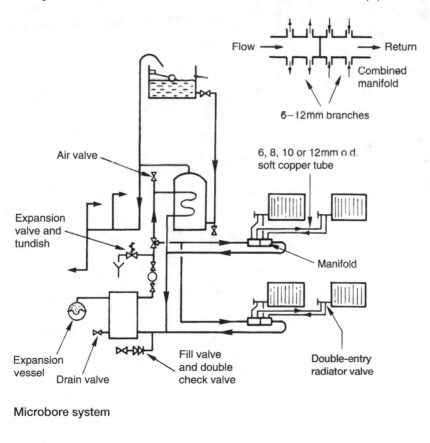

Microbore system

Double-Pump Heating and Hot Water Control

This is an alternative method for distributing hot water. It can be effected by using two separate pumps from the boiler flow: one to supply the hot water storage cylinder and the other the heating circuit. Grundfos Pumps Ltd have developed a purpose-made dual pump for this purpose, which is integrated into one body. This system conveniently replaces the conventional single pump and associated two- or three-port motorised distribution valves. Each pump is dedicated to hot water or heating and individually controlled by cylinder or room thermostat. The correct flow and pressure can be regulated to the characteristics of the specific circuit.

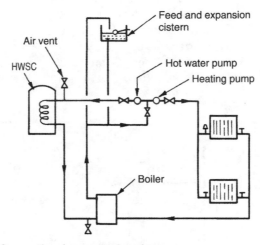

Conventional open-vent system

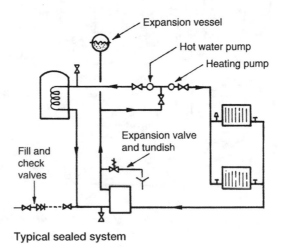

Typical sealed system

Air Elimination in Hot Water and Heating Systems

In conventional low-pressure systems, air and other gases produced by heating water should escape through the vent and expansion pipe. Air must be removed to prevent the possibility of airlocks, corrosion and noise. To assist air removal, a purpose-made device resembling a small canister may be used to concentrate the gases. This simple fitting is located on the boiler flow and vent pipe to contain the water velocity and ensure efficient concentration and release of air into the vent.

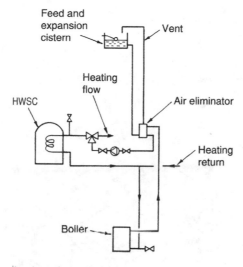

Application of air eliminator

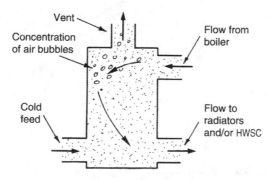

Air eliminator (approx. 100mm high × 75mm dia.
with standard 22mm o.d. copper tube connections)

Panel or Surface Heating

The system consists of 15mm or 22mm o.d. annealed copper pipes embedded in the floor, ceiling or walls. This has the benefit of avoiding unsightly pipes and radiators. Heat distribution is uniform, providing a high standard of thermal comfort as heat is emitted from the building fabric. However, thermal response is slow as the fabric takes time to heat up and to lose its heat. Thermostatic control is used to maintain the following surface temperatures:

Floors – 27°C
Ceilings – 49°C
Walls – 43°C

Joints on copper pipes must be made by capillary soldered fittings or by bronze welding. Unjointed purpose-made plastic pipes can also be used. Before embedding the pipes, they should be hydraulically tested, as described on page 212.

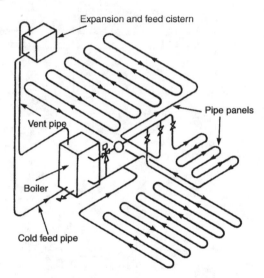

Installation of panel heating system

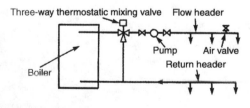

Detail of boiler and connections

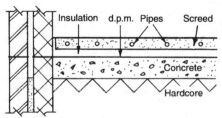

Method of embedding the panels

Current practice is to use jointless plastic piping in continuous coils. Pipes can be embedded in a 70mm cement and sand screed (50mm minimum cover to tube). In suspended timber floors, the pipe may be elevated by clipping tracks or brackets with metallic reflective support trays, prior to fixing the chipboard decking. Materials include

PEX: Cross-linked polyethylene.

PP: Co-polymer of polypropylene.

PB: Polybutylene.

These pipes are oxygen permeable; therefore, when specified for underfloor heating, they should include a diffusion barrier.

Alternative:
PEX/AL/PEX: Multi-layer pipe comprising cross-linked polyethylene with an aluminium core. Oxygen impermeable.

Boiler flow temperature for underfloor heating is about 50°C, while that for hot water storage and radiators is about 80°C. Therefore, where the same boiler supplies both hot water storage cylinder and/ or radiators and underfloor heating, a motorised thermostatic mixing valve is required to blend the boiler flow and underfloor heating return water to obtain the optimum flow temperature.

Extract from performance tables for a design room temperature of 21°C with a blended flow temperature of 50°C:

Solid floor – Pipe dia. (mm)	Pipe spacing (mm)	Output (W/m²)
15	100	82
15	200	67
18	300	55
Suspended floor –		
15	300*	47

*Assumes two pipe runs between floor joists spaced at 600mm centres.

For a room with a solid floor area of 13·5m² requiring a heating input of 779 watts (see page 199), the output required from the underfloor piping is:

$$779 \div 13 \cdot 5 = 57 \cdot 7 \text{ watts/m}^2$$

Therefore, 15mm diameter pipe at 200mm spacing (67W/m²) is more than adequate, while 18mm diameter pipe at 300mm spacing (55W/m²) is just below.

Manifold or header – manifolds are discretely located on a wall or within a boxed unit. Manifolds comprise:

- Flow ports (2–12).
- Return ports (2–12).
- Drain valve and hose connection (may be used for filling).
- Air ventilation valve.
- Isolating valve to each bank of ports.
- Visual flow meters to each flow port.
- Lockshield balancing valve on each return port.

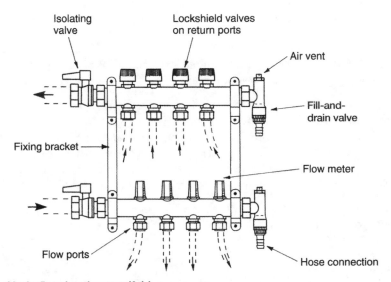

Underfloor heating manifold

Installation notes –

- One circulator per manifold.
- Combined radiator and panel systems, one circulator for each system.
- Screeded floor to have insulation turned up at edge to provide for expansion. Max. 40m² or 8m linear, without expansion joint.
- Timber floor to have 6–8mm expansion gap around periphery.

Refs. BS 5955-8: Plastics pipework (thermoplastics materials).

BS 7291-1, 2 and 3: Thermoplastic pipes.

BS EN 1264-4: Water-based surface embedded heating and cooling systems. Installation.

Underfloor installations

Suspended timber floor – 1

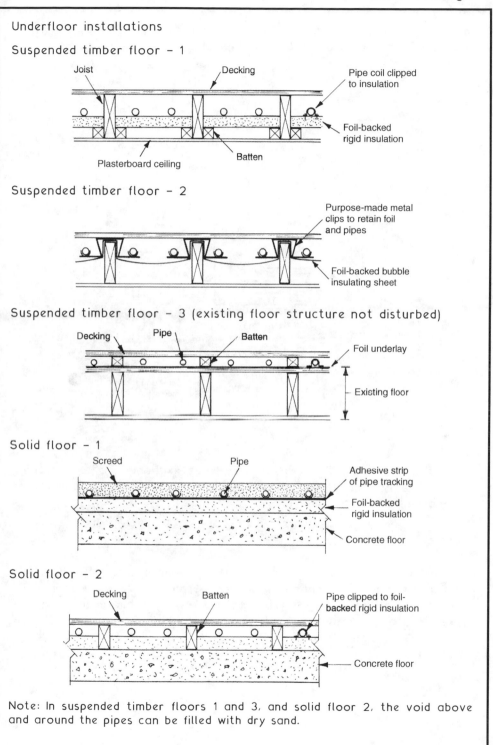

Solid floor – 1

Solid floor – 2

Note: In suspended timber floors 1 and 3, and solid floor 2, the void above and around the pipes can be filled with dry sand.

Schematic diagram of a solar energy panel in combination with a supplementary conventional boiler supply for hot water and underfloor heating –

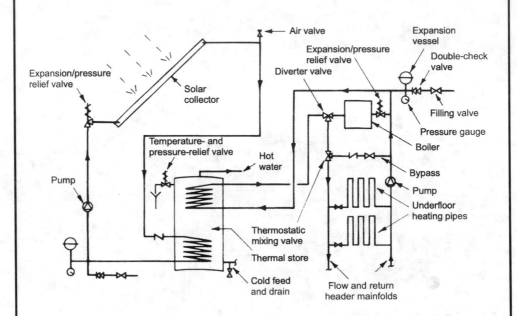

The storage cylinder has two primary heat exchange coils. The solar part of the system is connected to the lower coil and the boiler, and the heating circuit is connected to the upper coil.

For underfloor heating purposes, the thermostatic mixing valve maintains water at a constant delivery temperature by varying the proportions of hot flow water with cooler water returning via the bypass, a process known as modulating control.

The cost-effectiveness and efficiency of underfloor heating can be considerably improved when operating under steady-state modulated conditions regulated with a weather compensated circuit design (see pages 190 and 191). The system installation requires a motorised mixing valve and bypass, as shown on the previous page and on page 191. By computation of sensor-supplied data based on external temperature and the system heating curve, the control compensator regulates the flow temperature through the mixing valve.

Heating curve – a measure for comparing different system types, their design and application. Criteria used are based on the relationship between water flow temperature, room temperature and external design temperature. For underfloor heating, it is usual to select a flow temperature between 45°C and 55°C, with a return temperature of 35°C to 40°C. Typical heating curve values range between 0.85 and 1.50 for underfloor systems. Higher water temperature radiator systems have a value of about 2.50.

Formula –

$$\frac{\text{Design flow temperature} - \text{Design room temperature}}{\text{Design room temperature} - \text{Design external temperature}}$$

Examples using an underfloor design flow temperature of 50°C compared with a radiator design flow of 80°C, both applied to an internal design temperature of 21°C and an external design temperature of −1°C:

Underfloor –

$$\frac{50 - 21}{21 - -1} = \frac{29}{22} = 1.32$$

Radiators –

$$\frac{80 - 21}{21 - -1} = \frac{59}{22} = 2.68$$

Expansion Facilities in Heating Systems

In any water heating system, provision must be made for the expansion of water. A combined expansion and feed cistern is the traditional means. This will have normal expansion space under usual boiler-firing conditions of about 4% of the total volume of water in the system, plus a further third as additional expansion space for high boiler firing. Although the expansion can be accommodated up to the overflow level, there should be at least 25mm between overflow and the fully expanded water level.

Contemporary sealed systems have an expansion vessel connected close to the boiler. It contains a diaphragm and a volume of air or nitrogen to absorb the expansion. To conserve wear on the diaphragm, location is preferred on the cooler return pipe and on the negative side of the pump. System installation is simpler and quicker than with an expansion cistern. The air or nitrogen is pressurised to produce a minimum water pressure at the highest point on the heating system of 10kPa (approx. 1m head of water). In normal use, a pressure gauge with the expansion vessel will indicate about 1 bar (10m head or 100kPa).

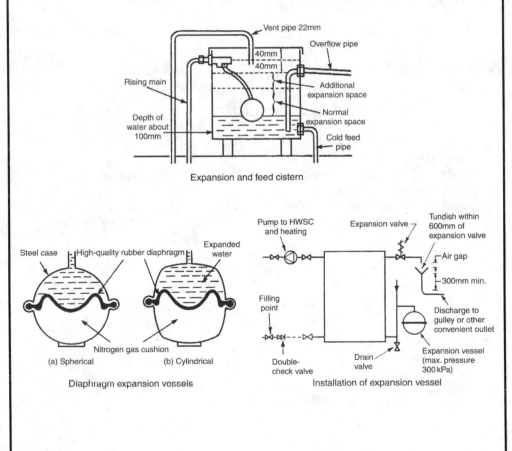

Expansion and feed cistern

Diaphragm expansion vessels

Installation of expansion vessel

Expansion vessels are produced to BS 6144. They must be correctly sized to accommodate the expansion of heated water without the system safety/pressure-relief valve operating. The capacity of an expansion vessel will depend on the static pressure (metres head from the top of the system to the expansion vessel), the system maximum working pressure (same setting as p.r.v.) obtained from the manufacturer's details and the volume of water in the system (approx. 15 litres per kW of boiler power).

Capacity can be calculated from the following formula:

$$V = \frac{e \times C}{1 - P_i/P_f}$$

where V = vessel size (litres)

   e = expansion factor (see table)

   C = capacity of system (litres)

   P_i = static pressure (absolute)*

   P_f = max. working pressure (absolute)*

* absolute pressure is 1 atmosphere (atm) of approx. 100kPa, plus system pressure.

E.g. C = 100 litres

   P_i = 1.5atm or 150kPa (5m head static pressure)

   P_f = 1.9atm or 190kPa (9m head static pressure)

Water temp. = 80°C

Temp.°C	Exp. factor
50	0·0121
60	0·0171
70	0·0227
80	0·0290
90	0·0359

$$V = \frac{0.029 \times 100}{1 - 150/190} = 13.80 \text{ litres}$$

Ref. BS 6144, Specification for expansion vessels using an internal diaphragm, for unvented hot water supply systems.

Solar Space Heating

Solar space heating must be complemented with a very high standard of thermal insulation to the building fabric. The solar panel shown on page 117 for hot water provision will need a much larger area, typically 40m² for a three- to four-bedroom detached estate house. A solar tank heat exchanger of about 40m³ water capacity is located in the ground. It is fitted with a pre-set safety-type valve which opens to discharge water to waste if it should overheat. The solar panel and associated pipework are mains filled and supplemented with a glycol or antifreeze additive.

With diminishing fossil fuel resources and inevitable rising fuel prices, solar heating is encouraged as a supplement or even an alternative to conventionally fuelled systems. For use as the sole energy for a heating system, there is still considerable scope for research and development. Technological developments are improving, particularly with the 'heat bank' or storage facility shown. In time, it may become viable even with the UK's limited solar energy in winter months.

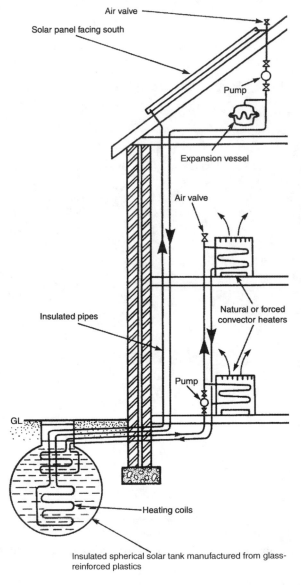

See also page 124, Properties of Heat – Hot Water. The following additional data have particular application to the design of hot water heating systems and components.

CHANGE OF STATE. Water has three basic characteristic states: solid (ice), liquid (fluid) or gas (steam). Water changes state at the specific temperatures of 0°C and 100°C.

LATENT HEAT is the heat energy absorbed or released at the point of change from ice to water and from water to steam, i.e. where there is no change in temperature. This is measured as specific latent heat, in units of joules per kilogram (J/kg).

$$\text{Specific latent heat of ice} = 335 \text{kJ/kg}$$

$$\text{Specific latent heat of water} = 2260 \text{kJ/kg}$$

SENSIBLE HEAT is the heat energy absorbed or released during change in temperature.

E.g. to convert 1kg of ice at 0°C to steam at 100°C:
Ice at 0°C to water at 0°C = 1kg × 335kJ/kg = 335kJ
Water at 0°C to water at 100°C = 1kg × Shc of water (approx. 4.2kJ/kg K) × 100K = 420kJ
Water at 100°C to steam at 100°C = 1kg × 2260kJ/kg = 2260kJ
The total heat energy will be 335 + 420 + 2260 = 3015kJ
Note: Total heat is also known as enthalpy.

HEAT ENERGY TRANSFER can be by:

• Conduction – heat travelling along or through a material without appreciable change in position of the material particles.

• Convection – heat travelling by movement of particles as they expand or contract.

• Radiation – heat transfer by electromagnetic waves through space from one material body to another.

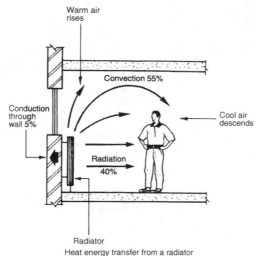

Heat energy transfer from a radiator
Note: Most heat energy is convected from a radiator, although the term radiator is preferred to differentiate from a convector, where about 90% of heat is convected.

High-Temperature, Pressurised Hot Water Heating Systems

Pressurisation allows water to be heated up to 200°C without the water changing state and converting to steam. This permits the use of relatively small diameter pipes and heat emitters, but for safety reasons, these systems are only suitable in commercial and industrial situations. Even then, convectors are the preferred emitter as there is less direct contact with the heating surface. Alternatively, radiators must be encased or provision made for overhead unit heaters and suspended radiant panels. All pipes and emitters must be specified to the highest standard.

Water can be pressurised by steam or nitrogen. Pressurised steam is contained in the upper part of the boiler. To prevent the possibility of the pressurised water from 'flashing' into steam, a mixing pipe is required between the heating flow and return. Nitrogen gas is contained in a pressure vessel separate from the boiler. It is more popular than steam as a pressurising medium, being easier to control, clean, less corrosive and less compatible with water. Air could be an alternative, but this is more corrosive than nitrogen and water soluble.

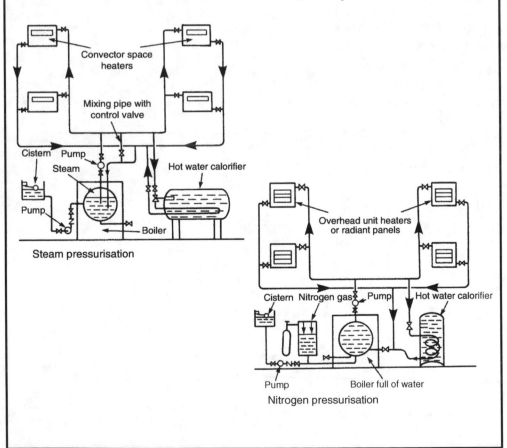

Steam pressurisation

Nitrogen pressurisation

When pressurising with nitrogen, it is important that the pressure increases in line with temperature. If it is allowed to deviate the water may 'flash', i.e. convert to steam, causing system malfunction and possible damage to equipment.

To commission the system:

1. Water is pumped from the feed and spill cistern.
2. Air is bled from high levels and emitters.
3. Air is bled from the pressure vessel until the water level is at one-third capacity.
4. Nitrogen is charged into the pressure vessel at half-design working pressure.
5. Boiler is fired and expansion of hot water causes the water volume and nitrogen pressure in the vessel to double.

Note: Pressure vessel must be carefully designed to accommodate expanded water – approximately 4% of its original volume.

Safety features include a pressure control relay. This opens a motorised valve which lets excess water spill into the feed cistern if the boiler malfunctions and overheats. It also detects low pressure, possibly from system leakage and engages the feed pump to replenish the water and pressure.

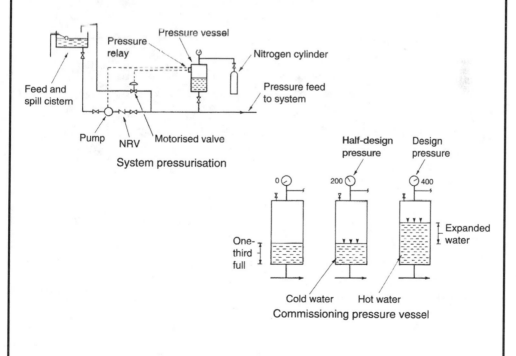

System pressurisation

Commissioning pressure vessel

Steam Heating Systems – 1

Steam was the energy source of the Victorian era. At this time, electricity and associated equipment that we now take for granted were in the early stages of development. Steam was generated in solid fuel boilers to power engines, drive machines and for a variety of other applications, not least as a medium for heat emitters. In this latter capacity, it functioned well, travelling over long distances at high velocity (24–36 m/s) without the need for a pump.

By contemporary standards, it is uneconomic to produce steam solely for heating purposes. However, it can be used for heating where steam is available from other processes. These include laundering, sterilising, kitchen work, manufacturing and electricity generation. Most of these applications require very high pressure; therefore pressure-reducing valves will be installed to regulate supply to heating circuits.

Steam systems maximise the latent heat properties of water when evaporating. This is approximately 2260 kJ/kg at boiling point, considerably more than the sensible heat property of water at this temperature of approximately 420 kJ/kg. Because of this high heat property, the size of heat emitters and associated pipework can be considerably less than those used for hot water systems.

Steam terminology:

Absolute pressure – gauge pressure + atmospheric pressure (101.325 kN/m^2 or kPa).

Latent heat – heat which produces a change of state without a change in temperature, i.e. heat which converts water to steam.

Sensible heat – heat which increases the temperature of a substance without changing its state.

Enthalpy – total heat of steam expressed as the sum of latent heat and sensible heat.

Dry steam – steam which has been completely evaporated, contains no droplets of liquid water.

Wet steam – steam with water droplets in suspension, present in the steam space, typically in pipes and emitters.

Flash steam – condensate re-evaporating into steam after passing through steam traps.

Saturated steam – steam associated with or in contact with the water in the boiler or steam drum over the boiler.

Superheated steam – steam which is reheated or has further heat added after it leaves the boiler.

Classification – low pressure, 35kPa–170kPa (108–130°C)
          medium pressure, 170kPa–550kPa (130–160°C)
          high pressure, over 550kPa (160°C and above).
          Note: Gauge pressures shown.

Systems can be categorised as gravity or mechanical. In both, the steam flows naturally from boiler to emitters without the need for a pump. In the mechanical system, a positive displacement pump is used to lift condensed steam (condensate) into the boiler. Steam pressure should be as low as possible as this will increase the latent heat capacity. A steam trap prevents energy loss at each emitter. These are fitted with a strainer or filter to contain debris and will require regular cleaning. A sight glass after each trap gives visual indication that the trap is functioning correctly, i.e. only condensate is passing. On long pipe runs a 'drip relay' containing steam valve, strainer, trap, sight glass, and a gate valve will be required to control condensing steam. This is represented by the strainer and trap in the mechanical system shown below. Expansion loops or bellows will also be required on long pipe runs to absorb thermal movement. All pipework and accessories must be insulated to a very high standard.

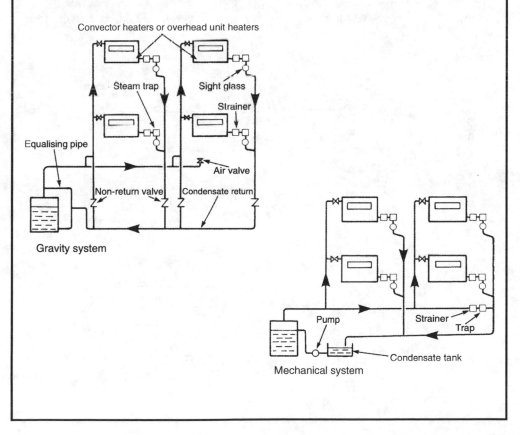

Convector heaters or overhead unit heaters

Steam trap

Sight glass

Strainer

Equalising pipe

Air valve

Non-return valve | Condensate return

Gravity system

Pump

Strainer

Trap

Condensate tank

Mechanical system

Steam Traps

The purpose of a steam trap is to separate steam from condensate, retaining the energy-efficient steam in distribution pipework and emitters. Traps are produced in various forms and sizes to suit all situations, some of which are shown below. The thermostatic and bimetallic types are for relatively small applications such as radiators and unit heaters. The bucket and ball-float types are more suited to separating larger volumes of condensate and steam at the end of long pipe runs and in calorifiers.

Thermostatic – bellows expand or contract in response to steam or condensate, respectively. Lower temperature condensate passes through.

Bimetallic – condensate flows through the trap until higher temperature steam bends the strip to close the valve.

Bucket – condensate sinks the bucket. This opens the valve, allowing steam pressure to force water out until the valve closes.

Ball-float – the copper ball rises in the presence of condensate opening the valve to discharge water until steam pressure closes the valve.

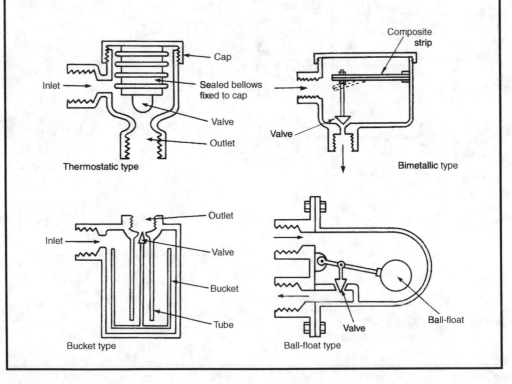

Non-storage type – used for providing instantaneous hot water for space heating. The steam tube bundle or battery occupies a relatively large area compared to the surrounding amount of water. To avoid temperature override and to control the steam flow, a thermostat and modulating valve must be fitted.

Storage type – used to store hot water for manufacturing processes and/or washing demands. Unlike non-storage calorifiers, these have a low steam-to-water ratio, i.e. a relatively small battery of steam pipes surrounded by a large volume of water.

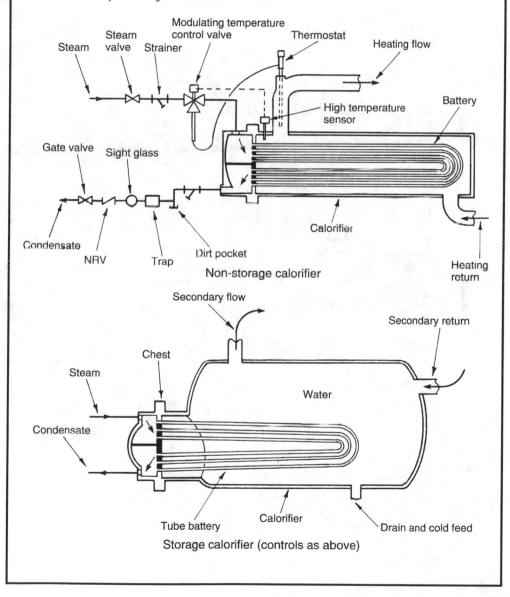

Non-storage calorifier

Storage calorifier (controls as above)

Steam Overhead Unit Heater

High-level fan-assisted unit heaters are often the preferred means of heat emission for use with steam heating systems. Unless housed, radiators and convectors can be dangerously hot to touch, and they take up useful floor space in industrial production and warehouse premises. A typical installation is shown below with a non-return type of check valve to control the flow of condensate.

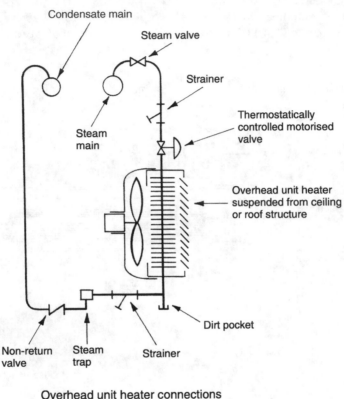

Overhead unit heater connections

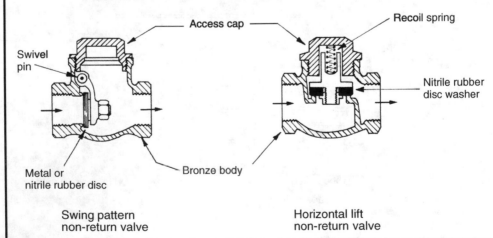

A district heating system is in principle an enlarged system of heating one building, extended to heat several buildings. It can be sufficiently large to heat a whole community or even a small town from one centralised boiler plant. Centralising plant and controls saves space in individual buildings. An effective plant management service will ensure the equipment is functioning to peak efficiency. Each building owner is required to pay a standing charge for the maintenance of plant and to subscribe for heat consumed through an energy-metered supply, similar to other utilities. An energy meter differs from a capacity or volume meter by monitoring the heat energy in the water flow, as this will vary in temperature depending on the location of buildings. The boiler and associated plant should be located in close proximity to buildings requiring a high heat load, e.g. an industrial estate. Long runs of heating pipes are required, and these must be well insulated. They are normally located below ground but may be elevated around factories. Systems can incorporate industrial waste incinerators operating in parallel with conventional boilers and may also use surplus hot water from turbine cooling processes in power stations or electricity generators. This is known as Combined Heat and Power.

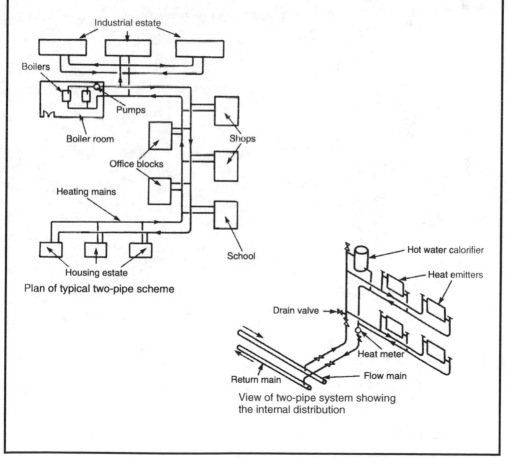

Plan of typical two-pipe scheme

View of two-pipe system showing the internal distribution

The three-pipe system is similar to the two-pipe system except for an additional small-diameter flow pipe connected to the boilers. This is laid alongside the larger diameter flow pipe and has a separate circulation pump. This smaller flow pipe is used during the summer months when space heating is not required, although in the intermediate seasons, it could supply both with limited application to heating. It should have enough capacity to supply the heating coils in the hot water storage cylinders plus a small reserve. It can be seen as an economic measure to reduce hot water heating volume, energy loss from the larger diameter pipe and pump running costs. A common large-diameter return pipe can be used.

Pipes must be at least 450mm below the surface as protection from vehicle loads. They must also be well insulated against heat loss and frost damage if water is not circulating. Insulation must be waterproof and the pipes protected from corrosion. Inevitably there will be some heat losses from the mains pipework. This will approximate to 15% of the system heating load.

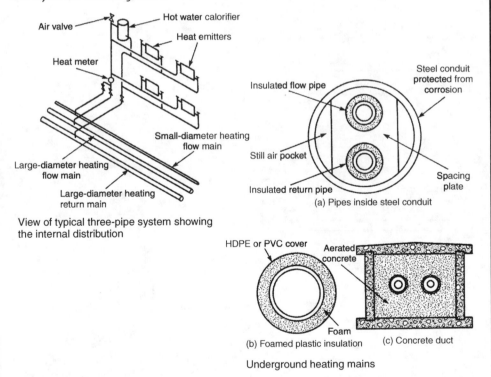

View of typical three-pipe system showing the internal distribution

(a) Pipes inside steel conduit

(b) Foamed plastic insulation

(c) Concrete duct

Underground heating mains

Note: Pre-insulated steel carrier pipes are used in many existing installations. Cross-linked polyethylene (PEX) is now the preferred material as unlike steel, will not corrode. PEX pipes are supplied rolled in long lengths, thereby reducing the amount of site joints.

The four-pipe system supplies both hot water and space heating as two separate systems. Individual hot water storage cylinders are not required, as large-capacity calorifiers are located in the boiler plant room and possibly at strategic locations around the district being served. This considerably simplifies the plumbing in each building as cold water storage cisterns are also unnecessary, provided all cold water outlets can be supplied direct from the main. However, the boiler plant room will be considerably larger to accommodate the additional components and controls. Excavation and installation costs will also be relatively expensive, but system flexibility and closure of the heating mains and associated boilers during the summer months should provide economies in use.

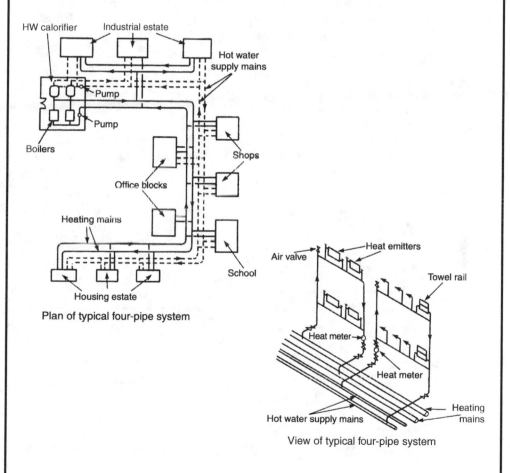

Plan of typical four-pipe system

View of typical four-pipe system

Combined Heat and Power (CHP)

Potential for more economic use of electricity-generating plant can be appreciated by observing the energy waste in the large plumes of condensing water above power-station cooling towers. Most power-stations are only about 50% efficient, leaving a considerable margin for reprocessing the surplus hot water.

Combining electricity generation with a supply of hot water (cogeneration) has become viable since the deregulation and privatisation of electricity supply. Prior to this, examples were limited to large factory complexes and remote buildings (e.g. prisons), which were independent of national power generation by special licence. Until recently, CHP has only been practical for large buildings or expansive collections of buildings such as university campuses and hospitals. The development of gas-fuelled micro-CHP for use in domestic situations is now viable, using units that are essentially condensing boiler with an electricity generator. See page 743.

Surplus energy from oil- or gas-fired engine-driven alternators occurs in hot water from the engine cooling system and the hot exhaust gases. In a CHP system, the rate of heat energy produced is directly related to the amount of electricity generated. There will be times when available hot water is insufficient. Therefore, a supplementary energy source from a conventional boiler will be required.

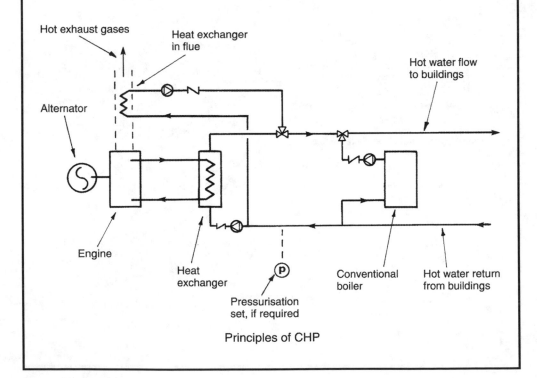

Principles of CHP

All pipe materials expand and contract when subject to temperature change. This linear change must be accommodated to prevent fatigue in the pipework, movement noise, dislocation of supports and damage to the adjacent structure.

Expansion devices:

- Natural changes in direction.
- Axial expansion bellows.
- Expansion loops.

Bellows and loops are not normally associated with domestic installations.

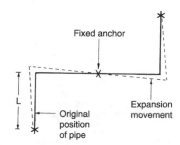

Natural changes in direction or offsets

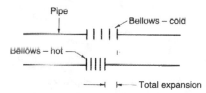

Axial expansion bellows responding to hot water

Bellows are factory-made fittings normally installed 'cold-drawn' to the total calculated expansion for hot water and steam services. The bellows can then absorb all anticipated movement by contraction. Where the pipe content is cold or refrigerated fluids, the bellows are compressed during installation.

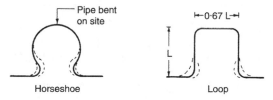

Site-made loops or horseshoe

Pipework Expansion – 2

Coefficients of linear expansion for common pipework materials:

Material	Coeff. of expansion $(m/mK \times 10^{-6})$
Cast iron	10·22
Copper	16·92
Mild steel	11·34
PVC (normal impact)	55·10
PVC (high impact)	75·10
Polyethylene (low density)	225·00
Polyethylene (high density)	140·20
ABS (acrylonitrile butadiene styrene)	110·20

E.g. An 80-mm diameter steel pipe of 20m fixed length is subject to a temperature increase from 20°C to 80°C (60K).

Formula:

Expansion = Original length × coeff. of expansion × Temp. diff.

$$= 20 \times 11·34 \times 10^{-6} \times 60$$

$$= 0·0136m \text{ or } 13·6mm$$

Single offset:

$L = 100 \sqrt{zd}$

L = see previous page

z = expansion (m)

d = pipe diameter (m)

$L = 100 \sqrt{0·0136} \times 0·080 = 3·30m$ minimum.

Loops:

$L = 50 \sqrt{zd}$

$L = 50\sqrt{0·0136} \times 0·080 = 1·65m$ minimum.

Top of loop $= 0·67 \times L = 1·10m$ minimum.

Notes:

- Provide access troughs or ducts for pipes in screeds (Part 15).
- Sleeve pipework through holes in walls, floors and ceilings (see page 476 for fire sealing).
- Pipework support between fixed anchors to permit movement, i.e. loose fit brackets and rollers.
- Place felt or similar pads between pipework and notched joists.
- Branches to fixtures to be sufficient length and unconstrained to prevent dislocation of connections.
- Allow adequate space between pipework and structure.

Thermostatic control of heating and hot water systems reduces consumers' fuel bills, regulates the thermal comfort of building occupants and improves the efficiency of heat-producing appliances. Approved Document L to the Building Regulations effects these provisions. This has the additional objective of limiting noxious fuel gases in the atmosphere and conserving finite natural fuel resources.

A room thermostat should be sited away from draughts, direct sunlight and heat emitters, at between 1·2 and 1·5m above floor level. Thermostatic radiator valves may also be fitted to each emitter to provide independent control in each room. A less expensive means of controlling the temperature in different areas is by the use of thermostatically activated zone valves to regulate the temperature of individual circuits.

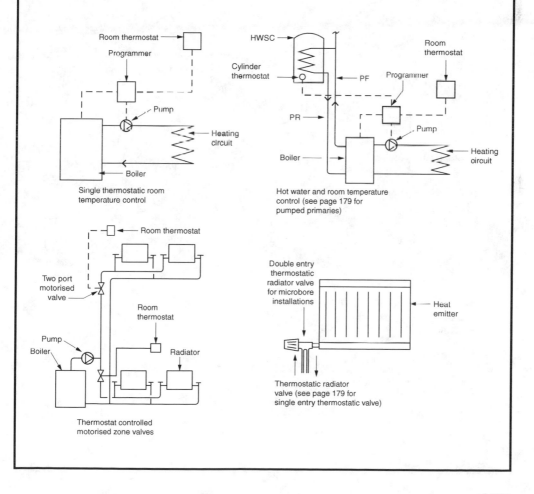

Single thermostatic room
temperature control

Hot water and room temperature
control (see page 179 for
pumped primaries)

Thermostat controlled
motorised zone valves

Thermostatic radiator
valve (see page 179 for
single entry thermostatic valve)

Thermostatic Control of Heating Systems – 2

Three port motorised valves direct hot water from the boiler to the primary and heating circuits. They can function as mixing or diverting depending on system design. The mixing valve has two inlets and one outlet. The diverting valve has one inlet and two outlets.

Valve movement is effected by a small electric motor or actuator that responds to thermostatic demand. When powered, the actuator rotates the valve stem through a simple geared drive against a spring resistance. When pre-set temperatures are satisfied, the air or water thermostat switches off electrical supply to the valve. The spring and geared mechanism return the valve to its resting position.

Although electrically powered, motorised valves incorporate a manual lever. In normal use, this is in the 'Auto' position. It can be moved to the 'Manual' position for draining down and filling the system.

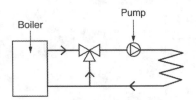

Mixing valve provides constant
flow rate and variable flow temperature

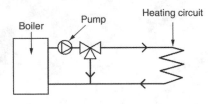

Diverting valve provides constant
flow temperature and variable flow

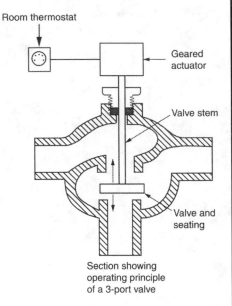

Section showing
operating principle
of a 3-port valve

The diverter valve may be used to close the heating circuit to direct hot water from the boiler to the hot water cylinder. The reverse is also possible, depending on whether hot water or heating is considered a priority. With either, when the thermostat on the priority circuit is satisfied, it effects a change in the motorised diverter valve to direct hot water to the other circuit.

A rod-type thermostat may be fitted into the hot water storage cylinder, or a surface contact thermostat applied below the insulation. At the pre-set temperature (about 60°C), a brass and invar steel strip expands to break contact with the electricity supply. A room thermostat also operates on the principle of differential expansion of brass and invar steel. Thermostatic radiator valves have a sensitive element which expands in response to a rise in air temperature to close the valve at a pre-set temperature, normally in range settings 5–27°C. Sensors are either a thermostatic coil, or a wax or liquid charged compartment which is insulated from the valve body.

A clock controller sets the time at which the heating and hot water supply will operate. Programmers are generally more sophisticated, possibly incorporating seven- or 28-day settings, bypass facilities and numerous on/off functions throughout the days.

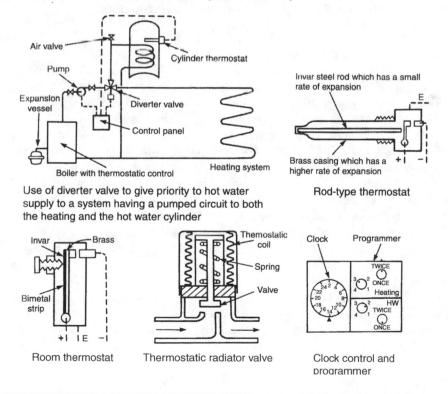

Air valve

Cylinder thermostat

Pump

Expansion vessel

Diverter valve

Control panel

Boiler with thermostatic control

Heating system

Use of diverter valve to give priority to hot water supply to a system having a pumped circuit to both the heating and the hot water cylinder

Invar steel rod which has a small rate of expansion

E

Brass casing which has a higher rate of expansion

Rod-type thermostat

Invar — Brass

Bimetal strip

+I IE −I

Room thermostat

Thermostatic coil

Spring

Valve

Thermostatic radiator valve

Clock   Programmer

TWICE
ONCE
Heating
HW
TWICE
ONCE

Clock control and programmer

Ref. Building Regulations, Approved Document L Conservation of fuel and power, Vol. 1 – Dwellings.

In 2002, revised mandatory standards of controls for hot water and heating installations were introduced, the objective being to limit consumption of finite fuel resources and to reduce the emission of atmospheric pollutants. Subsequently all new installations and existing systems undergoing replacement of components are affected.

Requirements for 'wet' systems –

• Only boilers of a minimum standard can be installed. See energy efficiency values on pages 132 and 133.

• Hot water storage cylinders must be to a minimum acceptable standard, i.e. BSs 1566 and 3198: Copper indirect cylinders and hot water storage combination units for domestic purposes, respectively, for vented systems. BS EN 12897: Water supply. Specification for indirectly heated unvented (closed) storage water heaters. Vessels for unvented systems may also be approved by the BBA, the WRC or other accredited European standards authority. See pages 754 and 755.

• New and replacement systems to be fully pumped. If it is impractical to convert an existing gravity (convection) hot water circulation system, the heating system must still be pumped, i.e. it becomes a semi-gravity system (see pages 177 and 182). Where a new boiler is installed, a fully pumped system is required. Existing system controls to be upgraded to include a cylinder thermostat and zone (motorised) valve to control the hot water circuit temperature and to provide a boiler interlock. Other controls are a programmer or clock controller, a room thermostat and thermostatic radiator valves (TRVs to BS EN 215) on all radiators except in rooms with a thermostat and in bathrooms.

Note: The boiler is said to be 'interlocked' when switched on or off by the room or cylinder thermostat (or boiler energy management system). The wiring circuit to and within the boiler and to the pump must ensure that both are switched off when there is no demand from the hot water or heating system, i.e. the boiler must not fire unnecessarily even though its working thermostat detects the water content temperature to be below its setting.

continued . . . .

Requirements for `wet' systems (continued) –

• Independent/separate time controls for hot water and space heating. The exceptions are

(1) combination boilers which produce instantaneous hot water and

(2) solid fuel systems.

• Boiler interlock to be included to prevent the boiler from firing when no demand for hot water or heating exists.

• Automatic bypass valve to be fitted where the boiler manufacturer specifies a bypass circuit.

Note: A circuit bypass and automatic control valve are specified by some boiler manufacturers to ensure a minimum flow rate while the boiler is firing. This is particularly useful where TRVs are used, as when these begin to close, a bypass valve opens to maintain a steady flow of water through the boiler. An uncontrolled open bypass or manually set bypass valve is not acceptable as this would allow the boiler to operate at a higher temperature, with less efficient use of fuel.

• Independent temperature control in living and sleeping areas (TRVs could be used for bedroom radiators).

• Installations to be inspected and commissioned to ensure efficient use by the local authority Building Control Department or self-certified by a `competent person', i.e. Gas Safe Registered, OFTEC or HETAS approved.

• System owners/users to be provided with equipment operating guides and maintenance instructions. This `log-book' must be completed by a `competent person'.

• Dwellings with less than 150 m^2 living space/floor area to have two space healing zones with independent temperature control, one dedicated to the general living area.

• Dwellings with over 150 m^2 living space/floor area to have the heating circuits divided into at least two zones. Each to have independent time and temperature control and to be included in the boiler interlock arrangement. A separate control system is also required for the hot water.

continued . . . .

Requirements for 'dry' systems –

• Warm air or dry systems (see page 196) should also benefit fully from central heating controls. Gas-fired air heaters should satisfy the following standard:

BS EN 778: Domestic gas-fired forced convection air heaters for space heating not exceeding a net heat input of 70kW, without a fan to assist transportation of combustion air and/or combustion products, or

• Replacement warm air heat exchanger units can only be fitted by a 'competent person'. All newly installed ducting should be fully insulated. See page 17 for definition of a 'competent person'.

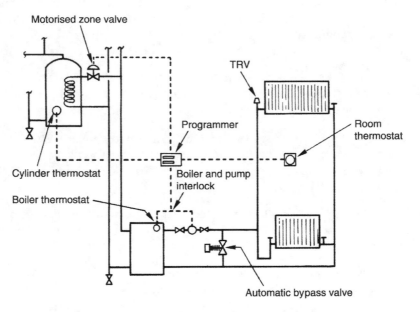

Typical semi-gravity system of hot water and heating controls

Note: Boiler and pump interlock is the wiring configuration as explained on the previous two pages.

Schematic of control systems –

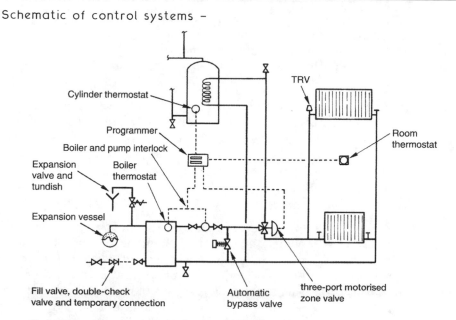

Cylinder thermostat

TRV

Programmer

Boiler and pump interlock

Room
thermostat

Expansion
valve and
tundish

Boiler
thermostat

Expansion vessel

Fill valve, double-check
valve and temporary connection

Automatic
bypass valve

three-port motorised
zone valve

Typical fully pumped system of hot water and heating

Note: Boiler and pump interlock is the wiring configuration as explained
on pages 180 and 181.

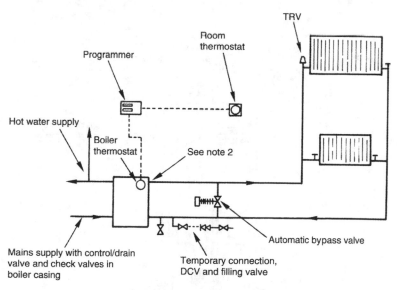

TRV

Room
thermostat

Programmer

Hot water supply

Boiler
thermostat

See note 2

Mains supply with control/drain
valve and check valves in
boiler casing

Temporary connection,
DCV and filling valve

Automatic bypass valve

Typical combination boiler (see also page 106)

Notes:
1. Hot water draw-off taps supplied direct from mains, through instantaneous water heater.
2. Heating water is sealed. Additional components include heating pump and expansion
   vessel in boiler casing, with expansion valve and tundish (see upper diagram).

Automatic Bypass Control

Modern boilers and heating systems are low water content to provide fuel efficiency and a rapid response. Therefore, to maintain a minimum flow through the boiler and to accommodate pump overrun, most boiler manufacturers will specify that a system bypass be used with their products.

An open bypass or bypass with a valve set in a fixed open position will satisfy the basic objectives, but with the boiler flow pipe feeding the return pipe at all operating times, the boiler will need to function at a higher temperature than necessary to fulfil system requirements. In addition, the heat energy transferred into the system will be limited, as a proportion of boiler flow water will be continually diverted through the bypass pipe.

Thermostatically controlled radiator valves and motorised zone and circuit valves are now standard installation. With these controls, parts of the system may be closed, leaving only a limited demand for heat. Selective demands will cause varying pump pressures, unless a bypass valve is in place to automatically adjust, regulate and respond to pressure changes from the pump. Some applications are shown on the previous two pages.

Typical automatic bypass valve –

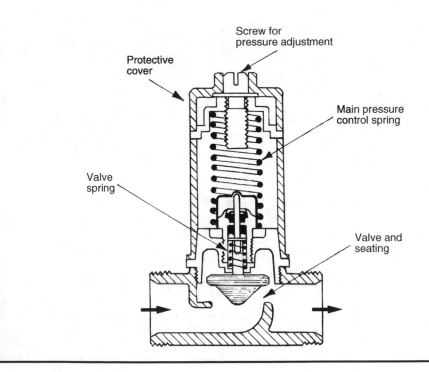

In addition to high-efficiency boilers, optimiser controls, thermostatic radiator valves and other fuel-saving measures considered elsewhere in this chapter, further economies and user comforts can be achieved by installing programmable thermostats with motorised valves dedicated to heat only a specific part or zone within a building.

Zone control or zoning provides fuel saving and user convenience by regulating heat/energy distribution to particular locations in response to occupancy. This prevents wasteful distribution of heat in a building that is not fully utilised.

Examples where zoning has greatest benefit:

- Unused upper-floor rooms, i.e. bedrooms, during daytime.
- Supplementary accommodation, bedsit or granny flat.
- Conservatories or other rooms with heating characteristics which are weather and seasonally variable.
- Office in the home, occupied while the remainder of the house is not.
- People with irregular working patterns (e.g. shift workers) may require heating downstairs when others will not.
- Insomniacs and people who get up regularly in the night (the elderly?) may require heating in a specific room at unusual times.

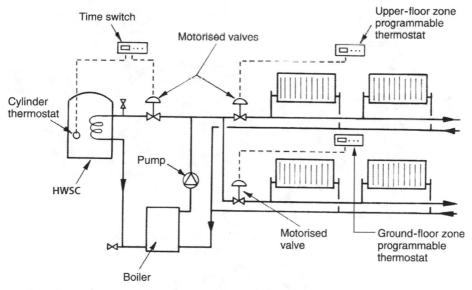

Example using zone control programmable thermostats with 'wireless' (radio frequency signals) regulation of the boiler

Note: See page 160 for boiler feed/fill and expansion facilities.

Time Proportional and Integral Programmable Control

Time proportional and integral (TPI) control matches boiler firing to system load demand. Boiler output is regulated proportionally with the temperature difference between varying room air temperature and the required set design temperature.

Standard room thermostats are effectively temperature-controlled switches, whereas TPIs incorporate programmable software. This provides accurate temperature control with minimal boiler firing, thereby maximising fuel use efficiency.

Graphical comparison between traditional on-off thermostat system operation with TPI programmable control —

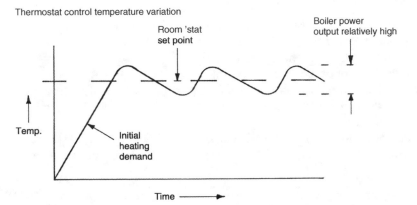

Thermostat control temperature variation

Room 'stat set point

Boiler power output relatively high

Temp.

Initial heating demand

Time

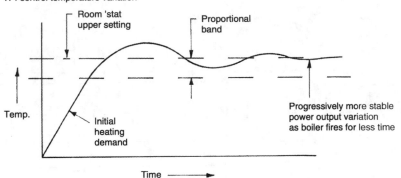

TPI control temperature variation

Room 'stat upper setting

Proportional band

Temp.

Initial heating demand

Progressively more stable power output variation as boiler fires for less time

Time

TPI operating principle – when room air temperature reaches the lower part of the proportional band, the TPI thermostat reduces the boiler firing time relative to heating demand (e.g. a proportional band set between 20°C and 22°C). When room temperature is at 21°C (i.e. half-way across the band), the boiler fires proportionally for half of the cycle period. Likewise, as room temperature increases above 21°C, the boiler will fire proportionally less. Control temperature will be smooth as shown in the graph, varying little relative to system performance with a traditional on-off thermostat. TPI complements boiler interlock requirements (see pages 180–181), optimising boiler firing time.

Piped water systems in modern, highly insulated buildings are unlikely to be affected by modest sub-zero external temperatures. Nevertheless, an automatic 24-hour frost damage fail-safe facility may be specified as a client requirement or to satisfy an insurer's standards. This is particularly appropriate for buildings located in very exposed parts of the country, and for buildings that are periodically unoccupied.

Frost thermostat – similar in appearance to a normal room thermostat but with a lower temperature range. Installed internally or externally on a north-facing wall and set to about 5°C.

Pipe thermostat – strapped to an exposed section of pipe to detect the temperature of the contents.

Both types of thermostat can be used independently or wired in series to the same installation as shown below. Whether used in combination or individually, they are installed to bypass the time control.

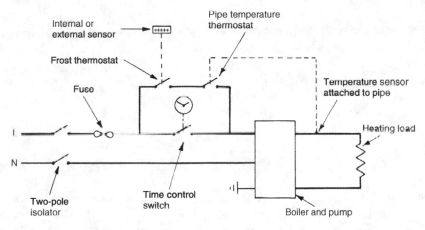

Thermostatic frost protection

Trace element frost protection – a low-voltage electric heating element taped to the pipe surface. Used mainly for external piped services.

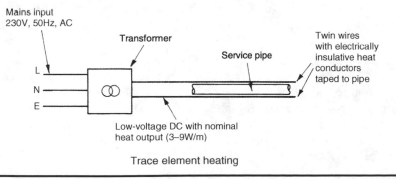

Trace element heating

Wireless Heating Controls

Wireless or radio frequency (RF) band communications are in common use (e.g. remote keyless entries, TV controls, portable telephones, burglar alarm systems, garage doors, estate gates and computer links). For heating system controls, this form of communications technology offers many benefits to both installer and property owner/end user, not least a saving in installation time, as hard-wiring between thermostatic controls, boiler controls, motorised valves and programmer is not required. There is also considerably less disruption to the structure and making good the superficial damage from channelling walls, lifting floorboards, drilling walls and holing joists. This is particularly beneficial where work is applied to existing buildings and refurbishment projects.

In principle, a battery cell power source is used to transmit a secure, unique radio signal from the hot water storage cylinder thermostat and each of the room thermostats. This signal is recognised by a receiver which is hard-wired to switching units placed next to the boiler, pump and motorised valves. Installation cabling is therefore reduced to an absolute minimum at localised receivers only. The appearance and location of thermostats are similar to conventional hard-wired units. The capital cost of components is significantly more, but the savings in installation time will justify this expenditure.

The use of radio frequencies for communications systems in modern society is strictly controlled and regulated by operator licensing regulations to prevent interference and cross-communications. For wireless domestic heating controls, this is not a problem as the unique low-power signals function at around 430MHz at a short range, typically up to 30 metres. At this specification, an operating licence is not required as it satisfies the recommendations of the European Telecommunications Standards Institute, European Standard EN 300-220 for equipment in the 25 to 1000Mhz frequency band at power levels up to 500mW.

To commission RF controls, each thermostat is digitally coded and programmed to the associated signal receiver. Therefore, the controls in one building will not interfere with similar controls in adjacent buildings, and vice versa. Siting of controls will require some care as large metal objects can inhibit the signalling function. Locations of the boiler and hot water storage cylinder are obvious examples that will need consideration.

There are a variety of wiring schemes depending on the degree of sophistication required and the extent of controls, i.e. thermostats, motorised valves, etc. Boiler and control equipment manufacturers provide installation manuals to complement their products. From these, the installer can select a control system and wiring diagram to suit their client's requirements.

The schematic diagrams shown relate to a gravity or convected primary flow and return and pumped heating system (see page 150) and a fully pumped hot water and heating system using a three-way motorised valve (see page 179).

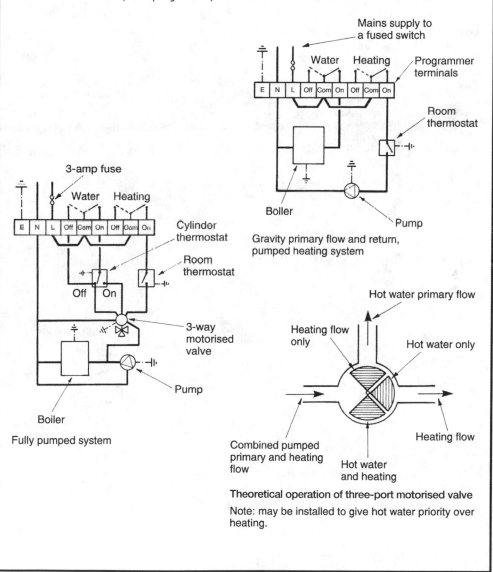

Gravity primary flow and return, pumped heating system

Fully pumped system

Theoretical operation of three-port motorised valve

Note: may be installed to give hot water priority over heating.

Energy Management Systems – 1

Optimum Start Controls – these have a control centre which computes the building's internal temperature and the external air temperature. This is used to programme the most fuel-efficient time for the boiler and associated plant to commence each morning and bring the building up to temperature ready for occupation. The system may also have the additional function of optimising the system shutdown time.

Compensated Circuit – this system also has a control centre to compute data. Information is processed from an external thermostat/sensor and a heating pipework immersion sensor. The principle is that the boiler water delivery temperature varied relative to outside air temperature. The warmer the external air, the cooler the system water and vice versa.

The capital cost of equipment for these systems can only be justified by substantial fuel savings. For large commercial and industrial buildings of variable occupancy, the expenditure is worthwhile, particularly in the intermediate seasons of autumn and spring, when temperatures can vary considerably from day to day.

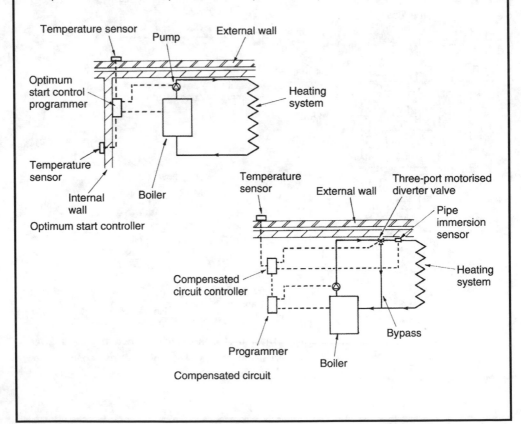

Weather compensated circuit – accurate control of indoor temperature depends on monitoring and modulating system heat input with the heat losses from a building. This differs considerably from the traditional heating system controlled solely by a thermostat. A thermostat functions relative to internal air temperature, switching on the boiler to supply water at a pre-set temperature.

Optimum comfort and economy are achieved if the heating system water is constantly circulated with temperature varied to suit occupancy needs. A balance is achieved by incorporating into the heating programme the external air temperature and internal heat gains from people, machinery, solar sources, etc. At the centre of the installation is a compensator-controlled three- or four-port motorised valve to blend the required amount of cool system return water with hot water supplied from the boiler. This ensures a continuous supply of water at the required temperature to satisfy ambient conditions. The motorised valve setting varies depending on the boiler water temperature, the system supply water temperature, internal air temperature and outdoor air temperature. The latter is measured by a thermostatic sensor fitted to a north-facing wall. Data from all four sources are computed in the compensator for positioning the motorised valve, activating the system circulator and to regulate the boiler functions.

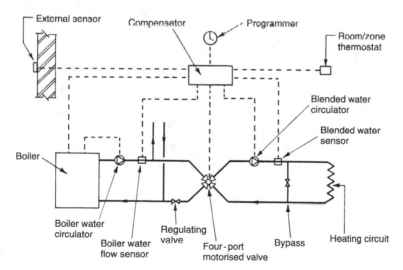

Weather compensated control system

Note: Variable water temperature systems are particularly suited to underfloor heating. The heating demand is more evenly controlled through the 'thermal' floor than by on–off thermostatic switching.

Energy Management Systems – 3

Energy management systems can vary considerably in complexity and degree of sophistication. The simplest timing mechanism to switch systems on and off at predetermined intervals on a routine basis could be considered as an energy management system. This progresses to include additional features such as programmers, thermostatic controls, motorised valves, zoning, optimum start controllers and compensated circuits. The most complex of energy management systems have a computerised central controller linked to numerous sensors and information sources. These could include the basic internal and external range shown schematically below, along with further processed data to include the time, the day of the week, the time of year, percentage occupancy of a building, meteorological data, system state feedback factors for plant efficiency at any one time and energy gain data from the sun, lighting, machinery and people.

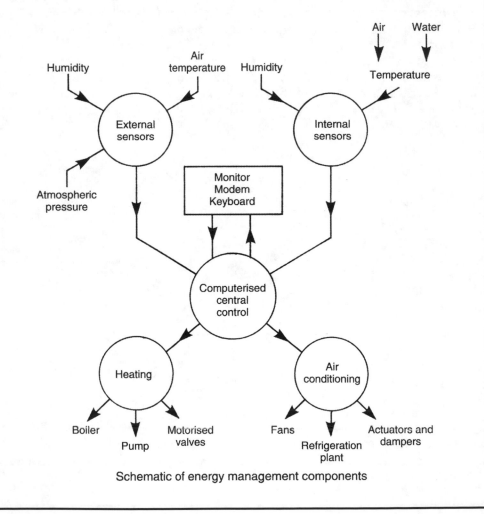

Schematic of energy management components

Operation – in principle, air is sourced through an air intake located in an externally housed plant room. An electrical compression unit pumps air to a storage vessel, otherwise known as a receiver. The compressor responds to upper and lower pressure sensors in the receiver. Most systems operate within a pressure range of 7 to 10 bars. Types of compressor vary, the most common being a dynamic centrifugal fan that draws in air and accelerates it through high speed impellors. The compressed air becomes very hot and requires cooling in an after cooler. In this state air contains a large amount of moisture, initially separated and drained at the cooler and thereafter at condensate receivers and a dryer. Contaminants, typically suspended particles, oil and odours can also be removed by installing selected filtration units before the air is finally delivered to process controls and other uses.

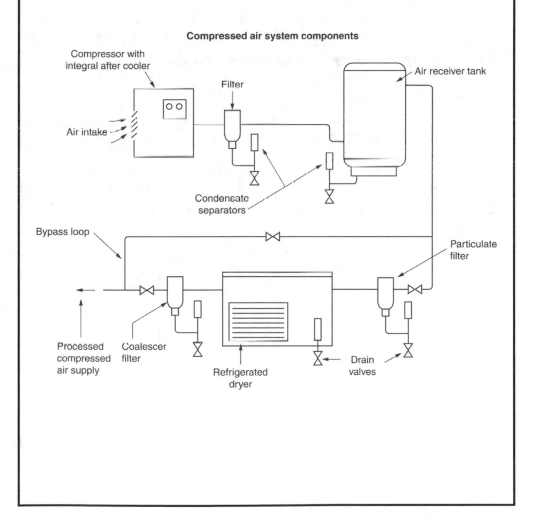

Compressed air system components

Compressed Air System Installation

Application – where building insurers and fire consultants require an alternative power source to electrical equipment when a production environment is deemed to be fire hazardous. Examples are piped services controls in industrial manufacturing processes that use or produce volatile chemicals and liquids including paints and plastics. Use of compressed air has other uses, and it is not just limited to process controls. Other applications include effluent aeration, glass blowing, vehicle tyre inflation and as a power source for hand tools.

Pneumatic pipelines – air delivery is conveyed from receiver to terminals in rigid metal or reinforced plastic pipes. Terminal transfer to controls, sensors and actuators is in microbore soft copper or flexible plastic tubing. Terminal tube fittings and connectors can be of brass or plastic push-fit or compression types. Plastic tubing is usually synthetic nylon or polyurethane, available in a variety of colour ranges. Standard outside diameters are 3, 4 and 6mm. Braided polyvinylchloride (PVC) and polytetrafluorethylene (PTFE) can also be used.

Linear actuators – a form of motorised valves where valve position is effected by converting compressed air or pneumatic energy into mechanical movement. Actuators can be single or double acting.

- Single acting requires compressed air to push a piston to close a valve. When the pressure is removed, the piston is retracted by a spring to open the valve.
- Double acting is more responsive but requires two separate compressed air supplies. One supply extends the piston, and the other retracts it.

Operating principle of pneumatic linear piston actuator

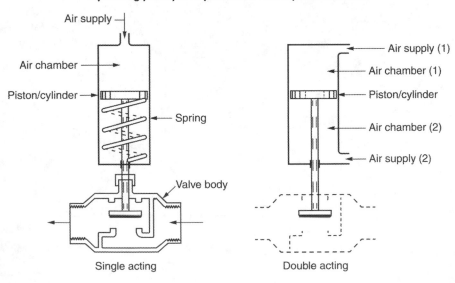

Single acting

Double acting

194

Linear diaphragm actuators are a variation of the single acting type shown in principle on the preceding page. A thin flexible membrane or diaphragm responds to a pneumatic supply. Air can only be supplied to one face of the diaphragm, hence the single action. These either function directly using a spring to retract or in reverse using a spring to extend.

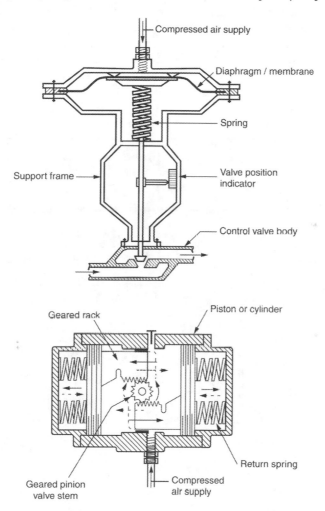

Rotary actuators are an alternative that comprise a pair of geared racks that drive a corresponding pinion attached to a valve stem. The racks attach to pistons contained within a sealed chamber. Air supply pushes the pistons and racks outwards to rotate the valve stem. Removal of air activates the valve stem in the opposing direction by force from return springs. Double acting rotary actuators are also available with two air supplies, the second supply replacing the need for return springs.

Warm Air Heating System

If there is sufficient space within floors and ceilings to accommodate ducting, warm air can be used as an alternative to hot water in pipes. There are no obtrusive emitters such as radiators. Air diffusers or grilles with adjustable louvres finish flush with the ceiling or floor. The heat source may be from a gas, oil or solid fuel boiler with a pumped supply of hot water to a heat exchanger within the air distribution unit. The same boiler can also be used for the domestic hot water supply. Alternatively, the unit may burn fuel directly, with air delivered around the burner casing. Control is simple, using a room thermostat to regulate heat exchanger and fan. The risk of water leakage or freezing is minimal, but air ducts should be well insulated to reduce heat losses. Positioning grilles in doors is an inexpensive means for returning air to the heater, but a return duct is preferred. Fresh air can be supplied to rooms through openable windows or trickle ventilators in the window frames. If rooms are completely sealed, fresh air should be drawn into the heating unit. The minimum ratio of fresh to recirculated air is 1:3.

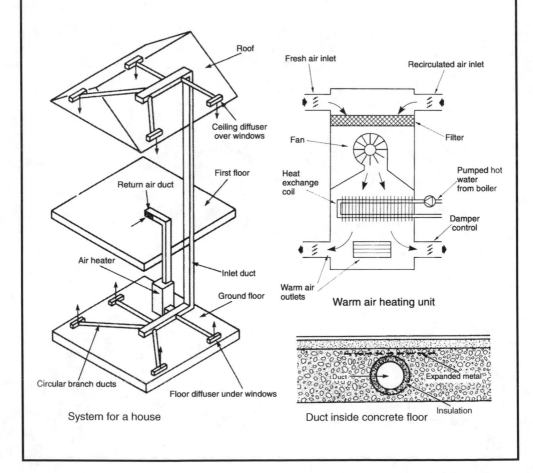

System for a house

Warm air heating unit

Duct inside concrete floor

The thermal transmittance rate from the inside to the outside of a building, through the intermediate elements of construction, is known as the 'U' value. It is defined as the energy in watts per square metre of construction for each degree kelvin temperature difference between the inside and outside of the building, i.e. $W/m^2 K$. 'U' values can vary depending on building type and construction method and with other criteria in the SAP calculations. Guidance is provided in Approved Document L to the Building Regulations.

Limiting area weighted average 'U' values for new and existing dwellings:

	New	Existing
External walls	0.26	0.18
Roof	0.16	0.15
External floor	0.18	0.18
Windows, doors	1.60	1.40
Party wall	0.20	0.00
Air permeability (see note)	8.0	–

Limiting 'U' values to existing buildings allows for the effect of, for example, replacement windows and/or extensions. For existing elements that are improved, e.g. installation of cavity wall insulation, improved 'U' values apply – see Table 4.3 in Vol. 1 of the Approved Document.

E.g. A room in a dwelling house constructed with limiting 'U' values has an external wall area of $30 m^2$ to include $3 m^2$ of double-glazed window. Given internal and external design temperatures of 22°C and –2°C, respectively, the heat loss through this wall will be

Area × 'U' × temperature difference

Wall: $27 \times 0.30 \times 24 = 194.40$

Window: $3 \times 2.00 \times 24 = 144.00$

338.40 watts

Notes: Area weighted average allows for interruption in the construction, e.g. meter cupboard voids.

Limiting refers to the worst acceptable.

Air permeability is the test reference pressure differential measured in m^3 per hour per m^2 of external envelope at 50Pa.

AD L Vol. 2 for 'U' values applicable to Buildings other than dwellings.

BRE Digest 443: Conventions for 'U' value calculations.

A heat emitter should be capable of providing sufficient warmth to maintain a room at a comfortable temperature. It would be uneconomical to specify radiators for the rare occasions when external temperatures are extremely low; therefore, an acceptable design external temperature for most of the UK is –1°C. Regional variations will occur, with a figure as low as –4°C in the north. The following internal design temperatures and air infiltration rates are generally acceptable:

Room	Temperature 0°C	Air changes per hour
Living	21	1·5
Dining	21	1·5
Bed/sitting	21	1·5
Bedroom	18	1·0
Hall/landing	18	1·5
Bathroom	22	2·0
Toilet	18	2·0
Kitchen	18	2·0

The study in the part plan shown below can be used to illustrate the procedure for determining heat losses from a room.

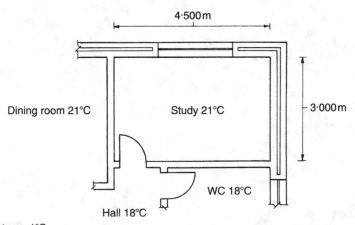

4·500m

Dining room 21°C   Study 21°C   3·000m

WC 18°C

Hall 18°C

External design temperature –1°C
Room height = 2·3m
Door area = 2m^2
Window area = 1·5m^2
Ventilation rate = 1·5a/c per hour
Bedrooms above at 18°C

To determine the total heat loss or heating requirement for a room, it is necessary to obtain the thermal insulation properties of construction. For the room shown on the previous page, the 'U' values can be taken as:

External wall . . . . 0·35W/m² K
Window . . . . 2·00
Internal wall . . . . 2·00
Door . . . . 4·00
Floor . . . . 0·25
Ceiling . . . . 2·50

Heat is also lost by air infiltration or ventilation. This can be calculated and added to the heat loss through the structure, to obtain an estimate of the total heating requirement.

Heat loss by ventilation may be calculated using the following formula:

$$\text{watts} = \frac{\text{room volume} \times \text{A/c per hour} \times \text{temp. diff. (int.-ext.)}}{3}$$

Note: The lower denomination, 3, is derived from density of air (1·2kg/m³) × s.h.c. of air (1000 J/kg K) divided by 3600 seconds.

For the study shown on the previous page:

(4·5 × 3 × 2·3) × 1·5 × (21 − −1) divided by 3 = 341·55 watts

Heat loss through the structure is obtained by summating the elemental losses:

Element	Area (m²)		'U' value		Temp. diff. (int.–ext.)		Watts
External wall	15·75	×	0·35	×	22	=	121·28
Window	1·5		2·00		22		66
Internal wall	8·35		2·00		3		50·10
Door	2		4·00		3		24
Floor	13·5		0·25		22		74·25
Ceiling	13·5		2·50		3		101·25
							436·88

Total heat loss from the study = 341·55 + 436·88 = 778·43, i.e.

779 watts

Heating Design – Radiator Sizing

Radiators are specified by length and height, number of sections, output in watts and number of panels. Sections refer to the number of columns or verticals in cast iron radiators and the number of corrugations in steel panel radiators. Panels can be single, double or triple. Design of radiators and corresponding output will vary between manufacturers. Their catalogues should be consulted to determine exact requirements. The following extract shows that a suitable single-panel radiator for the previous example of 779 watts, could be

450mm high × 1100mm long × 33 sections (832 watts), or
600mm high × 800mm long × 24 sections (784 watts).

Selection will depend on space available. Overrating is usual to allow for decrease in efficiency with age and effects of painting.

Height (mm)	Length (mm)	Sections	Watts (single)	Watts (double)
450	400	12	302	548
	500	15	378	686
	600	18	454	823
	700	21	529	960
	800	24	605	1097
	900	27	680	1234
	1000	30	756	1371
	1100	33	832	1508
	1200	36	907	1645
	1400	42	1058	1919
	1600	48	1210	2194
	1800	54	1361	2468
600	400	12	392	693
	500	15	490	866
	600	18	588	1039
	700	21	686	1212
	800	24	784	1386
	900	27	882	1559
	1000	30	980	1732
	1100	33	1078	1905
	1200	36	1176	2078
	1400	42	1372	2425
	1600	48	1568	2771
	1800	54	1764	3118

Note: Radiators are also manufactured in 300 and 700mm standard heights.

The tabulated heat output data in radiator manufacturers' catalogues should be read with regard to the stated mean water to air temperature difference. This is often referred to as Delta T or ΔT. Mean water temperature can be taken as the mid-point between system flow and return temperatures. Allowing for heat losses, this will be about 70°C. If room design air temperature is 20°C, ΔT will be 50K, the testing criteria as defined in BS EN 442. Typical heat outputs from radiators at 50K are shown on the preceding page. These figures are about 20% lower than outputs specified at 60K.

The BS EN testing process is based on pipe connections to the top and bottom, same ends (TBSE). Other conventions are the traditional gravity circulation connections at top and bottom, opposite ends (TBOE) and the UK standard for pumped circulation to domestic radiators of bottom, opposite ends (BOE).

Radiator connections –

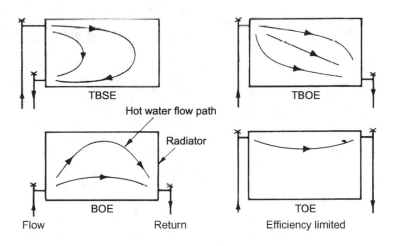

Ref. BS EN 442-1, and 2: Radiators and convectors. Technical specifications and requirements.

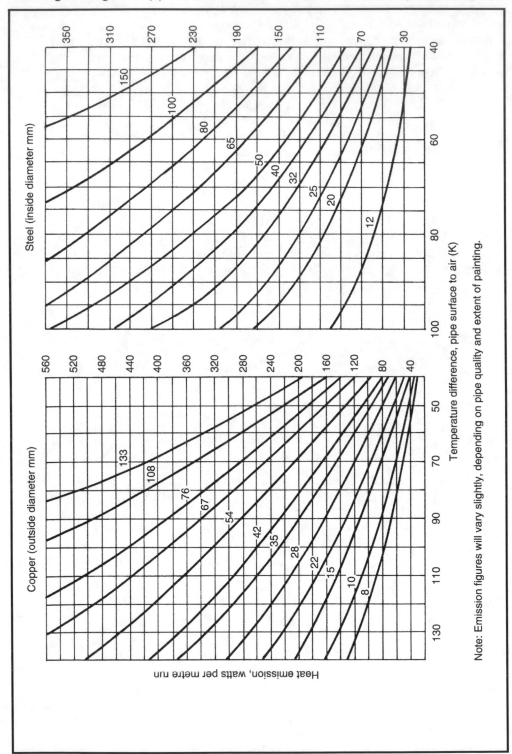

Steel (inside diameter mm)

Copper (outside diameter mm)

Heat emission, watts per metre run

Temperature difference, pipe surface to air (K)

Note: Emission figures will vary slightly, depending on pipe quality and extent of painting.

To determine the overall boiler rating, the requirement for hot water (see Part 3) is added to that necessary for heating. Heating requirements are established by summating the radiator specifications for each of the rooms. To this figure can be added a nominal percentage for pipework heat losses, the amount depending on the extent of insulation.

E.g. if the total radiator output in a house is 18kW and an additional 5% is added for pipework losses, the total heating requirement is:

$$18 + (18 \times 5/100) = 18 \cdot 9 kW.$$

Given the manufacturer's data of 80% boiler efficiency, the boiler gross heat input will be

$$18 \cdot 9 \times 100/80 = 23 \cdot 63 kW.$$

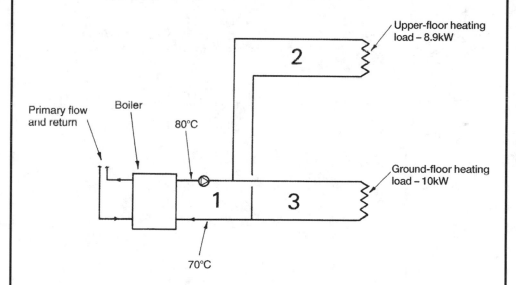

Pipes 1 – heating flow and return at boiler
Pipes 2 – to upper floor
Pipes 3 – to ground floor

Schematic illustration, assuming a heating load of 8·9kW on the upper floor and 10kW on the ground floor, i.e. 18·9kW total.

Heating Design – Pipe Sizes

The size of pipework can be calculated for each sub-circuit and for the branches to each emitter. Unless emitters are very large, 15mm o.d. copper tube or the equivalent is standard for connections to radiators in small bore installations. To illustrate the procedure, the drawing on the previous page allows for calculation of heating flow and return pipes at the boiler, and the supply pipes to each area of a house.

Pipes 1 supply the total heating requirement, 18·9kW.
Pipes 2 supply the upper-floor heating requirement, 8·9kW.
Pipes 3 supply the lower-floor heating requirement, 10kW.

For each pair of pipes (flow and return), the mass flow rate is calculated from:

$$kg/s = \frac{kW}{\text{S.h.c} \times \text{temp. diff. (flow} - \text{return)}}$$

Specific heat capacity (s.h.c.) can be taken as 4·2kJ/kg K. The temperature differential between pumped heating flow and return will be about 10K, i.e. 80°C–70°C.

Therefore, the mass flow rate for:

$$\text{Pipes 1} = \frac{18·9}{4·2 \times 10} = 0·45 kg/s$$

$$\text{Pipes 2} = \frac{8·9}{4·2 \times 10} = 0·21 kg/s$$

$$\text{Pipes 3} = \frac{10·0}{4·2 \times 10} = 0·24 kg/s$$

Selecting a pumped water velocity of 0·8m/s (see page 128) and copper tube, the design chart on page 206 indicates

Pipes 1 = 35mm o.d.
Pipes 2 = 22mm o.d.
Pipes 3 = 22mm o.d.

The specification for a pump is very much dependent on the total length of pipework, summated for each section within a system. In existing buildings, this can be established by taking site measurements. For new buildings at design stage, estimates can be taken from the architects' working drawings. Actual pipe lengths plus an allowance for resistance due to bends, tees and other fittings (see page 71) provides an effective length of pipework for calculation purposes.

Using the previous example, given that pipes 1, 2 and 3 are 6m, 10m and 12m effective lengths, respectively, the design chart shown on page 203 can be used to determine resistance to water flow in each of the three sections shown:

Pressure drop in pipes 1 = 200N/m² per metre (or pascals per metre).
Pressure drop in pipes 2 and 3 = 360N/m² per metre (Pa per m).

Therefore: Pipes 1 @   6m × 200Pa – 1200
           Pipes 2 @ 10m × 360Pa = 3600
           Pipes 3 @ 12m × 360Pa = <u>4320</u>
                              9120Pa or 9·12kPa

From this calculation, the pump specification is 0·45kg/s at 9·12kPa.

However, a higher figure for pump pressure will be necessary as the resistances in branch pipes to individual emitters will also need to be included. Pump selection is from manufacturers' pump performance charts similar to that shown on page 130.

Note: The smaller the pipe diameter, the greater the pressure drop or resistance to flow.

Water Flow Resistance through Copper Tube

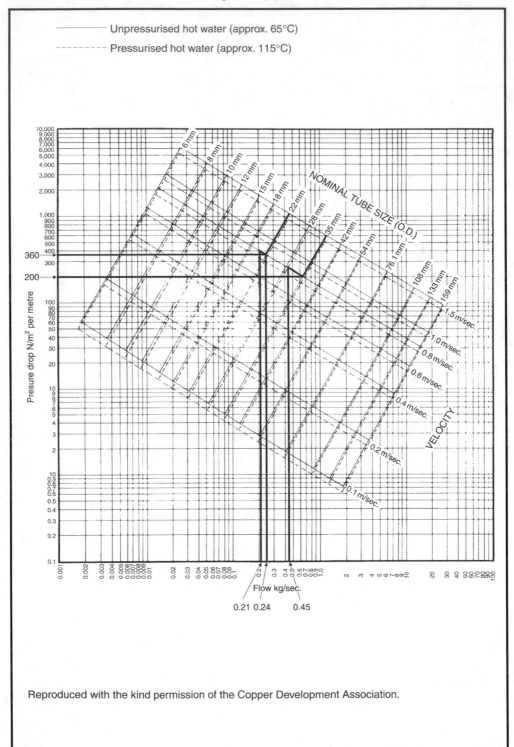

Reproduced with the kind permission of the Copper Development Association.

The name circulator as a component in domestic hot water and central heating systems is more realistic than pump, as the latter implies pressure of some intensity. For modest-sized installations designed for most dwellings, hot water is moved or circulated at a relatively low pressure. Nevertheless, the convenience and simplicity of the word pump has established it as preferred terminology.

Pumps with variable settings have been standard installation for domestic heating systems and remain functional in many existing premises. These permit some flexibility for adjustment to individual installations and adaptability for future system alterations or extensions. They also provide a 'one-fits-all' application, as one model will be suitable for a wide range of different situations. Modulating pumps are now preferred as they automatically adapt output to varying system demands typical of modern installations with thermostatic radiator valves and zone valves that can isolate parts of a heating circuit (see next page).

As indicated on pages 130 and 205, pump performance is specified by pressure output in kilo-Pascals (kPa or kN/m^2) or metres head (m) with a mass flow rate expressed in kilograms per second (kg/s) or litres per second (l/s).

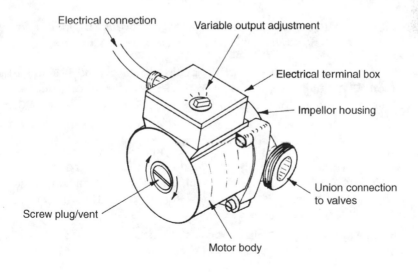

Electrical connection

Variable output adjustment

Electrical terminal box

Impellor housing

Union connection to valves

Screw plug/vent

Motor body

Typical power rating, 60 to 90 watts.

Modulating Circulator/Pump

Modulating pump – similar in appearance to a standard pump shown on the previous page. Internal rotation is from a glandless permanent magnet motor that has less potential to wear and relatively low energy consumption. Operation is demand controlled with a self-regulating output facility. This responds automatically to the varying load requirements of modern heating systems that incorporate numerous flow controls such as thermostats, motorised zone valves and thermostatic radiator valves. Motor speed is regulated to supply delivery pressure and flow rate in response to a heating demand that will change seasonally, daily and at different times of the day. A standard pump with a setting fixed to satisfy maximum demand is overrated for much of its running time, during which it consumes an excess of energy.

Energy label – European Directive 2009/125 introduced the following energy classification for pumps:

Class*	Energy efficiency index (EEI)	
A	< 0.4	Fixed speed pumps
B	0.4–0.6	are typically Class
C	0.6–0.8	D or E. Glandless
D	0.8–1.0	permanent magnet
E	1.0–1.2	motor modulating
F	1.2–1.4	pumps Class A.
G	≥ 1.4	

* Classes A to G correspond with the coloured energy efficiency labels commonly displayed on domestic appliances.

EEI specifies a pump's power input below a predetermined reference source. The lower the figure, the less the energy consumed.

EU Directive – new and replacement pump installations had an EEI of ≤ 0.27 from 2013 and from 2015 ≤ 0.23. Thereafter, most pumps were manufactured to an EEI of 0.20 with current indexing between 0.15 to 0.18.

Energy saving – older constant speed pumps are rated at about 60 watts, whereas modulating pumps have a power rating between 10 to 20 watts, achieving a significant reduction in individual consumers' electricity bills. On a wider scale, this represents a substantial reduction in CO_2 emissions.

Refs. Ecodesign Regulations for circulators – No. 641/2009. Requirements for glandless stand alone circulators – No. 622/2012.

Note: Compatibility with boiler must be determined, particularly where used as a replacement in an existing system.

Pump location is important, particularly with open-vent systems. The pump, cold feed and expansion/vent pipe positions should ensure that there is no positive or negative pump pressure where the cold feed connects to the circulatory system. This connection is known as the system-neutral point, where the only pressure is the head of water (h) from the feed cistern. If there is a significant imbalance at this point, water can pump over the expansion pipe and circulate through the feed cistern. Air may also be drawn into the system. The preferred relationship is illustrated below.

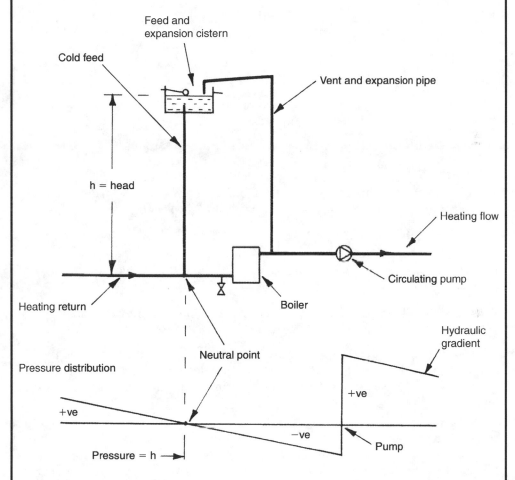

Some systems, notably older installations with higher water content heat exchangers, have the pump located on the return pipe. This should not present any problems with low circulating pressures and an adequate head of water from the feed cistern. An imbalance may occur if the system is partially closed by manual or automatic control, as pump pressure will increase in response to resistance.

Domestic Heating Circulator/Pump – Further Considerations

Water flow rates – the data on page 127 provides general guidance. For a more considerate design that has regard for noise that may be generated by water flowing, the following maximum velocities are recommended:

Pipe diameter (mm)	Water velocity/flow rate (m/s)
10	0·50
15	0·55
22	0·70
28	0·75
35	0·80
42	0·90
54	1·00

Note: Pipe diameter is expressed as copper outside diameter. For other materials, the nearest equivalent size is acceptable.

Pump position and installation –

• Low parts of a circuit are to be avoided as any sediment could accumulate in the pump body and contribute to wear.

• Accessible for maintenance, i.e. not secreted under floorboards or behind cupboards.

• Away from a wall or floor as pump vibration may generate noise through the structure.

• Isolating valves provided each side of the pump to avoid draining the whole system should the pump need to be removed.

• Preferably in a vertical pipe to ensure that the circulator shaft is horizontal. This reduces the load on the shaft bearings and allows air to purge itself from the rotor and impellor housing.

Note: Sometimes due to system restrictions, it may only be possible to place the pump in horizontal pipework. In these situations, the circulator shaft should not be less than horizontal. Just a few degrees higher at the vent plug end is better than allowing the shaft to suspend and possibly wear the shaft and bearing prematurely. If the motor is above the pump, its whole weight bears on the impellor and this too will cause premature wear. Also, any system air could become trapped at the top of the pump body.

Boiler Rating – Approximate Guide for Domestic Premises

A simple and reasonably accurate estimate for determining boiler size. Procedure:

- Establish dwelling dimensions and factor for location –

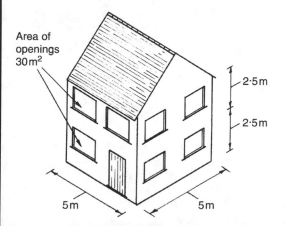

UK location	Factor
North & Midlands	29
Scotland	28·5
South east	27
Wales	27
Northern Ireland	26·5
South west	25

Detached house, location south east

- Approximate heat losses:

  Openings area (30 m²) × Openings 'U' value (2.00 ave.)* = 60 (A).

  Gross wall area (100 m²) – Openings area (30 m²) × Wall 'U' value (0.35)* = 24.5 (B).

  Roof length (5m) × Roof width (5m) × Roof 'U' value (0.25)* = 6.25 (C).

  Floor length (5m) × Floor width (5m) × Standard correction factor (0.7) = 17.5 (D).

  (For ceiling and floors in a mid-position flat, use zero where not exposed.)

- Summate fabric losses: A + B + C + D = 108.25.
- Multiply by location factor: 108.25 × 27 = 2922.75 watts.
- Calculate ventilation losses:

  Floor area (25 m²) × Room height (2.5m) × No. of floors (2) = Volume (125 m³) × Standard ventilation correction factor (0.25) × Location factor (27) = 843.75 watts.

- Boiler input (net) rating = 2922.75 + 843.75 + 2000 (watts for hot water) + calcs. for any extension to building = 5766.50 watts or 5.77 kW.

*See page 197 for current 'U' values.

211

Pressure Testing Installations

Testing medium – water is preferred to air, as water is virtually incompressible. In addition, about 200 times more energy would be stored in compressed air at the same pressure and volume as for water. This could have a damaging effect on personnel and property if a component leaked or failed.

Where premises are particularly sensitive to water leakage, a low-pressure air test can be undertaken before applying a hydraulic test.

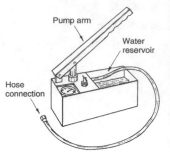

Hydraulic test pump

Procedure

- Disconnect ancillary equipment that may not be designed to withstand test pressures, e.g. shower, boiler, etc. Manufacturer's data should be consulted.
- Check all system high points for location of air vents.
- Blank or plug any open ends including float valves. Close valves where subsections only are being tested.
- Open all valves in the enclosed section under test.
- Attach test pump to a convenient point.
- Start filling the system by pump priming and replenishing the pump water reservoir.
- Ventilate air from high points until water shows.
- When the system is full, raise the pressure as required.
- If pressure falls, check joints, valves, etc., for leakage.
- When the test is satisfied, ensure the appropriate documentation is signed.

Test requirements

- Rigid pipes – provide an internal water pressure at the lowest point in the system at 50% above normal operating pressure. This should hold for one hour. For example, 1 bar (10m or 100kPa) operating pressure requires a 1.5 bar (15m or 150kPa) test pressure.
- Plastic pipes – elastic by nature will expand to some extent under pressure. Therefore, the test procedure for rigid pipes is inappropriate. Either of the following tests, A or B, is acceptable:

Test A – test pressure as for rigid pipes is applied and maintained for 30 minutes. After this time, pressure is reduced by one-third. For another 90 minutes, the test is satisfied if there is no further reduction in pressure. Test B – required test pressure is applied and maintained for 30 minutes. Test is satisfied if:

1. pressure drops <0.6 bar (60kPa) after a further 30 minutes, and
2. pressure drops <0.2 bar (20kPa) after a further 120 minutes, and
3. there is no visible leakage.

Application – underground and above-ground systems of water pipework.

Ref. Water Supply (Water Fittings) Regulations, Schedule 2, Paragraph 12.

Boilers with a cast iron heat exchanger used with an indirect sealed system are unlikely to corrode. However, some electrolytic reaction between copper pipes and steel heat exchangers in boilers and pressed steel radiators is possible. In addition, some corrosion of steel can occur where minute amounts of air enter the system. This may occur

- through undetected leakage at pipe joints;
- from air present in solution;
- from air dissolving into water contained in the feed and expansion cistern.

The initial indication of internal corrosion is one or more radiators failing to get hot and a need for frequent 'bleeding' through the air valve. Corrosion produces hydrogen gas. This may be detected by holding a lighted taper to the draught escaping at the air valve. Caution should be observed when effecting this test and, if the taper is seen to burn with a blue flame, hydrogen is present. Air will not burn.

Another characteristic of corrosion is black sludge accumulating in the bottom of radiators. This is known as magnetite, and it may also obstruct circulating pipes. Magnetite is the metallic breakdown of steel radiator walls. In addition to blockage and corrosion, magnetite is drawn to the magnetic field of the circulating pump where its abrasive presence may cause the impellor to fail. See page 141.

Corrosion in heating systems can be prevented or at least considerably reduced, by introducing a proprietary inhibitor to the feed and expansion cistern as the system is filled. With sealed systems, the inhibitor can be introduced with a funnel and hose temporarily connected to a high-level radiator (see page 140).

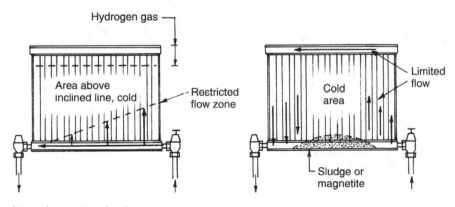

Internal corrosion of radiators

Ref. BS 7593: Code of practice for the preparation, commissioning and maintenance of domestic central heating and cooling water systems.

Overheating – New Residential Buildings

During high summer temperatures, the internal environment can become uncomfortably hot. Excessive heat removes moisture from the air, affecting the comfort and health of occupants.

A balance can be achieved by installing mechanical ventilation and/ or air conditioning. This is not the optimum solution as it involves capital expenditure, spatial requirements, increased energy use, fuel consumption bills and maintenance costs. An excess of internal heat along with equipment running costs to control it will significantly contribute to hazardous gas emissions impacting on global warming (see pages 135 and 136).

Preventative measures:

- External blinds/shutters/shading/awnings/overhangs.
- Limited depth of window reveal and area of glazing.
- Designing in appropriate amount of thermal mass to structure.
- Insulation to the external envelope. In winter retains warmth, in summer keeps heat out.
- Cross-ventilation – purge or background ventilators on opposing walls.
- Automated control – time and temperature – of mechanical ventilation (if provided).
- Consider orientation of glazing relative to solar exposure.
- Limit g-value.

Window g-value:

A measure of how much solar heat (infrared radiation) penetrates.

Zero (0%) = opaque, no solar heat enters.

1 (100%) = all solar heat enters, i.e. no glass at all.

Objective g-value is 0.4 (40%). Typical compliance is 0.7 (70%) with other preventative measures. Triple glazing is about 0.5 (50%).

Building Regulation A.D. O – Overheating, provides two methods that can be applied to the design of new residential buildings:

- Simplified method – practical provisions to minimise solar gain, some outlined under preventative measures above, along with compliance data and tables.
- Dynamic thermal modelling – analytical methodology with reference to CIBSE's TM 59 – Design methodology for the assessment of overheating risk in homes.

5 FUEL CHARACTERISTICS AND STORAGE

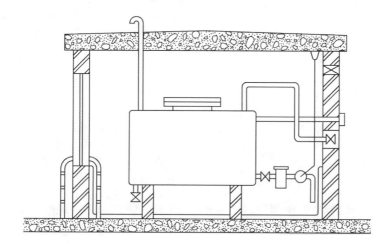

FUELS – FACTORS AFFECTING CHOICE

SOLID FUEL – PROPERTIES AND STORAGE

DOMESTIC SOLID FUEL BOILERS

SOLID FUEL – BIOMASS

SOLID FUEL – FLUES

OIL – PROPERTIES

OIL – STORAGE AND SUPPLY

OIL-FIRED BURNERS AND APPLIANCES

OIL – FLUES

NATURAL GAS – PROPERTIES

LIQUID PETROLEUM GAS – PROPERTIES AND STORAGE

ELECTRIC BOILER

ELECTRICITY – ELECTRODE BOILER

DOI: 10.1201/9781003434894-5

Fuels – Factors Affecting Choice

One of the most important considerations for providing an effective means of heating water is selection of an appropriate fuel. Choice and selection is a relatively new concept, as until the 1960s mains gas was rarely available outside large towns and cities. Also, the cost of fuel oil was prohibitive for most people. The majority of domestic premises were heated by solid fuel for open fires with a back boiler for hot water. Solid fuel boilers for hot water and central heating were available, but the associated technology of pumps and thermostatic controls was rudimentary by today's standards. Systems of the time required considerable attention, not least frequent replenishment of fuel and disposal of ash. The post-1960s era led to much higher expectations in domestic comfort and convenience standards. This coincided with considerable developments in fuel-burning appliances to complement the availability of new gas and oil resources from offshore sources.

Practical factors and amenity issues may still limit or simplify choice, e.g. in some areas, mains gas is not available and some buildings may have very limited space for fuel storage, or none at all. Personal preference as a result of previous experience, sales presentations or promotions may also have an important influence.

Amenity factors:

Facility to control the fuel, i.e. response to thermostatic and programmed automation.
Space for fuel storage.
Space for a boiler or special facilities to accommodate it.
Accessibility for fuel delivery.
Planning issues: chimneys and flue arrangements.
Location – conformity with Clean Air Act and exhaust emissions.
Maintenance requirements and after-care programme.
Availability.

Economic factors:

Capital cost of installation.
Cost of fuel storage facility.
Cost of special equipment.
Cost of equipment accommodation/plant room.
Cost of constructing a service area/access road.
Fuel costs – current and projected.
Flexibility of boiler, i.e. facility to change to another fuel.

Appropriate as logs of wood or as a coal product for open fires, stoves and boilers. A considerable amount of space is required for storage, and manual handling is very much a feature. Arrangements must be made for fuel deliveries and disposal of ashes. Although the combustion efficiency is generally lower than oil or gas, some degree of automation is possible with the more efficient slow-burning anthracites. Domestic boilers have several days' burning potential by gravity-fed integral hopper. Instantaneous control is not possible and skilful operation is required to maintain boilers at low output.

Chimney construction and flue requirements must comply with Approved Document J to the Building Regulations. These are generally much larger and more visual than that required for use with other fuels. The sulphur content from burnt coal products is corrosive to many materials; therefore, flue construction must not contain stainless steel linings or other materials which could be affected. The sulphur also contributes to atmospheric pollution.

Properties:

Fuel type	Calorific value MJ/kg	Sulphur content %	Bulk density* kg/m^3
Anthracite[†]	33	1·0	750–800
Coking coal	30	1·0	
Dry steam coal[†]	30	1·1	
Strong caking coal	29	1·9	600–800
Medium caking coal	27	1·9	
Weak caking coal	26	1·9	
Non-caking coal	24	1·8	
Manufactured coke[†]	28	N/A	400–500
Wood and wood products	16–19	N/A	300–800

Notes:
*Variation depending on granular size. Unit size and species for wood.
[†]Smokeless fuels.

Solid Fuel – Storage

When solid fuel is to be used, it is essential to consider accommodation for fuel storage and facilities available. For domestic and small buildings where requirements are minimal, a brick or concrete bunker of nominal size is adequate. Industrial and commercial premises will require a fuel bunker or hopper above the boiler to reduce manual handling. Motorised feed mechanisms can be used to regulate fuel delivery to the boilers and vacuum pumps can effect extraction of ashes.

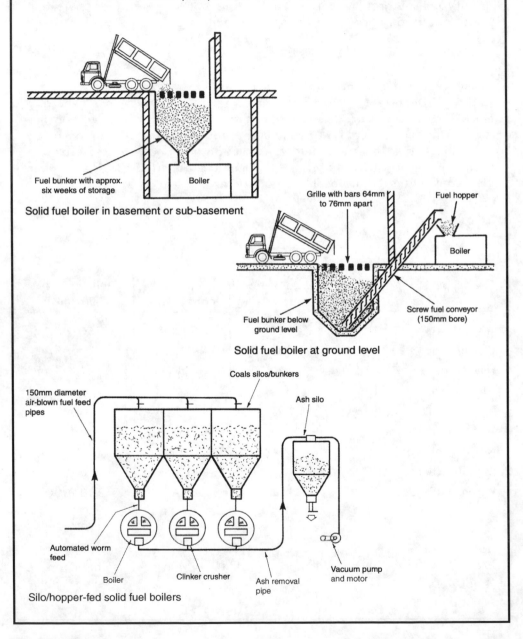

Fuel bunker with approx. six weeks of storage

Boiler

Solid fuel boiler in basement or sub-basement

Grille with bars 64mm to 76mm apart

Fuel hopper

Boiler

Fuel bunker below ground level

Screw fuel conveyor (150mm bore)

Solid fuel boiler at ground level

Coals silos/bunkers

150mm diameter air-blown fuel feed pipes

Ash silo

Automated worm feed

Boiler

Clinker crusher

Ash removal pipe

Vacuum pump and motor

Silo/hopper-fed solid fuel boilers

Back boilers situated behind a fireplace are limited to providing room heat from the fire, hot water by gravity circulation to a storage cylinder and perhaps a couple of radiators or a towel rail off the primary flow and return. They were standard installations in many 1930s houses but are now virtually obsolete. The combined room heater and boiler shown below is an improvement, having an enclosed fire and a convected outlet to heat the room in which it is installed. The water jacket is of sufficient capacity to provide hot water for storage and for several radiators. These appliances will require re-stoking every few hours.

Independent boilers are free-standing, automatically fed by hopper and require only a flue. A chimney structure is not necessary, provided the flue satisfies Approved Document J to the Building Regulations. The integral fuel store contains small granules or 'peas' of anthracite and will require minimal attention with a burning capacity of several days. Automatic control is by thermostat in the waterway to regulate a fan-assisted air supply for complete combustion. These boilers are designed with sufficient capacity to provide hot water and central heating for most domestic situations.

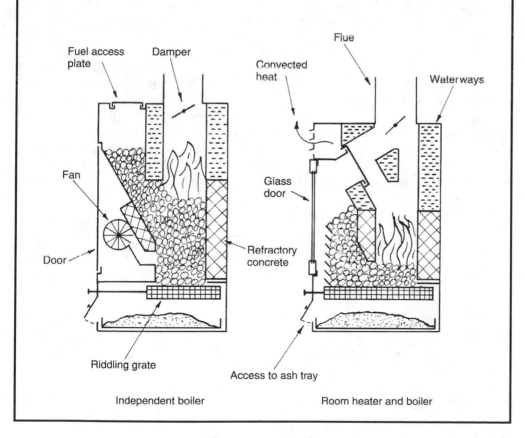

Independent boiler   Room heater and boiler

Solid Fuel – Biomass

Biomass – a biological material produced from living or recently living vegetable plant or animals. Carbon based and when burned generates heat energy.

Biofuel – a fuel derived from unfossilised material such as wood, energy crops, agricultural residue, food waste or industrial waste. Fossil fuels (coal, oil and gas) are also derived biologically but millions of years ago. The difference is the timescale.

Combustion – biomass products are composed of about 50% carbon, 40% oxygen and 5% hydrogen by weight. Combustion converts these constituents to carbon dioxide (CO_2) and water vapour (H_2O).

Emissions – CO_2 is a greenhouse gas (along with methane and nitrous oxide) that traps heat in the atmosphere keeping the Earth's surface warmer than it should be. CO_2 emissions from biofuels are relatively high compared to fossil fuels, due principally to a lower calorific value (see page 217), but the CO_2 is absorbed by biomass plant growth.

Carbon neutral – a reference to the concept of a closed carbon cycle with no net increase in levels of CO_2. The carbon used to produce a biomass product is taken out of the atmosphere as CO_2 using energy from the sun. In effect, wood is a store of solar energy. Using wood from sustained timber growth by management of regenerated forest farming produces an energy-friendly fuel. Therefore, atmospheric emissions of CO_2 from combustion of biomass fuel are balanced by the natural absorption of CO_2 by growing timber.

The assumption that biomass fuel is carbon neutral is not entirely true due to the energy input during the fuel's production. The amounts of nitrous oxides and sulphur dioxides in the combustion gases compare with those from burning fossil fuels. Suspended particulate emissions are higher.

Wood pellets – a by-product of wood-processing industries: 100% natural wood shavings and sawdust compressed into small cylindrical rolls and held together with lignin, a natural component of wood. One kilogram of pellets can provide about 3.5kWh of energy.

Solid Fuel – Biomass – Closed Carbon Cycle

The carbon cycle is a biogeochemical cycle that represents movement of elements and compounds* through living organisms and the non-living environment. During the plant growth of biomass and its subsequent combustion as a fuel, a process of carbon exchange occurs. Energy from sunlight contributes to the process of photosynthesis in plants which absorbs the carbon dioxide released into the atmosphere by biomass fuel combustion, thereby completing the cycle. As this cycle is closed, net carbon emissions are zero. Expressed another way, biomass is carbon neutral. A closed carbon cycle indicates that the world supports a fixed amount of carbon.

Simplified representation of the biomass closed carbon cycle –

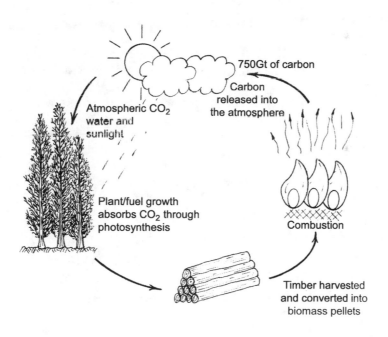

750Gt of carbon

Carbon released into the atmosphere

Atmospheric CO_2 water and sunlight

Plant/fuel growth absorbs CO_2 through photosynthesis

Combustion

Timber harvested and converted into biomass pellets

*Element – substance consisting of atoms of only one kind, e.g. carbon (C) and oxygen (O).

Compound – substance consisting of two or more elements, e.g. carbon dioxide (CO_2).

Solid Fuel – Biomass Appliances

Pellet boiler –

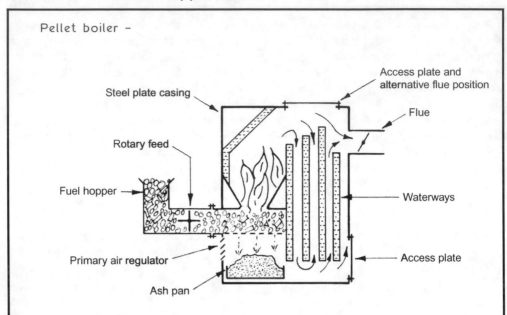

Log-burning stoves – more efficient than an open fire. Various shapes, sizes and designs are available. Useful as a supplementary heat source.

Log/pellet boiler – next step up from a log-burning stove. Effective for space heating workshops and similar open areas by warm air transfer, fan and ducting. A water jacket provides for stored hot water and some radiators.

Schematic layout of a combined biomass/solar hot water and heating system –

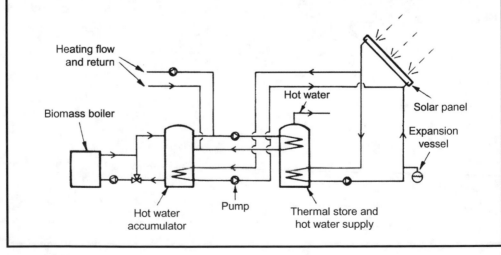

Flue pipes may be used to connect a solid fuel-burning appliance to a chimney. They must not pass through a roof space, partition, internal wall or floor. Acceptable connecting flue pipe materials are

- Cast iron – all but obsolete unless reclaimed. Formerly to BS 41 in spigot and socket format.
- Mild steel with a flue wall thickness of at least 3mm, complying with BS 1449-1: Steel plate, sheet and strip.
- Stainless steel with a flue wall thickness of at least 1mm, complying with BS EN 10088-1: Stainless steels (Grades 1·4401, 1·4404, 1·4432 or 1·4436).
- Vitreous enamelled steel pipe complying with BS EN 1856-2: Requirements for metal chimneys. Metal flue liners and connecting flue pipes.

All spigot and socket jointed pipes to be fitted socket uppermost and sealed with a non-combustible rope and fire cement or proprietory equivalent.

Any combustible material used in construction must be at least 200mm from the inside surface of the flue. Where any metal fixings are in contact with combustible materials, they must be at least 50mm from the inside surface of a flue.

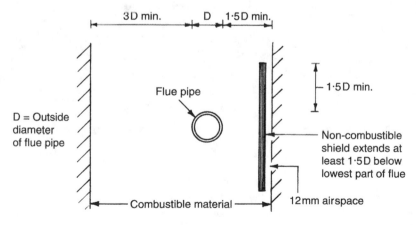

Other requirements for closeness of combustible material

Provisions for Solid Fuel Appliance Flues

Flue outlets should be above the roof line to effect clear, unhindered dispersal of combustion products without creating a fire hazard. This and other possibilities are shown on pages 227 and 228.

Flue length and height must be sufficient to encourage adequate draught and efflux (discharge) velocity at the terminal, with regard to limiting the possibility of condensation occurring in the flue. Flue gases cool relative to the flue pipe and surrounding structure temperature, until dew point of water occurs at about 60°C. Flue linings must therefore be impervious and resistant to corrosion. If condensation is a problem, a small-diameter vertical drain can be located at the base of the flue.

Flue direction should be straight and vertical wherever possible. Horizontal runs are to be avoided. If the appliance has a back outlet connection an exception is made, but the horizontal flue length must not exceed 150mm before connecting to a chimney or vertical flue. Bends should not exceed 45°C to the vertical to maintain a natural draught and to ease cleaning.

Flue size is never less than that provided on the appliance outlet.

Boiler, cooker or stove	Min. flue size
<20kW rated output	125mm-dia. or square/rectangular equivalent area, with a minimum dimension of 100mm in straight flues and 125mm for flues with bends
20–30kW rated output	150mm-dia. or square/rectangular equivalent area, with a minimum dimension of 125mm
30–50kW rated output	175mm-dia. or square/rectangular equivalent area, with a minimum dimension of 150mm

Flue size in chimneys varies between 125 and 200mm diameter (or square/rectangular equivalent) depending on application and appliance rating.

Ref. Building Regulations, Approved Document J: Combustion appliances and fuel storage systems. Sections 2.4 to 2.7.

Appliances require air (oxygen) for efficient combustion of fuel. This requires purpose-made ventilation openings in the structure, size depending on the appliance type and rating.

Appliance type	Permanently open ventilation (see note)
Boiler, cooker or stove with a flue draught stabiliser	$300\,mm^2/kW$ for the first 5kW of rated output, $850\,mm^2/kW$ thereafter
As above, without a flue draught stabiliser	$550\,mm^2/kW$ of rated output above 5kW

E.g. A 20kW boiler attached to a draught stabilised flue.

$$(300 \times 5) + (850 \times 15) = 14250mm^2$$

Taking the square root of 14250 indicates an open draught of at least $120 \times 120\,mm$.

Note: Dwellings with air permeability $>5m^3/h$ per m^2 at 50Pa $\leq 5m^3$, $850mm^2/kW$ (draught stabilised) and $550mm^2/kW$ (without stabiliser).

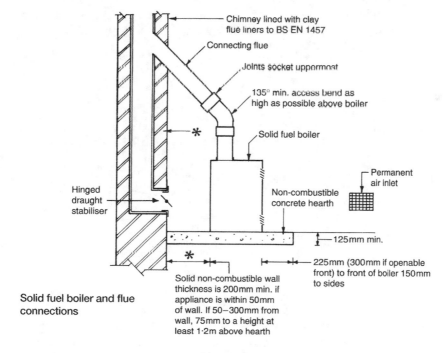

Chimney lined with clay flue liners to BS EN 1457
Connecting flue
Joints socket uppermost
135° min. access bend as high as possible above boiler
Solid fuel boiler
Permanent air inlet
Non-combustible concrete hearth
Hinged draught stabiliser
125mm min.
225mm (300mm if openable front) to front of boiler 150mm to sides
Solid non-combustible wall thickness is 200mm min. if appliance is within 50mm of wall. If 50−300mm from wall, 75mm to a height at least 1·2m above hearth

Solid fuel boiler and flue connections

Refs. Building Regulations, Approved Document J: Combustion appliances and fuel storage systems. Sections 2.1 to 2.3.

BS 5854: Code of practice for flues and flue structures in buildings.

Flue types –

- Flexible stainless steel.
  * A single-skin liner suitable for existing chimneys that is only for use with gas combustion appliances (see page 519).
  * A double-skin chimney liner manufactured with overlapping strips of high grade stainless steel. This provides a smooth sealed internal flue way that can be used with solid fuel appliances.
- Rigid stainless steel. Comprises two concentric walls with insulation in between. Specified to BS EN 1856-1. A useful application to existing buildings with no chimney. Used internally or externally.
- Clay/ceramic chimney liner to BS EN 1457-1. For use with new chimney construction. Surrounded with at least 100 mm masonry. See Part 8 of the 11th edition of the *Building Construction Handbook*.
- Pumice-based concrete to BS EN 1857. Construction as above.
- Chimney block systems.
  * Pumice. A double-wall (self-insulating double module) comprising square section pre-cast pumice concrete blocks for chimney construction with an inner void to accommodate pre-cast pumice concrete linings. BS EN 1858.
  * Clay/ceramic. A double wall system of square section pre-cast concrete chimney blocks with a circular void to accommodate a mineral wool insulated ceramic/clay lining. BS EN 13063-1.

Double wall chimney systems

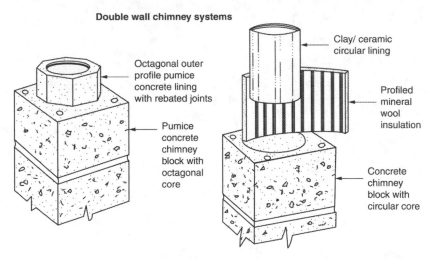

Octagonal outer profile pumice concrete lining with rebated joints

Pumice concrete chimney block with octagonal core

Clay/ ceramic circular lining

Profiled mineral wool insulation

Concrete chimney block with circular core

Refs.

BS EN 1856-1: Requirements for metal chimneys.

BS 1457-1: Clay/ceramic flue liners.

BS 1857: Concrete flue liners.

BS 1858: Concrete flue blocks.

BS EN 13063-1: System chimneys with clay/ceramic flue liners.

For safe dispersal of burnt solid fuels, flue outlets should discharge at a height and distance sufficient to clear combustible materials.

Flue outlets (also shown next page) –

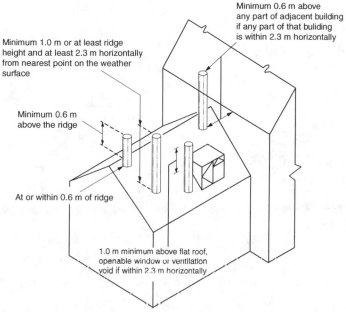

Minimum 0.6 m above any part of adjacent building if any part of that buliding is within 2.3 m horizontally

Minimum 1.0 m or at least ridge height and at least 2.3 m horizontally from nearest point on the weather surface

Minimum 0.6 m above the ridge

At or within 0.6 m of ridge

1.0 m minimum above flat roof, openable window or ventilation void if within 2.3 m horizontally

Note: Measurements for an external flue attached to an outside wall are as shown for internal flues, taken at the part of the roof nearest the flue

Flue datum measurements –

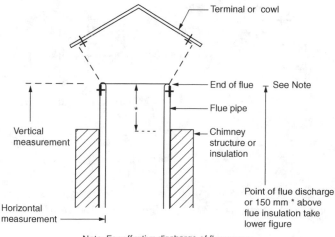

Terminal or cowl

End of flue — See Note

Flue pipe

Vertical measurement

Chimney structure or insulation

Point of flue discharge or 150 mm * above flue insulation take lower figure

Horizontal measurement

Note: For effective discharge of flue gases, 4.5 m minimum from top of appliance

The flue outlet must not terminate within the areas shown shaded.

Non-combustible roof coverings.

E.g. slates and tiles of concrete or clay –

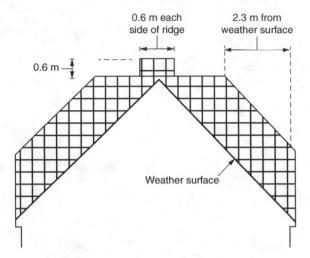

0.6 m each side of ridge

2.3 m from weather surface

0.6 m

Weather surface

Combustible roof coverings.

E.g. thatch and wood shingles –

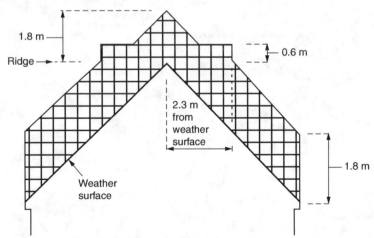

1.8 m

Ridge

0.6 m

2.3 m from weather surface

1.8 m

Weather surface

All dimensions minimum

Refs. Building Regulations, Approved Document J: Combustion appliances and fuel storage systems. Section 2.10 and 2.11.

BS EN 13384-1: Chimneys. Thermal and fluid dynamic calculation methods. Chimneys serving one combustion appliance.

228

Fuel for boilers is manufactured by processing crude oil. The crude is distilled and condensed to produce a variety of commercial brands including gasolenes, kerosenes and gas oils. Distillates are blended to create several grades suitable as boiler fuels.

Kerosene (known commercially as Class C2) is an unblended relatively expensive light distillate suitable for domestic vaporising or atomising oil-fired boilers. It is uncoloured or clear and has a viscosity of 28s as measured on the Redwood scale. Gas oil (Class D) is a heavier and less expensive distillate suitable for larger atomising burners in domestic and industrial applications. It is coloured red and has a viscosity of 35s. Fuel oils (Classes E, F, G and H) are a blend of residual oils with distillates that are considerably cheaper than the other classes. They are also heavier and generally require storage and handling plant with heating facilities. They require pre-heating before pumping and atomising for burning. These oils are limited to large-scale plant that has high-level chimneys to discharge the pollutants and dirty flue gases characteristic of their high sulphur content.

Characteristics:

	Kerosene	Gas oil	Residue-containing burner fuels			
Class	C2	D	E	F	G	H
Density	790	840	930	950	970	990 kg/m^3
Flashpoint	38	56	66	66	66	66°C
Calorific value	46·4	45·5	43·4	42·9	42·5	42·2 MJ/kg
Sulphur content	0·2	1·0	3·2	3·5	3·5	3·5%
Kinematic viscosity	20	5·5	8·2	20	40	56 *
Minimum storage temp.	N/A	N/A	10	25	40	45°C

Note: *Class C2 and D at 40°C.
Classes E, F, G and H at 100°C.

Ref. BS 2869: Fuel oils for agricultural, domestic and industrial engines and boilers. Specification.

Oil – Grading

Fuel oil for use in heating plant is graded by its viscosity or ability to flow. For purposes of applying a viscous grading to the different classifications of boiler fuels defined on the previous page, the British viscosity standard test adopted by the oil industry is known as the Redwood No. 1 viscosity test.

The test apparatus is simple in concept and is shown in principle below. It comprises a 50 ml (50 cm^3) oil container with a small orifice at its base through which the oil flows while being maintained at a constant temperature of 38°C. The Redwood viscosity is the time taken in seconds for the liquid to flow out of the container.

Redwood viscometer or viscosimeter

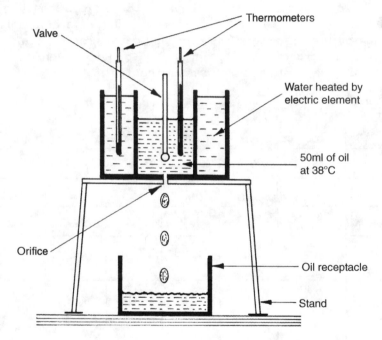

Comparison of oil classifications by Redwood grading

Class	Name	Redwood No. 1 (seconds)
C2	Kerosene	28
D	Gas oil	35
E	Light fuel oil	220
F	Medium fuel oil	950
G	Heavy fuel oil	3500
H	Residual tar oil	6000

An oil storage tank may be located externally. Unless the tank is underground or provided with a fire-resistant barrier, it must be sited at least 1·8m from the building. A plant room may be used if constructed in accordance with the local fire regulations. It must be built of fire-resistant materials, with the base and walls to flood level rendered with cement mortar to contain all the oil in the tank plus 10% in the event of a leakage. Where the oil storage room is within a building, it should be totally enclosed with walls and floors of at least four hours' fire resistance.

As a guide to tank capacity, it should equate to the normal delivery plus two weeks' supply at maximum consumption or three weeks' average supply – take the greater. Supply pipelines can be as little as 8 or 10mm o.d. annealed copper in coils to eliminate jointing. They can also be of steel for larger installations. Industrial supplies have the pipes insulated and trace wired to keep the oil warm. The tank should be elevated to provide at least 0·5m head to effect the level controller or metering valve. If this is impractical, the supply can be pumped. The maximum head is 4m.

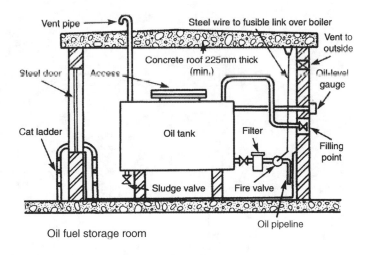

Oil fuel storage room

Refs: Environment Agency publication: Storing oil at your home or business.

BS 5410-1: Code of practice liquid fuel firing. Installations for space heating and hot water supply purposes for domestic buildings.

Building Regulations Approved Document J: Combustion appliances and fuel storage systems; Section 5, Provisions for liquid fuel storage and supply.

Installation of Oil Tank and Oil Supply

An oil storage tank is usually rectangular with a raised top designed to shed water. Tanks for domestic application have a standard capacity of 2275 litres (2·275m³) for economic deliveries of 2m³. A vertical sight glass attached to the side provides for easy visual indication of the level. Tanks are made from ungalvanised welded carbon steel or sectional pressed ungalvanised carbon steel with internal strutting to prevent deformity when full. They are also produced in plastic. Brick piers or a structural steel framework are used to raise the tank above the ground. This is necessary to avoid corrosion from ground contact and to create sufficient head or pressure (0.5m min.) from the outlet to the burner equipment. Location must be within 30m of the oil-tanker vehicle access point; otherwise, an extended fill line must be provided.

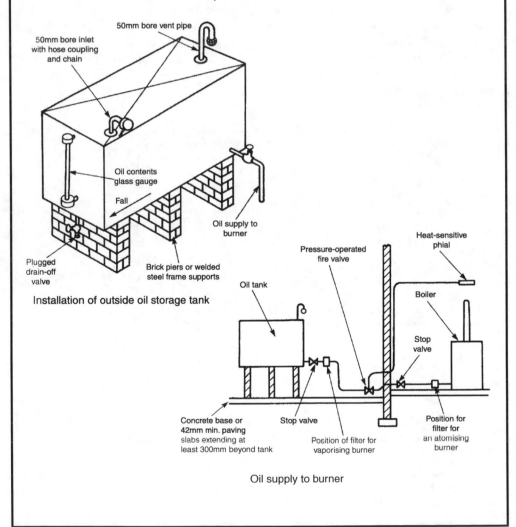

Installation of outside oil storage tank

Oil supply to burner

An oil tank is located for convenience of filling, maintenance, proximity to boiler and visual impact. More importantly, it should be positioned with regard to:

1. Protection of stored oil from an external fire.
2. Protection of nearby buildings if stored oil were to ignite.

Proximity to a building – the critical dimension is 1.8m from any part of a building. It can be less than this if

- a building's walls and projecting eaves are without permanent openings and have construction specified to at least 30 minutes' fire resistance (insulation, integrity and stability – BS 476–20, BS EN 1363 or BS EN 1364), OR
- a wall of at least 30 minutes' fire resistance is built between the building and the tank. This wall should extend at least 300mm beyond the tank height and width.

Further, to prevent vegetation growing over or under a storage tank, a dense concrete base or paving slabs of 42mm thickness is laid to extend at least 300mm beyond the tank perimeter.

Proximity to a boundary – if the tank is to be located less than 760mm from a boundary, a fire-resisting wall is required between tank and boundary. This wall is specified to 30 minutes' minimum fire resistance either side and extending to at least 300mm beyond the tank height and width.

Where a building and boundary limitations cannot be accommodated, a specially manufactured tank may be used. The tank construction must have an outer fire-resistant cladding over a fire-resistant insulating material to a fire-resistant impervious lining. A test certificate indicating 30 minutes' minimum fire resistance is required, and the base construction should be as described above.

Refs. BS 476: Fire tests on building materials and structures.
BS ENs 1363 and 1364: Fire-resistance tests.

Oil Tank Location – 2

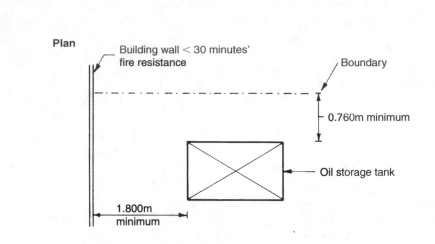

Plan

Building wall < 30 minutes' fire resistance

Boundary

0.760m minimum

Oil storage tank

1.800m minimum

Note: The dimensions given may be reduced if the building's wall has a fire resistance of 30 minutes or more, or if a 30 minutes' fire-resistant wall is constructed between tank and wall, and also between tank and boundary. A tank of 30 minutes' fire-resistant construction is an acceptable alternative.

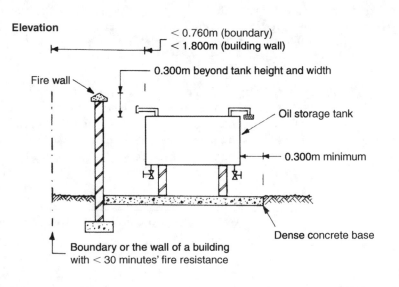

Elevation

< 0.760m (boundary)
< 1.800m (building wall)

0.300m beyond tank height and width

Fire wall

Oil storage tank

0.300m minimum

Dense concrete base

Boundary or the wall of a building with < 30 minutes' fire resistance

Bund – generally defined as an embankment or structure used for containment or retention purposes. In the context of oil storage and the interests of reducing the significance and expense of potential environmental damage due to oil seepage into the ground, a bund must be constructed around storage vessels. This is a precautionary measure to retain oil in the event of a leakage.

Objective – to prevent oil from percolating into the ground surrounding an oil tank and contaminating a watercourse.

Application – above-ground oil storage in excess of 200 litres' capacity for industrial, commercial and institutional (residential and non-residential) premises. Capacity exceeding 2500 litres in any situation.

Location – <10m from a watercourse (river, stream, land drain, etc.).

<50m from a well or borehole.

Where a spillage could access a drain opening.

Where the tank vent cannot be seen from the filling point.

Construction – of impermeable material, e.g. engineering brick wall and dense concrete base.

Capacity – minimum 110% of the storage tank volume.

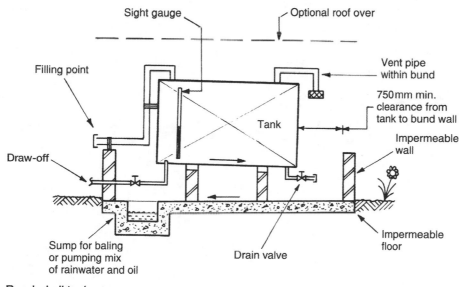

Bunded oil tank

Refs. Control of Pollution (Oil Storage) (England) Regulations. Water Resources Act, Sections 92 and 219.

Oil-Fired Burners

There are two types of oil burner: 1. vaporising; 2. atomising.

1. The natural draught vaporising burner consists of a cylindrical pot which is fed with oil at its base from a constant oil-level controller. When the burner is lit, a thin film of oil burns in the bottom. Heat is generated, and the oil is vaporised. When the vapour comes into contact with air entering the lowest holes, it mixes with the air and ignites. At full firing rate, more air and oil mix until a flame burns out of the top of the burner.

2. The pressure jet atomising burner has an atomising nozzle. This produces a fine spray of oil which is mixed with air forced into the burner by a fan. Ignition electrodes produce a spark to fire this air/oil mixture.

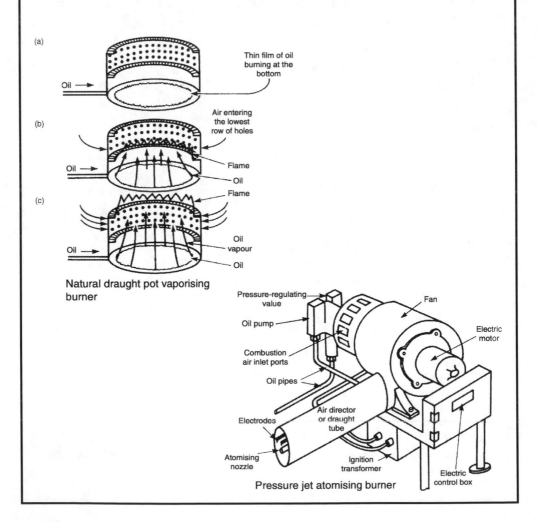

(a)

Oil →

Thin film of oil burning at the bottom

Air entering the lowest row of holes

(b)

Oil →

Flame

Oil

Flame

(c)

Oil →

Oil vapour

Oil

Natural draught pot vaporising burner

Pressure-regulating value

Fan

Oil pump

Electric motor

Combustion air inlet ports

Oil pipes

Electrodes

Air director or draught tube

Atomising nozzle

Ignition transformer

Electric control box

Pressure jet atomising burner

The wall-flame burner consists of a steel base plate securing a centrally placed electric motor. The armature of this motor is wound on a hollow metal shroud which dips into an oil well. A constant oil-level controller feeds the well, just covering the edge of the shroud. The shroud is circular with its internal diameter increasing towards the top, from which two holes connect with a pair of oil pipes. When the motor is engaged, oil is drawn up to the pipes and thrown onto the flame ring. Simultaneously, air is forced onto the rings by the fan. This air/oil mixture is ignited by the electrodes.

The constant oil-level controller is used to feed vaporising burners. If the inlet valve fails to close, oil flows into the trip chamber. The trip float rises and operates the trip mechanism, thus closing the valve.

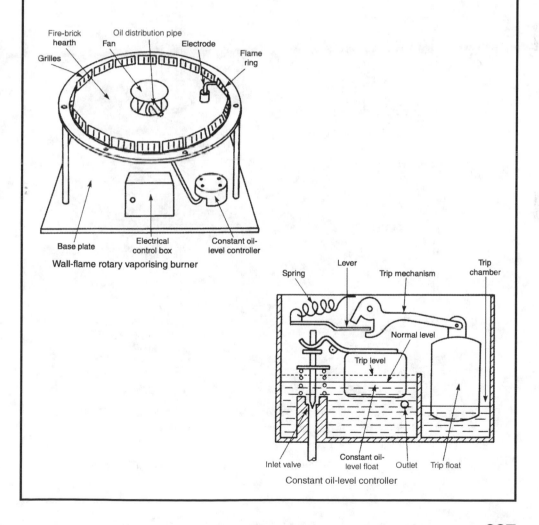

Wall-flame rotary vaporising burner

Constant oil-level controller

237

Ventilation for Oil-Fired Appliances < 45kW Output

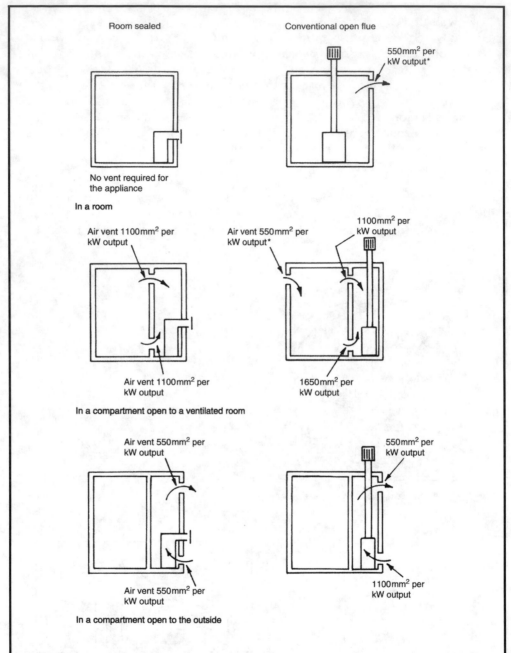

Room sealed

Conventional open flue

550mm² per kW output*

No vent required for the appliance

In a room

Air vent 1100mm² per kW output

Air vent 550mm² per kW output*

1100mm² per kW output

Air vent 1100mm² per kW output

1650mm² per kW output

In a compartment open to a ventilated room

Air vent 550mm² per kW output

550mm² per kW output

Air vent 550mm² per kW output

1100mm² per kW output

In a compartment open to the outside

* Ventilation should be increased by an additional 550mm² per kW output where the appliance has a draught break, i.e. a draught stabiliser or draught diverter.

Older dwellings (built pre-2009) having air permeability >5 m³/hour per m² at 50 Pa, the first 5 kW can be ignored.

Calculations relate to applications shown on the preceding page.

Example 1: A conventional open flue appliance of 12kW output rating in an older dwelling.

• Installed in a room.

No vent required up to 5kW, but 550mm^2 to be provided per kW thereafter:

12kW – 5kW = 7kW × 550mm^2 = 3850mm^2 air vent area.

• Installed in a cupboard compartment open to a ventilated room.

Air vent area is the same as above.

Vent area for cooling the appliance is 1100mm^2 for every kW rating:

12kW × 1100mm^2 = 13200mm^2.

Ventilation, cooling and combustion air area:

12kW × 1650mm^2 = 19800mm^2.

• Installed in a compartment open to the outside.

Air for cooling the appliance is 550mm^2 for every kW rating:

12kW × 550mm^2 = 6600mm^2.

Air for combustion:

12kW × 1100mm^2 = 13200mm^2.

Example 2: A room-sealed balanced flue appliance of 12kW output rating.

• In a cupboard compartment open to a ventilated room. Air for ventilation and cooling is 1100mm^2 per kW (twice):

12kW × 1100mm^2 = 13200mm^2 (twice).

• In a cupboard compartment open to the outside. Air for ventilation and cooling is 550mm^2 per kW (twice):

12kW × 550mm^2 = 6600mm^2 (twice).

Note: Provision for ventilation in walls may be partly by infiltration but is usually by purpose-made air-bricks built into the wall. These should not be covered over.

Outlets from flues serving oil-fired appliances, rated up to 45kW output, must be carefully located to ensure:

- natural draught for fuel combustion
- efficient and safe dispersal of combusted fuel products
- adequate air intake if combined with a balanced flue.
- A flue terminal facing a public access space, e.g. pavement, to be at least 2.1m above ground level.
- Flue terminal minimum 1.8m from oil tank.

In conjunction with the air inlet provisions shown on the previous page, the following guidance should ensure efficient combustion and burnt fuel gas dispersal.

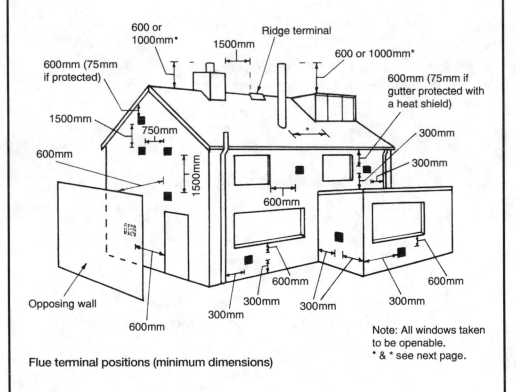

Flue terminal positions (minimum dimensions)

Note: All windows taken to be openable.
* & * see next page.

Ref. Building Regulations, Approved Document J: Combustion appliances and fuel storage systems. Section 4.

The following guidance provides minimum acceptable dimensions with regard to appliance efficiency, personnel and fire safety. The listing should be read with the illustration on the previous page. Local conditions such as wind patterns may also influence location of terminals. Flue terminal guards may be used as a protective barrier where direct contact could occur.

Location of terminal	Pressure jet atomising burner	Vaporising burner
Directly under an openable window or a ventilator	600	
Horizontally to an openable window or a ventilator	600	
Under eaves, guttering or drainage pipework	600	
As above, with a 750mm wide heat shield	75	
Horizontally from vertical drain or discharge pipes	300	
Horizontally from internal or external corners	300	
Horizontally from a boundary	300	
Above ground or balcony	300	
From an opposing wall or other surface	600	
Opposite another terminal	1200	
Vertically from a terminal on the same wall	1500	
Horizontally from a terminal on the same wall	750	
From a ridge terminal to a vertical structure	1500	
Above the intersection with a roof	600	1000 •
Horizontally to a vertical structure	750	2300 *
Above a vertical structure <750mm (pressure jet burner) or <2300mm (vaporising burner) horizontally from a terminal	600	1000 *

(Vaporising burner column annotated: ← Not to be used in these situations →)

Notes:
Dimensions in mm.
No terminal to be within 300mm of combustible material.
Where a vaporising burner is used, the terminal should be at least 2300mm horizontally from a roof.
See previous page for • and *.

Natural Gas – Properties

UK gas supplies originate from decaying organic matter found at depths up to 3km below the North Sea. Extraction is by drilling rigs and pipelines to the shore. On shore, it is pressurised to about 5kPa throughout a national pipe network.

Properties of natural gas:

Methane	89·5%
Ethane	4·5%
Propane	1·0%
Pentane	0·5%
Butane	0·5%
Nitrogen	3·5%
Carbon dioxide	0·5%

The composition shown will vary slightly according to source location. All the gases above are combustible except for nitrogen. Natural gas is not toxic, but incomplete combustion will produce carbon monoxide, hence the importance of correct burner and flue installations. A distinctive odour is added to the gas, as in its natural state it has no detectable smell. Natural gas is lighter than air with a specific gravity of about 0.6, relative to 1.0 for air.

Characteristics:

Calorific value	36–40MJ/m^3
Specific gravity	0.5–0·7
Wobbe No.	approx. 50%
Sulphur	approx. 20mg/m^3

Note: The Wobbe No. is sometimes used to represent the thermal input of an appliance for a given pressure and burner orifice. It is calculated from:

$$\frac{\text{Calorific value}}{\sqrt{\text{Specific gravity}}} \quad \text{e.g.} \quad \frac{40}{\sqrt{0.6}} = 51\%$$

Natural gas has many advantages over other fuels, including clean and efficient burning, no storage, less maintenance, relatively economic and a minimum of ancillaries.

Families of gases:

Family	Type of gas	Wobbe No. (%)
1	Manufactured, e.g. coal gas	22–30
2	Natural	39–55
3	Liquid petroleum	73–87

LPGs are a by-product of the oil refining process. They are also found naturally in the North Sea and other oilfields. These gases are liquefied in containers to about 1 of 200 of their volume as a gas by application of moderate pressure for convenience in transportation and storage. They are marketed as two grades, propane and butane, under various brand names. Both grades are heavier than air; therefore, periphery walls around storage containers are unacceptable. If there were a leakage, the vapour would be trapped at low level and be unable to disperse. Calorific values differ considerably from natural gas, therefore appliances are not interchangeable. Siting of storage vessels should be away from buildings, boundaries and fixed sources of ignition as a precaution in the event of fire.

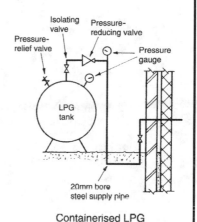

Containerised LPG

Storage tank capacity (m³)	Min. distance from building or boundary (m)
< 0·45	–
0·45–2·25	3.0
2·25–9·00	7·5
> 9·00	15·0

Characteristics:

Propane:

Calorific value	96 MJ/m³ (dry) 50 MJ/kg
Specific gravity	1·4–1·55
Sulphur content	0·02%
Air for combustion	24 m³ per m³ of gas

Butane:

Calorific value	122 MJ/m³ (dry) 50 MJ/kg
Specific gravity	1·9–2·1
Sulphur content	0·02%
Air for combustion	30 m³ per m³ of gas

Refs. UKLPG Code of Practice: Bulk LPG Storage at Fixed Installations, Part 1.

BS 6891: Specification for the installation and maintenance of low pressure gas installation pipework of up to 35mm (R1¼) on premises.

LPG – Cylinder Storage

Installation procedures for LPG are determined by:

- Capacity/volume of gas stored.
- Location of storage vessels, whether above or below ground.
- Function and purpose of the building served.

For most domestic and other small buildings, LPG is stored outdoors above ground in 47 kg cylinders. These are usually in a bank of four, two in service and two on standby. Transfer is effected with a simple manual or automatic change-over valve. Replacement cylinders are obtained from a contracted supplier.

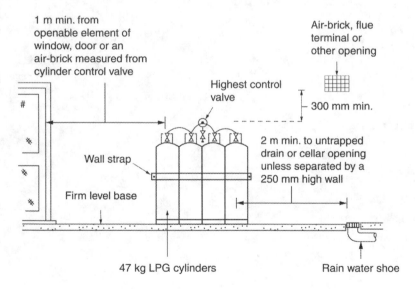

Cylinders to be padlocked and restrained in an upright position by straps, chains or a storage cage. They must also be protected from physical damage and be clear of exit routes.

LPG may be stored outdoors in tanks located below or above ground, but not within an open pit. Tanks are manufactured in standard capacities of 2 and 4m^3 (2000 and 4000 litres). They should be serviceable from an access road, positioned no more than 25 m from the replenishment vehicle.

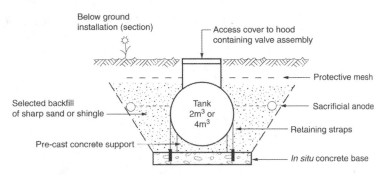

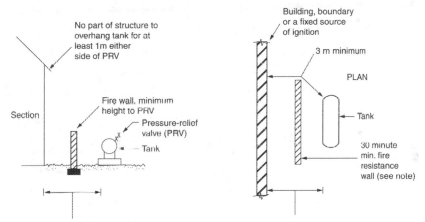

3 m normal separation may reduce to 1.5 m where fire wall intervenes

Note: Insulation, integrity and stability (resistance to collapse). Ref. Bldg. Regs. AD B and BS 476-22.

Refs. Building Regulations AD J6: Protection of liquid fuel storage systems. Section 5.

UKLPG Code of Practice 22: Design, Installation and Testing of LPG Piping Systems.

Institution of Gas Engineers and Managers Standard IGE/TD/4: Polyethylene (PE) and steel gas services and service pipework.

Electric Boiler

Electrically powered boilers have the advantage of no maintenance, no flue, over 99% efficiency* and no direct discharge of noxious gases.

* Energy loss is at the power-station where conversion of fuel energy into electricity can be as little as 50% efficient.

Primary thermal store (>15 litres capacity) – these use off-peak electricity, normally through a 3kW immersion heater as an economic means for creating a store of hot water. They have the option of supplementary power at standard tariff through higher rated immersion heaters to satisfy greater demand.

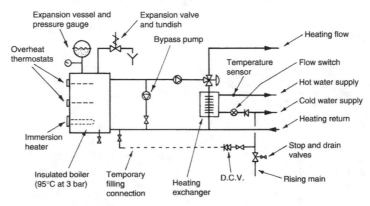

Primary store boiler

Instantaneous (<15 litres capacity) – these low water content, high-powered (6–12kW) units provide direct heat energy at standard tariff in response to programmed demand. They are very compact, generally about 100mm square × 1m in height. Integral controls include a thermal safety cut-out and 'soft' switching to regulate power supply as the unit is engaged.

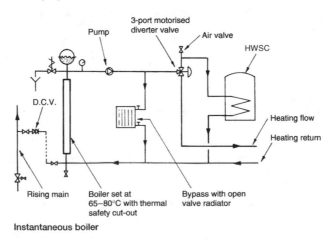

Instantaneous boiler

Electricity can be used directly in convectors, fan heaters, element fires, etc., or indirectly as shown below as hot water thermal storage heating. It is an alternative use of off-peak electricity to storage in concrete floors or thermal block space heaters and has the advantage of more effective thermostatic control.

Electricity is converted to heat energy in water by an electrode boiler and stored in a pressurised insulated cylinder at about 180°C. The water is circulated by a pump programmed for daytime use to heat emitters in the building. Careful design of the storage vessel is essential to maintain sufficient thermal capacity for the heating requirements. An assessment of demand will need to be presented to the supply authority and a reduced rate of electricity tariff may be negotiated, possibly between 1900 and 0700 hours.

Calorific value of electricity 3.6MJ/kWh

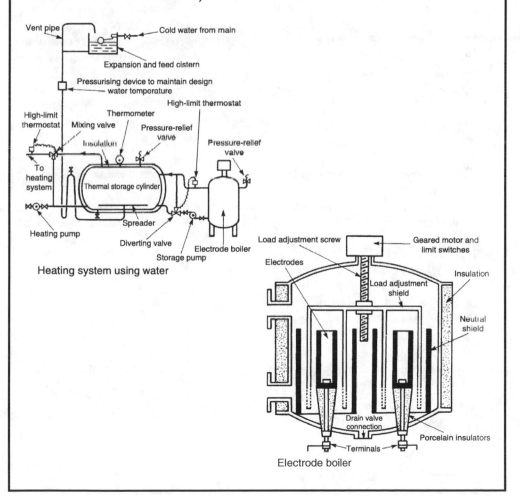

Heating system using water

Electrode boiler

Electrode Boiler

Principles – operates at a high voltage, typically between 6 and 24 kV AC with a capacity of 5–60 MW, both depending on application. Working pressure about 10 bar. Water within the boiler functions as both resistor and conductor to the electrical supply. A field of electricity flows between jet-type electrodes – usually 3 but may be more. This energises the molecules of water converting it to steam.

Background – originally developed some 70–80 years ago. During the 1950s and immediate decades thereafter, large-scale applications were widely used, especially in countries that had available cheap surplus hydro-electric power, notably Norway. In the latter part of the 20th century, its use declined as improvements in grid connections between countries made thermal energy generating systems less financially attractive.

Present-day applications – since the turn of the century, the technology has re-emerged, generally on a lesser scale than grid distribution, but as a supplement to intermittent/sporadic alternative energy generation such as wind turbines, and also, on a smaller scale still, where a back-up or a complementary energy facility is required. Examples being hospitals, CHP and district heating.

Advantages over alternative high temperature plant –

- Almost all (close to 100%) electrical energy input is converted to heat.
  Minimal energy loss is radiant heat from the insulated boiler casing.
- Rapid response time.
- Simpler and less expensive construction, fewer components than conventional tube heat exchanger boilers with combustion chambers and flues.
- Easier to control and maintain.
- Proven cost-effective efficient technology.
- Zero carbon footprint.
- Zero site atmospheric pollutants.
- Minimal thermal stress in boiler.
- Self-regulating as reduced water levels will limit current flow.

A disadvantage could be perceived as the amount of attention required to the necessary safety measures to avoid risks associated with very high-voltage equipment.

6 VENTILATION SYSTEMS

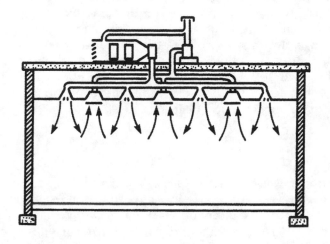

VENTILATION REQUIREMENTS

GUIDE TO VENTILATION RATES

DOMESTIC ACCOMMODATION

MECHANICAL VENTILATION

TYPES OF FAN

FAN LAWS

SOUND ATTENUATION IN DUCTING

AIR FILTRATION

LOW-VELOCITY AIR FLOW IN DUCTS

AIR DIFFUSION AND DISTRIBUTION PATTERNS

AIR CURTAIN

VENTILATION DESIGN

RESISTANCES TO AIR FLOW

DOI: 10.1201/9781003434894-6

Ventilation

Construction of modern homes is to a very high standard of energy-efficient insulation and airtightness. Standards have gradually improved over the past 50 years, initially in response to diminishing fossil fuel resources and more recently due to concerns about the effects of atmospheric pollution and global warming. Older homes had varying degrees of air leakage and natural ventilation that ensured a healthy internal environment. A high envelope air permeability allowed air infiltration and leakage through the building enclosure. Examples include single-glazed windows with ill-fitting sashes, door and window frames not sealed to adjacent walls, doors without draught proofing, fireplaces with open flues and air vents/bricks in pantries and boiler room walls (prebalanced flue boilers).

By comparison with past standards, contemporary construction has the potential to encourage an uncomfortable and unhealthy internal environment due to the discomfort of living within an energy-efficient sealed external envelope. To avoid the symptoms of sick building (house?) syndrome (see page 343), mould growths, asthma and dust mite allergies, provision of air circulation must be integrated with building design.

Air infiltration can be achieved by natural or mechanical means. The former is partly achieved by background trickle vents in window frames and by air gaps or undercutting to internal doors. Natural ventilation by these means is difficult to regulate in defined quantities; therefore, low-energy-use mechanical ventilation systems, particularly those with a heat recovery facility, are becoming quite common in new-build homes.

Building Regulations, Approved Document F1: Means of ventilation, supports provision of mechanical systems in homes that have an airtightness or air infiltration rate of less than 5m^3/h per m^2 envelope area at 50 Pascals (Pa or N/m^2) test pressure. Most dwellings are now constructed to be tighter than this criterion.

Ventilation – a means of changing the air in an enclosed space to:

- Provide fresh air for respiration – approx. 0·1 to 0·2l/s per person.
- Preserve the correct level of oxygen in the air – approx. 21%.
- Control carbon dioxide content to no more than 0.1%. Concentrations above 2% are unacceptable as carbon dioxide is poisonous to humans and can be fatal.
- Control moisture – relative humidity of 30% to 70% is acceptable.
- Remove excess heat from machinery, people, lighting, etc.
- Dispose of odours, smoke, dust and other atmospheric contaminants.
- Relieve stagnation and provide a sense of freshness – air movement of 0·15 to 0·5m/s is adequate.

Measures for control:

- Health and Safety at Work, etc., Act.
- Building Regulations (Approved Inspectors) Regulations.
- Air Quality Standards Regulations.
- Building Regulations, Approved Document F – Ventilation.
- BS 5925: Code of practice for ventilation principles and designing for natural ventilation.
- Health Act.

The statutes and regulations provide the Health and Safety Executive and approved building inspectors with authority to ensure buildings have suitably controlled internal environments. Building Regulation AD F and the British Standard provide measures for application.

Requirements for an acceptable amount of fresh air supply in buildings will vary depending on the nature of occupation and activity. As a guide, between 10l/s of outdoor air supply per person can be applied between the extremes of a non-contaminated environment, to an extract air rate of 36l/s per person in a room where the atmosphere is contaminated, e.g. by welding fumes from a factory manufacturing process. Converting this to m^3/h (divide by 1000, multiply by 3600), equates to $130m^3/h$ per person.

Air changes per hour or ventilation rate is the preferred criterion for system design. This is calculated by dividing the quantity of air by the room volume and multiplying by the occupancy.

E.g. $50m^3/h$, $100m^3$ office for five persons: $50/100 \times 5 = 2·5a/c$ per h.

Guide to Ventilation Rates

Room/building/accommodation	Air changes per hour
Assembly/entrance halls	3–6
Bathrooms (public)	6*
Boiler plant rooms	10–30†
Canteens	8–12
Cinema/theatre	6–10
Classrooms	3–4
Dance halls	10–12
Dining halls/restaurants	10–15
Domestic habitable rooms	1–4*
Factories/garages/industrial units	6–10
Factories – fabric processing	10–20
Factories (open plan/spacious)	1–4
Factories with unhealthy fumes	20–30
Foundries	10–15
Hospital wards	6–10
Hospital operating theatres	10–20
Kitchens (commercial)	20–60*
Laboratories	6–12
Laundries	10–15
Lavatories (public)	6–12*
Libraries	2–4
Lobbies/corridors	3–4
Offices	2–6
Smoking rooms	10–15
Warehousing	1–2

Notes:

*For domestic applications, see pages 253 and 254.

†18 air changes per hour is generally acceptable, plus an allowance of 0·5l/s (1·8m³/h) per kW boiler rating for combustion air. Double the combustion allowance for gas boilers with a diverter flue.

See also: BS 5925: Code of practice for ventilation principles and designing for natural ventilation.

Approved Document F (Ventilation) provides the minimum requirements for comfortable background ventilation and for preventing the occurrence of condensation. It is effected without significantly reducing the high standards of thermal insulation necessary in modern buildings.

Definitions:

- Habitable room – any room used for dwelling purposes, not solely a kitchen, utility room, bathroom or sanitary accommodation.
- Bathroom – any room with a bath and/or shower. Can include sanitary accommodation.
- Sanitary accommodation – any room with a WC.
- Wet room – kitchen, utility room, bathroom and sanitary accommodation.
- Ventilation opening – a means of ventilation, permanent or variable (open or closed) providing access to external air, e.g. door, window, louvre, air-brick or PSV.
- PSV – passive stack ventilation is a system of vertical ducting from room ceilings to roof outlets providing ventilation by stack effect and wind passing over the roof.
- Rapid or purge ventilation – openable window or mechanical fan extract system should be provided for each habitable room (see note).
- Background ventilation – permanent vents, usually trickle ventilators set in a window frame (see below). An air-brick with a sliding 'hit and miss' ventilator could also be used. Should provide air flow from outside to inside and inside to outside of a dwelling.
- Whole-building ventilation – continuous ventilation through background/trickle ventilators or other purpose-made vents.

Note: With background ventilation, some part of the ventilation opening should be at least 1.70m above the floor to avoid draughts.

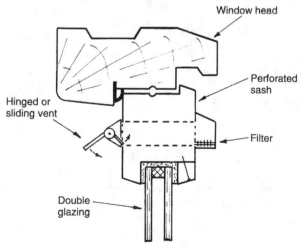

Window head
Perforated sash
Hinged or sliding vent
Filter
Double glazing

Window trickle ventilator

Note: Hinged or pivoted windows opening at least 30°, opening sash windows and external doors; min. 1/20 floor area of room served. 15° to 30°, 1/10 floor area. Less than 15° unsuitable.

Ventilation of Dwellings

Habitable rooms – rapid or purge ventilation should be capable of producing four air changes per hour for each room directly to outside, plus a whole dwelling ventilation rate of not less than:

Bedrooms	1	2	3	4	5
Ventilation rate (l/s)*	19	25	31	37	43

* For each additional bedroom, add 6 l/s.
* The minimum acceptable rate for any dwelling is 0.3l/s per m² total internal floor area.

Kitchen, utility room, bathroom and sanitary accommodation – local ventilation by intermittent or continuous mechanical means, i.e. an extractor fan capable of achieving the following minimum rates (l/s):

Room	Intermittent**	Continuous (high)	Continuous (low)
Kitchen	30 (adjacent to hob), or 60 (elsewhere)	13	Total extract for all rooms, not less than the whole-building ventilation rate.
Utility room	30	8	
Bathroom	15	8	
Sanitary accommodation	6	6	

** 15min. overrun where fitted to an internal room.
10mm ventilation gap cut under all internal doors above the floor finish.

Ventilation of dwellings can be provided by:

- For multiple floor dwellings, background/trickle ventilators of at least 8000mm² equivalent area in each habitable room and kitchen, 4000mm² in bathrooms (see next page). Purge or rapid ventilation by fan or openable window to every habitable room. Intermittent extractor fans with background ventilators for kitchen, utility room, bathroom and sanitary accommodation as table above.
- For single storey dwellings, min. 10,000mm² equivalent area in habitable rooms and kitchen, 4000mm² bathroom. For any dwelling type, no minimum equivalent area trickle ventilators for sanitary accommodation.
- Continuous mechanical extract (MEV) with background ventilators to all rooms. Purge ventilation to all habitable rooms (see page 261).
- Continuous mechanical supply and extract with heat recovery (MVHR). Purge ventilation to every habitable room (see page 262).
- Mechanical supply ventilation, also known as positive input ventilation (PIV). Purge ventilation to every habitable room (see page 264).

Note: For specific requirements relating to each of the above alternatives, see Building Regulations, Approved Document F – Ventilation, Section 1: New dwellings and Section 3: Existing dwellings.

The minimum equivalent ventilation area (EVA) requirements for background (trickle) ventilators are determined by specific room function as indicated below:

Function	Min. EVA – multiple floor dwellings (mm^2)	Min. EVA – single storey dwellings (mm^2)
Habitable rooms	8000	10,000
Kitchen	8000	10,000
Utility room	No minimum	No minimum
Bathroom	4000	4000
Sanitary accommodation	No minimum	No minimum

- Most trickle ventilators are purpose made with an EVA of 2500 mm^2.
- The above table does not apply where a dwelling has only one exposed façade, at least 70% of its openings in the same façade and where an internal kitchen has no windows or external wall available for a ventilator. Specialist consultancy required to determine alternatives.
- Open plan kitchen and living room situations require a minimum of three ventilators of the same EVA as provided for other habitable rooms in the same dwelling.
- Minimum of five ventilators in total required in habitable rooms and kitchens. Minimum of four in one-bedroom dwellings.
- Where a bathroom is without a window or external wall to accommodate a ventilator, the minimum EVA should be added to the EVA specified for other rooms.
- If a habitable room has no external walls, ventilation can be provided through an adjacent habitable room that has an external wall facility containing purge ventilation of fractional area (pg. 253) for both room floor areas. Background EVA min. 10,000 mm^2 and a permanent opening provided between the two rooms – min. 1/20 combined floor area.
- Cross-ventilation to be encouraged by locating ventilators on opposing walls.
- Purpose-made noise attenuating ventilators to be fitted where an external wall is close to a continuous noise source, e.g. busy road.
- Fans and ventilators in the same room at least 500 mm apart.

Ref. BS EN 13141-1: Externally and internally mounted air transfer devices.

Ventilation of Offices

Occupiable work rooms – will require a whole-building ventilation air supply rate of at least 10l/s per person, or 1 litre per second per sq. m floor area. Take higher value. Background or trickle ventilation can be used to satisfy this objective. As a guide, 4000mm² ventilation area per 10m² of floor area, with an additional 400mm² thereafter for every 1m² of floor.

Additional rapid or purge ventilation is also required for every unit of office accommodation. This may be satisfied with an openable window area at least equivalent to a percentage of the floor area as defined in BS 5925, or a mechanical air extract directly to outside, capable of at least 10l/s per person. For example, an office with an occupancy of six persons, floor area of 30m² and a room height of 3m (90m³ volume):

Background ventilation minimum = 10l/s per person

Purge/rapid ventilation minimum = 10l/s per person

$$\text{Total} = 20\text{l/s per person}$$

$$(20 \times 3600) \div 1000 = 72\text{m}^3/\text{h.}$$

$$(72 \div 90) \times 6 = 4\cdot8 \text{ air changes per hour (min.)}$$

Common space such as lift lobbies and corridors require either:

• Natural ventilation of at least 1/50 floor area,

or

• Mechanical ventilation supply of at least 0.5 l/s per m² common space floor area.

Kitchen (for food and beverage preparation), washrooms, sanitary accommodation, photocopy and print processing rooms – local extract ventilation by continuous or intermittent means as follows:

Room function	Local extract
Printing and photocopying for more than 30 minutes in every hour	20l/s per machine while in use if the room is permanently occupied, use greater value of extract and whole-building ventilation rate.
Sanitary accommodation and washrooms	Intermittent air extraction of: 15l/s per bath and shower. 6l/s per WC and urinal.
Food and beverage preparation areas (not commercial kitchens, see page 252)	Intermittent air extraction of: 15l/s for microwave and beverages only. 30l/s adjacent to hob with cooker(s). 60l/s elsewhere with cooker(s). Extract to engage automatically when food and beverage preparation equipment operates.

Note: Passive stack ventilation may be an acceptable alternative to use of local extract by mechanical means for sanitary accommodation and washrooms, and for food and beverage preparation areas.

Natural ventilation – non-mechanical air movement through intentional openings.

Refs.
CIBSE Application Manual 10.
Building Regulation Approved Document F, Vol. 2.

Natural ventilation is an economic means of providing air changes in a building. It uses components integral with construction such as air-bricks and louvres, or openable windows. The sources for natural ventilation are wind effect/pressure and stack effect/pressure.

Stack effect is an application of convected air currents. Cool air is encouraged to enter a building at low level. Here it is warmed by the occupancy, lighting, machinery and/or purposely located heat emitters. A column of warm air rises within the building to discharge through vents at high level, as shown on the following page. This can be very effective in tall office-type buildings and shopping malls, but has limited effect during the summer months due to warm external temperatures. A temperature differential of at least 10K is needed to effect movement of air; therefore, a supplementary system of mechanical air movement should be considered for use during the warmer seasons.

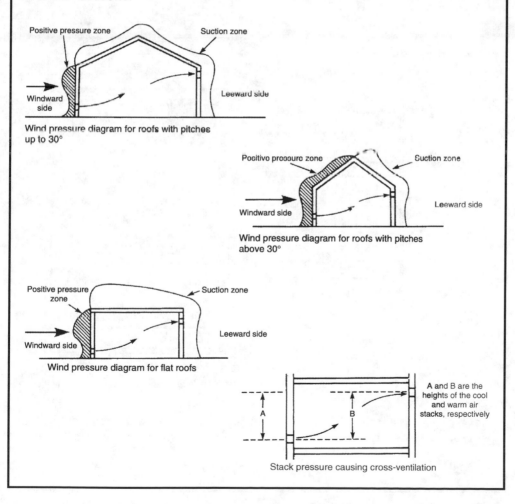

Wind pressure diagram for roofs with pitches up to 30°

Wind pressure diagram for roofs with pitches above 30°

Wind pressure diagram for flat roofs

Stack pressure causing cross-ventilation

The rates of air change are determined by the building purpose and occupancy, and local interpretation of public health legislation. Public buildings usually require a ventilation rate of 30m^3 per person per hour.

Wind passing the walls of a building creates a slight vacuum. With provision of controlled openings, this can be used to draw air from a room to effect air changes. In tall buildings, during the winter months, the cool, more dense, outside air will tend to displace the warmer, lighter, inside air through windows or louvres on the upper floors. This is known as stack effect. It must be regulated; otherwise, it can produce draughts at low levels and excessive warmth on the upper floors.

Ventilation and heating for an assembly hall or similar building may be achieved by admitting cool external air through low-level convectors. The warmed air rises to high-level extract ducts. The cool air intake is regulated through dampers integral with the convectors.

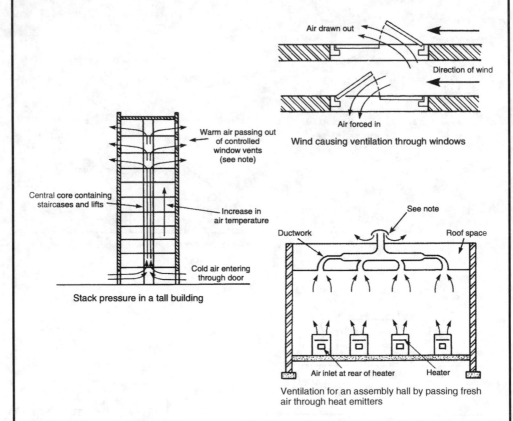

Air drawn out

Direction of wind

Air forced in

Wind causing ventilation through windows

Warm air passing out of controlled window vents (see note)

Central core containing staircases and lifts

Increase in air temperature

Cold air entering through door

Stack pressure in a tall building

See note

Ductwork

Roof space

Air inlet at rear of heater

Heater

Ventilation for an assembly hall by passing fresh air through heat emitters

Note: By current fuel conservation requirements, heat energy that would otherwise be lost by stack effect and convected air ventilation can be recovered. See pages 262 and 332 to 340.

Stack effect – a means for creating natural movement of air, particularly applicable to tall and narrow buildings. Established long ago for use in industrial buildings, notably mills and factories of the Victorian era. By current architectural designs, the principle shown on the preceding page has been applied to many open plan enclosed structures, notably atria and shopping malls.

Stack pressure – the following formula can be used as a guide to obtaining the pressure produced by stack effect ventilation:

$$P = 3462 \times h \, (1 \div t_o + 273) - (1 \div t_i + 273)$$

Where P = pressure (N/m^2 or Pascals)
  h = height between low air level inlet and high level outlet (m)
  t_o = temperature outside (°C)
  t_i = temperature inside (°C)

E.g. 1: h = 30m, t_o = 15°C, t_i = 20°C.
$P = 3462 \times 30 \, (1 \div 15 + 273) - (1 \div 20 + 273)$
$P = 103860 \, (1 \div 288) - (1 \div 293)$
$P = 103860 \, (0.0034722 - 0.0034129)$
$P = 103860 \times 0.0000593$
$P = 6.16$ N/m^2 or Pascals (Pa)

E.g. 2: h = 50m, t_o = 5°C, t_i = 25°C.
$P = 3462 \times 50 \, (1 \div 5 + 273) - (1 \div 25 + 273)$
$P = 173100 \, (1 \div 278) - (1 \div 298)$
$P = 173100 \, (0.0035971 - 0.0033557)$
$P = 173100 \times 0.0002414$
$P = 41.7$ N/m^2 or Pascals (Pa)

Approximate pressure (N/m^2 or Pa) due to stack effect –

$t_i - t_o$ (°C)	5	10	20	30	40	50	100m
–10	–2.2	–4.3	–8.6	–12.9	–17.2	–22	–43
0	0	0	0	0	0	0	0
5	1.1	2.2	4.3	6.5	8.6	11	22
10	2.2	4.3	8.6	12.9	17.2	22	43
20	4.3	8.6	17.2	25.8	34.4	43	86

Natural Ventilation – Passive Stack Ventilation (PSV)

PSV consists of vertical or near-vertical ducts of 100 to 150mm diameter, extending from grilles set at ceiling level to terminals above the ridge of a roof. Systems can be applied to kitchens, bathrooms, utility rooms and sometimes sanitary accommodation, in buildings up to four storeys requiring up to three stacks/ducts. More complex situations are better ventilated by a Mechanical Assisted Ventilation System (MAVS) (see next page).

PSV is energy efficient and environmentally friendly with no running costs. It works by combining stack effect with air movement and wind passing over the roof. It is limited to self-regulation, responding to temperature differentials where internal and external temperatures vary.

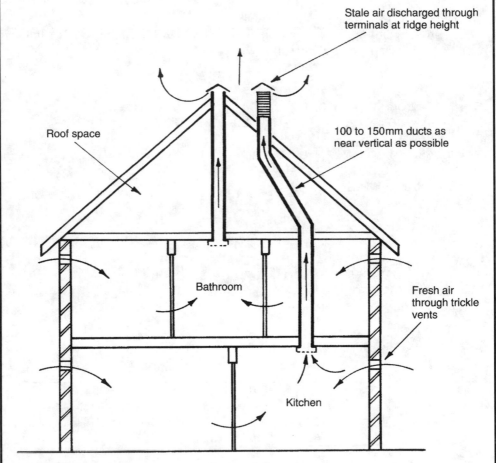

PSV to a dwelling house

Without an integral means of heat exchange/energy recovery (pg. 262) will contribute to air thermal losses.

Mechanically Assisted Extract Ventilation Systems (MAVS or MEV)

MAVS may be applied to dwellings and commercial premises where PSV is considered inadequate or impractical. This may be because the number of individual ducts would be excessive, i.e. too space-consuming and obtrusive with several roof terminals. A low-powered (40W) silent running fan is normally located within the roof structure. It runs continuously and may be boosted by manual control when the level of cooking or bathing activity increases. Humidity sensors can also be used to automatically increase air flow.

MAVS are acceptable to Approved Document F of the Building Regulations as an alternative to the use of mechanical fans in each room. However, both PSV and MAVS are subject to the spread of fire regulations (Approved Document B). Ducting passing through a fire-resistant wall, floor or ceiling must be fire protected with fire-resistant materials and be fitted with a fusible link automatic damper.

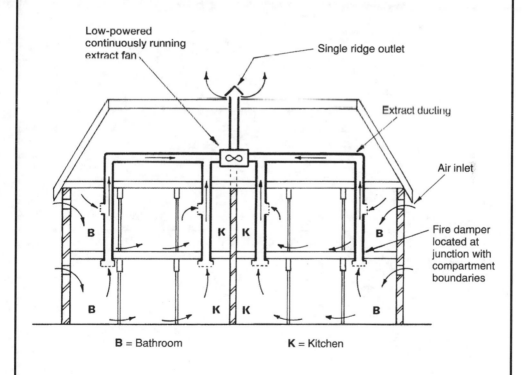

B = Bathroom   **K** = Kitchen

MAVS in a group of flats

Mechanical Ventilation with Heat Recovery (MVHR)

MVHR is a development of MAVS to include energy recovery from the warmth in fan-extracted moist air from bathrooms and kitchens. The heat recovery unit contains an extract fan for the stale air, a fresh air supply fan and a heat exchanger. This provides a balanced continuous ventilation system, obviating the need for ventilation openings such as trickle ventilators. Apart from natural leakage through the building and air movement from people opening and closing external doors, the building is sealed to maximise energy efficiency. Up to 70% of the heat energy in stale air can be recovered, but this system is not an alternative to central heating. A space heating system is required and MVHR can be expected to contribute significantly to its economic use. MVHR complies with the ʻalternative approachesʼ to ventilation of dwellings, as defined in Approved Document F to the Building Regulations.

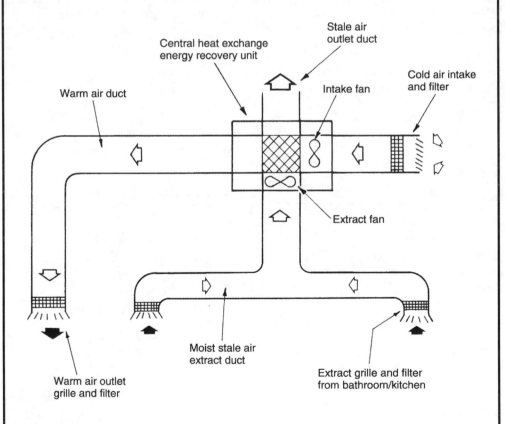

Schematic of an MVHR system of ventilation

Background trickle ventilators should not be used with MVHR systems as they could create unintended air pathways.

Mechanical ventilation systems are frequently applied to commercial buildings, workshops, factories, etc., where the air change requirements are defined for health and welfare provision. There are three categories of system:

1. Natural inlet and mechanical extract
2. Mechanical inlet and natural extract
3. Mechanical inlet and mechanical extract

The capital cost of installing mechanical systems is greater than natural systems of air movement, but whether using one or more fans, system design provides for more reliable air change and air movement. Some noise will be apparent from the fan and air turbulence in ducting. This can be reduced by fitting sound attenuators and splitters as shown on page 273. Page 288 provides guidance on acceptable noise levels.

Internal sanitary accommodation must be provided with a shunt duct to prevent smoke or smells from passing between rooms. In public buildings, duplicated fans with automatic change-over are also required in the event of failure of the duty fan.

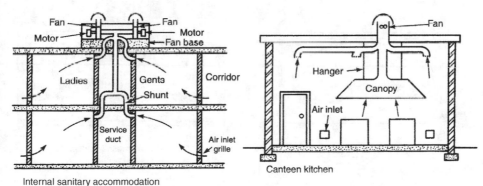

Internal sanitary accommodation

Canteen kitchen

Basement car parks require at least six air changes per hour and at exits and ramps where queuing occurs, local ventilation of at least ten air changes per hour. Duplicate fans should be provided with a fan failure automatic change-over.

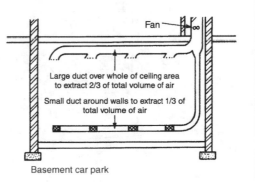

Basement car park

Fan-assisted ventilation systems supplying external air to habitable rooms must have a facility to pre-heat the air. They must also have control over the amount of air extracted; otherwise, there will be excessive heat loss. A mechanical inlet and mechanical extract system can be used to regulate and balance supply and emission of air by designing the duct size and fan rating specifically for the situation.

Air may be extracted through specially made light fittings. These permit the heat-enhanced air to be recirculated back to the heating unit. This not only provides a simple form of energy recovery, but also improves the light output by about 10%. With any form of recirculated air ventilation system, the ratio of fresh to recirculated air should be at least 1:3. i.e. min. 25% fresh, max. 75% recirculated.

In large buildings where smoking is not permitted, such as a theatre, a downward air distribution system may be used. This provides a uniform supply of warm filtered air.

Ductwork in all systems should be insulated to prevent heat losses from processed air and to prevent surface condensation.

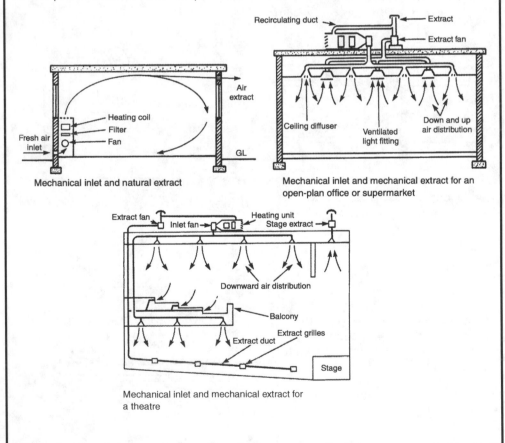

Mechanical inlet and natural extract

Mechanical inlet and mechanical extract for an open-plan office or supermarket

Mechanical inlet and mechanical extract for a theatre

Profile – generally circular, square or rectangular but may be oval. For efficient distribution of air, the uniformity of circular ducting is preferred for the following reasons:

• less opportunity for turbulence
• less resistance to friction
• inherent rigidity
• lower heat losses or gains
• sound transfer generally less
• less potential for air leakage

Where space is restricted under floors or in suspended ceilings, rectangular ducting of high-aspect ratio may be required for practical reasons (aspect ratio and conversion from circular to square or rectangular equivalent size are explained on pages 293 to 295). Square or rectangular ducting direction changes are more easily formed than with circular sections.

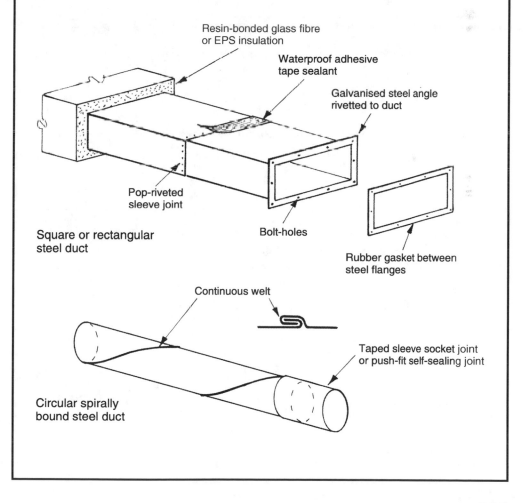

Resin-bonded glass fibre or EPS insulation

Waterproof adhesive tape sealant

Galvanised steel angle rivetted to duct

Pop-riveted sleeve joint

Bolt-holes

Square or rectangular steel duct

Rubber gasket between steel flanges

Continuous welt

Taped sleeve socket joint or push-fit self-sealing joint

Circular spirally bound steel duct

Ducting – Materials

Galvanised sheet steel is the most common material used for ventilation and air-conditioning ducting. Factory prefabricated sections are site jointed by bolted steel angle flanges with a rubber sealing gasket; the rigid angles can also function as suspended bracket fixings. Sleeve jointing with pop-rivets and tape sealant is also used with smaller profile sections.

In addition to galvanised steel, aluminium may be used in smaller profiles or externally in non-corrosive atmospheres. Copper or stainless steel is used where the ducting forms a feature, e.g. a cooker hood. Polypropylene and uPVC piping is suitable in short lengths and small diameters, mainly for domestic applications such as extract fan extensions. Plastic materials have limitations where performance in fire is a consideration.

Material	Sheet/wall thickness (mm)	Situation	Application
Galvanised steel	0·6	Low velocity < 10 m/s	Universal
		Low pressure < 500Pa	..
	0·8	Velocity > 10 m/s	..
		Pressure > 500Pa	..
Aluminium or copper	0·8	Low velocity	Features
Stainless steel	As galvanised steel		Features
UPVC	3·0	Low velocity	Domestic
Polypropylene	3·0	Low velocity	Domestic
Resin bonded glass fibre	3·0	Low velocity	Warm air heating

Apart from standard plastic pipe profiles (100 and 150mm nominal diameter drainage pipes), most ducting is factory produced to the designer's specification. It is unrealistic for sheet metal fabricators to produce standard sections due to unknown demand and the space requirement for storage.

Flexible ducts are useful for short connections from air distribution boxes or plenums to several diffusers within close proximity. They are also useful for correcting misalignments and for convenient connections to fan housings and terminals. Flexible connections to fans will help to reduce vibration and sound. Flexible ducting is produced in corrugations made up in a concertina format from thin sheet aluminium or from spirally wound steel-reinforced fabric. Lengths should be limited to as short as possible, as the concertina effect will impede air flow and create noise. Also, flexible ducting is more likely to suffer damage and leakage. Jointing is by taped sleeve and jubilee clip.

Propeller fan – several steel or plastic blades attached to a frame-mounted motor-driven shaft. Portable free-standing floor or desk units provide for local movement of air. Propeller fans can also be secured over a void in an external wall or ceiling/roof. Location within a ceiling or roof void is also possible with a short length of flexible ducting to discharge externally. Volumetric air flow can be very high with these fans, ideal for extracting smoke and other air suspended contaminants. Pressure potential is low at 15 to 20Pa and efficiency only about 40%. Therefore, propeller fans are not suitable for delivering air through ventilation duct systems.

Propeller fan

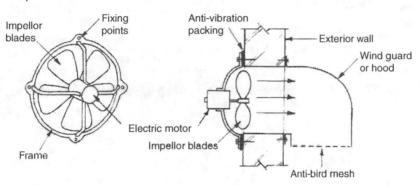

Cross flow or tangential fan – a cylindrical unit with peripheral forward curved impellors on a motor driven central axis. Air enters through one side of the fan casing and discharges in a series of vortices at the opposing side, sometimes deflected through 90°. Air velocity is high, although efficiency is only about 45%, therefore not for use with ducted systems. Mainly for portable fan heater units and fan-coil convectors, see pages 146 and 264.

Cross flow or tangential fan

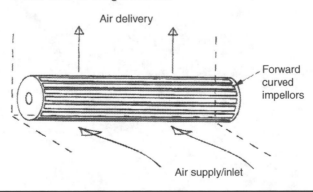

267

Types of Fan – Axial and Bifurcated Axial

Axial flow – several aerofoil section blades secured to a motor-driven central shaft. Located within a circular housing for in-line mounting to ventilation ducts. Large fans can be belt driven from a motor located externally. Close blade proximity to its housing is essential to generate high pressure, as is the aerodynamic design of the twist and pitch angle of the blades. Positioned within the ducting, the unit is compact; therefore, the only supplementary support requirement will be brackets to take the weight of the motor. High air pressures are possible, in excess of 2kPa not unusual. Where two fans are mounted in series, the pressure potential is doubled and so on proportionally. Fan efficiency is about 75%, but can be noisy at high speeds.

Axial flow fan

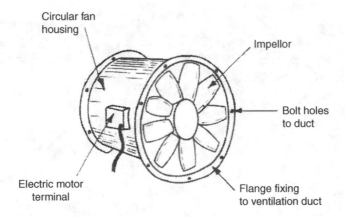

Circular fan housing

Impellor

Bolt holes to duct

Electric motor terminal

Flange fixing to ventilation duct

Bifurcated axial flow – a variation with an insulated inner housing to protect the fan cooled motor from greasy, hot and corrosive gases. Used in commercial kitchen hoods and boiler flue extracts.

Bifurcated axial flow fan (Section)

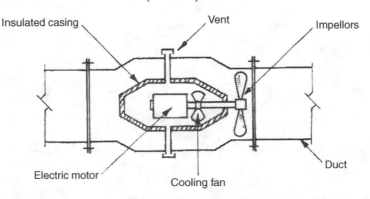

Insulated casing

Vent

Impellors

Electric motor

Cooling fan

Duct

Centrifugal fan – a cylindrical wheel with peripheral impellors. Blades comprising the impellor rotate on a shaft within an involute or scroll shaped casing. Air drawn in through the side(s) of the casing discharges radially under centrifugal force to the delivery ducting. Small fans have a motor integral to the impellor. For high-pressure air delivery through long ducted sections, the motor is external. This may be detached from the fan with a belt drive. Variable diameter pulley blocks regulate air flow conditions to seasonal requirements.

Centrifugal fan

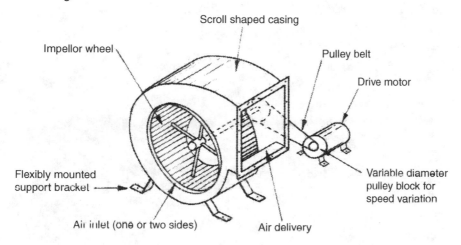

Characteristics – relatively quiet and capable of developing air pressures up to about 750Pa, possibly up to 1kPa with recent improvements in manufacturing technology. Efficiencies range between about 50% and 85% depending on configuration of impellors.

Centrifugal fan impellors

| Forward bladed for constant air pressure/volume | Backward bladed for variable air pressure/resistance | Radial or paddle bladed for air containing suspended particles |

Note : For all fan types, design and selection will depend on matching system characteristics with fan performance. See page 292.

Fan Laws

Fan performance depends very much on characteristics such as type and configuration of components. Given a standard set of criteria against which a fan's performance is measured (i.e. 20°C dry bulb temperature, 101·325kPa (1013mb) atmospheric pressure, 50% relative humidity and 1·2kg/m^3 air density), any variation in performance can be predicted according to the following fan laws:

• Discharge (volumetric air flow) varies directly with the fan speed.

$$Q2 = Q1(N2/N1)$$

• Fan pressure is proportional to the fan speed squared.

$$P2 = P1(N2/N1)^2$$

• Fan power is proportional to the fan speed cubed.

$$W2 = W1(N2/N1)^3$$

where Q = air volume in m^3/s
N = fan speed in rpm
P = pressure in pascals (Pa)
W = power in watts or kilowatts.

E.g. a mechanical ventilation system has the following fan characteristics:

Discharge (Q1) = 6m^3/s
Pressure (P1) = 400Pa
Power (W1)   = 3kW
Speed (N1)   = 1500rpm

If the fan speed is reduced to 1000rpm, the revised performance data will apply:

Discharge (Q2) = 6(1000/1500) = 4m^3/s
Pressure (P2) = 400(1000/1500)^2 = 178Pa
Power (W2)   = 3000(1000/1500)^3 = 890W

$$\text{Fan efficiency} = \frac{\text{Total fan pressure} \times \text{Air volume}}{\text{Power}} \times \frac{100}{1}$$

So, for this example: $\dfrac{178 \times 4}{890} \times \dfrac{100}{1} = 80\%$

Noise source – sound from fan blade rotation and air turbulence in ducting will be transmitted through air distribution systems and into the rooms served. In addition to the sound produced by the dynamics of air pressurisation and movement, there may also be vibration from mounting the fan on a rigid base.

Preventative measures –
• Limit the design air velocity and pressure drop or resistance to air movement, see page 288.
• Mount the fan on a base with an integral sound absorbing layer, or directly on rubber or spring fixings.
• Provide flexible synthetic fibre-reinforced PVC connections between the fan housing and adjacent ducting.
• Install attenuating measures in the ducting as shown on page 273.

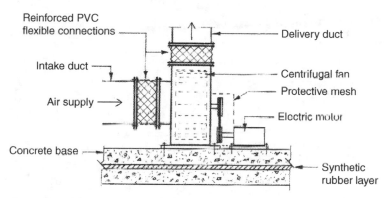

Centrifugal fan on sound absorbing base

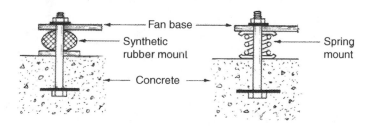

Alternative fan base bolt fixings

Fan Noise – Sound Levels

Fan manufacturers' catalogues and websites contain extensive details of the potential and application of their products, particularly in terms of output, delivery volume, etc. Also, there may be data relating to sound power level (Lw) under various operating conditions. In addition to this, or possibly in the absence of information, the following established empirical formulae may be used as an indication:

$$Lw = 67 + 10 \log (S) + 10 \log (p)$$
$$Lw = 40 + 10 \log (Q) + 20 \log (p)$$
$$Lw = 94 + 20 \log (S) - 10 \log (Q)$$

Where Lw = Sound power level (decibels, db)

log = Logarithm to the base 10

S = Rated motor power (kW)

p = Fan static pressure (Pa or N/m^2)

Q = Volume discharge (m^3/s)

E.g. Using data from the example on page 270:

S = 3kW

p = 400Pa

Q = 6m^3/s

$$Lw = 67 + (10 \log 3) + (10 \log 400)$$
$$Lw = 67 + (10 \times 0.4771) + (10 \times 2.6020)$$
$$Lw = 67 + 4.771 + 26.020 = 97.79 \text{ dB}$$

$$Lw = 40 + (10 \log 6) + (20 \log 400)$$
$$Lw = 40 + (10 \times 0.7781) + (20 \times 2.6020)$$
$$Lw = 40 + 7.781 + 52.040 = 99.82 \text{ dB}$$

$$Lw = 94 + (20 \log 3) - (10 \log 6)$$
$$Lw = 94 + (20 \times 0.4771) - (10 \times 0.7781)$$
$$Lw = 94 + 9.542 - 7.781 = 95.76 \text{ dB}$$

Therefore, expected sound power level will be in the region of the upper nineties decibels.

Note: The decibel (dB) is a measure for sound. It ranges from zero as the basis for the threshold of hearing, up to a sound level of 140 dB at the threashold of pain. Measurement is by hand-held sound level meter.

Sound transmission – noise generated by the fan and associated equipment can be conveyed through attached distribution ducts. Secondary noise within the air ducts can also be caused by sharp edges due to poor workmanship, restrictions through small radius bends and around flow regulating dampers. The length of a ventilation duct and perimeter will have some effect. Long ducts are more effective at damping sound pressure waves than those with a large area. Ducts of a large cross-section may need to be stiffened or braced to resist reverberation.

Methods for reducing and absorbing the transmission of sound –
• Ducting lined with a sound absorbent material such as low-density fibreboard.
• Sound absorbing attenuators located at strategic intervals.
• Vanes or splitters at bends to create a streamlined directional flow, thereby reducing impact and turbulence.
Note: Attenuation material to be fire proofed.

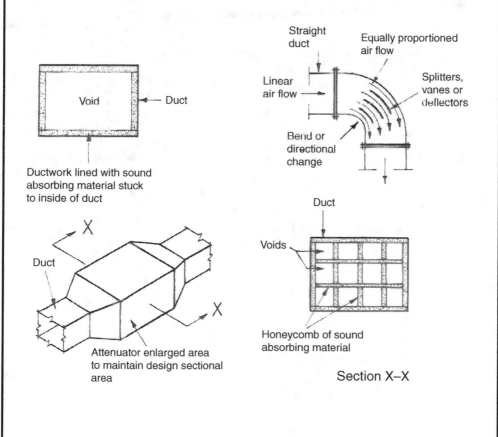

Void

Duct

Ductwork lined with sound absorbing material stuck to inside of duct

Straight duct

Equally proportioned air flow

Linear air flow

Splitters, vanes or deflectors

Bend or directional change

Duct

Voids

Honeycomb of sound absorbing material

Section X–X

X

Duct

X

Attenuator enlarged area to maintain design sectional area

Sound Transmission between Rooms

Corrective measures to prevent installation noise being transmitted through ventilation ducting are considered on the preceding page. Provisions may also be required to control airborne sound. This includes speech produced within a room being carried through a shared ventilation system to other rooms. In the interests of privacy and security, and to reduce the distraction of background noise from other rooms, each branch duct should be provided with an attenuator. This has the dual function of arresting system noise as well as unwanted airborne sound from adjacent rooms. Alternatively, the system could be designed with separation of ducting to individual rooms, with each separate duct being insulated.

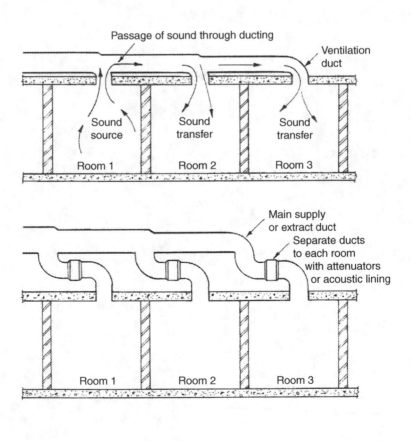

Ventilation ducts that convey air at room ambient temperature to remove stale and contaminated air such as smoke and dust, do not normally require thermal insulation. Ducts that transport pre-heated or pre-cooled air will be subjected to heat loss or gain, metal sheet ducts in particular, less so those of plastic sheet construction. Energy transfer through ducting must be avoided as uneconomic and environmentally unacceptable (Part L, Building Regulations). Combined with the moisture content in the atmosphere, the effect of energy transfer has potential to form surface condensation (sweating) on ducting. Prevention will require selection of appropriate measures for insulation. If surface moisture is not prevented, it can be damaging to both the duct material and components and the adjacent structure.

Thermal insulation applications –

Duct system type	Air in duct	Duct location	Potential problem
Cold air intake	Cold	Warm interior	Condensation on outer surface of duct
Warm air intake	Warm	Unheated interior	Condensation on inner surface of duct
Stale air extract	Warm	Warm interior	Heat energy loss through ducting and to external air*

*Energy recovery unit to be fitted (see pages 262, 339 and 340).

Useful references –
* BS 5422: Method for specifying thermal insulating materials for pipes, tanks, vessels, ductwork and equipment within temperature range minus 40°C to plus 700°C.
* BS 5970: Thermal insulation of pipework, ductwork, associated equipment and other industrial installations in the temperature range of minus 100°C to plus 870°C. Code of practice.
* Building Regulations:
  Part B – Fire safety.
  Part L – Conservation of fuel and power.

Thermal Insulation of Ventilation Ducts – Provisions

Pressure testing – ventilation ducts to be air pressure leak tested before applying insulation. Test pressure is usually set at the system design operating pressure plus 250 Pascals (1 Pa = $1N/m^2$).

Chilled and cold air supply – a fully sealed vapour barrier must be applied to the external warm surface of insulation. This is to prevent moisture in the atmosphere permeating the insulation and reducing its effectiveness. A situation known as 'wicking'. 'Wicking' is the term used where the voids in insulation become moisture saturated in the event of vapour barriers failing, thereby rendering the insulation useless.

Insulation format – purpose-made duct insulation is produced to suit circular, rectangular or square section profiles. Available types are rigid preformed slabs that can be cut to size on site, flexible rolls, blankets and matting. Most have a prepared vapour proof finish of reinforced aluminium foil.

Fixing – secured with an insulation bonding adhesive to the duct surface. Tie wires and clips may also be used. Joint abutments are staggered and sealed with self-adhesive reinforced aluminium foil tape to maintain the vapour barrier between foil surfaces.

Vapour barriers – necessary for all ducted air systems with temperatures below that of ambient air. To be fully effective, they must be continuous and undamaged. Examples include
- Solvent-based polymers, vinyl emulsions and bitumen emulsions, usually reinforced with fabric or glass fibre scrim.
- Thermoplastic elastomer (TPE) sheet with overlapping joints vapour sealed with adhesive.
- Polyvinylchloride (PVC), polyethylene (PE) or polyisobutylene (PIB) tape or sheet materials with sealed overlaps.
- Epoxy and polyester resins, also weather and chemical resistant.
- Polyester or glass fibre-reinforced aluminium foil.

Outdoor exposed finishes – these must be weather-proof to protect the insulation from saturation. Sheet materials include polyisobutylene (PIB) and bituminous roofing felt, both with lapped and sealed joints. Other possibilities are mastic solution painted over woven glass fibre matting and galvanised or stainless steel sheet metal cladding lapped with screw or riveted joints, mastic/silicone sealed.

Insulation materials – duct insulation materials should be both moisture resistant and of closed cell structure. Insulation that has open cells should not be specified for use with ventilation ducts as it is more likely to be prone to 'wicking'.

Insulation materials and properties –

Material type	Density (kg/m³)	Thermal conductivity (W/mK)
Expanded polystyrene	20	0.035
Mineral wool	25	0.040
Nitrile rubber	75	0.035
Phenolic foam	30	0.025
Polyethylene	30	0.040
Polyisocyanurate foam	40	0.020
Polyurethane foam	30	0.025

Figures given are typical and may vary slightly depending on manufacturing process.

See also page 56 for further data relating to insulating materials and their properties.

Typical application of thermal insulation –

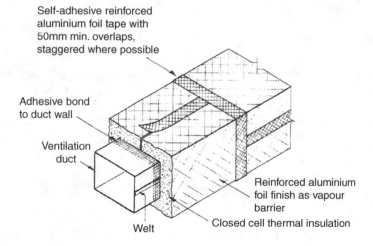

Self-adhesive reinforced aluminium foil tape with 50mm min. overlaps, staggered where possible

Adhesive bond to duct wall

Ventilation duct

Reinforced aluminium foil finish as vapour barrier

Closed cell thermal insulation

Welt

Air Filtration

Function – filters in ventilation and air-conditioning systems are usually located at the point of air entry, where they will have immediate effect in preventing the entry of dirty air. Here, they must be accessible for maintenance and replacement. Their purpose is to prevent suspended particles, contaminants and odours from entering a building and from interfering with the efficient operation of air movement plant.

Air suspended contaminants and particles – these can be solid, liquid, gas or organic in microbic dimensions mostly indiscernible to the eye. The tiniest particles that can be seen unaided by microscope or spectacles are about 10μm (ten microns or ten millionths of a metre, i.e. about one-tenth of a human hair of diameter 100μm).

Type	Typical size range (μm [m x 10^{-6}])
Bacteria	0.50–10.00
Dusts	1.00–50.00
Fungal spores	1.00–100.00
Human hair	30.00–200.00
Mist and fog	2.50–40.00
Pollen	10.00–100.00
Smoke and fumes	0.05–1.00
Viruses	< 0.50

Filter types – several types exist, all varying in composition and construction to suit different situations and applications. The most commonly used are

- Dry (cell or bag)
- Viscous
- Electrostatic
- Activated carbon

Examples are illustrated over the next few pages.

Dry element – material composition varies and may be paper, fine woven fabric, pleated glass paper or glass fibre in strands. Foamed plastics and kapok have also been used. Plastic and paper are not generally favoured unless provisions are made to protect them from exposure to fire. Ducted air systems are designed to accommodate standard sized filter elements in an open framed structure. Most of these filters are manufactured as inexpensive disposable units with a fairly short lifespan. They can be removed periodically for vacuum cleaning the upward dirty air side, although cleaning is limited as the element will eventually require replacement. Location is through an accessible compartment within the ducting.

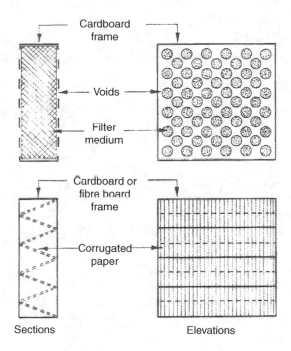

Types of disposable dry cell element filters

Air Filters – Roll and Bag

Dry roll – paper and fabric filters can be used in a roll format for suspension in the ducted air draught. Paper will have limited applications due to fire risk and other materials may require surface treatment with a fire retardant. The material can be rotated manually or by an electric motor on a timed control. Alternatively. the motor is activated by a pressure sensitive switch inside the ducting that detects resistance to air flow as the element efficiency reduces.

Dry bag – filter materials are produced in cotton or synthetic fabric and retained in an open frame within a ducted air system. While the fan is inactive, the bag hangs loosely inside the duct. When the fan engages the bag inflates to a horizontal profile providing a large surface area for retention of air suspended contaminants.

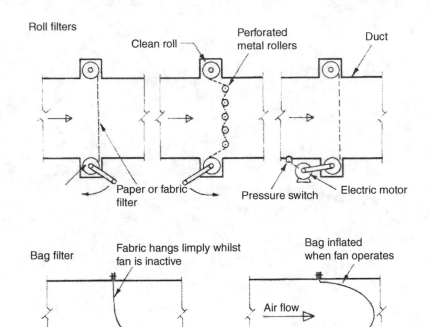

Viscous – produced as a panel containing either glass fibres, metal turnings, perforated or expanded metal sheet. The surface is coated with a non-toxic, non-flammable odourless oil. This type of filter has an excellent capacity for dust retention. It is therefore very suited to industrial applications where the density of air contamination by suspended particles is high. Periodic maintenance involves dipping and cleaning the framed elements in a solution of caustic soda or other oil diluting medium. An alternative contains a series of close-spaced metal plates (possibly corrugated to increase surface area) in a continuous oil spray.

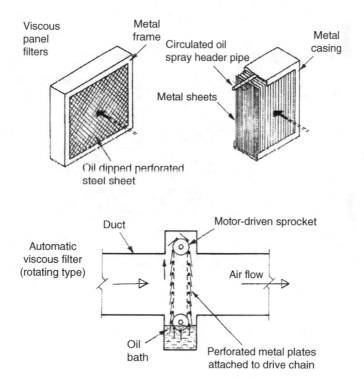

281

Air Filters – Electrostatic

Electric (electrostatic) – otherwise known as a precipitator. This type of filter is relatively expensive, somewhat compensated by being extremely effective in the removal of air suspended fine dust particles, pollen and smoke. They are usually installed with an insect screen or a dry cell pre-filter to remove larger suspended particles.

The unit has an ioniser producing some 12 to 15kV of direct current electricity. This is applied across electrodes that give any air suspended particles a positive charge, before they pass through opposingly charged metal plates. Negatively charged plates are earthed, and these attract and retain particles from the air stream. The negatively charged plates may be supplemented with a coating of oil or a gel to aid dust retention. Negative plates will require occasional cleaning and a replacement coating of oil. Very little electric current is used and resistance to air flow is minimal.

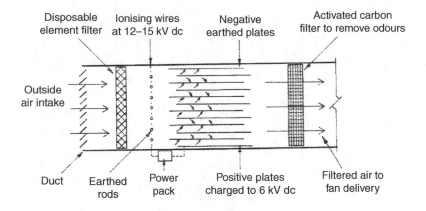

HEPA filter – High Efficiency Particulate Air filter. Developed by the US military during the 1940s to trap radioactive particles. Now applicable to any air filtration medium that is capable of retaining 99.95% (Euro. standard), 99.97% (US standard) particles of 0.0003mm (0.3 microns) or greater. Usually dense fibre strands or fabric bag. Often used with a less effective pre-filter to remove larger particles. Capable of removing dust, pollen, bacteria, moisture and micro-organisms.

Activated carbon – otherwise known as activated charcoal, this type of filter is the most effective for removing smells such as organic gases. A disposable filter composed of carbon particles resembling pieces of coconut shell and arranged to provide a large surface contact area. To increase its efficiency, a glass fibre pre-filtration matting is often used to contain the carbon shells. This type of filter is used specifically in commercial cooker hoods and in other greasy, odorous atmospheres, as the carbon is extremely absorbent. The attraction between hot, greasy fumes and carbon is termed adsorption. Activated carbon filters are disposable and must be easily accessible for inspection and replacement.

Typical application –

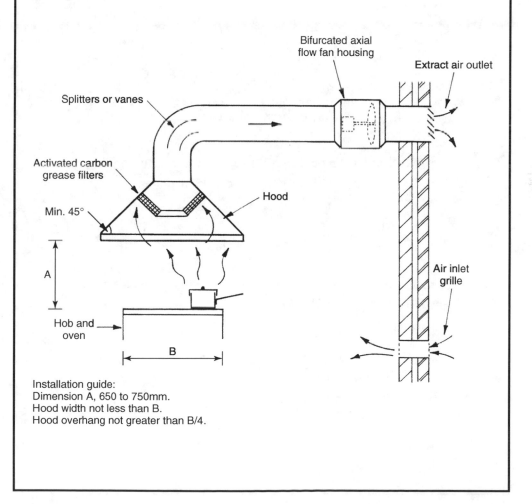

Installation guide:
Dimension A, 650 to 750mm.
Hood width not less than B.
Hood overhang not greater than B/4.

Low-Velocity Air Flow in Ducts

Simple ducted air systems, typical of those serving internal WCs and bathrooms, operate at relatively low air velocity with little frictional resistance or pressure drop. In these situations, the relationship between air flow and duct diameter can be expressed as:

$$Q = 6.3 \times 10^{-7} \times \sqrt{d^5 \times h \div L}$$

where Q = air flow rate in m³/sec.
       d = duct diameter in mm.
       h = pressure drop in mm water gauge.
       L = length of duct in metres.

To determine duct diameter from design input data, the formula is represented:

$$d = 305 \times \sqrt[5]{Q^2 \times L \div h}$$

E.g. A 10m long ventilation duct is required to provide air at 0.10 m³/sec at a pressure drop of 0.15 mm wg.

$$0.15mm = 1.5 \text{ pascals (Pa) (over 10m of ducting)}$$
$$= 0.015mm \text{ per m, or } 0.15Pa \text{ per m.}$$
$$d = 305 \times \sqrt[5]{(0.10)^2 \times 10 \div 0.15}$$
$$d = 305 \times \sqrt[5]{0.6667}$$
$$d = 305 \times 0.922 = 281mm \text{ diameter.}$$

To check that the calculated diameter of 281mm correlates with the given flow rate (Q) of 0.10 m³/sec:

$$Q = 6.3 \times 10^{-7} \times \sqrt{d^5 \times h \div L}$$
$$Q = 6.3 \times 10^{-7} \times \sqrt{(281)^5 \times 0.15 \div 10}$$
$$Q = 6.3 \times 10^{-7} \times 162110$$
$$Q = 0.102 \text{ m}^3/\text{sec}$$

Diffusers – these vary considerably in design from standard manufactured slatted grilles to purpose-made hi-tech profiled shapes and forms compatible with modern interiors. The principal objective of air distribution and throw must not be lost in these designs.

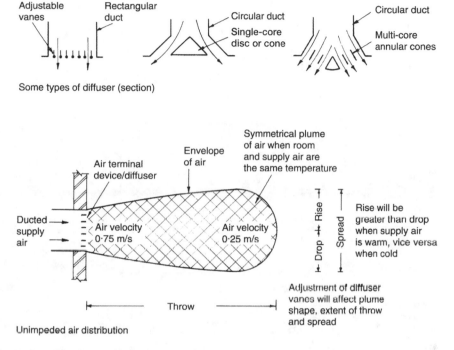

Some types of diffuser (section)

Unimpeded air distribution

Coanda effect – diffuser location must be selected to avoid unwanted draughts, air delivery impacting on beams, columns and other air deliveries. Where structural elements are adjacent, such as a wall and ceiling, the air delivery may become entrained and drawn to the adjacent surface. This can be advantageous as the plume of air throw, although distorted, may extend to run down the far wall as well.

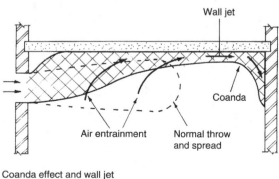

Coanda effect and wall jet

Other Air Diffusion Patterns

Location, type and size of supply air diffuser or grille will affect the way air is distributed within a room. Air temperature across the extremes of chilled to heated will also affect the air trajectory and pattern (see previous page).

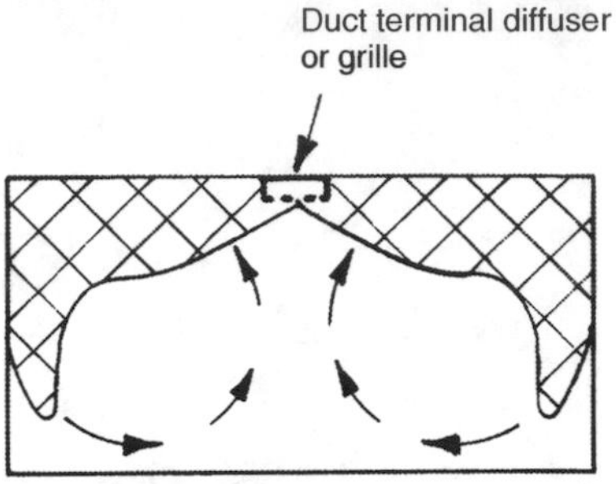

Duct terminal diffuser or grille

Centre ceiling with circular, square, rectangular or linear diffuser

Plume of supply air movement

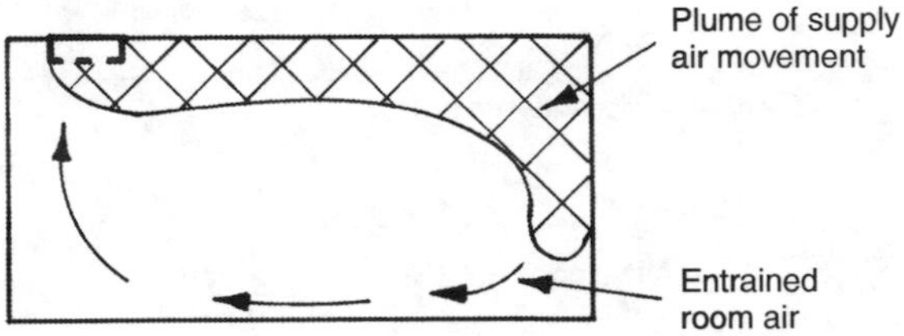

Ceiling periphery with square, rectangular or linear diffuser

Entrained room air

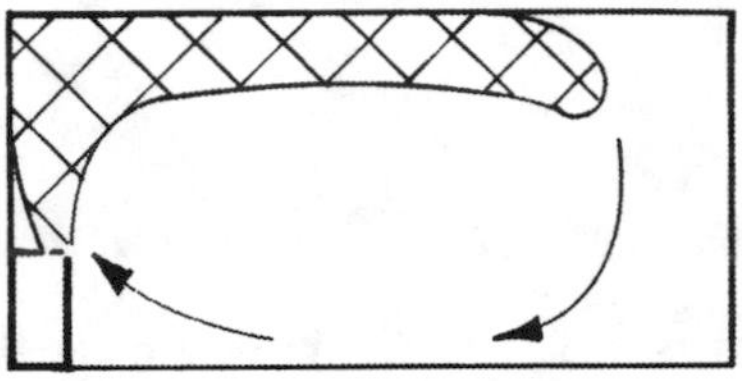

Sill or floor unit with modular or linear grilles

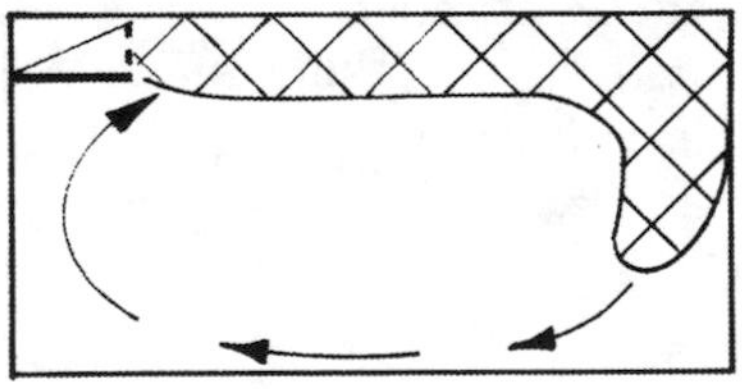

Ducted side grilles of modular or linear type

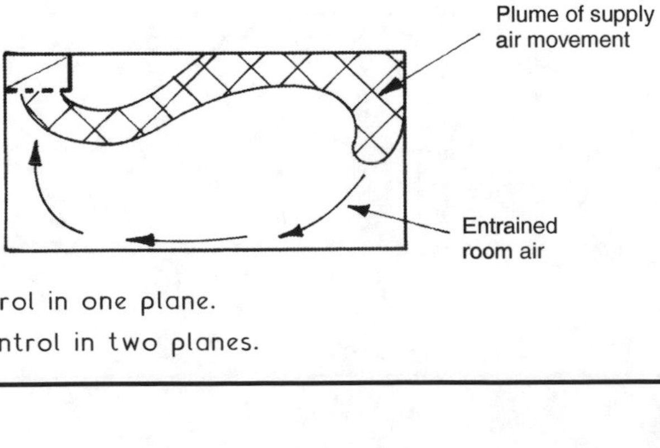

Plume of supply air movement

Ducted underside linear diffusers

Entrained room air

Linear grille - spread control in one plane.

Modular grille - spread control in two planes.

Principle – fan-directed warm air blown across an external door opening as a barrier against cold outside air entering. Fan power must be sufficient to direct air from the top of the opening to the floor, although air direction can be horizontal. An alternative specifically for food stores and restaurants, uses ambient air to create air turbulence to prevent flying insects from entering a building.

Applications include frequently accessed buildings, e.g.

* Shop openings/entrances
* Workshop entrances
* Public buildings
* Aeroplane hangars
* Restaurants

In less frequently used buildings, an air curtain can be effected by a micro-switch fitted between door and frame.

Typical design data:

* Discharge air temperature < 50°C
* Air velocity < 15m/s
* Air volume 3 to 6m^3/s per m of opening width
* Air jets angled 25° to 45° against external air

E.g.

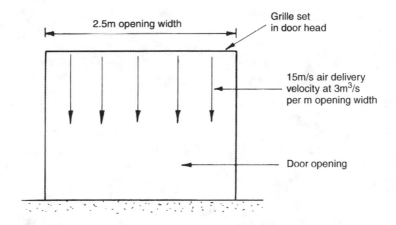

Air volume = 3m^3/s for a 2.5m-wide opening = 7.5m^3/s

Delivery ÷ velocity = 7.5m^3/s ÷ 15m/s = 0.5m^2 grille area

Grille will be 2.5m long (door-opening width) x 0.2m. i.e. 0.5m^2

Ventilation Design – Air Velocity and Resistance

Air velocity within a room or workplace should be between 0·15 and 0·50m/s, depending on the amount of activity. Sedentary tasks such as desk work will fall into the range of 0·15 to 0·30m/s, while more active assembly work, shop work and manufacturing, between 0·30 and 0·50m/s. These figures are designed to provide a feeling of freshness, to relieve stagnation without noise distraction from air movement equipment.

Conveyance of air and discharge through ducting and outlet diffusers will produce some noise. This should not be distracting and must be maintained at an unobtrusive level. As the extent of occupancy activity and/or machinery and equipment noise increases, so may the ducted air velocity, as background noise will render sound from air movement unnoticeable. For design purposes, the greater the ducted air velocity, the smaller the duct size and the less space-consuming the ducting. However, some regard must be made for acceptable ducted air noise levels and the following table provides some guidance:

Situation	Ducted air velocity (m/s)	Resistance or pressure drop (Pa/m)
Very quiet, e.g. sound studio, library, study, operating theatre	1·5–2·5	0.4
Fairly quiet, e.g. private office, habitable room, hospital ward	2·5–4·0	0.5
Less quiet, e.g. shop, restaurant, classroom, general office	4·0–5·5	0.6
Non-critical, e.g. gym, warehouse, factory, department store	5·5–7·5	0.7

Estimation of duct size and fan rating can be achieved by simple calculations and application to design charts. The example below is a graphical representation of the quantity of air (m^3/s), friction or pressure reduction (N/m^2 per m) or (Pa per m) and air velocity (m/s) in circular ductwork. Conversion to equivalent size square or rectangular ductwork is shown on pages 293–295.

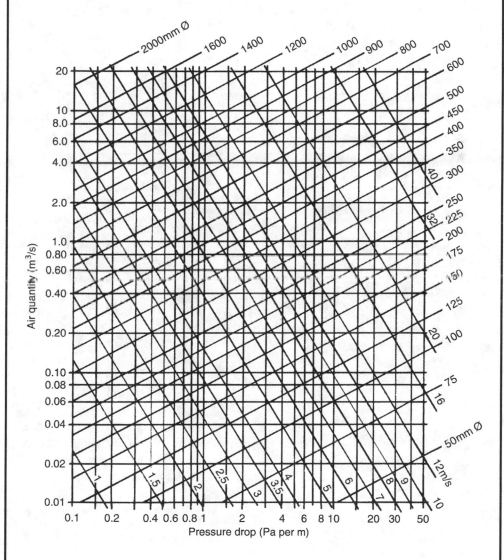

General air flow data for circular ducts

289

Ventilation Design – Air Quantity

For mechanical supply and extract systems, the air volume flow rate or quantity of air can be calculated from the following formula:

$$Q(m^3/s) = \frac{\text{Room volume} \times \text{Air changes per hour}}{\text{Time in seconds}}$$

Air changes per hour can be obtained from appropriate legislative standards for the situation or the guidance given on pages 251 and 252.

E.g.

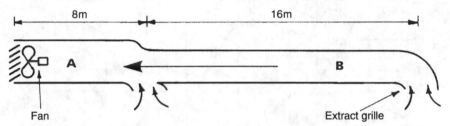

Room volume = 1800m³, requiring six air changes per hour

The ducted extract air system shown is a simple straight run, with duct A effectively 8m long and duct B effectively 16m long. Where additional bends, tees, offsets and other resistances to air flow occur, a nominal percentage increase should be added to the actual duct length. Some design manuals include `k´ factors for these deviations, and an example is shown on pages 296 and 297.

For the example given:

$$Q = \frac{1800 \times 6}{3600} = 3m^3/s$$

Disposition of extract grilles and room function will determine the quantity of air removed through each grille and associated duct. In this example, the grilles are taken to be equally disposed; therefore, each extracts 1·5m³/s. Duct A therefore must have capacity for 3m³/s and duct B, 1·5m³/s.

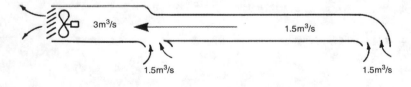

290

There are several methods which may be used to establish ventilation duct sizes, each having its own priority. The following shows three of the more popular, as applied to the design chart on page 289.

- Equal velocity – applied mainly to simple systems where the same air velocity is used throughout. For example, selected velocity is 7m/s (see page 288); therefore, the design chart indicates

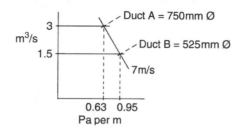

- Velocity reduction – air velocity is selected for the main section of ductwork and reduced for each branch. For example, selected air velocities for ducts A and B are 8m/s and 5m/s, respectively:

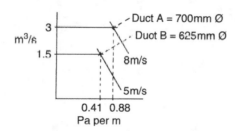

- Equal friction/constant pressure drop – air velocity is selected for the main section of ductwork. From this, the friction is determined and the same figure applied to all other sections. For example, selected air velocity through duct A is 7m/s:

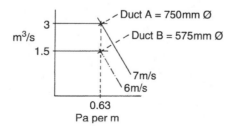

Ventilation Design – System and Fan Characteristics

Using the example on page 290 with the equal velocity method of duct sizing shown on page 291, the fan will be required to extract 3m³ of air per second at a pressure of:

$$\text{Duct (A)} = 8m \times 0.63\text{Pa per m} = 5.04\text{Pa}$$
$$\text{Duct (B)} = 16m \times 0.95\text{Pa per m} = \underline{15.20\text{Pa}}$$
$$20.24\text{Pa (i.e. 20.25)}$$

System pressure loss is calculated from: $k = P/Q^2$

$$\text{where } k = \text{pressure loss coefficient}$$
$$P = \text{pressure loss (Pa)}$$
$$Q = \text{air volume flow rate (m}^3\text{/s)}$$

Therefore, $k = 20.25/3^2 = 2.25$

Using this coefficient, the system characteristic curve may be drawn between the operating air volume flow rate of 3m³/s down to a nominal low operating figure of, say, 0·5m³/s. By substituting figures in this range in the above transposed formula, $P = k \times Q^2$, we have

$P = 2.25 \times (0.5)^2 = 0.56\text{Pa}$	[0·5m³/s @ 0·56Pa]
$P = 2.25 \times (1.0)^2 = 2.25\text{Pa}$	[1·0m³/s @ 2·25Pa]
$P = 2.25 \times (1.5)^2 = 5.06\text{Pa}$	[1·5m³/s @ 5·06Pa]
$P = 2.25 \times (2.0)^2 = 9.00\text{Pa}$	[2·0m³/s @ 9·00Pa]
$P = 2.25 \times (2.5)^2 = 14.06\text{Pa}$	[2·5m³/s @ 14·06Pa]
$P = 2.25 \times (3.0)^2 = 20.25\text{Pa}$	[3·0m³/s @ 20·25Pa]

Plotting these figures graphically against fan manufacturers' data will provide an indication of the most suitable fan for the situation:

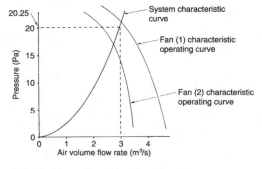

Select fan (1), as with variable settings this would adequately cover the system design characteristics.

Some ventilation design manuals limit data presentation to circular profile ductwork only. It is often more convenient for manufacturers and installers if square or rectangular ductwork can be used. This is particularly apparent where a high-aspect ratio profile will allow ducting to be accommodated in depth-restricted spaces such as suspended ceilings and raised floors.

Aspect ratio:

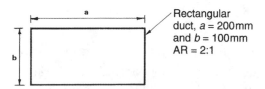

Rectangular duct, $a = 200$mm and $b = 100$mm
AR = 2:1

The numerical relationship between dimension a to b. Square = 1:1.

Conversion of circular ductwork to square or rectangular (or vice versa) using the equal velocity of flow formula:

$$d = \frac{2ab}{a + b}$$

where d = duct diameter
  a = longest dimension of rectangular duct
  b = shortest dimension of rectangular duct.

E.g. a 400mm diameter duct to be converted to a rectangular profile of aspect ratio 3:1.

$$a = 3b$$

Substituting in the above formula:

$$400 = \frac{2 \times 3b \times b}{3b + b} = \frac{6b^2}{4b} = \frac{6b}{4}$$

Therefore, $b = \dfrac{4 \times 400}{6} = 267$mm

  $a = 3b = 800$mm

For equal volume of flow and pressure drop, there are two possible formulae:

1. $$d = 1.265 \times \left| \frac{(a \times b)^3}{a + b} \right|^{0.2}$$

2. $$d = \left| \frac{32(a \times b)^3}{\pi^2(a + b)} \right|^{0.2}$$

Notes: 0.2 represents the fifth root of data in brackets.

Formulae assume identical coefficient of friction occurs between circular and rectangular ducts, i.e. same material used.

E.g. circular duct of 400mm diameter to be converted to rectangular having an aspect ratio of 3:1. Therefore, a = 3b.

Substituting in formula 1:

$$400 = 1.265 \times \left| \frac{(3b \times b)^3}{3b + b} \right|^{0.2}$$

$$400 = 1.265 \times \left| \frac{(3b^2)^3}{4b} \right|^{0.2}$$ From this, b = 216mm
a = 3b = 648mm

Substituting in formula 2:

$$400 = \left| \frac{32(3 \times b^2)^3}{\pi^2(3b + b)} \right|^{0.2}$$

$$400 = \left| \frac{3.242(27b^5)}{4} \right|^{0.2}$$ From this, b = 216mm
a = 3b = 648m

See next page for a simplified graphical conversion.

Note: A circular duct has diameter equivalent to the side of a square duct multiplied by 1.1.

Most ducting is sized using the same pressure drop or pressure loss per metre length. Larger ducting in a ventilation system will require a higher velocity to maintain a pressure drop equivalent to the smaller distribution ducting that it serves. The higher velocity will generate some increase in air movement noise, but this is not usually a problem as larger ducting is generally remote from occupied areas.

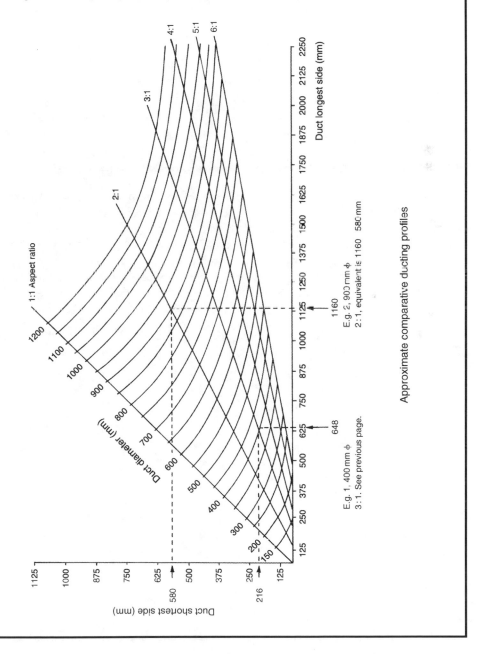

Approximate comparative ducting profiles

295

Resistances to Air Flow

There are many scientific applications to frictional or pressure losses created as air flows through ductwork. One of the most established is derived from Bernoulli's theorem of energy loss and gain as applied to fluid and air flow physics. Interpretation by formula:

$$h = k \left(\frac{V^2}{2g} \times \frac{\text{Density of air}}{\text{Density of water}} \right)$$

Where h = head or pressure loss (m)

k = velocity head loss factor

V = velocity of air flow (m/s)

g = gravity factor (9·81)

density of air = 1·2kg/m^3 @ 20°C and 1013mb

density of water = 1000kg/m^3

'k' factors have been calculated by experimentation using different ductwork materials. They will also vary depending on the nature of fittings (i.e. tees, bends, etc.), the profile, extent of direction change, effect of dampers and other restrictions to air flow. Lists of these factors are extensive and can be found in ventilation design manuals. The following is provided as a generalisation of some mid-range values for illustration purposes only:

Duct fitting	Typical 'k' factor
Radiused bend (90°)	0·30
Mitred bend (90°)	1·25
Branch (tee) piece (90°)	0·40–1·70*
Branch (tee) piece (45°)	0·12–0·80*
Reductions (abrupt)	0·25
Reductions (gradual)	0·04
Enlargements (abrupt)	0·35
Enlargements (gradual)	0·20
Obstructions (louvres/diffusers)	1·50
Obstructions (wire mesh)	0·40
Obstructions (dampers)	0·20–0·50[†]

Notes:
*Varies with area ratios of main duct to branch duct.
[†] Varies depending on extent of opening.

E.g. Calculate the pressure loss in a 10m length of 400mm-diameter ductwork containing four 90° radiused bends. Velocity of air flow is 5m/s.

$$k = \text{four No. bends} @ 0.30 = 1.20$$

Bernoulli's formula:

$$h = 1.2 \left[\frac{5^2}{2 \times 9.81} \times \frac{1.2}{1000} \right]$$

$$h = 0.00183m \text{ or } 1.83mm \text{ or approx. } 18Pa.$$

From the duct sizing chart on page 289, the pressure loss for a 400mm diameter duct at 5m/s is approximately 0.8Pa per metre.

For 10m of ductwork = 10 × 0.8 = 8Pa.*

Total pressure loss = 18Pa + 8Pa = 26Pa.

An alternative to the duct sizing chart for finding air flow resistance is the application of another established fluid and air flow theorem attributed to D'Arcy. This can be used for pipe sizing as well as for sizing small ducts.

D'Arcy's formula:

$$h = \frac{4fLV^2}{2gD} \times \frac{\text{Density of air}}{\text{Density of water}}$$

where f = friction coefficient, 0.005–0.007 depending on
         duct material
   L = length of duct (m)
   D = duct diameter (m).

Using the above example of a 10m length of 400mm (0.4m) ductwork conveying air at 5m/s:

$$h = \frac{4 \times 0.0052 \times 10 \times 5^2}{2 \times 9.81 \times 0.4} \times \frac{1.2}{1000}$$

h = 0.0008m or 0.8mm or approx. 8Pa.*

Notes:
*Varies with area ratios of main duct to branch duct.

Resistances to Air Flow – Directional Change

Resistance to air flowing through straight lengths of ventilation ducting is relatively less than the effect of air flowing around bends, through offsets and other changes in direction. The following formula with the list of coefficients for resistance provides comparative guidance:

$$H = F(V \div 4)^2$$

Where H = Head or pressure loss (mm water gauge)
       F = Coefficient of resistance
       V = Velocity of air flow (m/s)

Coefficients of resistance for directional change:

Profile	Type	F	Profile	Type	F
	90° sharp	1.5		90° radiused	0.5
	90° bend R = 2D	0.1		135° sharp	0.5
	135° long sweep	0.05		135° radiused R = 2D	0.2

	Increase in area: $\theta \leqslant 8°$	$0.15 \left[1 - \left(\dfrac{A_1}{A_2} \right) \right]^2$		A_1 and A_2 are cross-sectional areas
	$\theta > 8°$	$\left[1 - \left(\dfrac{A_1}{A_2} \right) \right]^2$		

Example of four 90° radiused bends, each having a coefficient of resistance (F) of 0.1, with air flowing at 5 m/s:

$H = F(V \div 4)^2$
$H = (0.1 \times 4)\,(5 \div 4)^2$
$H = 0.4 \times 1.56$
$H = 0.625$ mm water gauge or 6.25 Pa. The equivalent straight length resistance taken by the bends relative to their size/diameter is added.

When designing ventilation systems, provision must be made for the displacement of heat energy resulting from the movement of air. This is necessary for the maintenance of the building or room ambient temperature also to prevent cold draughts and condensation.

Cold supply air is pre-heated to discharge at the same temperature as the design air temperature for the room served. This will have no real effect on any separate heating system and can be regulated independently by a control thermostat. The following formula can be used to establish the ducted air heater rating in kW, relative to design temperature parameters:

Heater rating = m × Shc × Temp. diff. (int. − ext.)

where

m = mass air flow rate (kg/s)

Shc = Specific heat capacity of air (1.0 kJ/kg K)

Temp. diff. = Temperature differential between internal

room air and external supply air (K)

Air flow rate by volume (Q) is calculated in m^3/s. To convert this to mass air flow rate in kg/s, the volume rate is multiplied by air density (ρ) of 1.2kg/m^3.

Therefore,

Heater rating = Q × ρ × Shc × Temp. diff. (int. − ext.)

For example, a room with total fabric and infiltration heat losses of 3kW (see method of calculation on page 199), with air supply and temperature design factors as given below:

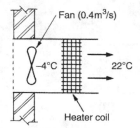

Fan (0.4m^3/s)

−4°C   22°C

Heater coil

Heater rating = 0·4 × 1·2 × 1·0 × (22 − −4)

= 12·48kw

Air duct heater calculation

Therefore, if the ducted air is required to supply all heating needs, then 12.48kW is added to the room losses of 3kW, bringing the total heat input to 15.48kW. If the ducted air system is to provide for the design room heat loss of 3kW, the discharge air temperature (T) can be found by rewriting the formula:

Room heat losses = Q × ρ × Shc × (T − int. air temp.)

Or: T = [Room heat losses ÷ (Q × ρ × Shc)] + 22

T = [3 ÷ (0·4 × 1·2 × 1·0)] + 22 = 28·25°C

Control of Mechanical Ventilation

Dwelling houses

- Ventilation of habitable rooms is generally manual through purge and background provisions. Automated controls with these are possible with the option of manual override.
- Systems with continuously running fans, e.g. MAVS (pg. 261) may have manual settings for speed and air volume. Automatic controls can have dedicated sensors that react to pollutants, e.g. smoke (kitchen or open fire), and/or humidity (moisture vapour in bathrooms and kitchen).

Buildings other than dwellings

- Air intake located away from sources of local pollution, e.g. traffic fumes and traffic noise.
- Intakes as high as possible (roof top).
- Central plant systems supplying several different control zones, i.e. rooms or areas within a building, each with differing occupancy functions to be designed with specific regard to solar exposure, pattern of occupancy and type of use.
- Controls in individual zones to have a dedicated facility for programming occupation (function), temperature (thermostatic control), rate of ventilation (fan speed and delivery volume) and air recirculation (proportion of fresh air to recirculated). Manual on/off override control to each zone/room.
- Where there is no demand or requirement for hot water and/or heating, the heat source, e.g. boiler and pump, to be programmed off.
- Heat exchange elements in warm air ventilation supplies and heat recovery ducted systems to be protected with a dedicated frost thermostat to engage heating element to prevent possibility of damage where sub-zero air temperature could enter at the intake.
- Air supply to intake ducting regulated relative to outdoor temperature by weather compensation (pgs. 190 and 191).
- Heat exchange energy recovery units (MVHR – pg. 262) to be installed where application permits.

Refs. Building Regulation A.D. F – Ventilation Vols.1 and 2.

Building Regulation A.D. L – Conservation of fuel and power Vol. 2.

7 AIR CONDITIONING

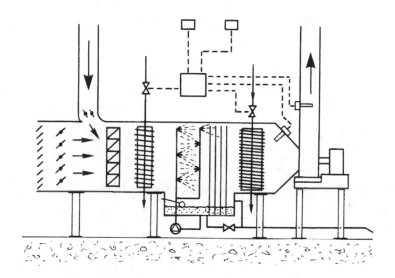

AIR CONDITIONING – PRINCIPLES AND APPLICATION
CENTRAL PLANT SYSTEM
AIR-PROCESSING UNIT
HUMIDIFIERS
VARIABLE AIR VOLUME
INDUCTION (AIR/WATER) SYSTEM
FAN-COIL (AIR/WATER) UNIT AND INDUCTION DIFFUSER
DUAL DUCT SYSTEM
CHILLED BEAMS AND CEILINGS
COOLING SYSTEMS
REFRIGERATION POWER UNITS
REFRIGERANT PROPERTIES
REFRIGERANT SYSTEM CHARACTERISTICS
PACKAGED AIR-CONDITIONING SYSTEMS
PSYCHROMETRICS – PROCESSES AND APPLICATIONS
HEAT PUMPS
HEAT RECOVERY DEVICES
LEGIONNAIRES' DISEASE
HEALTH CONSIDERATIONS AND BUILDING-RELATED
ILLNESSES

DOI: 10.1201/9781003434894-7

Air Conditioning – Principles

Air conditioning is achieved by developing the principles of moving air in ducted ventilation systems to include a number of physical and scientific processes which enhance the air quality. The objective is to provide and maintain internal air conditions at a predetermined state, regardless of the time of year, the season and the external atmospheric environment. For buildings with human occupancy, the design specification is likely to include an internal air temperature of 19–23°C and relative humidity between 40 and 60%.

The following is a glossary of some of the terminology used in air-conditioning design:

Dew point – temperature at which the air is saturated (100% RH) and further cooling manifests in condensation from water in the air.

Dry bulb temperature – temperature shown by a dry sensing element such as mercury in a glass tube thermometer (°C db).

Enthalpy – total heat energy, i.e. sensible heat + latent heat. Specific enthalpy (kJ/kg dry air).

Entropy – measure of total heat energy in a refrigerant for every degree of temperature (kJ/kg°C).

Latent heat – heat energy added or removed as a substance changes state, while temperature remains constant, e.g. water changing to steam at 100°C and atmospheric pressure (W).

Moisture content – amount of moisture present in a unit mass of air (kg/kg dry air).

Percentage saturation – ratio of the amount of moisture in the air compared with the moisture content of saturated air at the same dry bulb temperature. Almost the same as RH and often used in place of it.

Relative humidity (RH) – ratio of water contained in air at a given dry bulb temperature, as a percentage of the maximum amount of water that could be held in air at that temperature.

Saturated air – air at 100% RH.

Sensible heat – heat energy which causes the temperature of a substance to change without changing its state (W).

Specific volume – quantity of air per unit mass (m^3/kg).

Wet bulb temperature – depressed temperature measured on mercury in a glass thermometer with the sensing bulb kept wet by saturated muslin (°C wb).

This system is used where the air condition can be the same throughout the various parts of a building. It is also known as an all-air system and may be categorised as low velocity for use in buildings with large open spaces, e.g. supermarkets, theatres, factories, assembly halls, etc. A variation could incorporate a heating and cooling element in sub-branch ductwork to smaller rooms such as offices. Very large and high-rise buildings will require a high velocity and high pressure to overcome the resistances to air flow in long lengths of ductwork. Noise from the air velocity and pressure can be reduced just before the point of discharge, by incorporating an acoustic plenum chamber with low-velocity sub-ducts conveying air to room diffusers.

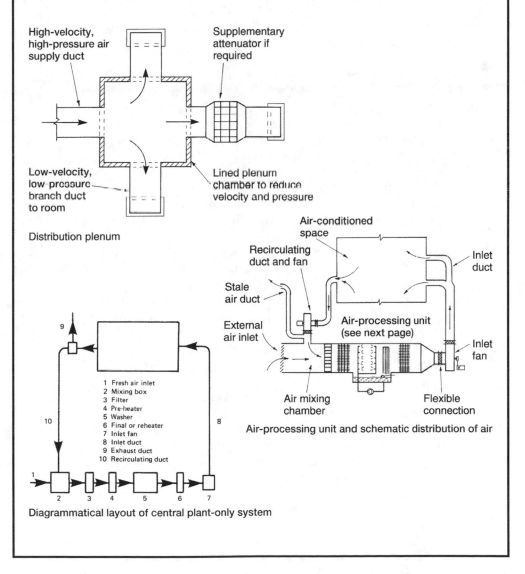

Distribution plenum

1 Fresh air inlet
2 Mixing box
3 Filter
4 Pre-heater
5 Washer
6 Final or reheater
7 Inlet fan
8 Inlet duct
9 Exhaust duct
10 Recirculating duct

Air-processing unit and schematic distribution of air

Diagrammatical layout of central plant-only system

Air-Processing Unit – Function

An air-processing unit, otherwise known as an air-handling unit (AHU), takes in air from the immediate external vicinity of a building and processes it for delivery through ducting to the interior.

- AHUs are usually located within a roof top plant room or if well insulated, positioned directly on a flat roof. At this uppermost part of a building, the air quality is likely to be better relative to street level.
- Fresh air enters through a louvred inlet or bird mesh to mix with recirculated air from the building's return duct. The ratio of recirculated to fresh air is a maximum of 3:1, i.e. 75% recirculated to 25% fresh.
- The mixed air is filtered to remove suspended dust and dirt particles. A dry filter is normally used for initial retention of suspended particles with a secondary viscous or electrostatic filter to eliminate finer suspended particles from the intake air.
- In winter, the air is pre-heated, usually by a hot water finned coil immersed in the air stream, then humidified with either a spray washer or by steam injection. A spray wash humidifier will cool the air to dew point temperature. These tend to be associated with older processing units, having been identified as a possible source of the bacteria associated with Legionnaires' disease (see page 342). If used, the water must be treated with a high concentration of biocide. With a steam humidifier, the air will gain slightly in temperature but from a safety perspective, bacteria are less likely to survive.
- In summer, the air can be cooled by a chilled water coil or a direct expansion (DX) coil, the latter being the evaporator coil in a refrigeration cycle. Condensation of the air will begin until saturation point when the air dehumidifies and reduces in temperature. Spray washing will also dehumidify the air.
- Air washers are rows of zigzag metal plates with water flowing over them. They effectively change the direction of air flow at every bend of the plates. They are usually in two parts. The first set, known as scrubber plates, wash away any remaining dust. The second set, known as eliminator plates or droplet eliminators, intercept the remaining moisture droplets so that only moisture absorbed in the air continues to final heating.
- The final heater or reheater is used to adjust the supply air temperature and relative humidity before fan delivery.

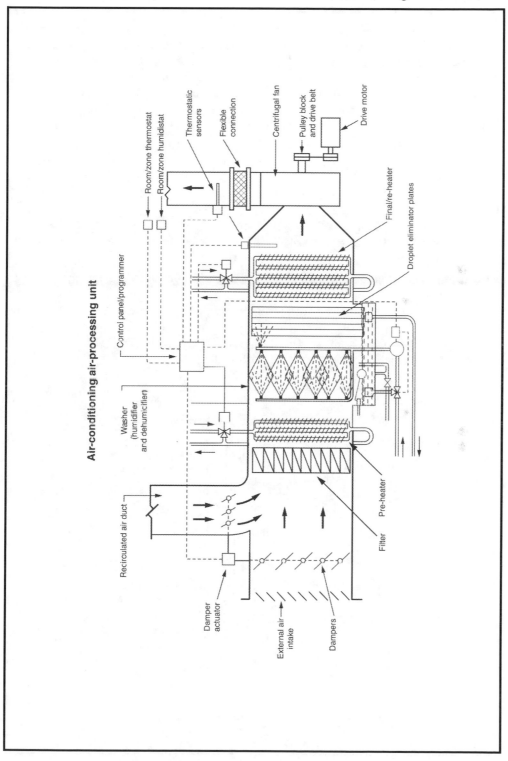

Air-conditioning air-processing unit

Room/zone thermostat
Room/zone humidistat

Thermostatic sensors

Flexible connection

Centrifugal fan

Pulley block and drive belt

Drive motor

Final/re-heater

Droplet eliminator plates

Control panel/programmer

Washer (humidifier and dehumidifier)

Recirculated air duct

Pre-heater

Filter

Damper actuator

External air intake

Dampers

Humidifiers

Depending on the state of the air on entering a spray washer, it can be humidified or dehumidified. Humidification in the presence of moisture is understandable, but dehumidification is less easy to comprehend. It occurs when the spray is at a lower temperature than the air and the dew point of the air. In this condition, the vapour pressure of the spray will be less than that of moisture in the air and some moisture from the air will transfer into the spray water. Hence, dehumidification.

Washers also remove some of the suspended dirt. Spray water pressure is usually between 200 and 300kPa. Air velocity through the washer is between 2 and 2·5m/s. Spray washers must be cleaned periodically and treated to neutralise any bacteria which could be living in the water. Water quality must also be monitored and findings documented. With numerous outbreaks of Legionnaires' disease originating from air-conditioning systems, the Health and Safety Executive have identified these spray washers as a possible health risk.

Contemporary air-processing units may incorporate steam injection humidifiers, but unlike washers, these should not be located immediately after the cooler coil. Here, the air will be close to saturation or even saturated (100% RH) and unable to accept further moisture. Therefore, dry saturated steam at over 200°C is better injected into the air close to its final discharge.

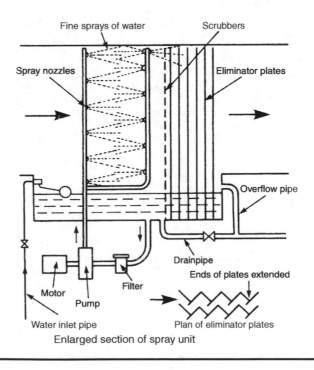

Enlarged section of spray unit

306

The VAV system has a central air-processing unit to produce air at a specified temperature and relative humidity. The conditioned air from the main unit is conveyed in ductwork to ceiling diffusers which incorporate thermostatically controlled actuators. These can change the air volume to suit each room load. In a large room, several of these VAV ceiling units may be controlled by one room thermostat.

Several rooms/zones may have separate thermostats to control the air flow to each room. The inlet fan may have variable pitched impellors operated by compressed air. A pressure switch controls the pitch angle. Air distribution is usually medium to high velocity. The air temperature in each zone can be varied with the heat energy in the delivery air volume, but the system is only suitable for buildings having a fairly evenly distributed cooling load.

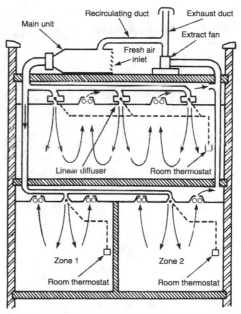

Layout of a typical variable air volume system

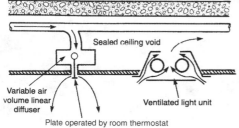

Note: The lighting fittings may require a fire damper

Section through plenum ceiling

Induction (Air/Water) System

Perimeter induction units – usually located under windows – blend primary air from the air-processing unit with secondary air from within the room. The high-velocity processed air delivery is induced into the unit through restrictive nozzles. This creates a negative pressure in its wake, drawing in the room secondary air for mixing and discharge. A damper regulates the volume of room air passing through a thermostatically controlled heating coil.

These coils may be used with chilled water as cooling coils in the summer months. If heating only is used, the system is known as the `two-pipe induction system`. With the additional two pipes for cooling water, the system is known as the `four-pipe change-over induction system`. The latter system gives excellent control of the air temperature in various zones but is very capital intensive, and therefore expensive to install.

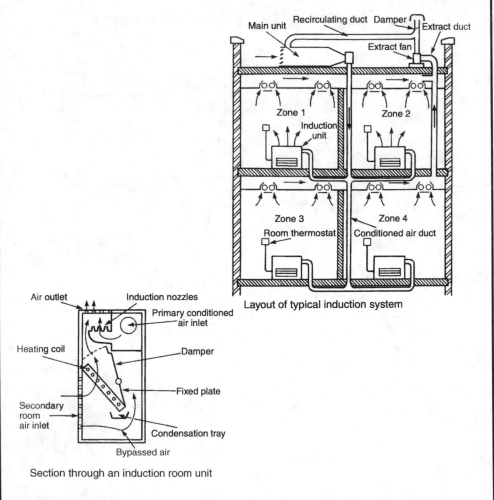

Layout of typical induction system

Section through an induction room unit

Fan-coil unit – an alternative discharge unit for application to the induction system shown on the previous page. Instead of nozzle injection of air, a low-powered fan is used to disperse a mixture of primary and secondary air after reheating or cooling from an energy exchanger within the unit.

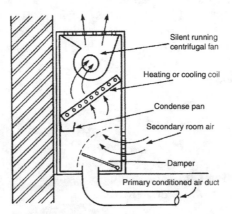

Section through a fan-coil room unit

Induction diffuser – another alternative which also uses a blend of recirculated room air with primary air. These locate at the end of branch ductwork and combine a diffuser with a simple primary and secondary air mixing chamber. The high-velocity primary air mixes with low-velocity secondary air drawn into a plenum ceiling from the room below. Light fitting extract grilles may be used to some advantage in this situation.

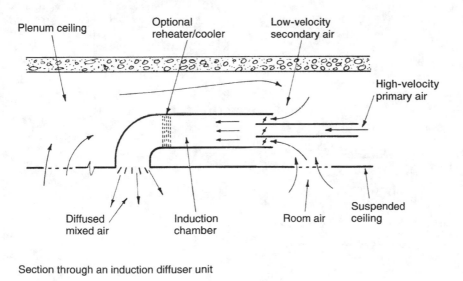

Section through an induction diffuser unit

Dual Duct System

The dual duct system is another means of providing varying air temperatures to different rooms in the same building. There is no water circulation to peripheral discharge units with terminal reheaters or coolers. This simplifies the plumbing installation as heating and cooling elements for each duct are located in the plant room. However, the system is space consuming and adequate provision must be made in suspended ceilings or raised flooring to accommodate both distribution ducts. The system is most energy economic when heating and cooling elements operate individually. For some of the year, this will not be practical and simultaneous delivery of cold and hot air is provided for blending at the point of discharge.

Delivery is at high velocity with hot and cold air regulated by a damper connected to a room thermostat. A control plate in the mixing unit maintains constant air volume. As with all systems of air conditioning, fire dampers are required where the ductwork passes through compartment walls and floors.

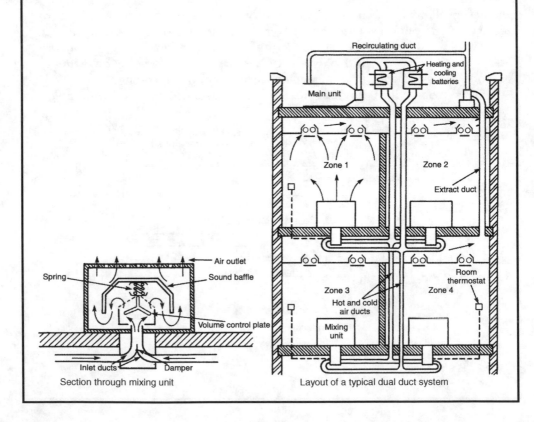

Section through mixing unit

Layout of a typical dual duct system

Chilled beams are usually formed as a bank of finned tubing, arranged in a square or rectangular profile. The tubing conveys chilled water and when encased and secured to the underside of a structural floor, the unit resembles a beam. An outer casing of sheet metal can be used to enclose the coiled pipes, and this may be perforated to encourage convection through the bank of finned tubing. A passive cooling effect is by natural convection, but active cooling can be achieved by using a fan-driven primary air supply. To conceal the installation, the underside of the box may be finished flush with a perforated suspended ceiling.

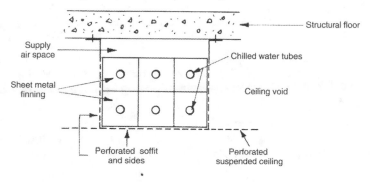

Chilled beam – typical output 150W/m^2 above a ceiling, 350W/m linear

Structural floor

Supply air space

Chilled water tubes

Sheet metal finning

Ceiling void

Perforated soffit and sides

Perforated suspended ceiling

Chilled ceilings were originally devised with chilled water pipes embedded within the underside of a concrete floor slab. The nominal increase in slab depth is justified by no visual intrusion of pipework. This form of radiant cooling has the disadvantage of creating a high thermal mass in the concrete slab, which is slow to respond to thermostatic control. These installations can also produce 'indoor rain' or condensation on the radiant underside of the slab. To prevent the ceiling from running wet, a suspended variation is preferred, with the option of an auxiliary or fan-driven primary air supply through perforations in the ceiling. These perforations will also increase the convective effect.

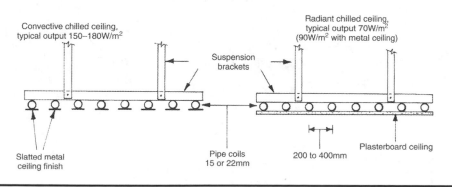

Convective chilled ceiling, typical output 150–180W/m^2

Radiant chilled ceiling, typical output 70W/m^2 (90W/m^2 with metal ceiling)

Suspension brackets

Slatted metal ceiling finish

Pipe coils 15 or 22mm

200 to 400mm

Plasterboard ceiling

311

Refrigeration Systems – Vapour Compression Cycle

Refrigeration systems are used to:

- Cool water for circulation through chiller coils. Brine may be used as a more efficient alternative to water.
- Directly chill air by suspending the cold evaporator coil in the air stream. When used in this way, the energy exchanger is known as a direct expansion (DX) coil.

The system most suited to air conditioning is the vapour compression cycle. It is a sealed pipe system containing refrigerant, compressor, condenser coil, expansion valve and evaporator coil, i.e. all the basic components of a domestic fridge. An alternative, the vapour absorption cooling cycle, is described on the next page.

Refrigerants are very volatile and boil at extremely low temperatures of –30 to –40°C. Some are capable of contributing to depletion of the ozone layer when released into the atmosphere. Dichlorodifluoromethane (R12), known as CFC, was used in many systems, but is now banned. Chlorodifluoromethane (R22), known as HCFC, is less ozone depleting. It is still used, while manufacturers research more environmentally friendly alternatives.

The refrigeration compression and evaporation cycle effects a change of temperature and state in the refrigerant, from liquid to gas and vice versa. Saturation pressure and temperature increase to emit heat at the condenser as heat energy is absorbed by the evaporator. As the liquid refrigerant changes to a gas through the expansion valve, it absorbs considerably more heat than during simple temperature change. This is known as the latent heat of vaporisation.

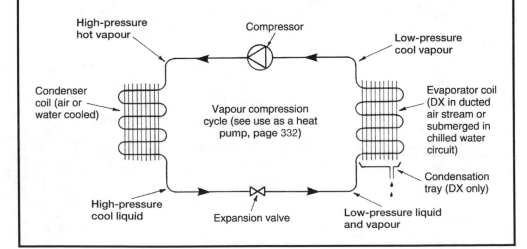

Concept – a chemically activated cycle with minimal moving parts relying on the dynamics of thermal and chemical reactions.

Refrigerant – ammonia or lithium bromide. Both are potentially hazardous. They are highly toxic, requiring associated equipment to be handled extremely carefully.

Application – have been used in small domestic appliances with a small gas flame energy source, but now only economically viable for large scale fixed load heat recovery systems. These may be waste heat energy sources from factory processes or from district heating plant. A solar heat source is also possible.

Evaporator: energy in heat transfers from a refrigerated space to vaporise the refrigerant.

Absorber: vapour is absorbed into a water/refrigerant solution and heat energy is transferred to water. Pump delivery to generator.

Generator: heater boils off refrigerant and weakened water/refrigerant solution returns to absorber.

Condenser: refrigerant gas loses heat to cooling water, returning to evaporator as a liquid.

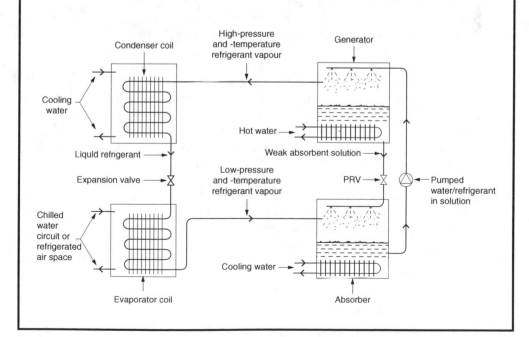

Refrigeration Power Units

Various types of refrigerant gas compressor power units can be used for chilling water in air-conditioning systems. Three are classified as positive displacement and the other dynamic.

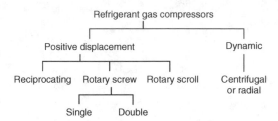

Reciprocating — contains several cylinders (up to 16) in which pistons compress an intake of gas before discharge as superheated vapour. Usually belt driven from an external electric motor. Up to 600kW of refrigeration capacity.

Rotary screw — single or double. A screw cut rotor, or more effectively two meshing screw cut rotors interacting by rotation in opposing directions to contain, reduce and compress the volume of refrigerant. Single 15 to 4500kW and double 120kW to 2MW.

Scroll — contains two spirals or interleaved scrolls, one stationary and the other rotating about it within a cylinder. The effect is to create reducing pockets and higher pressures that compress the refrigerant as it discharges through a restricted port, from 3 to 50kW.

Centrifugal — an aerodynamic lift compressor that raises pressure on the refrigerant by velocity or dynamic energy. Contains rotating impellors with aerofoil vanes, often gear driven to attain the necessary velocity. 300kW to 3MW, greater output possible if multi-staged.

Domestic fridge compressors have an electric motor of about 100W rating within the compressor housing. They may be either:

• Reciprocating — rotary motion changing to reciprocating through a crank and connecting rod to a piston.
• Rotary motor with blades attached directly to the shaft.

Properties — ideally, the following thermodynamic qualities:

- Low boiling point.
- Moderate density when liquid.
- High density in a gaseous state.
- High latent heat of vapourisation/evaporation.
- Easily liquified by compression.
- Chemically non-reactive.
- Safe environmentally, non-corrosive, non-toxic and non-flammable.

Natural refrigerants —

Water (R718) — inexpensive, readily available, non-toxic with no disposal problems. Requires a high vacuum to enable the water to boil at a low temperature of about 7°C. Equipment costs and size necessary to maintain a high vacuum limit its viability.

Air (R729) — readily available and free of environmental contamination issues. Requires extremely high pressures to liquefy, limiting its application to large and expensive plant.

Carbon dioxide (R744) — requires a fairly high power pressure input that limits its use to large-scale equipment. As a gas it is heavy, colourless, odourless, non-flammable and harmless in small quantities. In concentrations exceeding 5% it is toxic.

Ammonia (R717) — a highly efficient, inexpensive refrigerant that has been used in industrial and some domestic appliances since the late nineteenth century. Its energy efficiency is offset by high toxicity, irritation of the skin and potential for explosion.

Hydrocarbons (Propane R290 and Iso-butane R600a) — similar refrigerant properties to HCFC (R22) (see next page) with no ozone depletion potential and very low impact on global warming. Highly inflammable which limits its use.

R Numbering — a classification that identifies the molecular structure. Ninety is added to the number to give three digits representing the number of carbon, hydrogen and fluorine atoms, respectively.

E.g. Propane (R290)   290 + 90 = 380

i.e. 3 carbon, 8 hydrogen and 0 fluorine atoms.

A suffix a, b or c indicates progressively unsymmetric isomers.

Refrigerants – CFCs and HCFCs

Chlorofluorocarbon (CFC) – in the latter part of the nineteenth century and early part of the twentieth century, refrigerants in common use were ammonia (R717), methyl chloride (R40) and sulphur dioxide (R764). Leakage of these toxic and corrosive gases had very harmful, even deadly, effects, promoting research into the development and manufacture of less aggressive man-made chemical blends. The result was dichlorodifluoromethane (R12), marketed from 1930 under several trade names, one of the better known being Freon. This chlorine, fluorine and carbon (CFC) blend was colourless, odourless, non-flammable and non-toxic. For several decades, it was used in domestic appliances, commercial and automotive applications until the 1980s when it was found that CFC leakage contributed significantly to depletion of the ozone layer and global warming. These gases were then phased out of production and are now completely banned.

Hydrochlorofluorocarbon (HCFC) – refrigerant is a subclassification of CFC that has the addition of hydrogen. HCFCs have less impact on ozone depletion and global warming than CFCs due to a lower chlorine content. However, there is some environmental effect, so HCFCs are only regarded as transitional CFC replacements before being completely phased out in 2015. The HCFC chlorodifluoromethane (R22) has been widely used in domestic fridges and freezers as well as commercial heat pump, air-conditioning and refrigeration equipment.

Replacement ozone-friendly refrigerants known as zero depletion potential (ODPs), with very low contribution to global warming, have no chlorine content. These are known as hydrofluorocarbons (HFCs) and include the following:

Tetrafluoroethane (R134a)

          Trade names – Genetron 134a and Suva 134a.

Azeotropic blends: (R407c)   [23% R32, 25% R125, 52% R134a]

          Trade names – Klea and AC 9000.

          (R410a)   [50% R32, 50% R125]

          Trade names Puron and Suva 9100.

          (R417a)   [50% R134a, 46.5% R125, 3.5% R600]

          Trade name – Isceon 59.

The following is a small extract from hundreds of refrigerants. It is for comparison purposes and samples some of the refrigerants used in the past and some in current use.

Refrigerant	R No.	Molecular formula	Type	Boiling temp.°C	Critical temp.°C
Trichlorofluoromethane	11	CCl_3F	CFC	24	198
Dichlorofluoromethane	12	CCl_2F_2	CFC	-30	112
Chlorodifluoromethane	22	$CHClF_2$	HCFC	-41	96
Difluoromethane	32	CH_2F_2	HFC	-52	78
Chloromethane	40	CH_3Cl	HCC	-24	143
Trichlorotrifluoroethane	113	$C_2F_3Cl_3$	CFC	48	214
Dichlorotetrafluoroethane	114	$C_2F_4Cl_2$	CFC	4	146
Chloropentafluoroethane	115	C_2F_5Cl	CFC	-39	80
Pentafluoroethane	125	C_2HF_5	HFC	-48	66
Tetrafluoroethane	134a	$C_2H_2F_4$	HFC	-26	101
Propane	290	C_3H_8	HC	-4	97
Azeotropic	407c	R32 23% R125 25% R134a 52%	HFC	-44	86
Azeotropic	410a	R32 50% R125 50%	HFC	-51	104
Azeotropic	417a	R125 46% R134a 50% R600 4%	HFC	-41	90
Azeotropic	502	R22 49% R115 51%	HCFC	-45	81
Iso-butane	600a	C_4H_{10}	HC	-12	135
Ammonia	717	NH_3		-33	133
Water	718	H_2O		100	374
Air	729	N_2 78% O_2 21%		-193	-140
Carbon dioxide	744	CO_2		-78	31
Sulphur dioxide	764	SO_2		-10	158

Notes:

Azeotropic — a mix of two or more liquids.

Chloromethane — historically known as methylchloride.

Critical temp. — temperature above which it cannot exist as a liquid.

Refrigerant System Characteristics

Pressure enthalpy diagram – graphical representation of a refrigerant showing its total heat content (sensible + latent heat = enthalpy) during liquid, vapour and gaseous states at a given pressure.

Detailed charts are produced by refrigerant manufacturers such as ICI Plc and professional organisations such as the Chartered Institution of Building Services Engineers. The diagram below indicates the outline of these charts. The principal curved line divides the three states of a refrigerant during pressure, temperature and energy change.

For design purposes, the system operating characteristics can be superimposed on the chart to illustrate changes that occur during the refrigeration cycle. By comparing the system vapour compression cycle on various charts, it is possible to determine the most suitable refrigerant for the purpose.

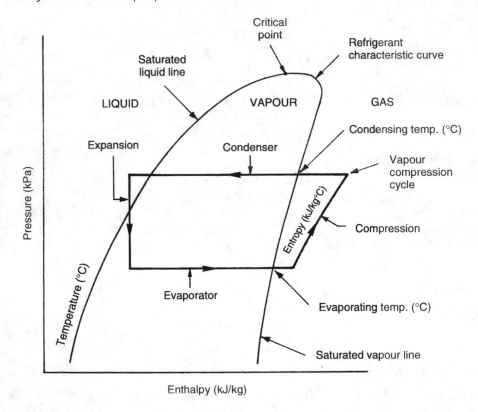

Typical pressure enthalpy diagram with a vapour compression cycle superimposed

318

Efficient operation of refrigeration systems depends to a large extent on maintaining condenser temperature at an optimum level. This is necessary for correct reaction of the refrigerant. The cooling medium can be water or air. Water is more effective, but for practical purposes and health issues (see page 321), air cooling is becoming more widely used.

The condenser coil on a domestic fridge is suspended at the back of the unit and exposed to ambient air to cool. This same principle can be applied to small packaged and portable air-conditioning units, possibly with the addition of a fan to enhance the cooling effect. Larger-scale air-conditioning installations have several high-powered fans to cool the condensers. These fans can be mounted horizontally or vertically to draw high-velocity air through the condenser coils.

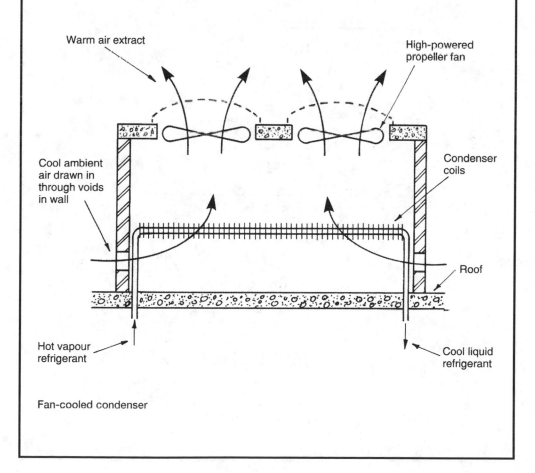

Fan-cooled condenser

Cooling Systems – Water-Cooled (Natural Draught) Condenser

Natural draught water cooling can take many forms. The simplest and most inexpensive is a pond. Cooled water is drawn from one end and warm return water pumped into the other. Spray ponds are more efficient and may incorporate ornamental fountains as part of the process. Both have a tendency to accumulate debris and will require regular attention.

More common are evaporative atmospheric cooling towers. These are usually located on the building roof or within the roof structure plant room. Wall construction is louvred to permit cross-flow of air. Internally the tower is either hollow or plastic baffled to increase the wetted contact area. Warm water from cooling the condenser is discharged through a bank of high-level sprays to cool as it descends through the air draught. It is then recirculated to the condenser.

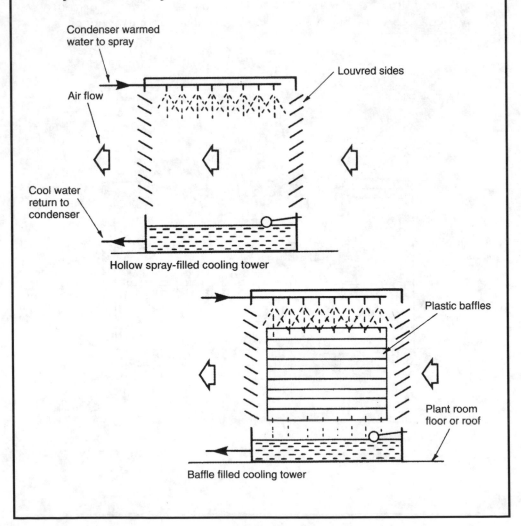

Condenser warmed water to spray

Louvred sides

Air flow

Cool water return to condenser

Hollow spray-filled cooling tower

Plastic baffles

Plant room floor or roof

Baffle filled cooling tower

Cooling Systems – Water-Cooled (Mechanical Draught) Condenser

Mechanical fan draught cooling provides absolute control over the air supply, operating independently of fickle weather and wind direction. Fan draught cooling towers are of two types:

1. Forced draught – similar in construction and operating principle to the natural draught tower, but with one or more low-level fans to force air through the tower.
2. Induced draught – a large high-level fan draws or induces air flow through the tower. The relatively large single fan is more economic in use and less likely to generate system noise and vibration.

Note: All water-cooling towers have become notorious as potential breeding areas for bacteria such as that associated with Legionnaires' disease. Therefore, towers must be maintained regularly and the water treated with a biocide, with regard to Workplace (Health, Safety and Welfare) Regulations 1992.

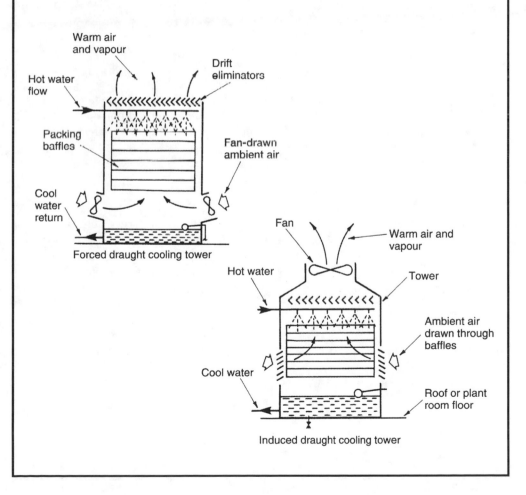

Forced draught cooling tower

Induced draught cooling tower

Packaged Air-Conditioning Systems – 1

Packaged air-conditioning systems are factory-manufactured units, delivered to site for direct installation. They contain a vapour compression cycle refrigeration system, using the evaporator for cooling and the condenser for heating, with fan delivery of the processed air. They are available in a wide range of power capacity, fan output, refrigeration and heating load for adaptation to various building types and situations.

Small- to medium-sized buildings are best suited to these systems as it would be too costly and impractical to provide numerous units for use in multi-roomed large buildings. The smallest units (1–3kW) are portable and free-standing, simply plugging into an electrical wall socket. Larger, fixed units (generally 10–60kW, but available up to 300kW) can be unsightly and difficult to accommodate. These may be located in a store-room and have short ductwork extensions to adjacent rooms.

Packages contain all the processes of conventional air-handling units, with the exception of a steam or water humidifier. Humidification is achieved with condensation from the direct expansion (DX) refrigeration coil suspended in the air intake.

For summer use, the cold (DX) coil cools incoming and recirculated air. The hot condenser coil is fan cooled externally. For winter use, the refrigeration cycle is reversed by a change-over valve to become a heat pump – see page 332. Now the cold incoming air is warmed or pre-heated through the hot condenser coil and may be further heated by an electric element or hot water coil at the point of discharge.

System types:

• Self-contained (single) package.
• Split (double) package.

Self-contained (single) package – suitable for relatively small rooms, e.g. shops, restaurants and classrooms. May be free-standing or attached to the structure.

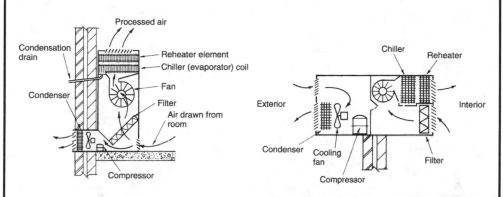

Single duct packaged unit   Single package wall-opening unit

Split (double) package – two separate units. One contains fan, filter, evaporator and expansion valve for interior location. The other contains condenser, fan and compressor for external location. The two link by refrigeration pipework. This has the advantage that one external unit can serve several interior units.

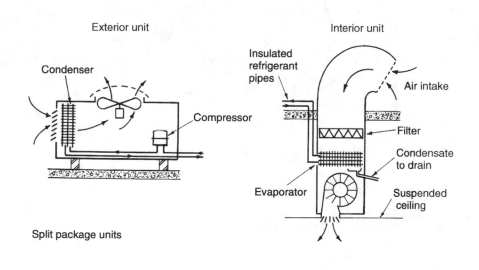

Split package units

Psychrometrics

Psychrometry – the science of moist air conditions, i.e. the characteristics of mixed air and water vapour. This can be calculated or design manuals consulted for tabulated information. Graphical psychrometric details are also available for simplified presentation of data. The chart outlined below is based on the calculated interrelationship of air properties at varying temperatures and conditions. In more detailed format, reasonably accurate design calculations can be applied. These are based on the processes shown plotted on the next page.

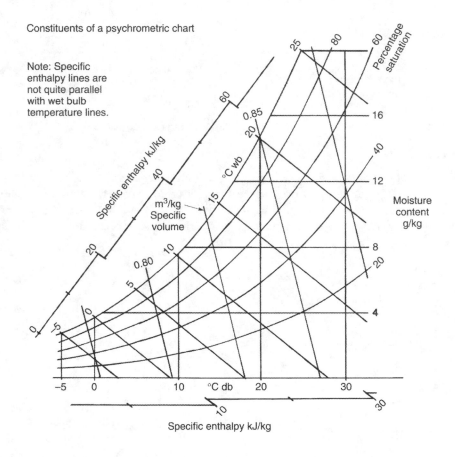

Constituents of a psychrometric chart

Note: Specific enthalpy lines are not quite parallel with wet bulb temperature lines.

Specific enthalpy kJ/kg

m³/kg Specific volume

°C wb

Percentage saturation

Moisture content g/kg

°C db

Specific enthalpy kJ/kg

The above diagram represents only the outline structure of a psychrometric chart. For accurate applications and calculations, detailed charts are available from the publications section of the Chartered Institution of Building Services Engineers. Also, subject to copyright, a variety of psychrometric chart applications can be accessed through computer downloads.

To locate a representative air condition on the psychrometric chart, two properties of the air must be known. The easiest coordinates to obtain are the dry and wet bulb temperatures. These can be measured from a sling psychrometer, also known as a whirling or sling hygrometer. Two mercury-in-glass thermometers are mounted in a frame for rotation about the handle axis. One thermometer bulb has a wetted muslin wick. After rotation, the wet bulb temperature will be lower than the dry bulb due to the evaporation effect of moisture from the muslin. The extent of evaporation will depend on the moisture content of the air.

For example, a sling psychrometer indicates 10°C db and 5°C wb temperatures. From the chart, the following can be determined:

Percentage saturation = 42%
Moisture content = 3·3g/kg dry air
Specific volume = 0·805m³/kg
Specific enthalpy = 18·5kJ/kg

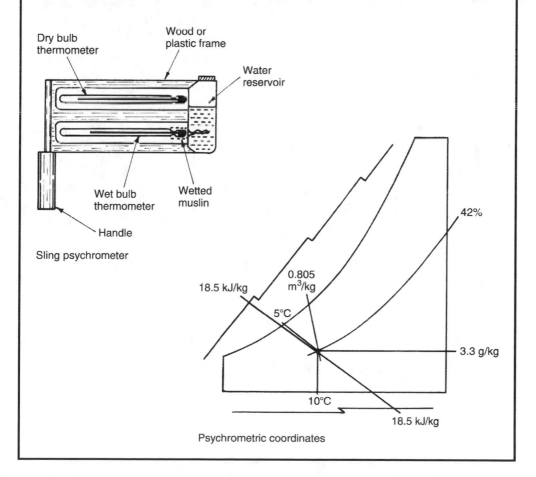

Sling psychrometer

Psychrometric coordinates

Psychrometric Processes – 2

Treatment of air is based on heating, cooling, humidification and dehumidification. These processes can be represented by lines drawn on the psychrometric chart.

- Heating (sensible) is depicted by a horizontal line drawn left to right. Dry bulb temperature increases with no change in moisture content, but there is a reduction in percentage saturation.

- Heating (latent) is the effect of steam humidification and is represented by a rising vertical line. Dry bulb temperature remains the same, moisture content and percentage saturation increase.

- Cooling (sensible) is depicted by a horizontal line drawn right to left. Dry bulb temperature decreases with no change in moisture content. Cooling by water spray humidifier is represented by an incline following the wet bulb temperature line. This is known as adiabatic humidification. Both cooling processes show an increase in percentage saturation.

- Dehumidification is shown with a descending vertical line. Moisture content and percentage saturation decrease.

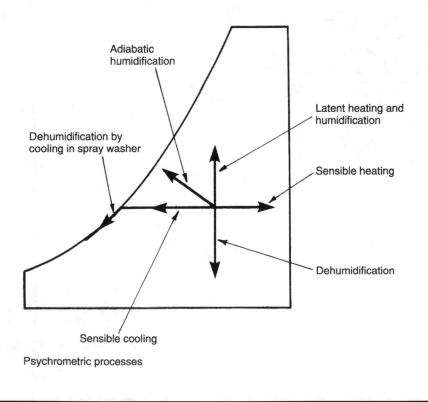

Psychrometric processes

Sensible heating of air may reduce its percentage saturation or relative humidity to an unacceptable level, i.e. <30%. Conversely, sensible cooling may increase the percentage saturation or humidity to an unacceptable level, i.e. >70%.

Applications:

1. Air enters the air-handling unit at 5°C db with an RH of 60%. Conditioned air is required at 20°C db with an RH of 50%. The air is pre-heated to 18·5°C db, cooled to 9°C dew point temperature (dry and wet bulb temperatures identical) and reheated to 20°C db (see lower diagram, centre).

2. Air enters the a.h.u. at 30°C db with an RH of 70%. Conditioned air is required at 20°C db with an RH of 50%. The air is cooled to 9°C dew point temperature and reheated to 20°C db (see lower diagram, right).

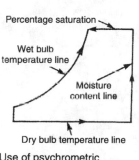

Use of psychrometric chart

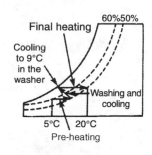

Condensation on room surfaces

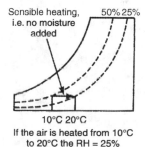

Heating of air without adding moisture

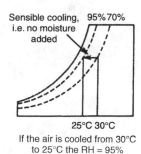

Cooling of air without dehumidification

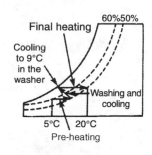

Humidifying by pre-heating, washing and final heating

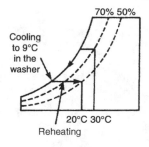

Dehumidifying by cooling, washing and reheating

327

Psychrometric Chart Applications – Air Mixing

Mixing of two air streams frequently occurs when combining fresh air with recirculated air from within the building. The process can be represented on a psychrometric chart by drawing a straight line between the two conditions and calculating a point relative to the proportions of mass flow rates.

Example 1:

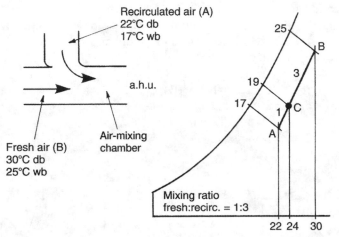

Mixed air condition (C) is found on straight line linking (A) and (B) proportioned 1:3, i.e. 24°C db, 19°C wb, 63% RH and 12 g/kg m.c.

Example 2:

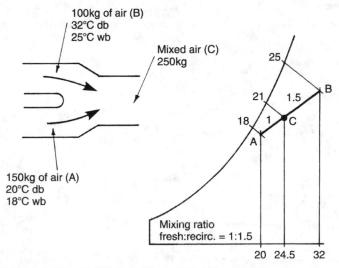

Mixed air (C) = 24.5°C db, 21°C wb, 72% RH and 14 g/kg m.c.

The calculation below relates to the example on page 327, where cool intake air at 5°C db, 60% RH is conditioned to 20°C db, 50% RH.

Applied to an office of 2400 m³ volume, requiring three air changes per hour, the quantity of air (Q) delivered will be

$$Q = \frac{\text{Volume} \times \text{Air changes per hour}}{3600} = \frac{2400 \times 3}{3600} = 2 \text{ m}^3/\text{s}$$

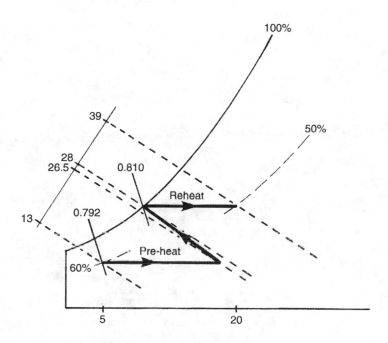

Humidifying by pre-heating, washing and final heating

Pre-heater enthalpy = 26·5 – 13 = 13·5 kJ/kg. Specific volume = 0·792 m³/kg

Reheater enthalpy = 39 – 28 = 11 kJ/kg. Specific volume = 0.810 m³/kg

Pre-heater

Specific volume converted to kg/s: 2·0 m³/s ÷ 0·792 m³/kg = 2·53 kg/s

Pre-heater rating: 2·53 kg/s × 13·5 kJ/kg = 34·2 kW

Reheater

Specific volume converted to kg/s: 2·0 m³/s ÷ 0·810 m³/kg = 2·47 kg/s

Reheater rating: 2·47 kg/s × 11 kJ/kg = 27·2 kW

Psychrometric Chart Applications – Plant Sizing (2)

The calculation below relates to the example on page 327, where warm intake air at 30°C db, 70% RH is conditioned to 20°C db, 50% RH.

With reference to the situation given on the previous page, the quantity of air delivered will be taken as 2m³/s.

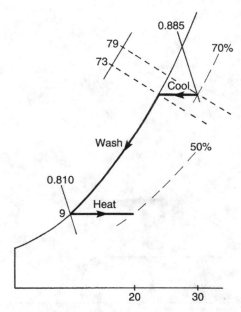

Dehumidifying by cooling, washing and reheating

Chiller enthalpy = 79 - 73 = 6kJ/kg. Specific volume = 0·885m³/kg

Specific volume converted to kg/s: 2·0m³/s ÷ 0·885m³/kg = 2·26kg/s

Chiller rating: 2·26kg/s × 6kJ/kg = 13·6kW

Note: Calculations on this and the preceding page assume 100% efficiency of plant. This is unrealistic; therefore, energy exchangers should be overrated to accommodate this.

E.g. If the chiller is 80% efficient, it will be rated: 13·6 × 100/80 = 17kW

Internal surface condensation can be minimised by providing a balance between heating, ventilation and insulation. Inadequate, intermittent or partial heating can produce a situation where the internal surfaces are colder than adjacent air temperatures. This will attract dampness to the surfaces from the moisture in the warmer air. A low rate of ventilation will also encourage a high level of humidity.

As shown in the diagram, external and internal environmental conditions can be plotted on a psychrometric chart to predict the risk of surface condensation.

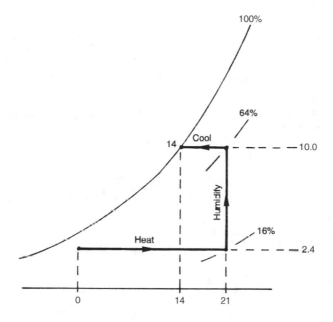

E.g.

External air conditions:

   0°C dry bulb temperature
   2·4 g/kg moisture content

Internal air conditions:

   Air warmed to 21°C dry bulb temperature
   Supply air moisture content remains at 2·4 g/kg
   RH or percentage saturation reduces to 16%
   Internal activities add 7·6 g/kg to moisture content (10 g/kg total)
   RH or percentage saturation increases to 64%

Condensation is shown to occur at 14°C or below, otherwise known as a dew point temperature of 14°C db and 14°C wb at 100% RH.

Heat Pumps – Principle

A heat pump is in principle a refrigeration cycle operating in reverse by extracting heat from a low-temperature source and upgrading it to a higher temperature for heat emission or water heating. The low-temperature heat source may be from water, air or soil which surrounds the evaporator.

A heat pump must be energy efficient; it must generate more power than that used to operate it. A measure of theoretical coefficient of performance (COP) can be expressed as:

$$COP = T_c/T_c - T_e$$

where T_c = condenser temperature based on degrees kelvin (0°C = 273K)

T_e = evaporator temperature based on degrees kelvin

E.g. T_c = 60°C, T_e = 2°C.

$$COP = \frac{60 + 273}{(60 + 273) - (2 + 273)} = 5\cdot74$$

i.e. 5·74kW of energy produced for every 1kW absorbed. Allowing for efficiency of equipment and installation, a COP of 2 to 3 is more likely.

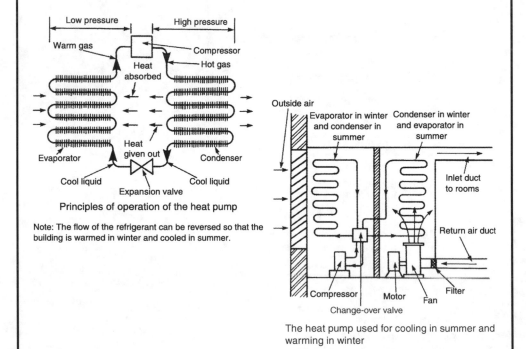

Principles of operation of the heat pump

Note: The flow of the refrigerant can be reversed so that the building is warmed in winter and cooled in summer.

The heat pump used for cooling in summer and warming in winter

Heat pump units are available as large items of plant that can be used to warm a whole building. However, small, self-contained units are more common. These are usually located under window openings for warm and cool air distribution in winter and summer respectively.

To transfer the warmth in stale extract duct air, water may be circulated through coils or energy exchangers in both the extract and cool air-intake ducts. This is known as a run-around coil and is shown in greater detail on page 339. Using water as the energy transfer medium is inexpensive but limited in efficiency. Use of a refrigerant is more effective, with an evaporator coil in the warm extract duct and a condenser coil in the cold air inlet duct.

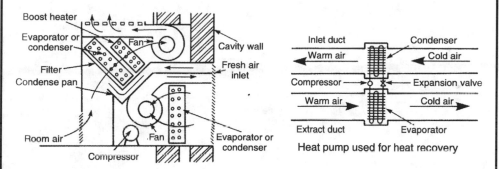

Unit heat pump fixed below window

Heat pump used for heat recovery

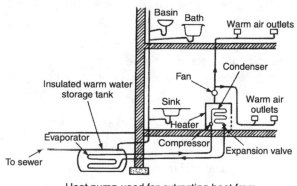

Heat pump used for extracting heat from warm waste water

Heat energy in warm waste water from sanitary fittings may be retrieved and used to supplement space heating by using a heat pump. An insulated tank buried below ground receives the waste water before it flows to the sewer. Heat energy is extracted through an evaporator inside the tank.

Heat Pumps – Energy Sources

The energy source for heat pumps can originate from the natural low heat in water, air and ground. The main energy-processing components are the source, the pump and the transfer.

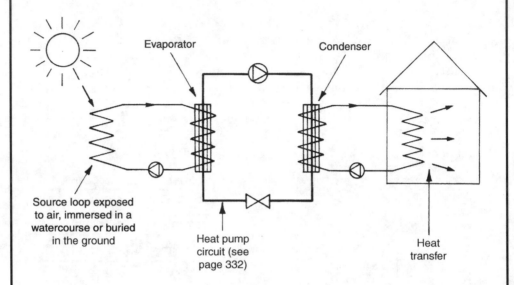

The principle is to absorb the heat from a low energy source, raise it in temperature and transfer it to storage or distribution.

Energy sources:

• Water in a standing body of some depth is preferred, typical of docklands or deep canals, although warm wastewater has potential as shown on the previous page. The source water can be used directly as an open-loop system in a run-around coil – see page 339

• Air is the least efficient in variable climate conditions due to its relatively low specific heat capacity. Application is comparably easy as groundwork and ground space are not required. Extractors can be installed on the inside or outside of an external wall.

• Ground loops can be horizontal within a few metres of the surface or vertical in boreholes of several metres' depth. Ground temperatures are fairly constant at 10°C, optimising equipment use. Application can be through polythene pipes containing a pumped distribution of water/antifreeze mixture between loop and evaporator.

Ground source heat pumps:

Objective – to extract heat from the thermal store of energy through pipes embedded in the ground. This energy store is absorbed from solar radiation, even during the winter. A pumped circuit between immersed pipes and heat pump evaporator contains a water/antifreeze mix or brine (salt dilution in water).

Ground temperature – 8 to 12°C within 15m of the surface.

Application – an immersed high-density polyethylene (HDPE) pipe loop (see previous page) is laid coiled in trenches 1.5 to 2.0m below the surface. Pipes may also be positioned vertically in a 'U' formation to greater depths. In these situations, higher quality cross-linked polyethylene (PE-Xa) is preferred.

Operation – relatively low-temperature water piped from the ground circuit to the heat pump evaporator has its heat energy exchanged into energy in a refrigerant before this is compressed to produce higher temperature energy at the condenser. Temperature at the condenser can be as high as 50°C. The condenser can be used to pre-heat boiler water for domestic use or it can be the source of medium-temperature water for underfloor heating.

Data – 0.414kg of carbon is produced for every kWh of electricity used. 0.194kg of carbon is produced for the same amount of gas used in a condensing boiler. If the boiler is 85% efficient, then $0.194 \times (100 \div 85) = 0.228$kg.

Example – given a GSHP operating at 300% efficiency, i.e. three times more energy output than that used to power the compressor and ground pump loop (COP of 3, see page 332).

0.414kg of carbon becomes 0.138kg, significantly less than that produced directly by electricity or the equivalent with gas at 0.228kg. Expressed another way, not only does this show carbon efficiency, but up to 3kW of energy is provided for every 1kW used, representing a substantial fuel cost saving.

Heat Pumps – Ground Source Applications

Types:

- Horizontal – pipes laid in trenches at 1.5 to 2.0m depth in straight lengths or coiled to maximise ground contact.

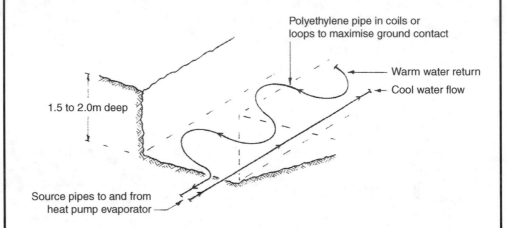

- Vertical – pipes positioned in boreholes, linked horizontally near the surface. Depths are usually at least 10m and may be as much as 200m depending on potential heat energy stored in the ground.

- Loop (closed) – consists of coils of pipework immersed in deep water lakes, rivers, docks, etc. The heat energy in surrounding water is transferred through the walls of submersed pipes. Similar in principle to the lower drawing on page 339.

- Loop (open) – heat energy stored in underground aquifers or in very deep rivers is extracted and pumped directly around the heat pump evaporator and then back to its source. Similar in principle to the upper drawing on page 339.

- Thermal piles – helical pipes incorporated within deep piled concrete foundations to large buildings.

Ref. Ground Source Heat Pump Association (GSHPA) Standards.

Unit profile –

Dimensions vary
depending on situation
and heat load,
typically about
1.0m x 0.5m
x 0.2 to 0.3m deep

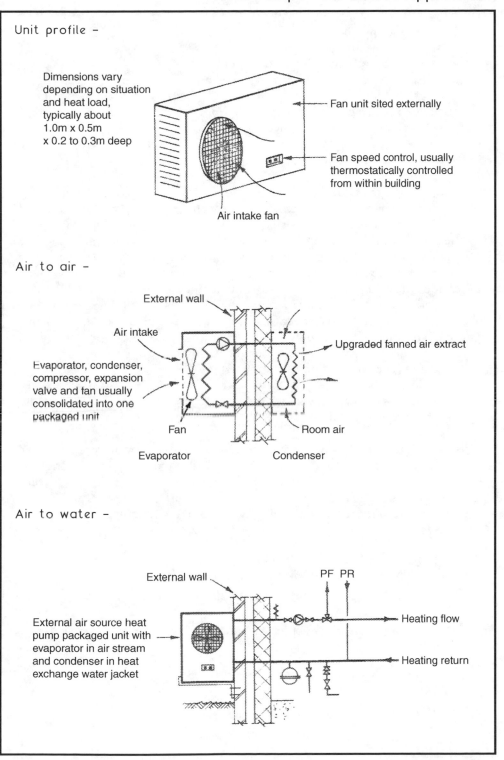

Fan unit sited externally

Fan speed control, usually
thermostatically controlled
from within building

Air intake fan

Air to air –

External wall

Air intake

Evaporator, condenser,
compressor, expansion
valve and fan usually
consolidated into one
packaged unit

Upgraded fanned air extract

Fan

Room air

Evaporator

Condenser

Air to water –

External wall

PF PR

External air source heat
pump packaged unit with
evaporator in air stream
and condenser in heat
exchange water jacket

Heating flow

Heating return

Heat Pumps – Air Source

Air source heat pumps:

Objective – to extract the heat energy contained in air outside a building using a fan unit to transfer its potential into heat energy in stored hot water or into heat emitters. The heat energy in air alone is insufficient, but it can be upgraded by applying the principle of a heat pump refrigeration cycle. The heat energy in outside air is extracted in the same way that a domestic refrigerator extracts heat energy from the air within its storage compartment.

Function – the outside fan unit contains the evaporator component of a refrigeration cycle which contains refrigerant in a vapour/liquid state. As the refrigerant absorbs heat from the air, it evaporates and is compressed/pressurised to increase its temperature to about 100°C at the condenser. The condenser coil containing this high-temperature vaporised refrigerant is immersed in water and functions as a heat exchanger serving a hot water storage facility or a heating circuit. As the water absorbs heat energy from the condenser coil, the refrigerant within the coil loses temperature and condenses back to a liquid. Thereafter, it passes through the restriction of an expansion valve to reduce refrigerant pressure and temperature further, before vaporising in the evaporator to continue the cycle.

Refrigerant – the type used is in the azeotropic category, typically R407c, see pages 316 and 317.

Efficiency – dependent on outside air temperature relative to heated water design temperature. As air temperatures are limited in the UK, air to water heat pumps are most suited to reduced temperature applications such as underfloor heating circuits, i.e. 40°C to 50°C. Radiators will need to be significantly oversized to compensate unless there is a supplementary heat source from a boiler.

Theoretical COP –

$$\frac{\text{Heat energy output (say, 10kW)}}{\text{Compressor power + system circulating pump power (say, 4kW)}}$$

Therefore, COP = 2.5

Ref.
BS EN 16147: Heat pumps with electrically driven compressors.

The run-around coil can be used as a direct energy transfer system or as a system of heat recovery.

Direct, open-loop system

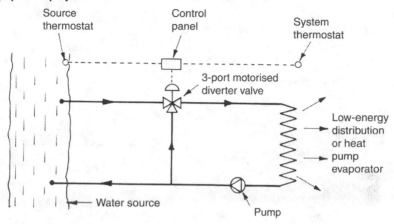

Indirect, closed-loop system

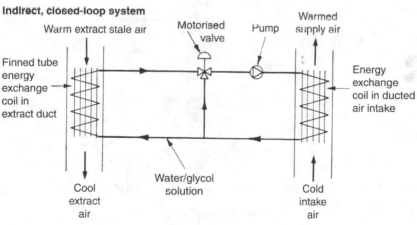

Note: Long and uninsulated pipe runs will limit efficiency.

Further Heat Recovery Devices

The concept of a thermal or heat wheel was devised about 50 years ago by Carl Munter, a Swedish engineer. Wheels range from 600 mm to 4 m in diameter; therefore, sufficient space must be allowed for their accommodation. They have an extended surface of wire mesh or fibrous paper impregnated with lithium chloride. Lithium chloride is an effective absorbent of latent heat energy in the moisture contained in stale air. A low-power (700 W) electric motor rotates the wheel at an angular velocity of 10–20 rpm. Heat from the exhaust air transfers to the inlet air and the purging section extracts the contaminants. Efficiency can be up to 90%.

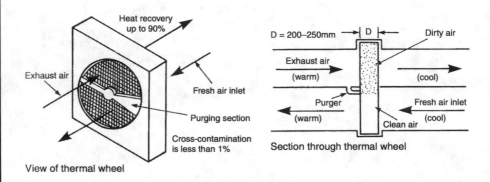

View of thermal wheel

Section through thermal wheel

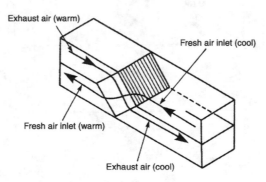

Heat recovery duct

The heat recovery duct or plate heat exchanger has warm exhaust air separated from the cool inlet air by metal or glass vanes. Heat from the exhaust vanes is transferred to the inlet vanes to warm the incoming air. Ducts must be well insulated to conserve energy and to reduce condensation. Condensation should be drained at the base of the unit. Efficiency is unlikely to exceed 50%.

Buildings are designed with the intention of providing a comfortable internal environment. To achieve this efficiently, many incorporate air-conditioning and ventilation systems. Misuse of some of the system equipment may cause health hazards to the occupants.

Legionnaires' disease – obtained its name from the first significant outbreak that occurred during an American Legionnaires' convention at a hotel in Philadelphia, USA in 1976. The bacterial infection was contracted by 221 people; it has similar symptoms to pneumonia. Of these, 34 died. Subsequently, numerous outbreaks have been identified worldwide, not least in the UK. They are generally associated with hot water systems (see page 131) and air-conditioning water cooling towers.

The organisms responsible occur naturally in swamps and similar humid conditions. In limited numbers, they are harmless, but when concentrated, they contaminate the water in which they live. If this water is suspended in the air as an aerosol spray, it can be inhaled to establish lung disease in susceptible persons.

Some other worst-known outbreaks –

Date	Place	No. affected (see note)	No. of deaths	Source
1999	Bovenkarspel, Holland. Exhibition centre	318	32	Hot tub
1985	Staffordshire District Hospital, UK	180	28	Cooling tower (pages. 320–321)
2005	Toronto, Canada. Old people's home	127	21	Cooling tower
2005	Fredrikstad, Norway. Factory	56	10	Spray washer (page 306)
2002	Barrow-in-Furness, UK. Arts centre	180	7	Cooling tower
2014	Lisbon, Portugal. Fertiliser plant.	302	7	Cooling tower
2017	Lisbon, Portugal. Hospital.	53	5	Water supply
2019	N. Carolina, USA. Exhibition centre.	141	4	Hot tub

Note: Figures given for number affected vary between different information data resources, as those exposed and showing pneumonia-type symptoms were not definitely affected by the LD bacteria. Also, some with milder respiratory illnesses may have been affected by the same bacteria but suffered only flu-like symptoms. This is known as Pontiac Fever.

Areas for concern –

- Water systems with a temperature between 20°C and 60°C. The optimum breeding temperature of the bacteria is about 40°C.
- Water cooling towers, particularly the older type with coarse timber packing. Where located in dirty and dusty atmospheres typical of that found close to building sites in busy city centres, contaminated spray dispersed in the air may be inhaled by people in the vicinity. The spray may also be drawn into ventilation inlets of adjacent buildings and be distributed through the ductwork.
- Water in spray humidifiers can be affected by contaminated air drawn into air-processing units. Humidifiers are a possible breeding area for bacteria unless the water is strongly dosed with biocide. They should be replaced with steam humidifiers.

People most at risk –

The elderly, those with existing respiratory problems, heavy smokers and those in a generally poor state of health. Nevertheless, there have been many incidents of fit, healthy young people being infected.

Solution –

- Abolition of wet cooling towers and replacement with air-cooled condensers.
- Use of packaged air conditioning with air cooling.
- Documented maintenance of plant and equipment, particularly wet cooling towers to involve regular draining, cleaning, water replacement and treatment with a biocide.

Outlook –

In the UK, there are around 300 new occurrences of LD every year. Due to public awareness, cure is successful if detection and diagnosis are early enough. However, the bacteria are becoming more resilient and concern is that they can resist and breed at higher temperatures. Solar heating can produce stored water at temperatures compatible with the bacteria flourishing; therefore these systems must be supported with hot water-temperature control capable of eliminating the bacteria.

Refs. Workplace (Health, Safety and Welfare) Regulations.

HSE Approved Code of Practice and Guidance (L8): The Control of Legionella Bacteria in Water Systems.

Humidifier fever – this is not an infection, but an allergic reaction producing flu-like symptoms such as headaches, aches, pains and shivering. It is caused by micro-organisms which breed in the water reservoirs of humidifiers while they are shut down, i.e. weekends or holidays. When the plant restarts, concentrations of the micro-organisms and their dead husks are drawn into the air stream and inhaled. After a few days' use of the plant, the reaction diminishes and recommences again after the next shutdown. Water treatment with a biocide is a possible treatment or replacement with a steam humidifier.

Sick building syndrome – this is something of a mystery as no particular cause has been identified for the discomfort generally attributed to this disorder. The symptoms vary and can include headaches, throat irritations, dry or running nose, aches, pains and loss of concentration. All or some may be responsible for personnel inefficiency and absenteeism from work. While symptoms are apparent, the causes are the subject of continued research. Some may be attributed to physical factors such as

- Noise from computers, machinery, lighting or ducted air movement.
- Strobing from fluorescent strip lights.
- Static electricity from computer screens, copiers, etc.
- Fumes from cleaning agents.
- Glare from lighting and monitors.
- Unsympathetic internal colour schemes.
- Carpet mites.

Other factors are psychological:

- Lack of personal control over an air-conditioned environment.
- No direct link with the outside world, i.e. no openable windows.
- Disorientation caused by tinted windows.
- Working in rooms with no windows.
- Dissatisfaction with air conditioning does not provide the ideal environment.

More apparent may be lack of maintenance and misuse of air-conditioning plant. Energy economising by continually recirculating the same air is known to cause discomfort for building occupants. The research continues and, as a result of sick building syndrome, new building designs often favour more individual control of the workplace environment or application of traditional air movement principles such as stack effect.

Other Causes of Unhealthy Buildings

Modern public buildings, offices, sports centres, hospitals, hotels, etc., are usually highly serviced to provide optimum atmospheric conditions. Less so in domestic dwellings, but whatever the building type, the objective is to achieve a cost viable comfortable temperature, air cleanliness and an acceptable level of humidity. Successfully attained, a pleasant environment complements our general health with a positive perspective on work and daily routines.

As indicated on the preceding three pages, poor standards of installation and maintenance of building services can have very serious consequences for the wellbeing of occupants.

Further considerations:

Lead – in buildings constructed prior to the 1950s, lead piping was commonly used for water mains, supply and service pipes in addition to internal distribution. This may have no impact at all on personal health, but in some instances, long-term sickness, mental disorders and even death may be attributed to lead poisoning. Long since banned, including use in paint and petrol. Lead solders, see pages 38 and 40.

Asbestos – in fibrous form mixed with water and applied with wire reinforcement as insulation to commercial boilers, associated plant and pipes. Suspected as a possible cause of respiratory disorders from the early part of the 20th century, it wasn't until the 1960s that it became linked to lung cancer (asbestosis). Now banned, but still prevalent in some old buildings.

Electro-magnetic fields – invisible waves of electricity or radiation, apparent, e.g., in wireless communication devices, intruder alarm systems and central heating controls. Research continues into whether EMFs are a health hazard. It is yet to be defined.

Particulates – fine particles, classified as < 2.5 micrometres, associated with mould and fungi, bacteria, aerosol products and smoke. Air pollution from wood-burning, especially stoves, is an established source which may manifest respiratory problems.

8 DRAINAGE SYSTEMS, SEWAGE TREATMENT AND REFUSE DISPOSAL

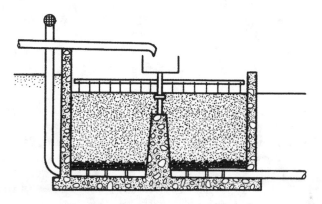

COMBINED AND SEPARATE SYSTEMS
PARTIALLY SEPARATE SYSTEM
RODDING POINT SYSTEM
SEWER CONNECTION
DRAINAGE VENTILATION
UNVENTILATED STACKS
DRAIN LAYING
MEANS OF ACCESS
BEDDING OF DRAINS
DRAINS UNDER OR NEAR BUILDINGS
DRAINPIPE MATERIALS
JOINTS USED ON DRAINPIPES
ANTI-FLOOD DEVICES
GARAGE DRAINAGE
DRAINAGE PUMPING
SUBSOIL DRAINAGE
TESTS ON DRAINS
SOAKAWAYS
CESSPOOLS AND SEPTIC TANKS
DRAINAGE FIELDS AND MOUNDS
RAINWATER MANAGEMENT
DRAINAGE DESIGN AND FORMULAE
WASTE AND REFUSE PROCESSING

DOI: 10.1201/9781003434894-8

Drainage Systems – Combined and Separate Systems

The type of drainage system selected for a building will be determined by the local water authority's established sewer arrangements. These will be installed with regard to foul water processing and the possibility of disposing of surface water via a sewer into a local watercourse or directly into a soakaway.

Combined system – this uses a single drain to convey both foul water from sanitary appliances and rainwater from roofs and other surfaces to a shared sewer. The system is economical to install, but the processing costs at the sewage treatment plant are high.

Separate system – this has foul water from the sanitary appliances conveyed in a foul water drain to a foul water sewer. The rainwater from roofs and other surfaces is conveyed in a surface water drain into a surface water sewer or a soakaway. This system is relatively expensive to install, particularly if the ground has poor drainage qualities and soakaways cannot be used. However, the benefit is reduced volume and treatment costs at the processing plant.

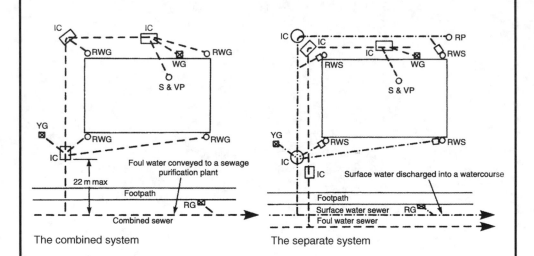

The combined system   The separate system

Key:

IC	= Inspection chamber	RWG	= Rainwater gully
WG	= Waste gully	RG	= Road gully
YG	= Yard gully	RWS	= Rainwater shoe
RP	= Rodding point	S & VP	= Soil and vent pipe (discharge stack)

Partially separate system – most of the rainwater is conveyed by the surface water drain into the surface water sewer. For convenience and to reduce site costs, the local water authority may permit an isolated rainwater inlet to be connected to the foul water drain. This is shown with the rainwater inlet at A connected to the foul water inspection chamber. Also, a rodding point is shown at B. These are often used at the head of a drain, as an alternative to a more costly inspection chamber.

A back inlet gully can be used for connecting a rainwater downpipe or a waste pipe to a drain. The bend or trap provides a useful reservoir to trap leaves. When used with a foul water drain, the seal prevents air contamination. A yard gully is solely for collecting surface water and connecting this with a drain. It is similar to a road gully but smaller. A rainwater shoe is only for connecting a rainwater pipe to a surface water drain. The soil and vent pipe or discharge stack is connected to the foul water drain with a rest bend at its base. This can be purposely made or produced with two 135° bends. It must have a centre-line radius of at least 200 mm.

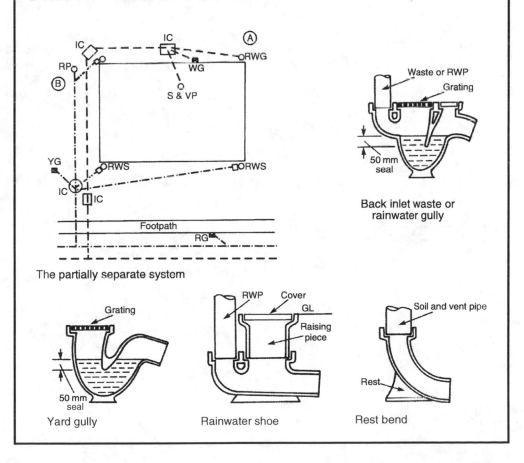

The partially separate system

Back inlet waste or rainwater gully

Yard gully

Rainwater shoe

Rest bend

Rodding Point System

Rodding points or rodding eyes provide a simple and inexpensive means of access at the head of a drain or on shallow drain runs for rodding in the direction of flow. They eliminate isolated loads that manholes and inspection chambers can impose on the ground, thus reducing the possibility of uneven settlement. The system is also neater, with less surface interruptions. Prior to installation, it is essential to consult with the local authority to determine whether the system is acceptable and, if so, to determine the maximum depth of application and any other limitations on use. As rodding is only practical in one direction, an inspection chamber or manhole is usually required before connection to a sewer.

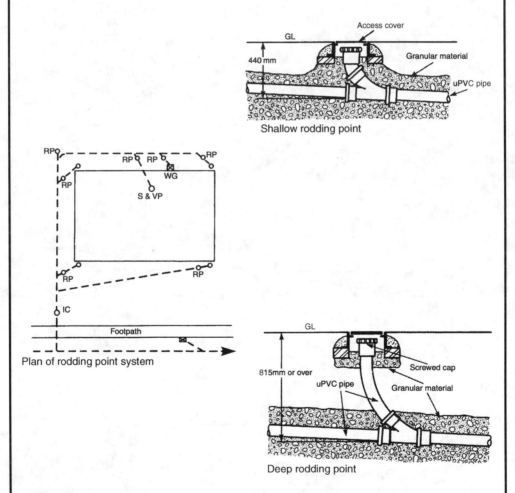

Shallow rodding point

Plan of rodding point system

Deep rodding point

Refs. Building Regulations, Approved Documents H1: Foul water drainage and H3: Rainwater drainage.
BS EN 752: Drain and sewer systems outside buildings.

Connections between drains and sewers must be obliquely in the direction of flow. Drains may be connected independently to the public sewer so that each building owner is responsible for the maintenance of the drainage system for that building up to the property boundary. Thereafter, ownership and responsibility for the drain are with the water and sewerage authority. In situations where there would be long drain runs, it may be more economical to connect each drain to a shared sewer. This requires only one sewer connection for several buildings. Responsibility for maintenance of the shared sewer is the local water and sewerage authority's.

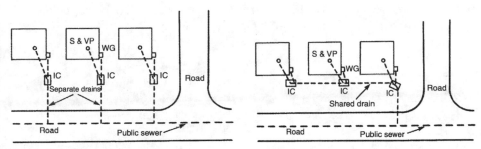

Use of separate drains | Use of private sewer

Connection of a drain to the public sewer can be made with a manhole. If one of these is used at every connection, the road surface is unnecessarily disrupted. Therefore, a saddle is preferred, but manhole access is still required at no more than 90m intervals. Saddles are bedded in cement mortar in a hole made in the top of the sewer.

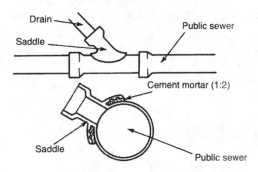

Use of saddle connection

Note: The term lateral drain may be used to describe a shared drain or sewer.

Venting of foul water drains is necessary to prevent a concentration of gases and to retain the air inside the drain at atmospheric pressure. This is essential to prevent the loss of trap water seals by siphonage or compression. The current practice of direct connection of the discharge stack and drain to the public sewer provides a simple means of ventilation through every stack. In older systems, generally pre-1950s, an interceptor trap with a 65mm water seal separates the drain from the sewer. The sewer is independently vented by infrequently spaced high-level vent stacks. Through ventilation of the drain is by fresh air inlet at the lowest means of access and the discharge stack. It may still be necessary to use this system where new buildings are constructed where it exists. It is also a useful means of controlling rodent penetration from the sewer.

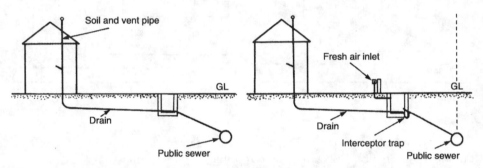

Without the use of an interceptor trap   With the use of an interceptor trap

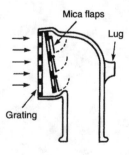

Fresh air inlet

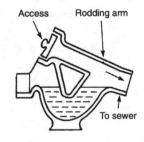

Interceptor trap

To reduce installation costs and to eliminate roof penetration of ventilating stacks, discharge stacks can terminate inside a building. This is normally within the roof space, i.e. above the highest water level of an appliance connected to the stack, provided the top of the stack is fitted with an air admittance valve (AAV). An AAV prevents the emission of foul air, but admits air into the stack under conditions of reduced atmospheric pressure. AAVs are limited in use to dwellings of no more than three storeys, in up to four adjacent buildings. The fifth building must have a conventional vent stack to ventilate the sewer.

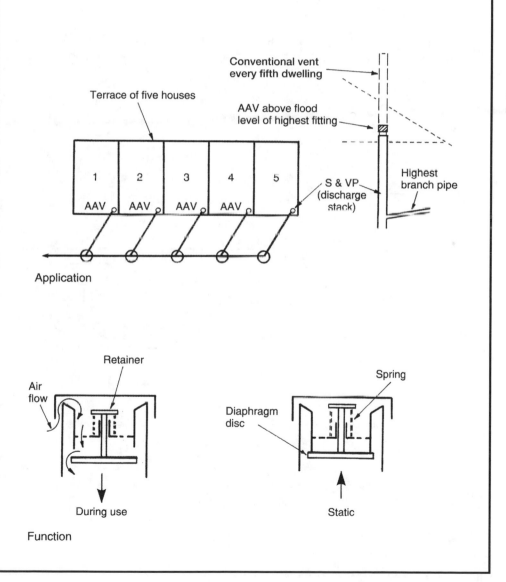

Conventional vent every fifth dwelling

AAV above flood level of highest fitting

Terrace of five houses

1 2 3 4 5

AAV AAV AAV AAV

S & VP (discharge stack)

Highest branch pipe

Application

Retainer

Air flow

Spring

Diaphragm disc

During use

Static

Function

Unventilated Stacks – Ground Floor Only

Direct connection – a WC may discharge directly into a drain, without connection to a soil and ventilating stack. Application is limited to a maximum distance between the centre line of the WC trap outlet and the drain invert of 1·5m.

Stub stack – this is an extension of the above requirement and may apply to a group of sanitary fittings. In addition to the WC requirement, no branch pipes to other fittings may be higher than 2m above a connection to a ventilated stack or the drain invert.

The maximum length of branch drain from a single appliance to a means of drain access is 6m. For a group of appliances, it is 12m.

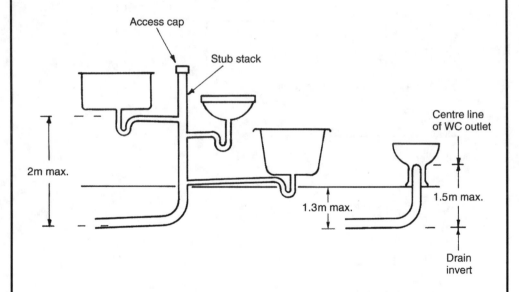

Stub stack for a group of fittings   WC direct connection

Refs. Building Regulations, Approved Document H1, Section 1: Sanitary pipework.

BS EN 12056-2: Gravity drainage systems inside buildings.

Sanitary pipework, layout and calculation.

The bottom of a drain trench must be excavated to a gradient. This is achieved by setting up sight rails, suitably marked to show the centre of the drain. These are located above the trench and aligned to the gradient required. At least three sight rails should be used. A boning rod (rather like a long 'T' square) is sighted between the rails to establish the level and gradient of the trench bottom. Wooden pegs are driven into the trench bottom at about 1m intervals. The required level is achieved by placing the bottom of the boning rod on each peg and checking top alignment with the sight rails. Pegs are adjusted accordingly and removed before laying the drains. For safe working in a trench, it is essential to provide temporary support to the excavation.

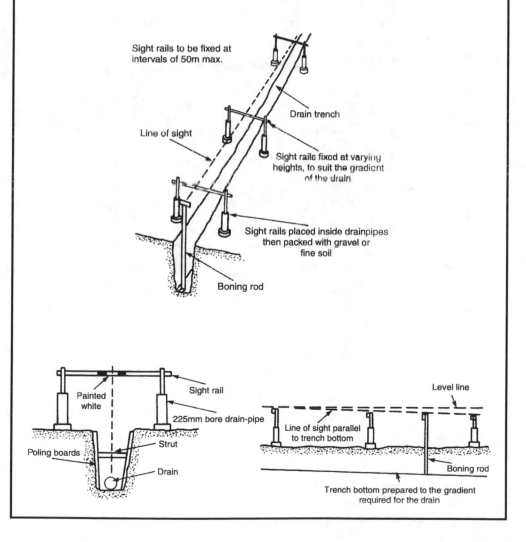

Sight rails to be fixed at intervals of 50m max.

Line of sight

Drain trench

Sight rails fixed at varying heights, to suit the gradient of the drain

Sight rails placed inside drainpipes then packed with gravel or fine soil

Boning rod

Painted white

Sight rail

225mm bore drain-pipe

Poling boards

Strut

Drain

Level line

Line of sight parallel to trench bottom

Boning rod

Trench bottom prepared to the gradient required for the drain

Drain access may be obtained through rodding points (page 348), shallow access chambers, inspection chambers and manholes. Pipe runs should be straight and access provided only where needed, i.e.:

- at significant changes in direction
- at significant changes in gradient
- near to or at the head of a drain
- where the drain changes in size
- at junctions
- on long, straight runs.

Maximum spacing (m) of access points based on Table 13 of Approved Document H1 to the Building Regulations:

From \ To	Access fitting Small	Access fitting Large	Junction	Inspection chamber	Manhole
Start of drain	12	12	–	22	45
Rodding eye	22	22	22	45	45
Access fitting:					
150 dia.	–	–	12	22	22
150 × 100	–	–	12	22	22
225 × 100	–	–	22	45	45
Inspection chamber	22	45	22	45	45
Manhole	–	–	–	45	90

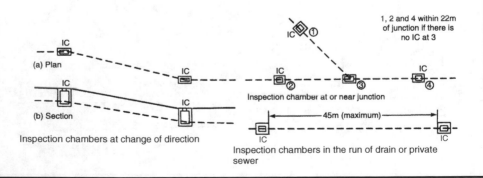

(a) Plan

(b) Section

Inspection chambers at change of direction

1, 2 and 4 within 22m of junction if there is no IC at 3

Inspection chamber at or near junction

45m (maximum)

Inspection chambers in the run of drain or private sewer

Shallow access chambers or access fittings are small compartments similar in size and concept to rodding points, but providing drain access in both directions and possibly into a branch. They are an inexpensive application for accessing shallow depths up to 600mm to invert. Within this classification manufacturers have created a variety of fittings to suit their drain products. The uPVC bowl variation shown combines the facility of an inspection chamber and a rodding point.

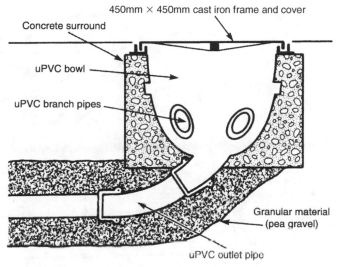

The Marscar access bowl

Note: Small lightweight cover plates should be secured with screws, to prevent unauthorised access, e.g. children.

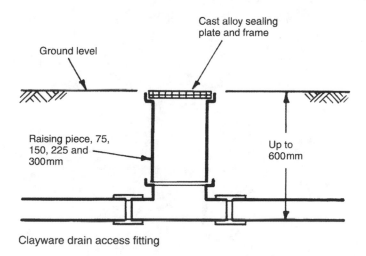

Clayware drain access fitting

355

Means of Access – 3

Inspection chambers are larger than access chambers, having an open channel and space for several branches. They may be circular or rectangular on plan and preformed from uPVC, precast in concrete sections or traditionally constructed with dense bricks from a concrete base. The purpose of an inspection chamber is to provide surface access only; therefore, the depth to invert level does not exceed 1m.

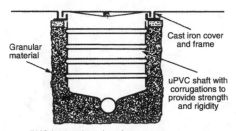

uPVC inspection chamber

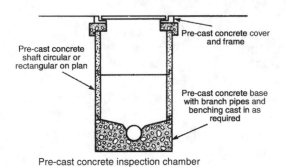

Pre-cast concrete inspection chamber

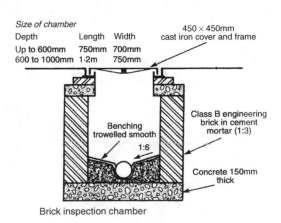

Brick inspection chamber

Size of chamber		
Depth	Length	Width
Up to 600mm	750mm	700mm
600 to 1000mm	1·2m	750mm

The term manhole is used generally to describe drain and sewer access. By comparison, manholes are large chambers with sufficient space for a person to gain access at drain level. Where the depth to invert exceeds 1m, step irons should be provided at 300mm vertical and horizontal spacing. A built-in ladder may be used for very deep chambers. Chambers in excess of 2·7m may have a reduced area of access known as a shaft (min. 1050 × 840mm or 1050mm diameter), otherwise the following applies:

Depth (m)	Internal dimensions (mm) l × b	Cover size
<1·5	1200 × 750 or 1050 dia.	Min. dimension 600 mm
1·5–2·7	1200 × 750 or 1200 dia.	Min. dimension 600 mm
>2·7	1200 × 840 or 1200 dia.	Min. dimension 600 mm

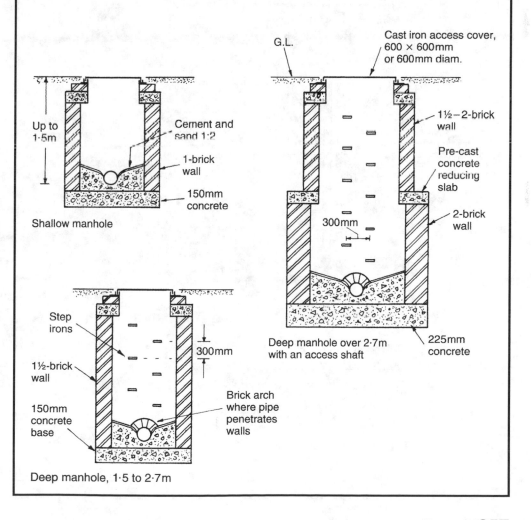

Shallow manhole

Deep manhole, 1·5 to 2·7m

Deep manhole over 2·7m with an access shaft

357

Means of Access – Backdrop Manhole

Where there is a significant difference in level between a drain and a private or public sewer, a backdrop may be used to reduce excavation costs. Backdrops have also been used on sloping sites to limit the drain gradient, as at one time it was thought necessary to regulate the velocity of flow. This is now considered unnecessary, and the drain may be laid to the same slope as the ground surface. For use with cast iron and uPVC pipes up to 150mm bore, the backdrop may be secured inside the manhole. For other situations, the backdrop is located outside the manhole and surrounded with concrete.

The access shaft should be 1050 × 840mm minimum and the working area at least 1·2m × 840mm.

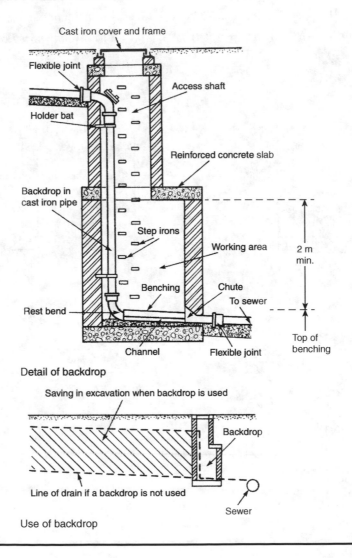

Detail of backdrop

Use of backdrop

Cover plates – produced from durable materials; traditionally cast iron, also cast or pressed steel (galvanised), aluminium alloy, reinforced concrete and plastics. Selection will depend on application. GRP, other plastic composites and light metal alloys are suitable for relatively small covers to rodding eyes and access fittings. Plastics have the advantage of zero value to metal thieves. Small lightweight access covers are screw fixed to deter unauthorised access and for safety purposes.

BS grouping and classification by application –

Group	Class	Static test load	Typical application
1	A15	1.5 tonnes	Pedestrian areas
2	B125	12.5 ..	Occasional vehicular access
3	C250	25.0 ..	Car parks and forecourts
4	D400	40.0 ..	Lorry access and highways
5	E600	60.0 ..	Docks and aircraft standings
6	F900	90.0 ..	Extremely high wheel loads

Types – generally rectangular or circular on plan, in single or double seal format. Double seals are used where drain access is required within a building. Here, the cover is sealed to its frame with grease to maintain an airtight fit.

Manhole covers and frames –

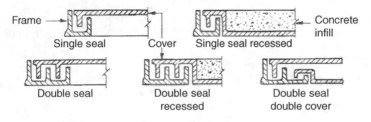

Refs. BS EN 124: Gully tops and manhole tops for vehicular and pedestrian areas. Parts 1 to 6.
Building Regulations Approved Document H1 (Section 2.54) '. . . removable non-ventilating covers . . .'

Drains must be laid with due regard for the subsoil condition and the imposed loading. The term bedding factor is applied to laying rigid drainpipes. This describes the ratio of the pipe strength when bedded to the pipe test strength as given in the relevant British Standard.

Class A bedding gives a bedding factor of 2·6, which means that a rigid drainpipe laid in this manner could support up to 2·6 times the quoted BS strength. This is due to the cradling effect of concrete, with a facility for movement at every pipe joint. This method may be used where extra pipe strength is required or great accuracy in pipe gradient is necessary. Class B bedding is more practical, considerably less expensive and quicker to use. This has a more than adequate bedding factor of 1·9. If used with plastic pipes, it is essential to bed and completely surround the pipe with granular material to prevent the pipe from distortion.

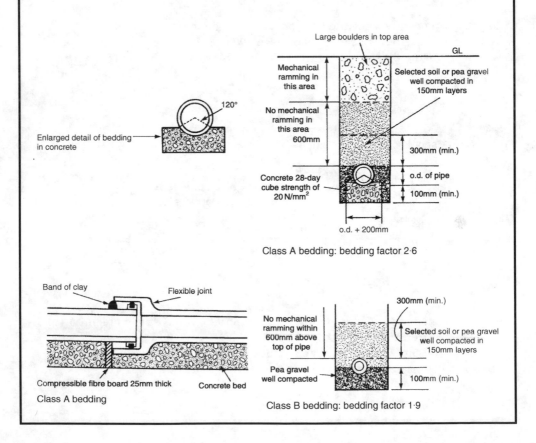

Enlarged detail of bedding in concrete

120°

Large boulders in top area

GL

Mechanical ramming in this area

Selected soil or pea gravel well compacted in 150mm layers

No mechanical ramming in this area 600mm

300mm (min.)

Concrete 28-day cube strength of 20 N/mm²

o.d. of pipe

100mm (min.)

o.d. + 200mm

Class A bedding: bedding factor 2·6

Band of clay

Flexible joint

Compressible fibre board 25mm thick

Concrete bed

Class A bedding

No mechanical ramming within 600mm above top of pipe

300mm (min.)

Selected soil or pea gravel well compacted in 150mm layers

Pea gravel well compacted

100mm (min.)

Class B bedding: bedding factor 1·9

Approved Document H to the Building Regulations provides many methods which will support, protect and allow limited angular and lineal movement to flexibly jointed clay drainpipes. Those shown below include three further classifications and corresponding bedding factors. Also shown is a suitable method of bedding flexible plastic pipes. In waterlogged trenches, it may be necessary to temporarily fill plastic pipes with water to prevent them from floating upwards while being laid. In all examples shown, space to the sides of pipes should be at least 150mm.

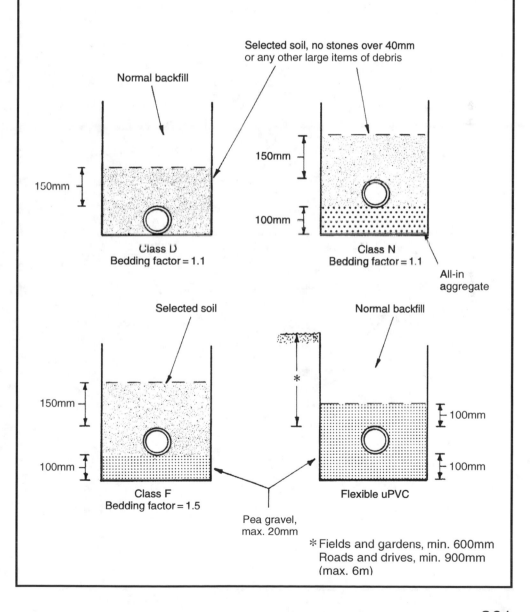

Normal backfill

Selected soil, no stones over 40mm or any other large items of debris

150mm

Class D
Bedding factor = 1.1

150mm

100mm

Class N
Bedding factor = 1.1

All-in aggregate

Selected soil

150mm

100mm

Class F
Bedding factor = 1.5

Pea gravel, max. 20mm

Normal backfill

*

100mm

100mm

Flexible uPVC

*Fields and gardens, min. 600mm
Roads and drives, min. 900mm
(max. 6m)

Drains Under Buildings

Wherever possible, siting drain runs beneath buildings should be avoided. This is because accessibility for maintenance would be restricted and there could be damage or movement to the drain from building settlement. Where ground floors are raised and ventilated, suspended drainage is possible. Otherwise, a drain may be protected by a complete surround of concrete integrated with the floor slab, or with a reinforced concrete floor slab cast above the drain. Drainpipes below the reinforced concrete to be flexibly jointed in a granular bedding beneath a compressible cover. Drains below a floor slab that are over 300mm to the pipe crown should be protected with a 100mm surround of granular fill (pea shingle).

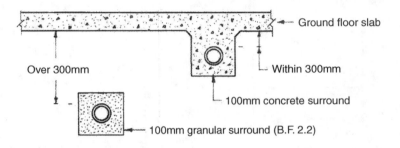

Ground floor slab

Over 300mm

Within 300mm

100mm concrete surround

100mm granular surround (B.F. 2.2)

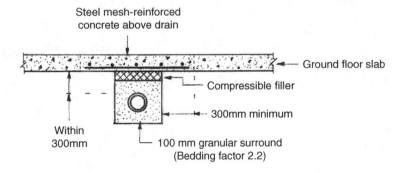

Steel mesh-reinforced concrete above drain

Ground floor slab

Compressible filler

300mm minimum

Within 300mm

100 mm granular surround (Bedding factor 2.2)

Ref. Building Regulations Approved Document H1 (Section 2.23).

Drainage pipes that penetrate a wall below ground level and above the building foundations should be installed in anticipation of some building settlement. Provision is achieved by either:

• Positioning the pipe through a purpose-made void in the wall, or
• Rigidly securing the pipe to the wall and providing flexible pipe joints and surround no more than 150mm from each end of the retained pipe.

Drain passing through sub-surface wall

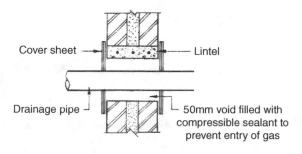

Cover sheet — Lintel

Drainage pipe — 50mm void filled with compressible sealant to prevent entry of gas

Alternative

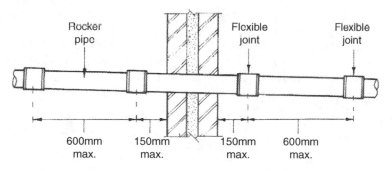

Rocker pipe   Flexible joint   Flexible joint

600mm max.   150mm max.   150mm max.   600mm max.

Drains and sewers should be located as far away as practicable from a new building. This is particularly important where there is a possibility for a future side or rear extension to the building. However, boundary limitations often determine location of drains and sewers; therefore, they should be installed with due regard to the settlement potential that any adjacent wall and foundations may have.

Ref. Building Regulations Approved Document H1 (Section 2.24).

Approved Document H4 (Section 1.12) to the Building Regulations has provision for construction over a drain and sewer where it is greater than 2m deep to invert level. In this situation, the design of foundations should incorporate a lintel or beam to over-span the pipe by at least 1.5m either side. Where less than 2m, the wall and foundation should be deeper locally to accommodate the pipe as shown on the previous page. AD H4 (Section 1.10).

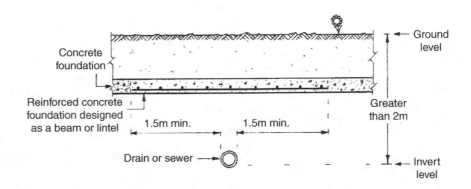

Approved Document H1 (Section 2.25) to the Building Regulations provides guidance on location of drain trenches close to a building's foundations. Where the trench is within 1m of a foundation, the pipe should be surrounded in concrete and the trench similarly filled to the underside of the foundation. If the trench distance exceeds 1m from the foundation, concrete fill is to a point below the lowest level of the building equal to the trench distance less 150mm.

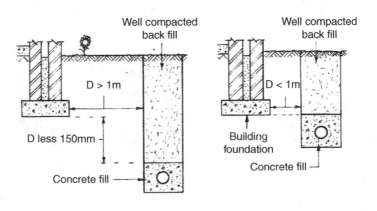

The critical dimension for new building work in the proximity of an existing drain or sewer is 3 metres. Building Regulations Approved Document H4 gives specific guidance where proposed building work would be constructed above or within 3m horizontal distance of drains or sewers, along with some exempting measures. If practicable, diversion of the sewer should be considered. The objective of avoiding sewers with new building work is that some existing sewers may be prone to groundwater leakage, thereby allowing fine subsoil materials such as silt and loam to percolate into the sewer undermining the structural integrity of the soil beneath a foundation. Also, the imposed load from a new building with the consolidated subsoil beneath it could fracture an existing sewer pipe. This will have an obvious disruption to the drainage facility and may cause the building to suffer structural settlement.

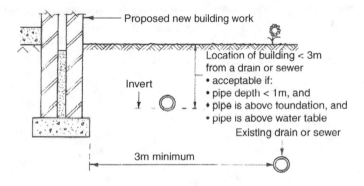

Proposed new building work

Location of building < 3m
from a drain or sewer
• acceptable if:
• pipe depth < 1m, and
• pipe is above foundation, and
• pipe is above water table

Invert

Existing drain or sewer

3m minimum

Ref. Building Regulations Approved Document H4: Building over sewers. The Water Industry Act.

Drainpipe Materials

- Vitrified clayware – 100, 150, 200, 225, 250 and 300 mm diameter nominal bore in 1.75 m lengths (1.60 m for 100 mm nom. bore). Also produced in a range of sizes from 400 to 600 mm diameter nominal bore for communal sewer pipes.
  Ref. BS EN 295-1.

- uPVC – standard outside diameters of 110 and 160 mm corresponding with nominal bore diameters of 100 and 150 mm in 3.00 and 6.00 m lengths. Other nominal bore diameters are 200, 250 and 315 mm.
  Refs. BS 4660, BS EN 1401-1 and BS EN 13598-1.
  Also produced in a structured or profiled twin wall format with a smooth bore and a ribbed external surface to provide for axial rigidity and radial strength.

- Cast iron – available in standard nominal bore diameters of 100, 150 and 225 mm. Larger nominal bore diameters are available in several sizes up to 600 mm. Suitable for suspending under raised floors and in false ceilings, for bridge drainage and use in unstable ground and ground known to contain methane gas.
  Ref. BS 437.

- Concrete – sewer pipes in nominal bore diameters of 300, 375, 450, 525, 600 and 675 mm. Above this, in a range of diameters up to 2.10 m nominal bore. Standard lengths are 2.50 m.
  Ref. BS 5911-1 and BS EN 1916.

- Asbestos cement – obsolete due to being a potential health hazard and superseded by lightweight uPVC. Used as drainage pipe material until the early 1970s; therefore, it may be found when existing drainage systems are exposed. Where located, must only be handled by specialist contractors (see page 7).

- Pitch fibre – also obsolete since the introduction of lightweight uPVC, but will be found in existing drainage systems particularly those installed between about 1950 and the late 1970s. Made from wood fibre impregnated with coal tar, pitch or bitumen. Design life is about 40 years, after which delamination may occur causing loss of strength and subsequent collapse.

Rigid jointing of clay drainpipes is now rarely specified, as flexible joints have significant advantages:

- They are quicker and simpler to make.
- The pipeline can be tested immediately.
- There is no delay in joint setting due to the weather.
- They absorb ground movement and vibration without fracturing the pipe.

Existing clay drains will be found with cement and sand mortar joints between spigot and socket. Modern pipe manufacturers have produced their own variations on flexible jointing, most using plain-ended pipes with a polypropylene sleeve coupling containing a sealing ring. Cast iron pipes can have spigot and sockets cold caulked with lead wool. Alternatively, the pipe can be produced with plain ends and jointed by rubber sleeve and two bolted couplings. Spigot and socket uPVC pipes may be jointed by solvent cement or with a push-fit rubber 'O' ring seal. They may also have plain ends jointed with a uPVC sleeve coupling containing a sealing ring.

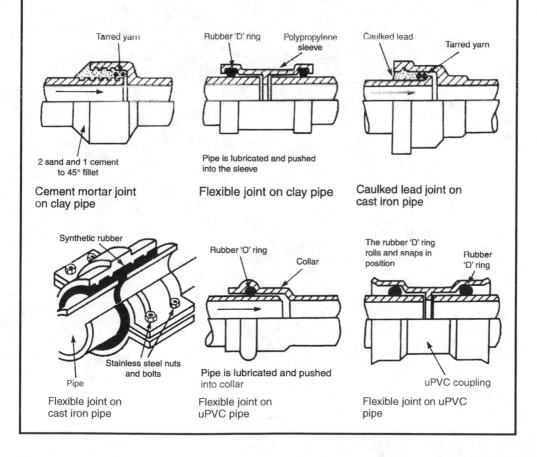

Tarred yarn

2 sand and 1 cement
to 45° fillet

Cement mortar joint on clay pipe

Rubber 'D' ring   Polypropylene sleeve

Pipe is lubricated and pushed into the sleeve

Flexible joint on clay pipe

Caulked lead   Tarred yarn

Caulked lead joint on cast iron pipe

Synthetic rubber

Stainless steel nuts and bolts

Pipe

Flexible joint on cast iron pipe

Rubber 'O' ring   Collar

Pipe is lubricated and pushed into collar

Flexible joint on uPVC pipe

The rubber 'D' ring rolls and snaps in position   Rubber 'D' ring

uPVC coupling

Flexible joint on uPVC pipe

Anti-Flood Devices – Interceptor Trap and Gully

Blockage in a public sewer or a period of prolonged heavy rainfall can cause surcharging and back flooding in drains. An anti-flooding facility must be installed to buildings on relatively low-lying sites, cellars, basements and where rainwater discharges directly into watercourses that may overflow. A risk assessment survey will determine whether a supplementary pump installation is required.

In situations where an interceptor trap is installed, an anti-flooding type should be specified in place of a standard interceptor. An anti-flood gully is simpler and quite effective if regularly maintained to remove silt and other deposits.

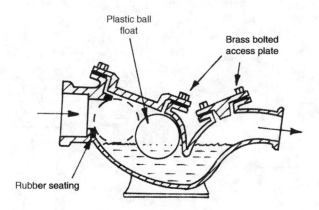

Anti-flooding interceptor trap

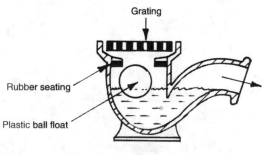

Anti-flooding gully trap

Anti-flood valves are an alternative to anti-flood gullies. Located in the drain run, they require less maintenance and attention than floor- or ground-level gullies. Anti-flood trunk valves have been in use for a considerable time. These are contained within a purpose-built access chamber or manhole close to the sewer or main drain connection.

Modern variations are supplied fitted within a factory-made access chamber with external push-fit pipe connectors. Inside the chamber, plastic floats positioned either side of the drainage pipe link to a valve or stopper. If a reverse surcharge occurs, the floats rise to close the valve until normal conditions return. Double valve units are also available and these are generally specified as a preference.

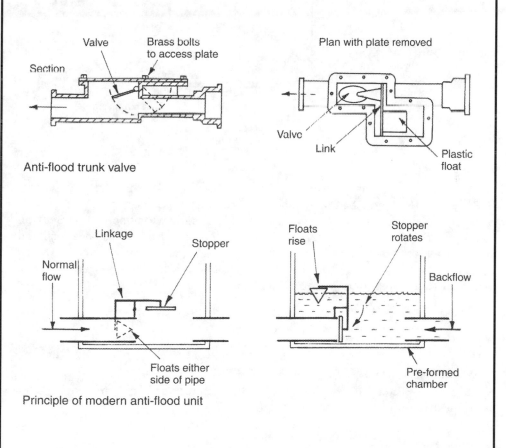

Anti-flood trunk valve

Principle of modern anti-flood unit

Grease Trap

Waste water from canteen sinks, commercial restaurants and dishwashers contains a considerable amount of fat, oil and grease. If not removed it can build up and block a drain. A grease trap functions by allowing the grease to cool in a relatively large volume of water. It congeals, solidifies and floats to the surface. At regular servicing intervals a tray is lifted out for cleaning and removal of the grease, usually by controlled burning. The trap outlet functions as a water seal to retain grease, preventing it from contaminating the drain or sewer.

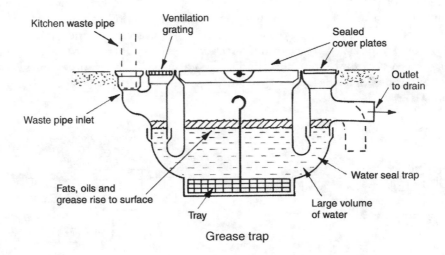

Grease trap

Correctly sized, a grease trap will retain about 50% of fats, etc. Greater effect can be achieved by installing more than one in series. It is a Building Regulation requirement (Approved Document H1, Section 2.21) for all new and refurbished commercial hot food preparation premises to have an effective means of grease removal from waste water before it discharges to a public drain or sewer. Grease separators complying with BS EN 1825 will satisfy this.

Refs. BS EN 1825-1: Grease separators. Principles of design, performance and testing, marking and quality control.

BS EN 1825-2: Grease separators. Selection of nominal size, installation, operation and maintenance.

The Public Health Act prohibits discharge of petroleum and oil into a sewer. Garage floor washings will contain petrochemicals, and these must be prevented from entering a sewer. The floor layout should be arranged so that one garage gully serves up to 50m² of floor area. The gully will retain some oil and other debris, which can be removed by emptying the inner bucket. A petrol interceptor will remove both petrol and oil. Both rise to the surface with some evaporation through the vent pipes. The remaining oil is removed when the tanks are emptied and cleaned. The first chamber will also intercept debris, and this compartment will require more regular cleaning. Contemporary petrol interceptors are manufactured from reinforced plastics for simple installation in a prepared excavation.

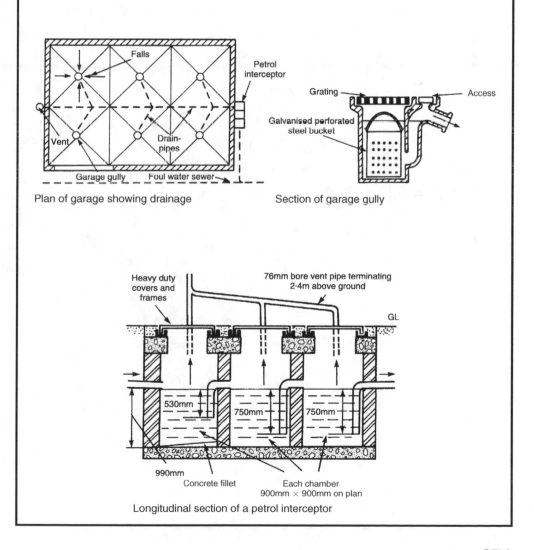

Plan of garage showing drainage

Section of garage gully

Longitudinal section of a petrol interceptor

The contents of drainage pipelines should gravitate to the sewer and sewage processing plant. In some situations, site levels or basement sanitary facilities will be lower than adjacent sewers and it becomes necessary to pump the drainage flows. A pumping station or plant room can be arranged with a motor room above or below surface level. Fluid movement is by centrifugal pump, usually immersed and therefore fully primed. For large schemes, two pumps should be installed with one on standby in the event of the duty pump failing. The pump impellor is curved on plan to complement movement of sewage and to reduce the possibility of blockage. The high-level discharge should pass through a manhole before connecting to the sewer.

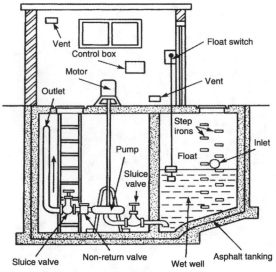

Section through pumping station

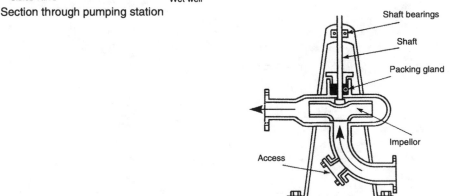

Section through centrifugal pump

Refs. BS EN 12056-4: Gravity drainage systems inside buildings.

Waste water lifting plants. Layout and calculation.

BS EN 12050: Waste water lifting plants for buildings and sites.

A sewage ejector may be used as an alternative to a centrifugal pump for lifting foul water. The advantages of an ejector are

- Less risk of blockage.
- Fewer moving parts and less maintenance.
- A wet well is not required.
- One compressor unit can supply air to several ejectors.

Operation:

- Incoming sewage flows through inlet pipe A into ejector body B.
- Float rises to the top collar.
- Rod is forced upwards, opening an air inlet valve and closing an exhaust valve.
- Compressed air enters the ejector body, forcing sewage out through pipe C.
- The float falls to the bottom collar and its weight plus the rocking weight closes the air inlet valve and opens the exhaust valve.

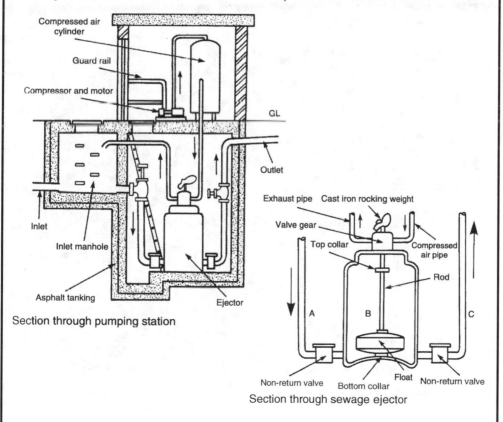

Section through pumping station

Section through sewage ejector

Ref. BS EN 16932: Drain and sewer systems outside buildings. Pumping systems.

When considering methods of drainage pumping, equipment manufacturers should be consulted with the following details:

• Drainage medium – foul or surface water, or both.
• Maximum quantity – anticipated flow in m³/h.
• Height to which the sewage has to be elevated.
• Length of delivery pipe.
• Availability of electricity – mains or generated.
• Planning constraints, regarding appearance and siting of pump station.

In the interests of visual impact, it is preferable to construct the motor room below ground. This will also absorb some of the operating noise.

In basements, there may be some infiltration of ground water. This can be drained to a sump and pumped out as the level rises. In plant rooms, a sump pump may be installed to collect and remove water from any leakage that may occur. It is also useful for water extraction when draining down boilers for servicing.

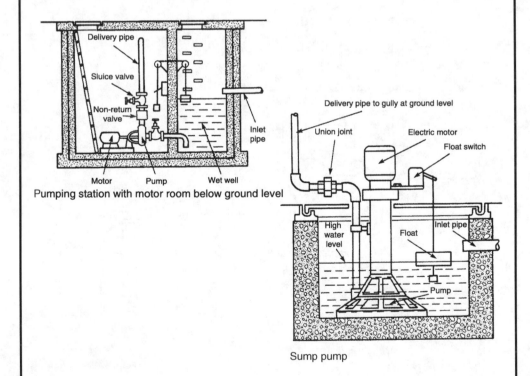

Pumping station with motor room below ground level

Sump pump

Ref. Design guidance for external pumped installations may be found in BS ENs 12050-1 to 4: Wastewater lifting plants for buildings and sites.

A means for providing drainage from existing basement space where gravity discharge to adjacent drain and sewer is impractical. This is quite common when older dwellings are converted and adapted into several smaller habitable units and floor space is maximised to include basement accommodation. Also applicable to purpose built basements.

Typical application to an existing basement –

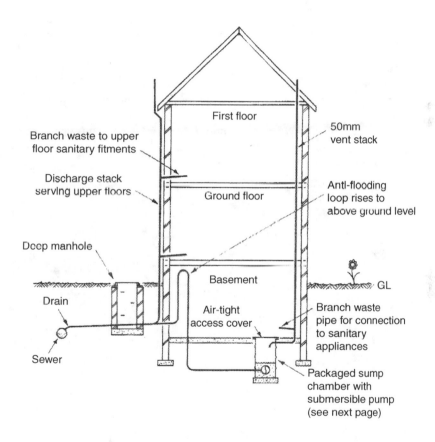

Note: Any gullies connected to be of the anti-flood type (page 368).

375

Basement pumping, protection against backflow flooding –
- Non-return valve on pump outlet pipe.
- Anti-flood loop.
- Gullies of the anti-flood type.

Building Regulations Approved Document H1 Foul drainage, Section 2.36 – Pumping installations.

Minimum sump requirements:
- Sized to contain 24-hour inflow capacity in event of power equipment failure; unless alternative provision is made, e.g. from back-up electricity generation.
- 150 litres/person per day design requirement for domestic use. Other building types based on water intake demand.
- Design capacity pro rata where only part of the foul water system is pumped.
- Controls designed to optimise pump operation, i.e. time programmed as well as level/volume float control.

Factory manufactured packaged foul water sump and pump

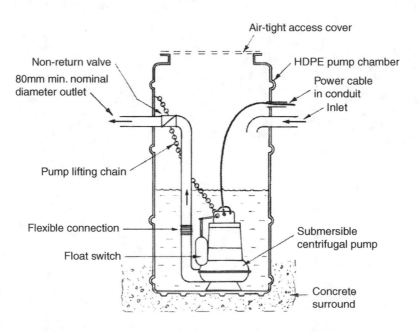

Note: Also suitable for rainwater applications, e.g. basements subject to occasional surface water ingress.

Ideally, buildings should be constructed with foundations above the subsoil water table. Where this is unavoidable or it is considered necessary to generally control the ground water, a subsoil drainage system is installed to permanently lower the natural water table. Various ground drainage systems are available; the type selected will depend on site conditions. The simplest is a French drain. It comprises a series of strategically located rubble-filled trenches excavated to a fall and to a depth below high water table. This is best undertaken after the summer, when the water table is at its lowest. Flow can be directed to a ditch, stream or other convenient outfall. In time the rubble will become silted up and need replacing.

An improvement uses a polyethylene/polypropylene filament fabric membrane to line the trench. This is permeable in one direction only and will also function as a silt filter. This type of drain is often used at the sides of highways with an open rubble surface.

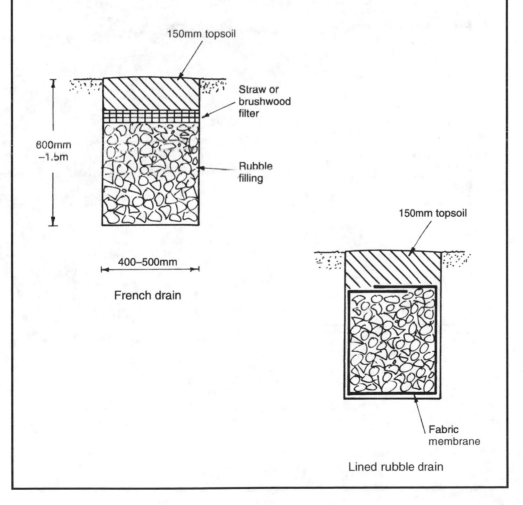

150mm topsoil

Straw or brushwood filter

600mm –1.5m

Rubble filling

400–500mm

French drain

150mm topsoil

Fabric membrane

Lined rubble drain

The layout and spacing of subsoil drainage systems depends on the composition and drainage qualities of the subsoil and the disposition of buildings. For construction sites, the depth of drainage trench will be between 600mm and 1·5m. Shallower depths may be used in agricultural situations and for draining surface water from playing fields. Installation of pipes within the rubble drainage medium has the advantage of creating a permanent void to assist water flow. Suitable pipes are produced in a variety of materials, including clay (open jointed, porous or perforated), concrete (porous (no-fine aggregate) or perforated) and uPVC (perforated). The pipe void can be accessed for cleaning, and the system may incorporate silt traps at appropriate intervals. Piped outlets may connect to a surface water sewer with a reverse-acting interceptor trap at the junction.

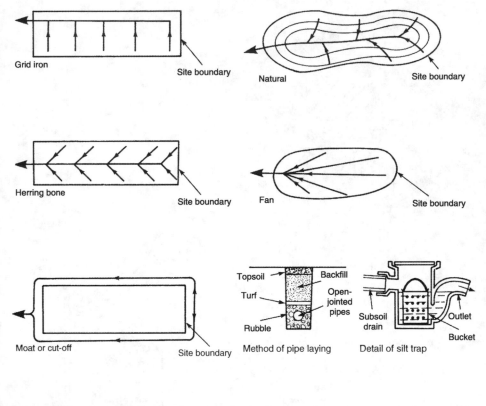

Note: The installation of subsoil drainage may be necessary under Building Regulation provisions. The purpose of this is to prevent the passage of ground moisture into a building and the possibility of damage to a building.

Ref. Approved Document C, Section 3: Subsoil drainage.

British Standard pipes commonly used for subsoil drainage:

- Perforated clay, BS EN 295-5.
- Porous clay, BS 1196.
- Profiled and slotted plastics, BS 4962.
- Perforated uPVC, BS 4660.
- Porous concrete, BS withdrawn no manufacturing interest.

Silt and other suspended particles will eventually block the drain unless purpose-made traps are strategically located for regular cleaning. The example shown on the previous page is adequate for short drain runs, but complete systems will require a pit which can be physically accessed. This is an essential requirement if the drain is to connect to a public surface water sewer. In order to protect flow conditions in the sewer, the local water authority may only permit connection via a reverse-acting interceptor trap. This item does not have the capacity to function as a silt trap.

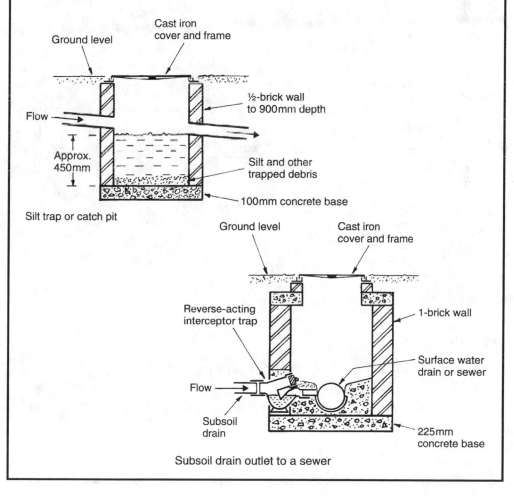

Silt trap or catch pit

Subsoil drain outlet to a sewer

Tests on Drains

Drains must be tested before and after backfilling trenches.

Air test – the drain is sealed between access chambers and pressure tested to 100mm water gauge with hand bellows and a 'U' gauge (manometer). The pressure must not fall below 75mm during the first five minutes.

Smoke test – may be used to detect leakage. The length of drain to be tested is sealed and smoke pumped into the pipes from the lower end. The pipes should then be inspected for any trace of smoke. Smoke pellets may be used in the smoke machine, or with clay and concrete pipes they may be applied directly to the pipeline.

Water test – effected by stopping the lower part of the drain and filling the pipe run with water from the upper end. This requires a purpose-made test bend with an extension pipe to produce a 1·5m head of water. This should stand for two hours and if necessary topped up to allow for limited porosity. For the next 30 minutes, maximum leakage for 100mm and 150mm pipes is 0·05 and 0·08 litres per metre run, respectively.

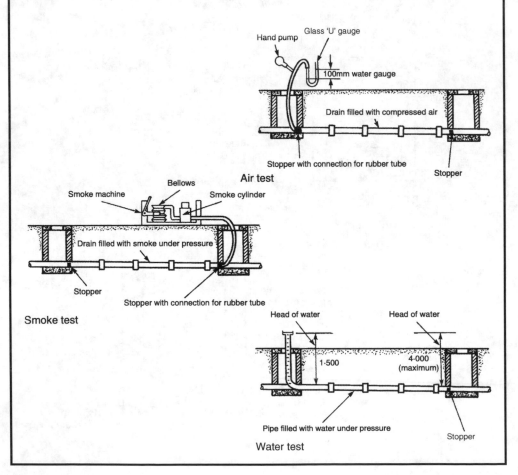

Hand pump   Glass 'U' gauge

100mm water gauge

Drain filled with compressed air

Stopper with connection for rubber tube   Stopper

Air test

Smoke machine   Bellows   Smoke cylinder

Drain filled with smoke under pressure

Stopper

Stopper with connection for rubber tube

Smoke test

Head of water   Head of water

1·500   4·000 (maximum)

Pipe filled with water under pressure

Stopper

Water test

Where a surface water sewer is not available, it may be possible to dispose of rainwater into a soakaway. A soakaway will only be effective in porous soils and above the water table. Water must not be allowed to flow under a building and soakaways should be positioned at least 3m away (most local authorities require 5m). A filled soakaway is inexpensive to construct, but it will have limited capacity. Unfilled or hollow soakaways can be built of pre-cast concrete or masonry.

Soakaway capacity can be determined by applying a rainfall intensity of at least 50mm per hour to the following formula:

$$C \times A \times R \div 3$$

where C = capacity in m³
      A = area to be drained in m²
      R = rainfall in metres per hour.
E.g. a drained area of 150m²
      C — 150 × 0·050 ÷ 3 = 2·5 m³

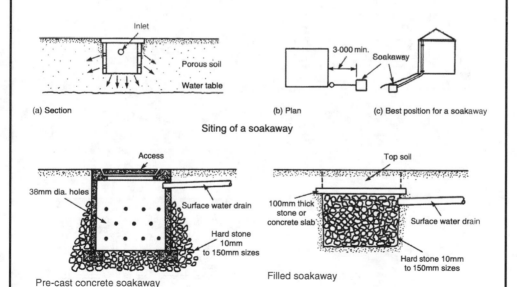

(a) Section

Inlet
Porous soil
Water table

(b) Plan

3·000 min.
Soakaway

(c) Best position for a soakaway

Siting of a soakaway

Pre-cast concrete soakaway

Access
38mm dia. holes
Surface water drain
Hard stone 10mm to 150mm sizes

Filled soakaway

Top soil
100mm thick stone or concrete slab
Surface water drain
Hard stone 10mm to 150mm sizes

Note: BRE Digest 365: Soakaways, provides a more detailed approach to capacity calculation. BRE also produce the soakaway design software, BRESOAK.

Cesspools

A cesspool is an acceptable method of foul water containment where main drainage is not available. It is an impervious chamber requiring periodic emptying, sited below ground level. Traditional cesspools were constructed of brickwork rendered inside with waterproof cement mortar. Pre-cast concrete rings supported on a concrete base have also been used, but factory-manufactured glass-reinforced plastic units are now preferred. The Building Regulations require a minimum capacity below inlet level of 18000 litres. A cesspool must be impervious to rainwater, well ventilated and have no outlets or overflows. It should be sited at least 15m from a dwelling.

Capacity is based on 150 litres per person per day at 45-day emptying cycles, e.g. a four-person house:

$$= 4 \times 150 \times 45 = 27\,000 \text{ litres } (27m^3)$$

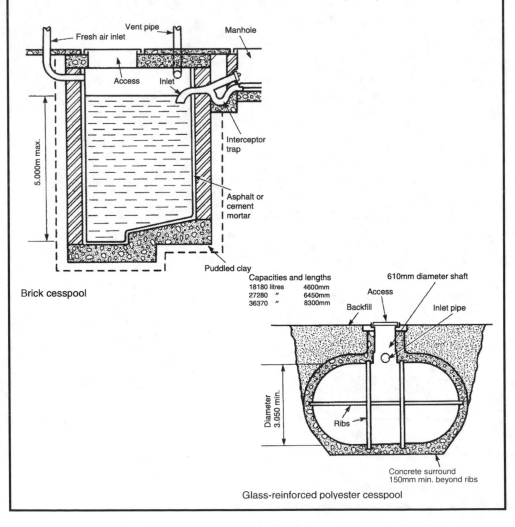

Brick cesspool

Capacities and lengths
18180 litres   4600mm
27280   "   6450mm
36370   "   8300mm

Glass-reinforced polyester cesspool

Where main drainage is not available, a septic tank is preferable to a cesspool. A septic tank is self-cleansing and will only require annual desludging. It is in effect a private sewage disposal plant, which is quite common for buildings in rural areas. The tank is a watertight chamber in which the sewage is liquefied by anaerobic bacterial activity. This type of bacteria lives in the absence of oxygen which is ensured by a sealed cover and the natural occurrence of a surface scum or crust. Traditionally built tanks are divided into two compartments with an overall length of three times the breadth. Final processing of sewage is achieved by conveying it through subsoil drainage pipes or a biological filter. Capacity is determined from the simple formula:

$$C = (180 \times P) + 2000$$

where C = capacity in litres
        P = no. of persons served

E.g. 10 persons; C = $(180 \times 10) + 2000 = 3800$ litres $(3.8\,m^3)$.

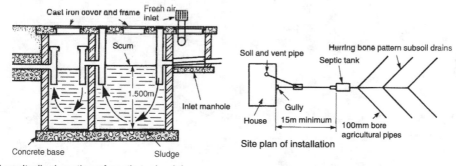

Longitudinal section of septic tank minimum volume under Building Regulations = 2.7 m³

Site plan of installation

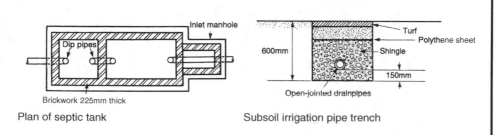

Plan of septic tank

Subsoil irrigation pipe trench

Klargester Settlement/Septic Tank

The Klargester settlement tank is a simple, reliable and cost-effective sewage disposal system manufactured from glass-reinforced plastics for location in a site-prepared excavation. A standard range of tanks are produced in capacities ranging from 2700 to 10000 litres, to suit a variety of applications from individual houses to modest developments including factories and commercial premises. The sewage flows through three compartments (1,2,3) on illustration where it is liquefied by anaerobic bacterial activity. In similarity with traditionally built tanks, sludge settlement at the base of the unit must be removed annually. This is achieved by pushing away the floating ball to give extraction tube access into the lowest chamber. Processed sewage may be dispersed by subsoil irrigation or a biological filter.

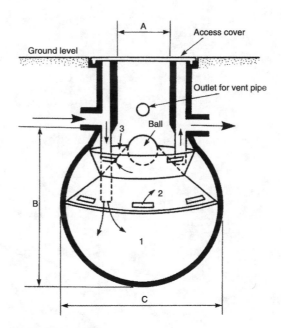

Section through tank

Capacity of tank in litres	Number of users with flow rate per head per day		Nominal dimensions in mm.		
	180 litres	250 litres	A	B	C
2700	4	3	610	1850	1800
3750	9	7	610	2060	2000
4500	14	10	610	2150	2100
6000	22	16	610	2400	2300
7500	30	22	610	2630	2500
10000	44	32	610	2800	2740

Ref. Building Regulations, Approved Document H2: Waste water treatment systems and cesspools.

The biological disc has many successful applications to modest-sized buildings such as schools, prisons, country clubs, etc. It is capable of treating relatively large volumes of sewage by an accelerated process. Crude sewage enters the biozone chamber via a deflector box which slows down the flow. The heavier solids sink to the bottom of the compartment and disperse into the main sludge zone. Lighter solids remain suspended in the biozone chamber. Within this chamber, micro-organisms present in the sewage adhere to the partially immersed slowly rotating discs to form a biologically active film feeding on impurities and rendering them inoffensive. Baffles separate the series of rotating fluted discs to direct sewage through each disc in turn. The sludge from the primary settlement zone must be removed every six months.

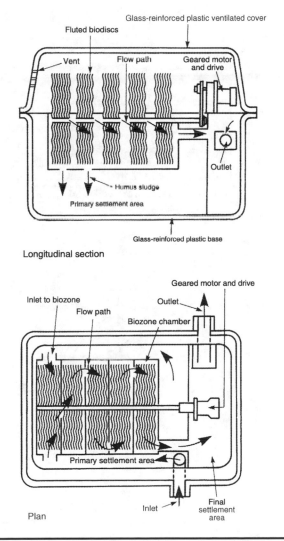

Longitudinal section

Plan

Biological Filter

Treatment of septic tank effluent – liquid effluent from a septic tank is dispersed from a rotating sprinkler pipe over a filter of broken stone, clinker, coke or polythene shingle. The filter surfaces become coated with an organic film which assimilates and oxidises the pollutants by aerobic bacterial activity. This type of bacteria lives in the presence of oxygen, encouraged by ventilation through under-drains leading to a vertical vent pipe. An alternative process is conveyance and dispersal of septic tank effluent through a system of subsoil drains or a drainage field. To succeed, the subsoil must be porous and the pipes laid above the highest water table level. Alternatively, the primary treated effluent can be naturally processed in constructed wetland phragmite or reed beds (see page 389). Whatever method of sewage containment and processing is preferred, the local water authority will have to be consulted for approval.

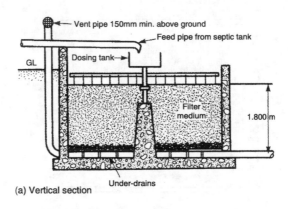

(a) Vertical section

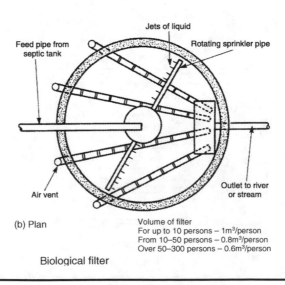

(b) Plan

Volume of filter
For up to 10 persons – 1m³/person
From 10–50 persons – 0.8m³/person
Over 50–300 persons – 0.6m³/person

Biological filter

Drainage fields and mounds are a less conspicuous alternative to the use of a biological filter for secondary processing of sewage. Disposal and dispersal is through a system of perforated pipes laid in a suitable drainage medium.

Location:

- Min. 10m from any watercourse or permeable drain.
- Min. 50m from any underground water supply.
- Min. distance from a building:

<5 people	15m
6–30 people	25m
31–100 people,	40m
>100 people	70m

- Downslope of any water source.
- Unencroached by any other services.
- Unencroached by access roads or paved areas.

Ground quality:

- Preferably granular, with good percolation qualities. Subsoils of clay composition are unlikely to be suited.
- Natural water table should not rise to within 1m of distribution pipes invert level.
- Ground percolation test:

1. Dig several holes 300 × 300mm, 300mm below the expected distribution pipe location.
2. Fill holes to a 300mm depth of water and allow to seep away overnight.
3. Next day refill holes to 300mm depth and observe time in seconds for the water to fall from 225mm depth to 75mm. Divide time by 150mm to ascertain average time (Vp) for water to drop 1mm.
4. Apply floor area formula for drainage field:

$$At = p \times Vp \times 0.25$$

where, At = floor area (m^2)

p = no. of persons served

e.g. 40min. (2400 secs) soil percolation test time in a system serving six persons.

$$Vp = 2400 \div 150 = 16$$
$$At = 6 \times 16 \times 0.25 = 24 \text{ m}^2$$

Note: Vp should be between 12 and 100. Less than 12 indicates that untreated effluent would percolate into the ground too rapidly. A figure greater than 100 suggests that the field may become saturated.

Drainage Fields and Mounds – 2

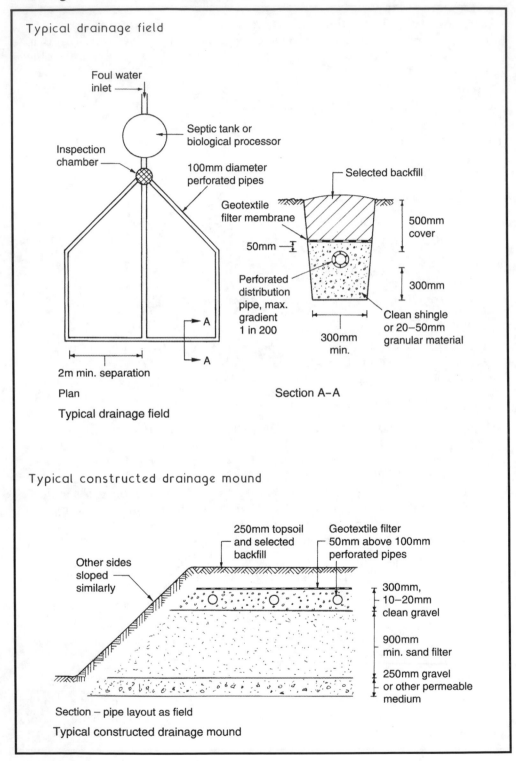

Typical drainage field

Foul water inlet

Septic tank or biological processor

Inspection chamber

100mm diameter perforated pipes

2m min. separation

Plan

Perforated distribution pipe, max. gradient 1 in 200

Typical drainage field

Selected backfill

Geotextile filter membrane

50mm

500mm cover

300mm

300mm min.

Clean shingle or 20–50mm granular material

Section A–A

Typical constructed drainage mound

250mm topsoil and selected backfill

Geotextile filter 50mm above 100mm perforated pipes

Other sides sloped similarly

300mm, 10–20mm clean gravel

900mm min. sand filter

250mm gravel or other permeable medium

Section – pipe layout as field

Typical constructed drainage mound

Reed Beds and Constructed Wetlands

These provide a natural method for secondary treatment of sewage from septic tanks or biological processing equipment.

Common reeds (*Phragmites australis*) are located in prepared beds of selected soil or fine gravel. A minimum bed area of $20\,m^2$ is considered adequate for up to four users; $5\,m^2$ should be added for each additional person. Reeds should be spaced about every 600mm and planted between May and September. For practical purposes, application is limited to about 30 people, due to the large area of land occupied. Regular maintenance is necessary to reduce unwanted weed growth which could restrict fluid percolation and natural processing. The site owners have a legal responsibility to ensure that the beds are not a source of pollution, a danger to health or a nuisance.

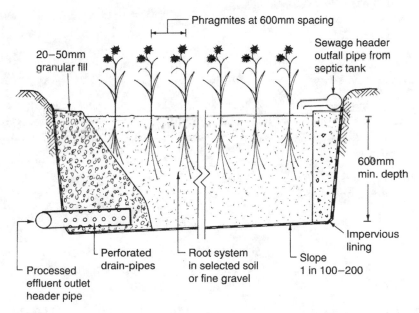

Section through a typical reed bed

Ref. Building Regulations, Approved Document H2: Waste water treatment systems and cesspools.

Sustainable Urban Drainage Systems (SUDS)

Extreme weather situations in the UK have led to serious property damage from flooding, as drains, rivers and other watercourses are unable to cope with the unexpected volumes of surface water. A possible means of alleviating this and moderating the flow of surface water is the construction of SUDS between the drainage system and its outfall.

Objectives are to:

- decrease the volume of water discharging or running off from a site or building
- reduce the run-off rate
- filter and cleanse the debris from the water flow.

Formats:

- soakaways
- swales
- infiltration basins and permeable surfaces
- filter drains
- retention or detention ponds
- reed beds.

Soakaways – See page 381. For application to larger areas, see BS EN 752: Drain and sewer systems outside buildings.

Swales – Channels lined with grass. These slow the flow of water, allowing some to disperse into the ground as they convey water to an infiltration device or watercourse. They are best suited to housing, car parks and roads.

Infiltration basins and permeable surfaces – Purposely located depressions lined with grass and positioned to concentrate surface water into the ground. Permeable surfaces such as porous asphalt or paving can also be used to the same effect.

Filter drains – Otherwise known as French drains (see page 377). Note that drainage may be assisted by locating a perforated pipe in the centre of the gravel or rubble filling.

Retention or detention ponds – These are man-made catchments to contain water temporarily, for controlled release later.

Reed beds – These are not restricted to processing septic tank effluent, as shown on page 389. They are also a useful filter mechanism for surface water, breaking down pollutants and settlement of solids.

Refs. Sustainable Urban Drainage Systems – Design manual for England and Wales – (C 522) and for Scotland and N. Ireland (C 521) – CIRIA. SUDS Manual (C753) – CIRIA.

In terms of demand, fresh water is becoming a relatively scarce resource, resulting in an ongoing programme by the water authorities to meter water consumption to all buildings. Rainwater harvesting is an economic means for supplementing the use of processed water, thereby reducing utilities bills and impact on the environment. The process involves intercepting, storing and filtering the surface water run-off from roofs and hard landscaping. Some applications to rainwater drainage systems are shown below.

Typical domestic garden application –

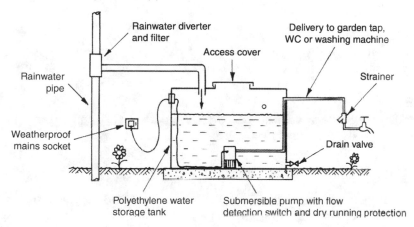

Typical commercial/industrial site application –

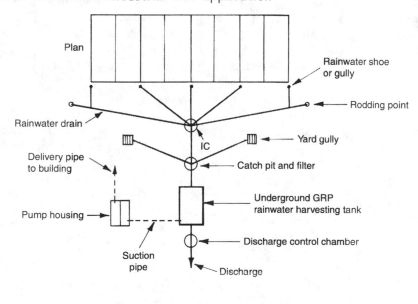

Objectives for domestic water consumption are set at various levels, the requirement defined in Building Regulation AD G2 – Water Efficiency. This sets a whole building standard of water consumption at 125 litres/day per person in new dwellings, somewhat less than the generally accepted average of 150 litre/day. Water authorities objectives are well below this, with the simple technology of rainwater harvesting seen as a major contributor to attaining water consumption economies.

Anticipated benefits –

- Domestic mains water consumption reduced by up to 50%.
- Water consumption within commercial premises with large roof areas reduced by up to 75% where there is high usage of sanitary fitments.

Water captured from rainfall run-offs is directed through filters, stored and recycled as illustrated on the previous page.

Requirements –

- Separate pipework from mains-supplied potable/wholesome water.
- No direct connection between harvested water and mains-supplied water.
- Pipework conveying harvested water to be marked and labelled with clear identification '… so as to be easily distinguished from any supply pipe or distributing pipe'. Some examples of identification could be RECLAIMED WATER, UNWHOLESOME WATER, NON-POTABLE WATER or WATER NOT FOR HUMAN CONSUMPTION. Extract from Schedule 2, Para. 14 The Water Supply (Water Fittings) Regulations.
- Non-potable, harvested rainwater must be fully serviced/supplied (topped up) even during periods of dry weather. Therefore, a non-potable water storage vessel must have a mains water top-up facility via an air gap or header tank as a backflow prevention installation.
- A harvested water storage tank overflow pipe is required to provide controlled discharge during heavy rainfall. This should run off via a drain into an adequately sized soakaway or storm water drainage system.

Refs. BS 8595: Code of practice for the selection of water reuse systems.

BS EN 16941-1: On-site non-potable water systems. Systems for the use of rainwater.

Separation of water supplies – there must be no possibility of cross-connection between a pipe conveying wholesome water and a pipe conveying harvested rainwater or any other type of unwholesome water. Where a mains supply of wholesome water is used to top up a vessel containing harvested rainwater, backflow prevention is essential. Mechanical pipeline fittings such as double-check valves and variable backflow preventers with reduced pressure zone (RPZ valve) are not sufficient (see pages 54 and 55). Backflow prevention is by air gap separation, as shown below.

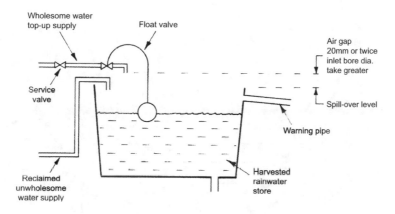

Marking and labelling – pipes conveying reclaimed water must be labelled (see previous page) and identified with a BS 1710 colour coding to distinguish them from other pipeline services. Storage vessels and valves should also be identified. The examples show green as the basic colour for water. See pages 761 and 762 for others.

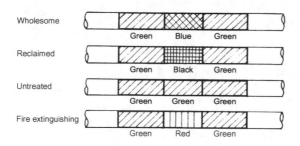

Ref. BS 1710: Specification for identification of pipelines and services.

Rainwater Attenuation

Rainwater attenuation applies to controlling and managing rainfall run-off from buildings and hard landscaping. In many situations, this can be achieved with soakaways or with retention/detention ponds. An alternative is an underground retention and discharge process that uses a system of fabricated plastic modular cells. The individual units, similar in appearance to milk crates, are tied or clipped together to create a matrix. These can be made up to an overall size large enough to accommodate the run-off demands for numerous buildings. The completed matrix is wrapped in an impermeable membrane, but this can be partially omitted if a soakaway facility is required.

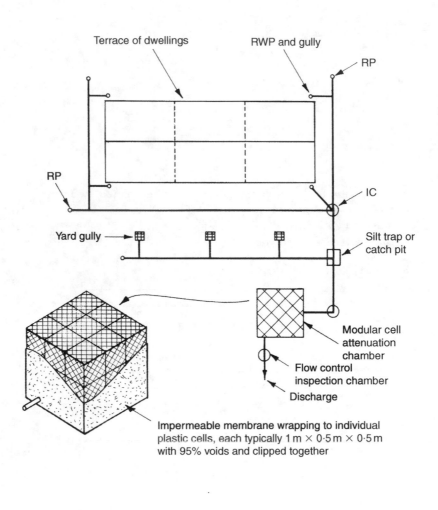

Terrace of dwellings

RWP and gully

RP

RP

IC

Yard gully

Silt trap or catch pit

Modular cell attenuation chamber

Flow control inspection chamber

Discharge

Impermeable membrane wrapping to individual plastic cells, each typically 1 m × 0·5 m × 0·5 m with 95% voids and clipped together

The size of gutters and downpipes will depend on the effective surface area to be drained. For flat roofs, this is the plan area, while pitched roof effective area (A_e) can be calculated from:

$$\text{Roof plan area} \div \text{Cosine pitch angle}$$

Roofs over 70° pitch are treated as walls, with the effective area taken as:

Elevational area × 0·5.

Actual rainfall varies throughout the world. For UK purposes, a rate of 75mm/h (R) is suitable for all but the most extreme conditions. Rainfall run-off (Q) can be calculated from:

$$Q = (Ae \times R) \div 3600 = l/s$$

E.g. a 45° pitched roof of 40m² plan area.

$$Q = ([40 \div \text{Cos } 45°] \times 75) \div 3600$$
$$Q = ([40 \div 0·707] \times 75) \div 3600$$
$$Q = 1·18 \ l/s$$

Size of gutter and downpipe will depend on profile selected, i.e. half round, ogee, box, etc. Manufacturers' catalogues should be consulted to determine a suitable size. For guidance only, the following is generally appropriate for half-round eaves gutters with one end outlet:

Half-round gutter (mm)	Outlet dia. (mm)	Flow capacity (l/s)
75	50	0·38
100	65	0·78
115	65	1·11
125	75	1·37
150	90	2·16

Therefore, the example of a roof with a flow rate of 1·18l/s would be adequately served by a 125mm gutter and 75mm downpipe.

- Where an outlet is not at the end, the gutter should be sized to the larger area draining into it.
- The distance between a stopped end and an outlet should not exceed 50 times the flow depth.
- The distance between two or more outlets should not exceed 100 times the flow depth (see example below).
- For design purposes, gutter slope is taken as less than 1 in 350.

E.g. a 100mm half-round gutter has a 50mm depth of flow, therefore:

100 × 50 = 5000mm or 5m spacing of downpipes.

See next page and page 482 for alternative guides to estimating size of rainwater goods.

Drainage Design – Surface Areas (2)

Another method of rainwater downpipe sizing is provided in BS EN 12056-3: Gravity drainage systems inside buildings. Roof drainage, layout and calculations. This Standard provides tabulated data calculated from the Wyly-Eaton equation to determine the capacity of rainwater pipes:

$$Q = 2.5 \times 10^{-4} \times k^{-0.167} \times d^{2.667} \times f^{1.667}$$

where Q = capacity of the rainwater pipe (l/s)

   k = pipe roughness factor (usually taken as 0.25mm)

   d = inside diameter of the rainwater downpipe (mm)

   f = filling degree or proportion of the rainwater pipe cross section filled with water (dimensionless)

d (mm)	Q (l/s)	f
65	1·5	0·20
65	3·4	0·33
75	2·2	0·20 (see calculation below)
75	5·0	0·33
90	3·5	0·20
90	8·1	0·33
100	4·6	0·20
100	10·7	0·33

Example by calculation for a 75mm diameter rainwater pipe with a filling degree of 0·2:

$$Q = 2.5 \times 10^{-4} \times 0.25^{-0.167} \times 75^{2.667} \times 0.20^{1.667}$$
$$Q = 2.5 \times 10^{-4} \times 1.26 \times 100181.69 \times 0.07$$
$$Q = 2.21, \text{ i.e. } 2.2 \text{ l/s}$$

To calculate rainwater pipe diameter, the formula is rearranged:

$$d^{2.667} = \frac{Q}{2.5 \times 10^{-4} \times k^{-0.167} \times f^{1.667}} \quad \text{Using } k = 0.25\text{mm and } f = 0.20$$

$$d^{2.667} = \frac{2.2}{2.5 \times 10^{-4} \times 0.25^{-0.167} \times 0.20^{1.667}} \quad \text{Then } d = 74.89, \text{ i.e. } 75\text{mm}$$

This alternative procedure can be seen to allow a greater amount of flow capacity than that indicated on the previous page.

When designing rainfall run-off calculations for car parks, playgrounds, roads and other made-up areas, a rainfall intensity of 50mm/h is considered adequate. An allowance for surface permeability (P) should be included, to slightly modify the formula from page 395:

$$Q = (A \times R \times P) \div 3600 = l/s$$

Permeability factors:

Asphalt	0·85–0.95
Concrete	0·85–0.95
Concrete blocks (open joints)	0·40–0.50
Gravel drives	0·15–0.30
Grass	0·05–0.25
Paving (sealed joints)	0.75–0.85
Paving (open joints)	0.50–0.70

E.g. a paved area (P = 0.75) 50m × 24m (1200m^2)

$$Q = (1200 \times 50 \times 0·75) \div 3600$$

$$Q = 12·5 \; l/s \; or \; 0·0125 \; m^3/s$$

The paved area will be served by several gullies (at 1 per 300m^2 = 4) with subdrains flowing into a main surface water drain. Each drain can be sized according to the area served, but for illustration purposes, only the main drain is considered here. The pipe sizing formula is:

$$Q = V \times A$$

where Q = quantity of water (m^3/s)
      V = velocity of flow (min. 0·75 m/s) – see next page
      A = area of water flowing (m^2)

Drains should not be designed to flow full bore as this leaves no spare capacity for future additions. Also, fluid flow is eased by the presence of air space. Assuming half full bore, using the above figure of 0.0125m^3/s, and the minimum velocity of flow of 0.75m/s:

$$Q = V \times A$$
$$0·0125 = 0·75 \times A$$

Transposing,

$$A = 0·0125 \div 0·75$$

$$A = 0·017 \; m^2$$

This represents the area of half the pipe bore, so the total pipe area is double, i.e. 0·034m^2.

Area of a circle (pipe) = πr^2 where r = radius of pipe (m).
Transposing,

$$r = \sqrt{Area \div \pi}$$
$$r = \sqrt{0.034 \div \pi}$$
$$r = 0.104m \; or \; 104mm$$

Therefore, the pipe diameter = 2 × 104 = 208mm.

The nearest commercial size is 225mm nominal inside diameter.

397

Drainage Design – Velocities and Hydraulic Mean Depth

Velocity of flow – 0·75m/s is the accepted minimum to achieve self-cleansing. It is recognised that an upper limit is required to prevent separation of liquids from solids. A reasonable limit is 1·8m/s for both surface and foul water drainage, although figures up to 3m/s can be used especially if grit is present. The selected flow rate will have a direct effect on drain gradient; therefore, to moderate excavation costs a figure nearer the lower limit is preferred. Also, if there is a natural land slope and excavation is a constant depth, this will determine the gradient and velocity of flow.

Hydraulic mean depth (HMD) – otherwise known as hydraulic radius represents the proportion or depth of flow in a drain. It will have an effect on velocity and can be calculated by dividing the area of water flowing in a drain by the contact or wetted perimeter. Thus for half full bore:

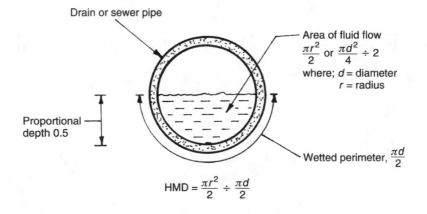

Drain or sewer pipe

Area of fluid flow
$\dfrac{\pi r^2}{2}$ or $\dfrac{\pi d^2}{4} \div 2$
where; d = diameter
r = radius

Proportional depth 0.5

Wetted perimeter, $\dfrac{\pi d}{2}$

$$HMD = \frac{\pi r^2}{2} \div \frac{\pi d}{2}$$

This table summarises HMD for proportional flows:

Depth of flow	HMD
0·25	Pipe dia. (m) ÷ 6·67
0·33	Pipe dia. (m) ÷ 5·26
0·50	Pipe dia. (m) ÷ 4·00
0·66	Pipe dia. (m) ÷ 3·45
0·75	Pipe dia. (m) ÷ 3·33
Full	Pipe dia. (m) ÷ 4·00

E.g. a 225mm (0·225m) drain flowing half bore:

$$HMD = 0·225 \div 4 = 0·05625$$

Drains are usually designed with a maximum flow condition of three-quarters full bore, i.e. depth of flow or proportional depth 0·75. It is essential to maintain some air space within a drain to prevent pressure variations. Half full bore is a more conservative design, allowing ample space for future connections and extensions to the system.

The relationship between drain capacity or proportional depth of flow, velocity of flow (m/s) and discharge (m³/s) is represented in graphical format:

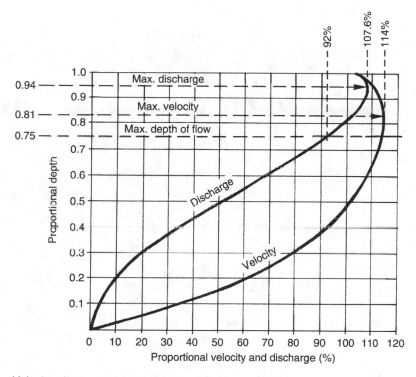

Velocity, discharge and proportional depth

Taking the example on page 403, the drain is designed to flow at half full bore with a flow velocity of 0·8m/s and discharge of 0·052m³/s. If at some later date, additional buildings are connected to the drainage system to produce an anticipated flow of up to 0·75 proportional depth, the graph indicates revised relative velocity and discharge rates of 114% or 0·912m/s and 92% or 0·048m³/s, respectively.

Drainage Design – Gradient (1)

The fall, slope or inclination of a drain or sewer will relate to the velocity of flow and the pipe diameter. The minimum diameter for surface water and foul water drains is 75 mm and 100 mm, respectively.

Maguire's rule of thumb is an established measure of adequate fall on drains and small sewers. Expressing the fall as 1 in x, where 1 is the vertical proportion to horizontal distance x, then:

$$x = \text{pipe diameter in mm} \div 2.5$$

E.g. a 150 mm nominal bore drainpipe:

$$x = 150 \div 2.5 = 60, \text{ i.e. 1 in 60 minimum gradient.}$$

Pipe dia.(mm)	Gradient
100	1 in 40
150	1 in 60
225	1 in 90
300	1 in 120

For full and half bore situations, these gradients produce a velocity of flow of about 1·4 m/s.

The Building Regulations, Approved Documents H1 and H3, provide guidance on discharge capacities for surface water drains running full and foul water drains running 0·75 proportional depth. The chart below is derived from these data:

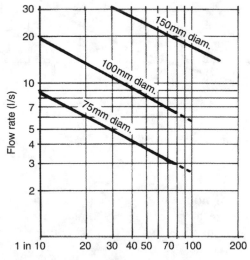

Notes: (1) – – – = Rainwater only.
(2) 75 mm is for rainwater or waste water (no soil) only.
(3) To convert l/s to m³/s, divide by 1000.

An alternative approach to drainage design is attributed to the established fluid flow research of Antoine Chezy and Robert Manning. This can provide lower gradients:

Chezy's formula: $V = C\sqrt{m \times i}$

where V = velocity of flow (min. 0.75m/s)

C = Chezy coefficient

m = HMD (see page 400)

i = inclination or gradient as 1/X or 1 ÷ X.

Manning's formula: $C = (1 \div n) \times (m)^{\frac{1}{6}}$

where

C = Chezy coefficient

n = coefficient for pipe roughness 0.010*

m = HMD

$\frac{1}{6}$ = sixth root

*A figure of 0.010 is appropriate for modern high-quality uPVC and clay drainware – for comparison purposes, it could increase to 0.015 for a cast concrete surface.

E.g. A 300mm (0.3m) nominal bore drainpipe flowing 0.5 proportional depth (half full bore). The Chezy coefficient can be calculated from Manning's formula:

$$HMD = 0.3 \div 4 = 0.075 \text{ (see page 400)}$$

$$C = (1 \div n) \times (m)^{\frac{1}{6}}$$

$$C = (1 \div 0.010) \times (0.075)^{\frac{1}{6}} = 65$$

Using a velocity of flow shown on the previous page of 1.4m/s, the minimum gradient can be calculated from Chezy's formula:

$$V = C\sqrt{m \times i}$$

$$1.4 = 65\sqrt{0.075 \times i}$$

$$(1.4 \div 65)^2 = 0.075 \times i$$

$$0.00046 \div 0.075 = i$$

$$i = 0.00617$$

$$i = 1 \div X$$

So, X = 1 ÷ 0.00617 = 162, i.e. 1 in 162

Drainage Design – Derivation of Formulae

Chezy formula – attributed to experiments that determined the relationship between fluid flow in open channels relative to the velocity of flow. In 1775, the Frenchman, Antoine Chezy, published his formula which has since become the practical basis for drainage design calculations:

$$V = C\sqrt{m \times i}$$

Chezy's C cannot be evaluated as a pure number or as a constant. It has the dimensions of acceleration, i.e. $\sqrt{L \times (T^{-1})}$ where L is the length of drain run and T is time. Chezy's C is therefore a coefficient affected by the hydraulic mean depth (m) and the pipe surface roughness (n).

Kutter and Ganguillet formula – these Swiss engineers determined a factor for channel and pipe surface roughness (n) for a variety of materials ranging from glass (0·009) to rough timber (0·160). For purposes of modern pipe materials, these values or coefficients range from 0·010 to 0·015 as qualified on the previous page. On the basis of their research, in 1869 Kutter and Ganguillet produced a formula for evaluating Chezy's C:

$$C = \frac{23 + (0 \cdot 00155 \div s) + (1 \div n)}{1 + (23 + (0 \cdot 00155 \div s)) \times (n \div \sqrt{m})}$$

where s is expressed as the sine of the bed slope or drain inclination.

Manning formula – following the earlier work of Kutter and Ganguillet, in 1888 the Irish engineer Robert Manning produced his much simpler formula for the Chezy coefficient:

$$C = (1 \div n) \times (m)^{\frac{1}{6}}$$

The value of pipe surface roughness (n) is attributed to Kutter and Ganguillet. It is often referred to as Kutter's n.

Although extremely dated and some being quite empirical in their formulation, these formulae have stood the test of time. They are still favoured by engineers and drainage designers and continue to feature in research and product development.

Small drainage schemes:

< 20 dwellings, 100mm nom. bore pipe, min. gradient 1 in 80.

20–150 dwellings, 150mm nom. bore pipe, min. gradient 1 in 150.

Minimum size for a public sewer is 150mm. Most water authorities will require a pipe of at least 225mm to allow for future developments and additions to the system.

For other situations, estimates of foul water flow may be based on water consumption of 225 litres per person per day. A suitable formula for average flow would be:

$$l/s = \frac{\text{Half consumption per person per day}}{6 \text{ hours} \times 3600 \text{ seconds}}$$

Note: 6 hours is assumed for half daily flow.

E.g. A sewer for an estate of 500, four-person dwellings:

$$l/s = \frac{112 \times 4 \times 500}{6 \times 3600} = 10.4$$

Assuming maximum of five times average flow – 52l/s or 0.052m³/s.

Using the formula Q = V × A (see page 397) with a velocity of flow of, say, 0.8m/s flowing half full bore (0.5 proportional depth):

$$Q = 0.052 \text{m}^3/\text{s}$$
$$V = 0.8 \text{m/s}$$
$$A = \text{half bore (m}^2)$$

Transposing the formula:

$$A = Q \div V$$
$$A = 0.052 \div 0.8 = 0.065 \text{m}^2$$

A represents half the bore, therefore the full bore area = 0.130m².

Area of a circle (pipe) = πr^2, therefore πr^2 = 0.130

$$\text{Transposing: } r = \sqrt{0.130 \div \pi}$$
$$r = 0.203 \text{m radius}$$
$$\text{Therefore diameter} = 0.406 \text{m or } 406 \text{mm}$$
$$\text{Nearest commercial size is } 450 \text{mm nominal bore.}$$

Drainage Design – Foul Water (2)

An alternative approach to estimating drain and sewer flows is by summation of discharge units and converting these to a suitable pipe size. Discharge units represent frequency of use and load producing properties of sanitary appliances. They are derived from data in BS EN 12056-2 and BS EN 752, standards for drainage systems inside and outside buildings, respectively. Although intended primarily for sizing discharge stacks, they are equally well applied to drains and sewers.

Appliance	Situation	No. of units
WC	Domestic	7
	Commercial	14
	Public	28
Basin	Domestic	1
	Commercial	3
	Public	6
Bath	Domestic	7
	Commercial	18
Sink	Domestic	6
	Commercial	14
	Public	27
Shower	Domestic	1
	Commercial	2
Urinal		0·3
Washing machine		4–7
Dishwasher		4–7
Waste disposal unit		7
Group of WC, bath and one or two basins		14
Other fittings with an outlet of:		
	50mm nom. i.d.	7
	65mm nom. i.d.	7
	75mm nom. i.d.	10
	90mm nom. i.d.	10
	100mm nom. i.d.	14

Note:

Domestic = houses and flats.

Commercial = offices, factories, hotels, schools, hospitals, etc.

Public or peak = cinemas, theatres, stadia, sports centres, etc.

Using the example from page 403, i.e. 500, four-person dwellings. Assuming one WC, one shower, two basins, two sinks, one group of appliances, washing machine and dishwasher per dwelling.

WC	7	discharge units
Shower	1	discharge unit
Basins	2	discharge units
Sinks	12	discharge units
Group	14	discharge units
Washing machine	4	discharge units
Dishwasher	4	discharge units

Total = 44 discharge units × 500 dwellings = 22000 discharge units.

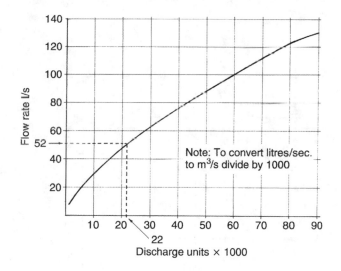

Sewer size can be calculated for a 0·052m³/s flow at half full bore using the formula Q = V × A as shown on page 403. Gradient can be calculated using the Chezy and Manning formulas as shown on page 401.

Combined surface and foul water drains will require separate calculations for both flow conditions. Drain size can be based on the greater flow, not the total flow, as the chance of the peak flows of both coinciding is remote.

See pages 480 and 481 for alternative 'K' factor method of drainage design.

Drainage Design – Gradient and Velocity

Adaptation of Kutter and Ganguillet´s formulations for pipe sizes, gradients and velocity where n = 0.013 and HMD is diameter ÷ 4, i.e. half and full bore.

Example shown is for a flow velocity of 1.4m/s in a 300mm pipe laid to a gradient of 1 in 90. The same flow velocity in a 150mm pipe requires a gradient of 1 in 29. A 150mm pipe at 1 in 90 has a velocity of 0.8m/s.

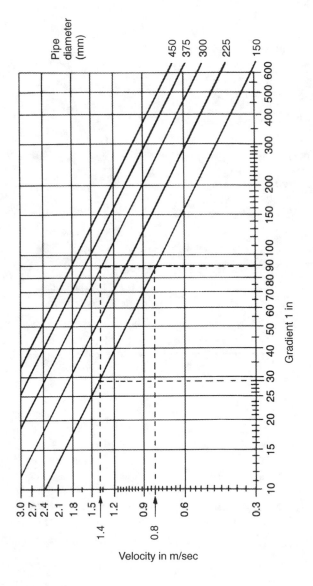

Adaptation of Kutter and Ganguillet's formulations for pipe sizes, gradients and discharge rates where n = 0.013 and HMD is diameter ÷ 4, i.e. half and full bore.

Examples shown are from the previous page, with pipes of 150mm dia. graded at 1 in 29 and 300mm dia. at 1 in 90.

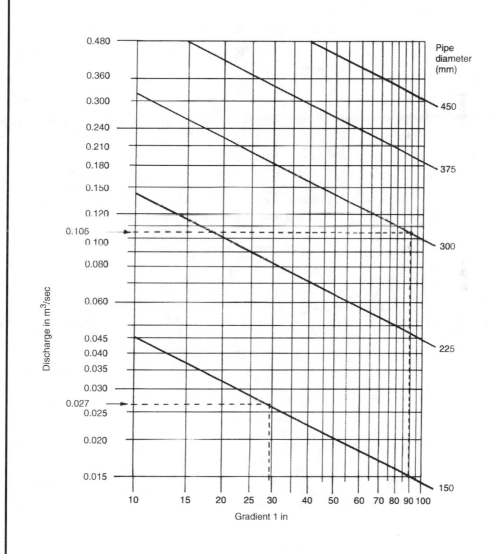

Crimp and Bruges' formula – during the mid-1890s, William Crimp and Charles Bruges formulated an adaptation of Kutter and Ganguillet's earlier work on fluid flow analysis. They produced a simplified evaluation of flow velocity in drains and sewers based on the following:

$$v = C\ m^{0.67}\sqrt{i}$$

where
v = velocity of flow in m/s

C = constant; 84 is used in the following calculations as this complements the HMD proportional flows indicated on page 398

m = Hydraulic Mean Depth

i = inclination or gradient as 1 in X

Note: C, not compared with or confused with Chezy's constant.

$HMD^{0.67}$ or $m^{0.67}$ can also be expressed as $\sqrt[3]{(m)^2}$

Therefore, $\quad v = C\ \sqrt[3]{(m)^2}\ \sqrt{i}$

E.g. 1. Taking the example on page 401 of a 300mm drain flowing half bore at a minimum gradient of 1 in 162 with a flow velocity of 1.4m/s. Using Crimp and Bruge's formula, the velocity of flow can be compared:

$$v = 84\ \sqrt[3]{(0.075)^2}\ \sqrt{1 \div 162}$$
$$v = 84 \times 0.1778 \times 0.0786 = 1.18\text{m/s (slightly less)}$$

E.g. 2. Using the formula to calculate the minimum gradient for a 300mm pipe flowing half bore discharging at a flow rate of 0.8m/s.

$$0.8 = 84\ \sqrt[3]{(0.075)^2}\ \sqrt{1 \div X}$$
$$(0.8 \div 84 \times 0.1778)^2 = 1 \div X$$
$$X = 1 \div 0.00267 = 348$$

Therefore, a minimum gradient of 1 in 348.

E.g. 3. A 150mm pipe flowing 0.75 depth of flow at 0.8m/s:

$$0.8 = 84\ \sqrt[3]{(0.045)^2}\ \sqrt{1 \div X} \qquad \text{[Note: m = pipe dia. } \div 3.33]$$
$$(0.8 \div 84 \times 0.126)^2 = 1 \div X$$
$$X = 1 \div 0.00567 = 176$$

Therefore, a minimum gradient of 1 in 176.

C.F. Colebrook and C.M. White formula – an equation devised in the late 1930s based on theory and experimentation of partial flows in drains and sewers:

$$V = -2\sqrt{(2gs\psi D)} \log [(K + 3.7\psi D) + (\sqrt{2.51v} + \psi D\sqrt{(2gs\psi D)})]$$

where

V = velocity of flow (m/s)

g = gravitational acceleration ($9.81m/s^2$)

s = slope or pipe gradient (1 in X)

ψ = correction factor for flows less than half full

D = diameter of pipe

K = roughness coefficient 0.6 (rainfall) 1.5 (foul water)

v = kinematic viscosity

As can be seen, the Colebrook/White calculation is quite extensive and very time-consuming to use. Fortunately, their formulations are available graphically and the reader is referred to the Thomas Telford publication *Gravity Flow Pipe Design Charts* by D. Butler and B.R.C. Pinkerton, where numerous charts are available for a wide range of pipe sizes and applications.

Since the mid-twentieth century, there have been many more theories and research applications into fluid flow and hydraulic conditions in drainage pipes. Consistency of results remains somewhat elusive as can be seen from the outcome of various calculations over the preceding pages. Therefore, no new theory is propounded here, nor any of the formulae used suggested as best. They are for guidance and reference only as a basis for further study.

Factors outside calculation and laboratory experimentation will inevitably affect flow conditions in drains. These may include the quality of workmanship during installation, irregular settlement of pipes, leakage, possible ingress of tree roots and later additions to the system. With regard to these, it is for drainage consultants and engineers to determine their own design criteria based on theoretical guidance and personal experience, and to produce specifications with regard to the practicalities of site characteristics.

Refuse Chute

The quantity and location of refuse chutes depends upon:

- layout of the building
- number of dwellings served – max. six per hopper
- type of material stored
- frequency of collection
- volume of refuse
- refuse vehicle access – within 25m.

The chute should be sited away from habitable rooms, but not more than 30m horizontal distance from each dwelling. It is more economical to provide space for additional storage beneath the chute, than to provide additional chutes. Chute linings are prefabricated from refractory or Portland cement concrete with a smooth and impervious internal surface. The structure containing the chute void should have a fire resistance of one hour. The refuse chamber should also have a one hour fire resistance and be constructed with a dense impervious surface for ease of cleaning.

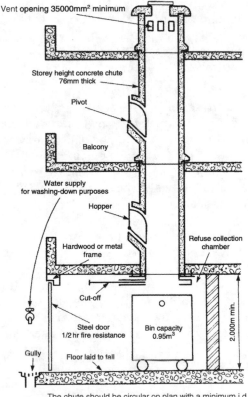

Vent opening 35000mm² minimum

Storey height concrete chute 76mm thick

Pivot

Balcony

Water supply for washing-down purposes

Hopper

Refuse collection chamber

Hardwood or metal frame

Cut-off

Steel door 1/2 hr fire resistance

Bin capacity 0.95m³

2.000m min.

Gully

Floor laid to fall

The chute should be circular on plan with a minimum i.d. of 450mm

Ref. BS 5906: Waste management in buildings. Code of practice.

410

This system has a flue to discharge the incinerated gaseous products of combustion above roof level. A fan ensures negative pressure in the discharge chute to prevent smoke and fumes from being misdirected. A large combustion chamber receives and stores the refuse until it is ignited by an automatic burner. Duration of burning is thermostatically and time controlled. Waste gases are washed and cleaned before discharging into the flue. There is no restriction on wet or dry materials, and glass, metal or plastics may be processed.

Health risks associated with storing putrefying rubbish are entirely eliminated as the residue from combustion is odourless and sterile. Refuse removal costs are reduced because the residual waste is only about 10% of the initial volume.

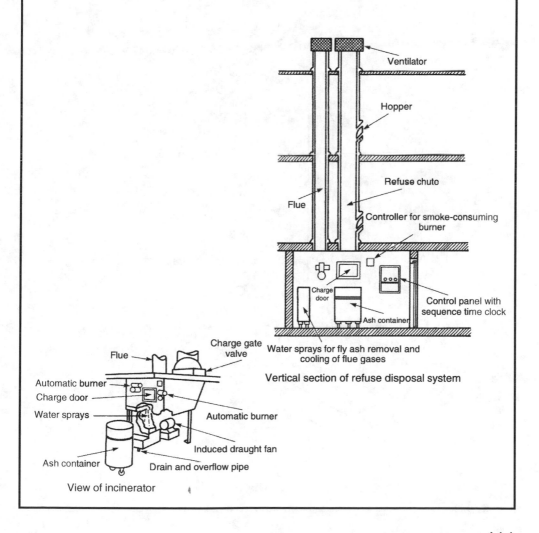

Ventilator

Hopper

Refuse chute

Flue

Controller for smoke-consuming burner

Charge door

Control panel with sequence time clock

Ash container

Charge gate valve

Water sprays for fly ash removal and cooling of flue gases

Vertical section of refuse disposal system

Flue

Automatic burner

Charge door

Water sprays

Automatic burner

Induced draught fan

Ash container

Drain and overflow pipe

View of incinerator

411

Sanitary Incineration

Incinerators are the quickest, easiest and most hygienic method for disposing of dressings, swabs and sanitary towels. They are usually installed in office lavatories, hospitals and hotels. When the incinerator door is opened, gas burners automatically ignite and burn the contents. After a predetermined time, the gas supply is cut off by a time switch. Each time the door is opened, the time switch reverts to its original position to commence another burning cycle. Incinerators have a removable ash pan and a fan-assisted flue to ensure efficient extraction of the gaseous combustion products. In event of fan failure, a sensor ensures that gas burners cannot function. The gas pilot light has a thermocoupled flame failure device.

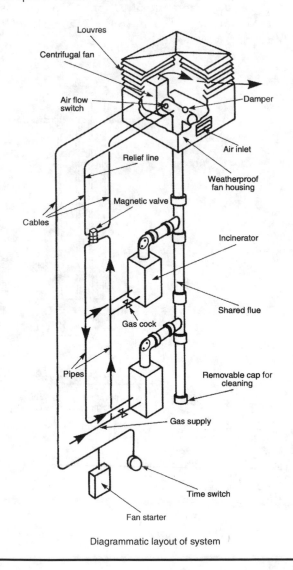

Diagrammatic layout of system

Food waste, bottles, cans and cartons are disposed of at source, without the need to grind or crush the refuse. A bowl beneath the sink retains the normal waste water. Refuse is placed inside a central tube in the sink. When the tube is raised the waste-water and the refuse are carried away down a stack or discharge pipe to a chamber at the base of the building. Refuse from the chamber is collected at weekly intervals by a specially equipped tanker in which the refuse is compacted into a damp, semi-solid mass that is easy to tip. One tanker has sufficient capacity to contain the refuse from up to 200 dwellings. Waste-water from the tanker is discharged into a foul water sewer.

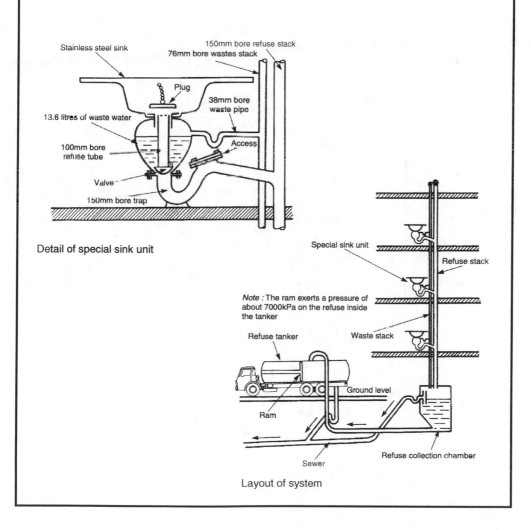

Detail of special sink unit

Layout of system

413

Pneumatic Transport of Refuse

Refuse from conventional chutes is collected in a pulveriser and disintegrated by grinder into pieces of about 10mm across. The refuse is then blown a short distance down a 75mm bore pipe in which it is retained, until at predetermined intervals a flat disc valve opens. This allows the small pieces of refuse to be conveyed by vacuum or air stream at 75 to 90km/h through a common underground service pipe of 150–300mm bore. The refuse collection silo may be up to 2·5km from the source of refuse. At the collection point, the refuse is transferred by a positive pressure pneumatic system to a treatment plant where dust and other suspended debris is separated from bulk rubbish. The process can be adapted to segregate salvagable materials such as metals, glass and paper.

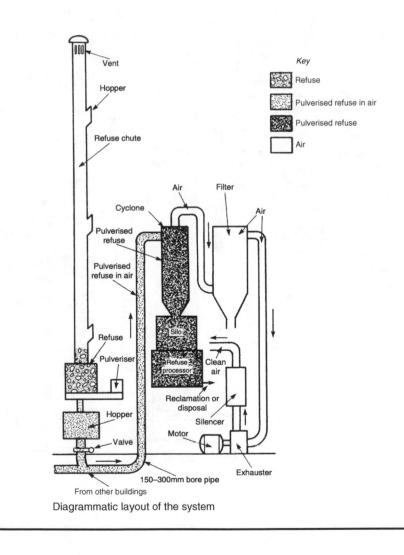

Diagrammatic layout of the system

Food waste disposal units are designed for application to domestic and commercial kitchen sinks. They are specifically for processing organic food waste and do not have the facility to dispose of glass, metals, rags or plastics. Where a chute or Garchey system is not installed, these units may be used to reduce the volume otherwise deposited in dustbins or refuse bags.

Food waste is fed through the sink waste outlet to the unit. A grinder powered by a small electric motor cuts the food into fine particles which is then washed away with the waste water from the sink. The partially liquefied food particles discharge through a standard 40mm nominal bore waste pipe into a back inlet gully. As with all electrical appliances and extraneous metalwork, it is essential that the unit and the sink are earthed.

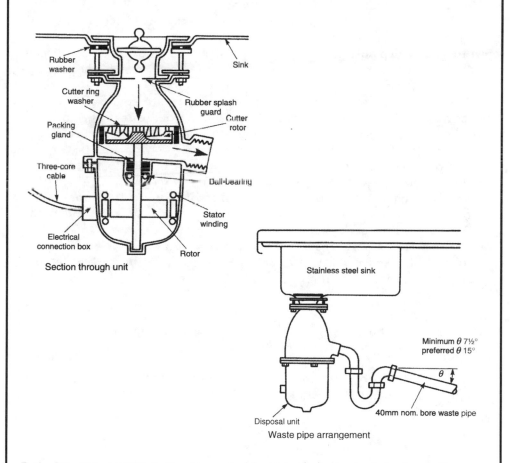

Section through unit

Waste pipe arrangement

Ref. BS EN 60335-2-16: Household and similar electrical appliances. Safety, Particular requirements for food waste disposers.Garage Drainage

415

Food Waste Recycling

An alternative to pre-disposal bulk collection within buildings and to individual food waste disposal units connected to the sanitation and drainage system. Mainly intended for use by occupants of dwellings to separate their food waste into small-sealed containers. Commercial premises are serviced similarly but with much larger waste receptive containers. Collected weekly and removed for recycling through bio-collection and processing plant.

Process sequence:

- Food waste is separated from general waste and deposited in purpose-made sealed boxes for weekly collection.
- Recycling contractors empty waste boxes into purpose-made mobile units that collect community food waste.
- Delivery of accumulated food waste to a local depot with an anaerobic digestion facility.
- Plastic wrappings, packaging and other non-processable items removed by pre-filtering, although these items should have been removed previously.
- Remaining volume of size reduced food waste pumped through temperature regulated collection tanks where micro-organisms/ anaerobic bacteria break down the waste by digestion. See pages 383–385, 731–734 and 752.
- Approximately 3 days allowed for the waste to change state into a biogas (mainly methane).
- Gas is cleaned by filtration and monitored for quality before being introduced as a supplement to the national gas supply though the gas distribution grid.
- The remaining non-gaseous effluent is dried as a digestate or compost for soil conditioning. Sold bulk for agricultural purposes or bagged as a fertiliser for garden use.

Ref. BS PAS 110: Specification for whole digestate, separated liquor and separated fibre derived from the anaerobic digestion of source-segregated biodegradable materials.

Page 752 illustrates the principle of food waste processing into a biogas and associated plant.

9 SANITARY FITMENTS AND APPLIANCES: DISCHARGE AND WASTE SYSTEMS

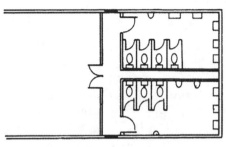

DOI: 10.1201/9781003434894-9

Flushing Cisterns

Bell type – this form of flushing cistern is now virtually obsolete, although some reproductions are available for use in keeping with the refurbishment of historic premises. Cast iron originals may still be found in use in old factories, schools and similar established buildings. It is activated by the chain being pulled which also lifts the bell. As the chain is released, the bell falls to displace water down the stand pipe, effecting a siphon which empties the cistern. The whole process is relatively noisy.

Disc type – manufactured in a variety of materials including plastics and ceramics for application to all categories of building. Depressing the lever raises the piston and water is displaced over the siphon. A siphonic action is created to empty the cistern. Some cisterns incorporate an economy or dual-flush siphon. When the lever is depressed and released promptly, air passing through the vent pipe breaks the siphonic action to give a 4.5-litre flush. When the lever is held down, a 7.5-litre flush is obtained. Since 2001, the maximum permitted single flush to a WC pan is 6 litres.

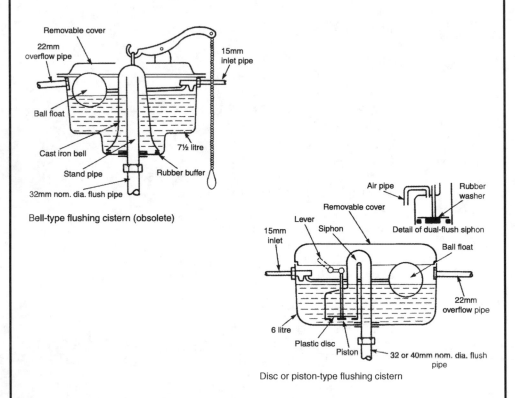

Bell-type flushing cistern (obsolete)

Disc or piston-type flushing cistern

Refs. BS 1125 and 7357: Specifications for WC flushing cisterns.
The Water Supply (Water Fittings) Regulations 1999.

418

A flushing trough may be used as an alternative to several separate flushing cisterns where a range of WCs are installed. They are particularly applicable to school, factory and office sanitary accommodation. Trough installation is economic in equipment and time. It is also more efficient in use as there is no waiting between consecutive flushes. The disadvantage is that if it needs maintenance or repair, the whole range of WCs are unusable. The trough may be bracketed from the rear wall and hidden from view by a false wall or ceiling.

The siphon operates in the same manner as in a separate cistern, except that as water flows through the siphon, air is drawn out of the air pipe. Water is therefore siphoned out of the anti-siphon device, the flush terminated and the device refilled through the small hole.

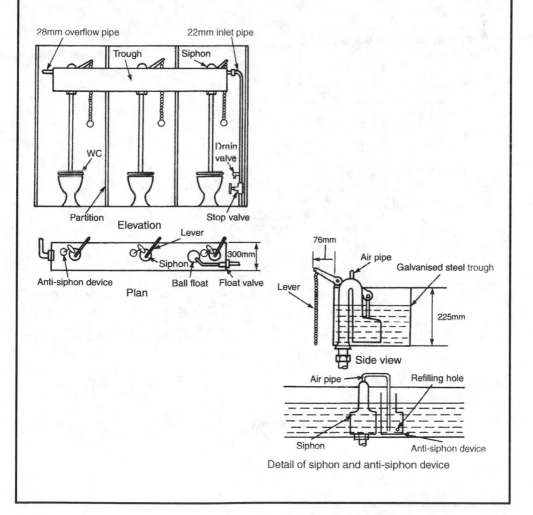

Detail of siphon and anti-siphon device

Automatic Flushing Cisterns

Roger Field's flushing cistern is used for automatically flushing WCs. It has application to children's lavatories and other situations where the users are unable to operate a manual flush device. As the cistern fills, air in the stand pipe is gradually compressed. When the head of water 'H' is slightly above the head of water 'h', water in the trap is forced out. Siphonic action is established, and the cistern flushes the WC until air enters under the dome to break the siphon.

With the smaller urinal flush cistern, water rises inside the cistern until it reaches an air hole. Air inside the dome is trapped and compressed as the water rises. When water rises above the dome, compressed air forces water out of the 'U' tube. This lowers the air pressure in the stand pipe, creating a siphon to empty the cistern. Water in the reserve chamber is siphoned through the siphon tube to the lower well.

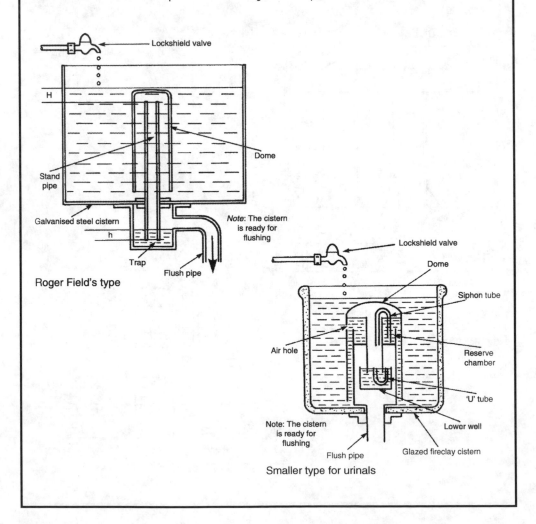

Roger Field's type

Smaller type for urinals

Flushing valves are a more compact alternative to flushing cisterns, often used in marine applications, but may only be used in buildings with approval of the local water authority. The device is a large equilibrium valve that can be flushed at any time without delay, provided there is a constant source of water from a storage cistern. The minimum and maximum head of water above valves is 2·2 m and 36 m, respectively. When the flushing handle is operated, the release valve is tilted and water is displaced from the upper chamber. The greater force of water under piston 'A' lifts valve 'B' from its seating and water flows through the outlet. Water flows through the bypass and refills the upper chamber to cancel out the upward force acting under piston 'A'. Valve 'B' closes under its own weight.

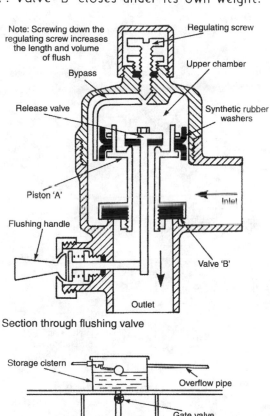

Note: Screwing down the regulating screw increases the length and volume of flush

Regulating screw

Upper chamber

Bypass

Release valve

Synthetic rubber washers

Piston 'A'

Inlet

Flushing handle

Valve 'B'

Outlet

Section through flushing valve

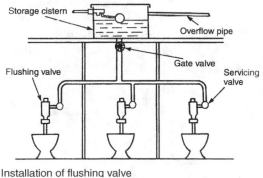

Storage cistern

Overflow pipe

Gate valve

Flushing valve

Servicing valve

Installation of flushing valve

421

Flushing Valve – Installation

- The minimum flow rate at an appliance is 1·2 litres per second. By domestic standards, this is unrealistically high; therefore, pressure flushing valves are not permitted in houses.
- Where connected to a mains supply pipe or a cistern distributing pipe, a flushing valve must include a backflow prevention device having a permanently vented pipe interrupter situated at least 300 mm above the spill-over level of the served WC.
- If a permanently vented pipe interrupter is not fitted, the water supply to a flushing valve must be from a dedicated cistern with an air gap (see page 53) below its float valve delivery.
- The maximum flush in a single operation is 6 litres.
- Flushing valves may be used to flush urinals. In this situation, they should deliver no more than 1·5 litres of water to each bowl or position per operation. See page 447.

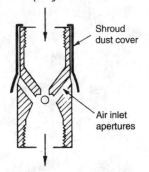

Shroud
dust cover

Air inlet
apertures

Pipe interrupter with a permanent
atmospheric vent

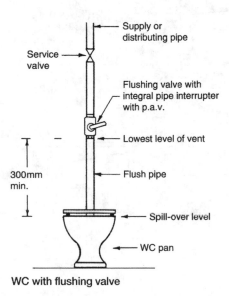

Supply or
distributing pipe

Service
valve

Flushing valve with
integral pipe interrupter
with p.a.v.

Lowest level of vent

300mm
min.

Flush pipe

Spill-over level

WC pan

WC with flushing valve

The washdown WC pan is economic, simple and efficient. It rarely becomes blocked and can be used in all types of buildings with colour variations to suit internal decor. Manufacture is primarily from vitreous china, although glazed fireclay and stoneware have been used. Stainless steel WCs can be specified for use in certain public areas and prisons. Pan outlet may be horizontal, 'P', 'S', left- or right-handed. Horizontal outlet pans are now standard, with push-fit adaptors to convert the pan to whatever configuration is required. Plastic connectors are commonly used for joining the outlet to the soil branch pipe. The flush pipe joint is usually made with a rubber cone connector which fits tightly between WC and pipe.

WC pan outlet <80mm, trap diameter = 75mm

WC pan outlet >80mm, trap diameter = 100mm

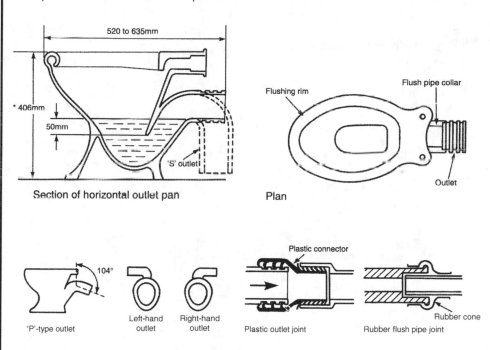

Section of horizontal outlet pan   Plan

'P'-type outlet   Left-hand outlet   Right-hand outlet   Plastic outlet joint   Rubber flush pipe joint

*Note: Add approximately 25mm to the top of the WC to allow for seat height. Overall height for the disabled is 480mm, junior schoolchildren 355mm and infants 305mm.

Refs. BS EN 997: WC pans and WC suites with integral trap.
BS EN 33: WC pans and WC suites. Connecting dimensions.

Siphonic Water Closets

Siphonic WCs are much quieter in operation than washdown WCs and they require less flush action to effect an efficient discharge. They are not suitable for schools, factories and public buildings as they are more readily blocked if not used carefully.

The double-trap type may be found in house and hotel bathrooms. When flushed, water flows through the pressure-reducing fitting 'A'. This reduces the air pressure in chamber 'B'. Siphonic action is established, and the contents of the first trap are removed. This is replenished from reserve chamber 'C'.

The single-trap variant is simpler and has limited application to domestic bathrooms. When the cistern is flushed, the content is discharged through the trap and restricted in flow by the specially shaped pan outlet pipe. The pipe fills with water which causes a siphonic effect. Sufficient water remains in the reserve chamber to replenish the seal.

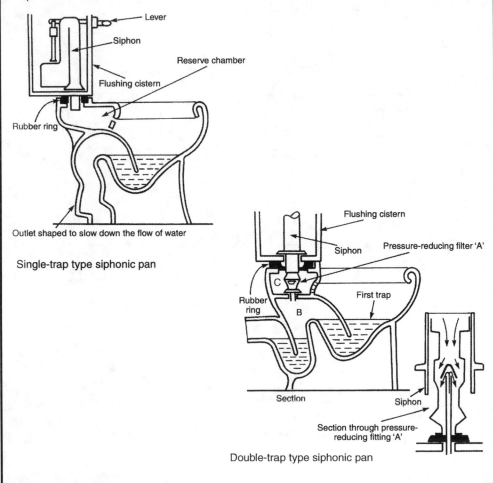

Single-trap type siphonic pan

Double-trap type siphonic pan

A bidet is classified as a waste fitting. The requirements for a discharge pipe from a bidet may therefore be treated in the same manner as a basin waste of the same diameter – nominally 32mm. It is an ablutionary fitting used for washing the excretory organs, but may also be used as a foot bath. Hot and cold water supplies are mixed to the required temperature for the ascending spray. For greater comfort, the rim of the fitting may be heated from the warm water. Ascending spray-type bidets are not favoured by the water authorities because the spray nozzle is below the spill level, risking water being back-siphoned into other draw-off points. This is prevented by having independent supply pipes to the bidet which are not connected to any other fittings. A further precaution would be installation of check valves on the bidet supply pipes or a thermostatic regulator with integral check valves. Over-the-rim hot and cold supplies are preferred. These have an air gap requirement between rim and tap outlets, (see next page and page 53).

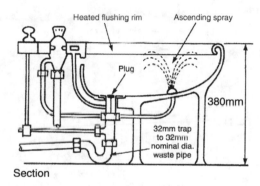

Section

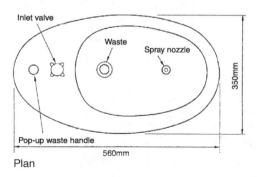

Plan

Note: Size and profile vary between manufacturers. Dimensions shown are representative.

Ref. BS EN 14528: Bidets. Functional requirements and test methods.

Bidets – Installation

Over-the-rim – bidets in domestic plumbing systems may have cold and hot water supplied through individual taps, or a combined mixer tap. There must be a distinct air gap between the tap water outlet and the spillover level (rim) of the appliance. This gap requirement is 20mm for standard 12mm (1/2") taps (see also page 53).

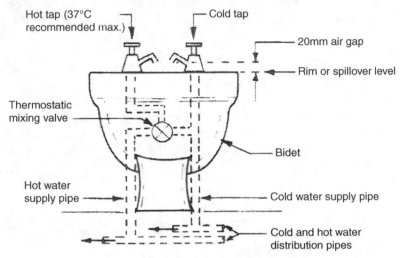

Ascending spray –

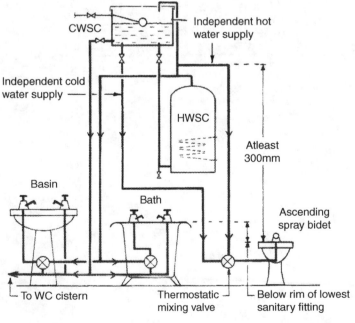

Ref. BS EN 35: Pedestal and wall-hung bidets with over-rim supply. Connecting dimensions.

A shower is more economical to use than a bath as it takes less hot water (about one-third), it is arguably more hygienic and it takes up less space. The mixing valve should be thermostatic (see pages 440–443) to avoid the risk of scalding. A minimum 1m head of water should be allowed above the shower outlet. If this is impractical, a pumped delivery could be considered (see next page). The shower outlet (rose) should also be at least 2m above the floor of the shower tray. Supply pipes to individual showers are normally 15mm o.d. copper or equivalent. These should incorporate double-check valves if there is a flexible hose to the rose, as this could be left in dirty tray water which could back-siphon. An exception to check valves is where the shower head is fixed and therefore well above the air gap requirements and spill-over level of the tray.

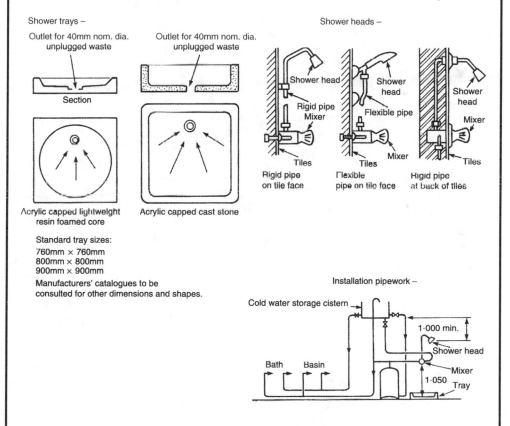

Shower trays –

Outlet for 40mm nom. dia. unplugged waste

Outlet for 40mm nom. dia. unplugged waste

Section

Acrylic capped lightweight resin foamed core

Acrylic capped cast stone

Standard tray sizes:
760mm × 760mm
800mm × 800mm
900mm × 900mm
Manufacturers' catalogues to be consulted for other dimensions and shapes.

Shower heads –

Shower head
Rigid pipe
Mixer
Tiles
Rigid pipe on tile face

Shower head
Flexible pipe
Mixer
Tiles
Flexible pipe on tile face

Shower head
Mixer
Tiles
Rigid pipe at back of tiles

Installation pipework –

Cold water storage cistern
1·000 min.
Shower head
Bath   Basin
Mixer
1·050   Tray

Refs. BS EN 251: Shower trays. Connecting dimensions.

BS EN 263: Sanitary appliances. Cross-linked cast acrylic sheets for baths and shower trays for domestic purposes.

BS EN 14428: Shower enclosures.

Pumped Showers – 1

Where the 1m minimum head of water above the shower outlet is not available and it is impractical to raise the level of the supply cistern, a pump can be fitted to the mixer outlet pipe or on the supply pipes to the mixer. The pump is relatively compact and small enough to be installed on the floor of an airing cupboard or under the bath. It must be accessible for maintenance, as the pump should be installed with filters or strainers which will require periodic attention, particularly in hard water areas. The pump will operate automatically in response to the shower mixer being opened. A pressure sensor and flow switch detect water movement to activate the pump and vice versa. Electricity supply can be from an isolating switch or pull cord switch with a 3 amp fuse overload protection spurred off the power socket ring main.

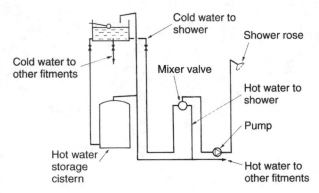

Pump on mixed water outlet

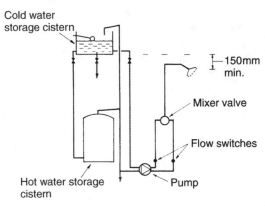

Pumped supply to mixer

Note: Double-check valves may be required on the supply pipes as described on the previous page. The mixing valve and pump may incorporate check valves – refer to manufacturer's information.

Minimum cold water storage 230 litres per bathroom, 365 litres for one bathroom and an en-suite shower room. Where two or more bathrooms are provided, a 28mm o.d. min. cold feed pipe to the HWSC should be used. Water supplies to the pump can be the first tee-branch connection, but as shown below an independent arrangement is preferable.

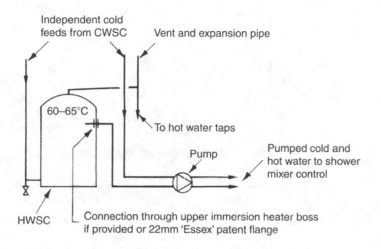

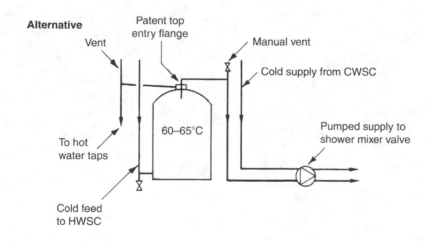

Note: Water supply to a shower pump is not to be taken direct from the mains.

Other applications –

Pump located above the hot water connection to the hot water storage cylinder, e.g. in a roof void.

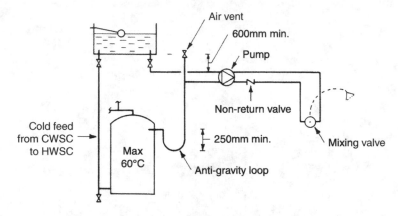

Note: With anti-gravity loop hot water supply, the stored water temperature is limited to reduce the possibility of gravity circulation and aeration.

Pumps provided for a multiple shower situation, e.g. sports complex.

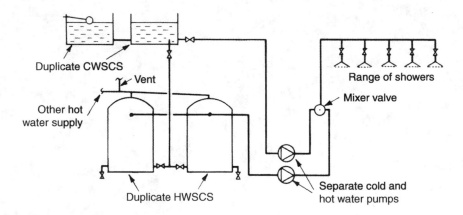

Note: For all types of installation, the pump manufacturer's data should be consulted for recommended pipe sizes. For most domestic situations, the gravity feed is 22mm o.d. copper and the pumped supply 15mm o.d. copper.

Instantaneous electric water heating for showers is an economical, simple-to-install alternative to a pumped shower. This is particularly apparent where there would otherwise be a long secondary flowpipe from the hot water storage cylinder to the shower outlet, possibly requiring additional secondary return pipework to avoid a long 'dead leg'. Cold water supply is taken from the rising main in 15mm o.d. copper tube. This will provide a regulated delivery through the shower unit of up to 3 litres/min. The unit contains an electric element, usually of 7·2 or 8·4kW rating. It also has a number of built-in safety features:

- Automatic low-pressure switch to isolate the element if water pressure falls significantly or the supply is cut off.
- Thermal cut-off. This is set by the manufacturer at approximately 50°C to prevent the water from overheating and scalding the user.
- Non-return or check valve on the outlet to prevent back-siphonage.

Electricity supply is an independent radial circuit, originating at the consumer unit with a miniature circuit breaker (MCB) appropriately rated. Alternatively a suitable rated fuseway may be used in the consumer unit and added protection provided with an in line residual current device (RCD) trip switch. All this, of course, is dependent on there being a spare way in the consumer unit. If there is not, there will be additional expenditure in providing a new consumer unit or a supplementary fuse box. A doublepole cord-operated pull switch is located in the shower room to isolate supply.

Shower rating (kW)	Cable length (m)	Fuse of MCB rating (amps)	Cable size (mm^2)
7·2	<13	30 or 32	4
7·2	13–20	30 or 32	6
7·2	20–35	30 or 32	10
8·4	<17	40 or 45	6
8·4	17–28	40 or 45	10

Mains Fed, Electric Shower – 2

Unit detail and installation:

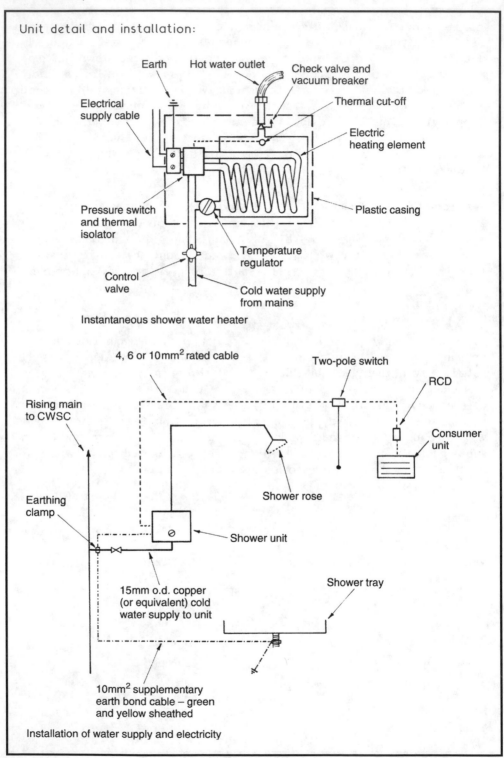

Earth

Hot water outlet

Check valve and vacuum breaker

Electrical supply cable

Thermal cut-off

Electric heating element

Pressure switch and thermal isolator

Plastic casing

Control valve

Temperature regulator

Cold water supply from mains

Instantaneous shower water heater

4, 6 or 10mm² rated cable

Two-pole switch

RCD

Rising main to CWSC

Consumer unit

Shower rose

Earthing clamp

Shower unit

15mm o.d. copper (or equivalent) cold water supply to unit

Shower tray

10mm² supplementary earth bond cable – green and yellow sheathed

Installation of water supply and electricity

432

Baths are manufactured in acrylic sheet, reinforced glass fibre, enamelled pressed steel and enamelled cast iron. The acrylic sheet bath has the advantage of light weight to ease installation, it is comparatively inexpensive and is available in a wide range of colours. However, special cleaning agents must be used, otherwise the surface can become scratched. It will require a timber support framework, normally laid across metal cradles. Traditional cast iron baths are produced with ornate feet and other features. Less elaborate, standard baths in all materials can be panelled in a variety of materials including plastic, veneered chipboard and plywood.

The corner bath is something of a luxury. It may have taps located to one side to ease accessibility. A Sitz bath is stepped to form a seat. It has particular application to nursing homes and hospitals for use with the elderly and infirm.

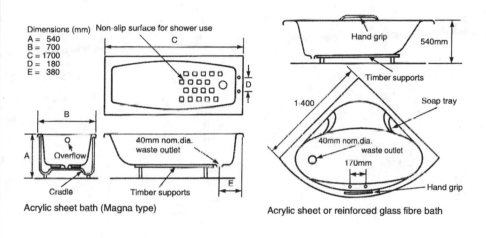

Acrylic sheet bath (Magna type)

Acrylic sheet or reinforced glass fibre bath

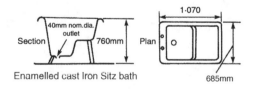

Enamelled cast Iron Sitz bath

Refs. BS 1189 and 1390: Specifications for baths made from porcelain enamelled cast iron and vitreous enamelled sheet steel, respectively.

BS 4305-2: Baths for domestic purposes made of acrylic material.

BS EN 232: Baths. Connecting dimensions.

BS EN 198: Sanitary appliances. Baths made from cross-linked cast acrylic sheets. Requirements and test methods.

Sinks

Sinks are designed for culinary and other domestic uses. They may be made from glazed fireclay, enamelled cast iron or steel, stainless steel or from glass fibre-reinforced polyester.

The Belfast sink has an integral weir overflow, and water may pass through this to the waste pipe via a slotted waste fitting. It may have a hardwood or dense plastic draining board fitted at one end only or a draining board fitted at each end. Alternatively, the sink may be provided with a fluted drainer of fireclay. The London sink has similar features, but it does not have an integral overflow. In recent years, sinks of this type have lost favour to surface built-in metal and plastic materials, but there is now something of a resurgence of interest in these traditional fittings. Stainless steel sinks may have single or double bowls, with left- or right-hand drainers or double drainers. These can be built into a work surface or be provided as a sink unit with cupboards under. The waste outlet is a standard 40 mm nominal diameter.

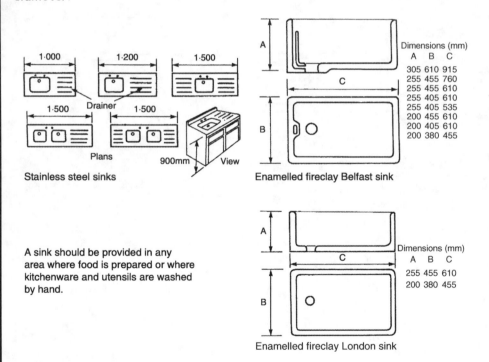

Stainless steel sinks

Enamelled fireclay Belfast sink

Dimensions (mm)

A	B	C
305	610	915
255	455	760
255	455	610
255	405	610
255	405	535
200	455	610
200	405	610
200	380	455

A sink should be provided in any area where food is prepared or where kitchenware and utensils are washed by hand.

Enamelled fireclay London sink

Dimensions (mm)

A	B	C
255	455	610
200	380	455

Refs. BS 1206: Specification for fireclay sinks, dimensions and workmanship.

BS EN 13310: Kitchen sinks. Functional requirements and test methods.

Building Regulations, Approved Document G6: Kitchens and food preparation areas.

These are rarely necessary in domestic situations, but have an application to commercial premises, schools, hospitals and similar public buildings. They are usually located inside the cleaning contractor's cubicle and are fitted at quite a low level to facilitate ease of use with a bucket. They are normally supported by built-in cantilever brackets and are additionally screwed direct to the wall to prevent forward movement. 13mm bore (half-inch) hot and cold water draw-off bib-taps may be fitted over the sink, at sufficient height for a bucket to clear below them. 19mm bore (three-quarter-inch) taps may be used for more rapid flow. A hinged stainless steel grating is fitted to the sink as a support for the bucket. The grating rests on a hardwood pad fitted to the front edge of the sink to protect the glazed finish. A 40mm nominal diameter waste pipe is adequate for this type of sink.

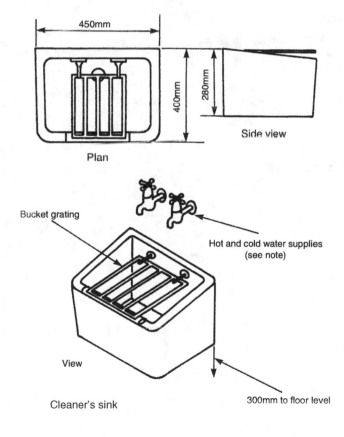

450mm

400mm

280mm

Side view

Plan

Bucket grating

Hot and cold water supplies
(see note)

View

Cleaner's sink

300mm to floor level

Note: For all sanitary fitments, the convention of hot tap to the left and cold tap to the right applies.

Wash Basins

There are various types of basin, ranging in size and function from hand rinsing to surgical use. A standard basin for domestic application to bathrooms and cloakrooms consists of a bowl, soap tray, weir overflow and holes for taps and outlet. It may be supported by cast iron brackets screwed to the wall, a corbel which is an integral part of the basin built into the wall or a floor pedestal which conceals the pipework. Water supply is through 13mm (half-inch) pillar taps for both hot and cold. A standard 32mm nominal diameter waste outlet with a slot to receive the integral overflow connects to a trap and waste pipe of the same diameter. A plug and chain normally controls outflow, but some fittings have a pop-up waste facility.

Most basins are made from coloured ceramic ware or glazed fireclay. There are also metal basins produced from stainless steel, porcelain enamelled sheet steel or cast iron.

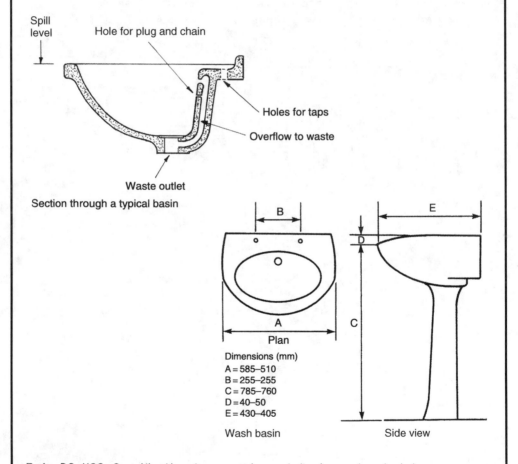

Spill level

Hole for plug and chain

Holes for taps

Overflow to waste

Waste outlet

Section through a typical basin

Plan

Dimensions (mm)
A = 585–510
B = 255–255
C = 785–760
D = 40–50
E = 430–405

Wash basin

Side view

Refs. BS 1188: Specification for ceramic wash basins and pedestals.

BS 5506-3: Specification for wash basins.

BS EN 31: Wash basins. Connecting dimensions.

436

Washing troughs are manufactured circular or rectangular on plan in ceramic materials or stainless steel. They are an economical and space-saving alternative to a range of basins, for use in factory, school and public lavatories. Some variations have an overall umbrella spray or fountain effect operated by a foot pedal. These are no longer favoured by the water supply undertakings as a trough must have a separate draw-off tap for every placement. In common with other sanitary fitments, there must be provision to prevent the possibility of back-siphonage, i.e. an adequate air gap between tap outlet and spill level of the trough. Hot and cold water supply to the taps is thermostatically blended to 40°C maximum.

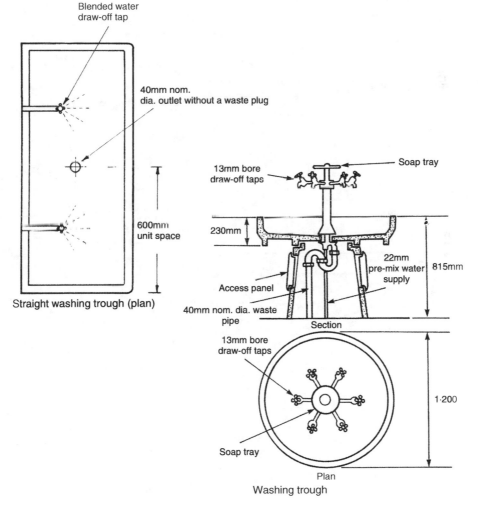

Blended water draw-off tap

40mm nom. dia. outlet without a waste plug

600mm unit space

Straight washing trough (plan)

13mm bore draw-off taps

Soap tray

230mm

Access panel

22mm pre-mix water supply

815mm

40mm nom. dia. waste pipe

Section

13mm bore draw-off taps

Soap tray

1·200

Plan

Washing trough

Note: Draw-off taps adapted or purposely designed to deliver a maximum of 0.06 l/s (3.6 l/min.)

437

Unplugged Appliances

To prevent an excess of water use and unnecessary waste, most sanitary appliances are fitted with a means for plugging the waste outlet. Baths, basins, sinks and similar use appliances intended to contain a volume of water for washing should have attached or be fitted with a readily accessible plug or other device such as a pop-up waste closer.

Exceptions –

- Appliances fitted with spray taps only.
- Washing troughs, as shown on the preceding page.
- Washing appliances (basins or troughs) with self-closing taps.
- Shower trays or baths used with only a shower supply.
- Drinking water fountains.
- Purpose-made appliances for use in medical, dentistry and veterinary situations, e.g. dentist's mouthwash bowl.
- Wash basins with water supply restricted to a 0.06 litres/second maximum flow.

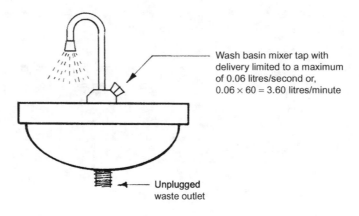

Wash basin mixer tap with delivery limited to a maximum of 0.06 litres/second or, $0.06 \times 60 = 3.60$ litres/minute

Unplugged waste outlet

Ref. The Water Supply (Water Fittings) Regulations, Schedule 2, Para. 28.

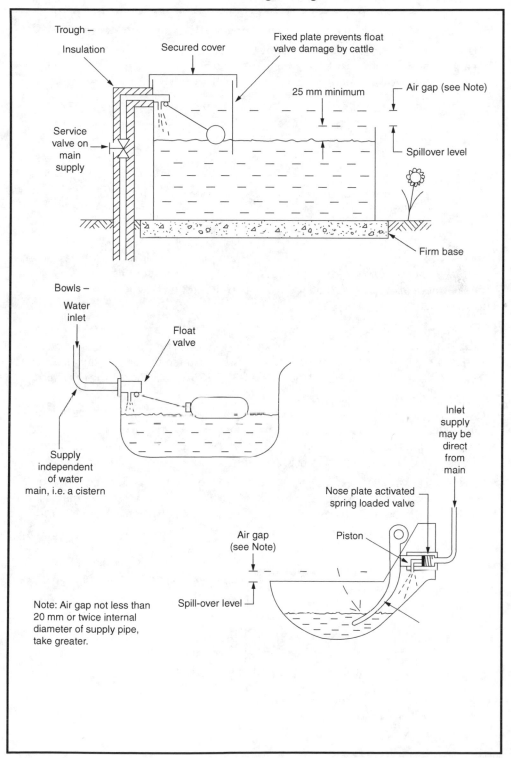

Trough –

Insulation

Secured cover

Fixed plate prevents float valve damage by cattle

25 mm minimum

Air gap (see Note)

Service valve on main supply

Spillover level

Firm base

Bowls –

Water inlet

Float valve

Supply independent of water main, i.e. a cistern

Inlet supply may be direct from main

Nose plate activated spring loaded valve

Piston

Air gap (see Note)

Spill-over level

Note: Air gap not less than 20 mm or twice internal diameter of supply pipe, take greater.

439

Thermostatic Mixing Valve – 1

Safe water temperature – safety is considered from two perspectives:

1. Legionella or Legionnaires' disease – 60 to 65°C stored hot water requirement to prevent development of the bacteria.
2. Scalding – water temperatures above 45°C can cause injury.

Recommended maximum temperatures at draw-off points are

Appliance	Temperature °C
Bath	43
Shower	40
Hand basin	40
Bidet	37
Sink	48

As can be seen, there is a conflicting temperature differential between the two perspectives of about 20°C. Therefore, scalding is possible unless measures are introduced to blend the hot water supply with cooler water.

Mixing valve types:

Type 1 (TMV 1) – mechanical mixing with an over-temperature stop to BS EN 1286: Sanitary tapware. Low-pressure mechanical mixing valves. General technical specification. Or, BS 5779: Specification for spray mixing taps.

Type 2 (TMV2) – thermostatic mixing to BS EN 1287: Sanitary tapware. Low-pressure thermostatic mixing valves. General technical specification. Or, BS EN 1111: Sanitary tapware. Thermostatic mixing valves (PN 10). General technical specification.

Type 3 (TMV 3) – thermostatic mixing with enhanced thermal performance. This should comply with NHS Estates Model Engineering Specification DO8, thermostatic mixing valves (healthcare premises).

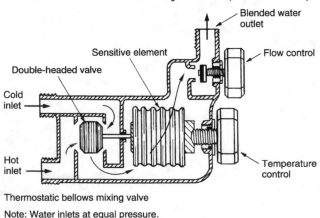

Thermostatic bellows mixing valve

Note: Water inlets at equal pressure.

Any newly built property or an existing property subject to alterations or refurbishment that include bathroom facilities must incorporate a device to prevent the hot water supply to a bath or a bidet from exceeding 48°C. This device will usually be a thermostatic mixing valve required to be set between 44°C and 46°C with a +/– tolerance of 2°C. In due course, these Building Regulation requirements may be applied to hot water temperatures at the outlets of other sanitary fitments. The need for these controls is in response to the unacceptably high number of scalding incidents through hot water supplied at normal storage temperature of about 60°C, as accounted for at the top of the preceding page.

This objective of safe hot water outlet temperatures can be achieved by installing a thermostatic mixing valve to blend cold and hot water supplies before they discharge through a terminal fitting, i.e. a tap or a shower rose. The basic mixing valve shown on the previous page has a temperature-sensitive element that responds to inlet water temperature changes to ensure a stable outlet temperature. In addition to this blending facility, the outlet must also have an over-temperature detection element and valve to close the delivery if the cold water supply fails or if the sensitive element within the mixing unit malfunctions.

Thermostatic mixing valves to TMV1 specification have largely been superseded by the higher expectations of TMV2 standards. However, the TMV1 specification is still acceptable in situations where persons using the blended water supply are not considered to be at any risk. This generally excludes use in premises occupied by children, elderly or infirm people. Valves to TMV2 specifications will normally satisfy the standards expected for domestic use, i.e. installation in dwellings, housing association properties, hotels and hostels. Valves to TMV3 specification are for higher risk situations, such as nursing/ convalescent homes and hospitals.

Refs. Building Regulations Part G, AD G3: Hot Water Supply and Systems.

Building Research Establishment, Information Paper IP 14/03, Preventing hot water scalding in bathrooms: using TMVs.

BS 7942: Thermostatic mixing valves for use in care establishments.

Typical applications

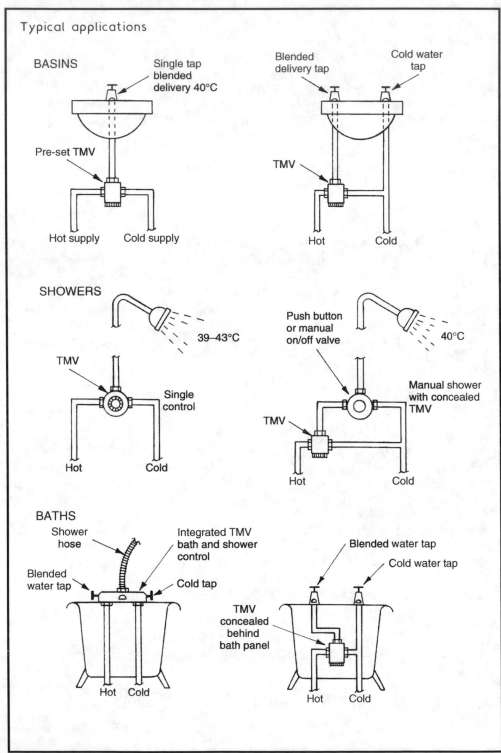

BASINS

Single tap
blended
delivery 40°C

Pre-set TMV

Hot supply   Cold supply

Blended
delivery tap

Cold water
tap

TMV

Hot   Cold

SHOWERS

39–43°C

TMV

Single
control

Hot   Cold

Push button
or manual
on/off valve

40°C

Manual shower
with concealed
TMV

TMV

Hot   Cold

BATHS

Shower
hose

Integrated TMV
bath and shower
control

Blended
water tap

Cold tap

Hot   Cold

Blended water tap

Cold water tap

TMV
concealed
behind
bath panel

Hot   Cold

Typical TMV unit

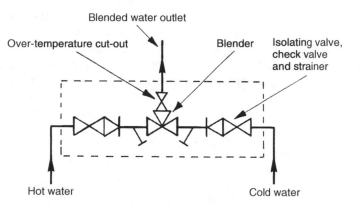

The stored hot water temperature must be at least 60°C to prevent the development of bacteria. Therefore, simply setting the hot water storage cylinder thermostat to provide a general supply of hot water at about 48°C is prohibited. Subject to the local water supply authority's requirements and approval, it may be possible to install just one central or master thermostatic mixing valve, to regulate domestic hot water requirements to all fitments as shown in principle below.

Master mixing valve installation

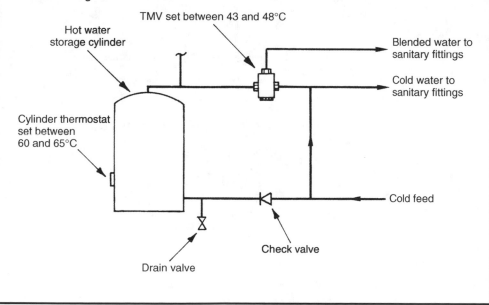

Tempering Valve

A tempering valve can be used to provide a simple means for temperature control of potable hot water distribution to several sanitary fitments. It therefore has application to large-scale domestic and commercial sanitary plumbing installations. It can also be used for industrial process water supplies. Unlike TMVs that are normally provided as dedicated point-of-use water temperature controls to every sanitary fitting, only one tempering valve is used for a whole system and this should be fitted as close as practical to the hot water source. Therefore, the installation costs with a tempering valve are relatively economical for large systems of sanitation where good overall temperature control is adequate.

A tempering valve functions by blending cold and hot water in proportional volumes without significantly impeding the flow rate. The valve can be set manually where precise temperature control is not a requirement, but where used to supply sanitary appliances an anti-scald protection measure by automatic thermostatic control set to a maximum of 48°C will be necessary. Thermostatic tempering valves are produced for this purpose.

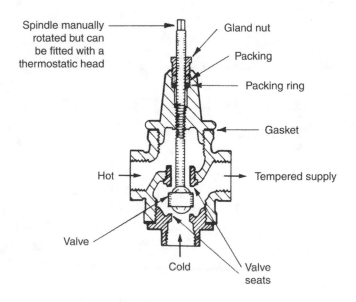

Ref. BS EN 15092: Building valves. In-line hot water supply tempering valves. Tests and requirements.

When mixing hot water with cold water to provide a blended supply, the quantities and temperatures of each can be estimated relative to the required water temperature objective. Factors such as lengths of individual supply pipes, effect of pipe insulation if and where provided and slight variances in water density at different temperatures will make a nominal contribution and may be included for academic reasons. However, for practical purposes, the formula procedure shown below will provide an adequate approximation:

Heat lost by hot water supply = Heat gained by cold water supply

Heat lost by hot water supply = $q_1(t_h - t)$

$$\text{where } q_1 = \text{quantity of hot water (litres or m}^3\text{)}$$
$$t_h = \text{temperature of hot water supply (°C)}$$
$$t = \text{temperature of mixed water (°C)}$$

Heat gained by cold water supply = $q_2(t - t_c)$

$$\text{where } q_2 = \text{quantity of cold water (litres or m}^3\text{)}$$
$$t_c = \text{temperature of cold water supply (°C)}$$
$$t = \text{temperature of mixed water (°C)}$$

Therefore,

$$q_1(t_h - t) = q_2(t - t_c)$$
$$(q_1 \times t_h) - (q_1 \times t) = (q_2 \times t) - (q_2 \times t_c)$$
$$(q_1 \times t_h) + (q_2 \times t_c) = (q_2 \times t) + (q_1 \times t)$$
$$(q_1 \times t_h) + (q_2 \times t_c) = (q_2 \times q_1)t$$
$$\frac{(q_1 \times t_h) + (q_2 \times t_c)}{(q_2 + q_1)} = t$$

Example: A thermostatic mixing valve is set to blend hot water at 60°C with cold water at 8°C in the proportion of 2·5:1, i.e. 2·5 litres of hot water for every 1 litre of cold water. The resultant delivery temperature will be

$$t = \frac{(2 \cdot 5 \times 60) + (1 \times 8)}{(2 \cdot 5 + 1)}$$

$$t = \frac{150 + 8}{3 \cdot 5} = 45 \cdot 14 \text{ approximately } 45°C$$

Urinals

These are used in virtually all buildings and public lavatories containing common facilities for male conveniences. They reduce the need for a large number of WCs. Three formats are available in ceramic ware or stainless steel:

- Bowl – secured to the wall and provided with division pieces where more than one is installed.
- Flat slab – fixed against the wall with projecting return end slabs and a low-level channel.
- Stall – contains curved stalls, dividing pieces and low-level channel.

Urinals are washed at intervals of 20 minutes by means of an automatic flushing cistern discharging 4·5 litres of water per bowl of 610mm of slab/stall width. The water supply to the cistern should be isolated by a motorised valve on a time control, to shut off when the building is not occupied. A hydraulically operated inlet valve to the automatic flushing cistern can be fitted. This closes when the building is unoccupied and other fittings not used.

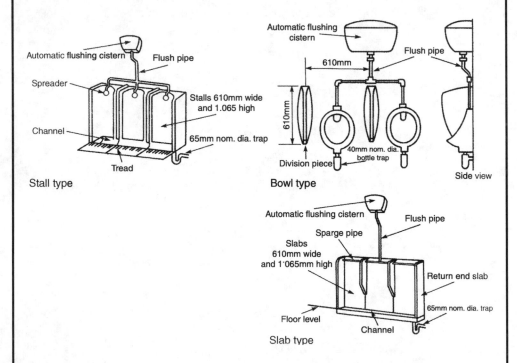

Stall type

Bowl type

Slab type

Refs. BS 4880-1: Specification for urinals. Stainless steel slab urinals.

BS 5520: Specification for vitreous china bowl urinals.

BS EN 13407: Wall hung urinals. Functional requirements and test methods.

See page 420 and preceding page for automatic devices.

Urinals usually have automatically operated flushing mechanisms. However, manual operation is also acceptable by use of:

- Flushing cistern.
- Flushing valve.
- Wash basin tap and hydraulic valve (combination of manual and automatic). See next page.

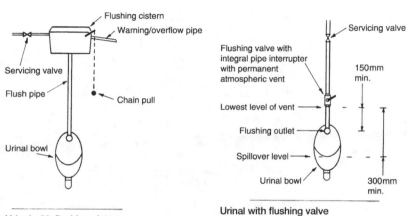

Urinal with flushing cistern

Urinal with flushing valve

Urinal with hydraulic valve

Urinals – Automatic Flushing Device

Function – the automatic flushing device is effectively a hydraulic valve operated by variable water pressure in the pipe supplying an automatic flushing cistern.

Hydraulic valve

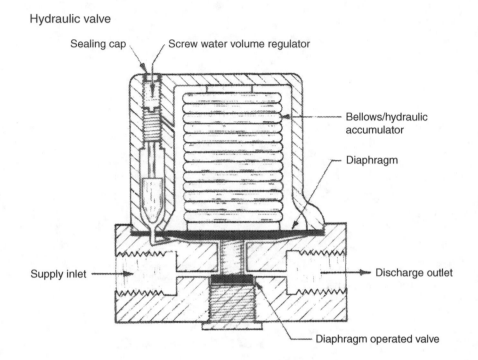

Operation – under normal static supply conditions, the hydraulic valve contains an equilibrium of water pressure and remains stable in a closed position. The equilibrium is disturbed when a basin tap connected to the same water supply is opened. A WC flush will also have the same effect where similarly connected. The pressure impulse causes a diaphragm within the valve to fluctuate. This moves a valve from its seating to allow a regulated flow of water to discharge into an automatic flushing cistern (see previous page and page 420). After several repeated uses of a basin tap and/or WC, sufficient water will have discharged for the automatic flushing cistern to function.

An hydraulic valve is fitted on the automatic flushing cistern supply pipe. A servicing valve is positioned before the inlet. The outlet is preferably unrestricted with an open end, but may have a lockshield valve or pet cock permanently open to control water discharge to the cistern.

Special types of sanitary appliances are required for hospital sluice rooms. The slop hopper is used for the efficient disposal of bedpan excrement and general waste. It is similar in design to the washdown WC pan, but has a hinged stainless steel grating for a bucket rest. Another grating inside the pan prevents the entry of large objects which could cause a blockage.

The bedpan washer has a spray nozzle for cleaning bedpans and urine bottles. To prevent possible contamination of water supply, it is essential that the water supplying the nozzle is taken from an interposed cold water storage cistern used solely to supply the bedpan washer. Alternatively, the design of the bedpan washer must allow for an air gap (min. 20mm) between spray outlet nozzle and water spill level. A 90-mm nominal diameter outlet is provided for the pan.

See pages 421 and 422 for flushing valve alternatives.

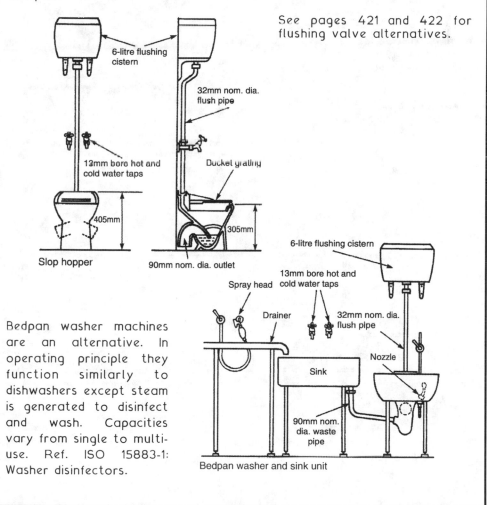

6-litre flushing cistern

32mm nom. dia. flush pipe

13mm bore hot and cold water taps

Bucket grating

405mm

305mm

Slop hopper

90mm nom. dia. outlet

6-litre flushing cistern

13mm bore hot and cold water taps

Spray head

Drainer

32mm nom. dia. flush pipe

Nozzle

Sink

90mm nom. dia. waste pipe

Bedpan washer and sink unit

Bedpan washer machines are an alternative. In operating principle they function similarly to dishwashers except steam is generated to disinfect and wash. Capacities vary from single to multi-use. Ref. ISO 15883-1: Washer disinfectors.

Sanitary Conveniences – Building Regulations

Approved Document G4 provides for minimum quantity, use and disposition of sanitary conveniences. These should contain sufficient appliances relative to a building's purpose and be separated by a door from places where food is stored or prepared. Layout of appliances and installation should allow for access and cleaning. The diagrams illustrate various locations for sanitary conveniences, with an intermediate lobby or ventilated space as required. En-suite facilities are acceptable direct from a bedroom, provided another sanitary convenience is available in the building. All dwellings must have at least one WC with a wash basin facility. The wash basin should be located in the room containing the WC or in a room or space giving direct access to the WC room (provided that it is not used for the preparation of food). A dwelling occupyied by more than one family should have the sanitary facilities available to all occupants.

Approved Document M1, Section 10 requires a WC in the principal/entrance storey of a dwelling (see page 454).

Any dwelling should have a bathroom containing a fixed bath or a shower, plus a wash basin.

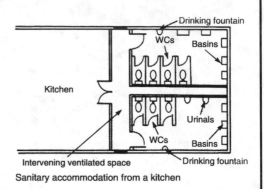

Sanitary accommodation from a kitchen

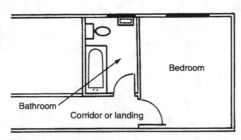

Entry to a bathroom via a corridor or landing

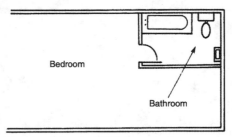

Entry to a bathroom directly from a bedroom

Refs. Building Regulations, Approved Document G4: Sanitary conveniences and washing facilities.
Building Regulations, Approved Document F1 – Ventilation.

The British Standard recommends that every new dwelling is fitted with at least one WC, one bath or shower, one wash basin and one sink. In dwellings accommodating five or more people, there should be two WCs, one of which may be in a bathroom. Any WC compartment not adjoining a bathroom shall also contain a wash basin. Where two or more WCs are provided, it is preferable to site them on different floors.

The number of appliances recommended for non-domestic premises such as offices, factories, shops, etc. varies considerably. BS 6465-1 should be consulted for specific situations. A general guide is provided on the next page.

Bathroom arrangements are detailed in BS 6465-2. Some simple domestic layouts are shown below, with minimum dimensions to suit standard appliances and activity space.

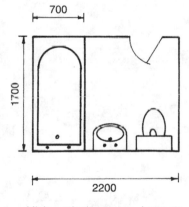

Minimum bathroom requirements

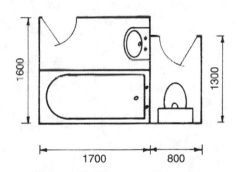

Alternative with adjacent WC compartment

Design of appliances should be such that they are smooth, impervious and manufactured from non-corrosive materials. They should be self-cleansing in operation and easily accessible for manual cleaning. Simplicity in design and a regard to satisfactory appearance are also important criteria.

Refs. BS 6465-1: Sanitary installations. Code of practice for the design of sanitary facilities and scales of provision of sanitary and associated appliances.

BS 6465-2: Sanitary installations. Space recommendations. Code of practice.

Sanitary Conveniences – Washrooms

The Offices, Shops and Railway Premises Act, now largely superseded by the Health and Safety at Work etc. Act and associated regulations, requires occupied buildings to have suitably located accommodation for sanitary appliances. This can be achieved by complying with the various regulations and other published guidance listed at the bottom of the page.

In general, the following minimum provisions apply:

- Adequate ventilation
- Regular cleaning schedule
- Cold and hot running water, or mixed warm water
- Means for cleaning (soap) and drying (towels or warm air)
- Showers if the type of work justifies it
- Toilet paper and coat hook in the WC cubicle
- Privacy, preferably with separate male and female accommodation unless each facility is separated with a lockable door for use by one person at a time
- Accessibility – not necessarily in the workplace but within the vicinity.

Minimum facilities:

Mixed use or female use only –

Persons	WCs	Washbasins
1–5	1	1
6–25	2	2

Thereafter, one additional WC and one additional wash basin per 25 persons

Male use only –

Persons	WCs	Urinals	Wash basins
1–15	1	1	As above
16–30	2	1	.. ..
31–45	2	2	.. ..
46–60	3	2	.. ..
61–75	3	3	.. ..
76–90	4	3	.. ..

Thereafter, allocated on the same proportional basis.

Refs. Building Regulations, Approved Document G4: Sanitary conveniences and washing facilities.

BS 6465-1 to 4: Sanitary installations.
Workplace (Health, Safety and Welfare) Regulations.
Food Hygiene Regulations.

Sufficient space for comfort and accessibility should be provided within WC compartments. The following guidance accepts overlap of adjacent activities and door opening:

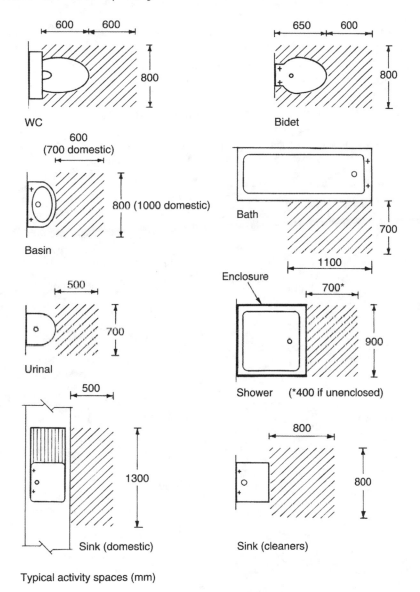

Typical activity spaces (mm)

See also the following three pages for spatial and access requirements for disabled persons.

Sanitary Conveniences for Disabled People (Dwellings)

Objectives for WC provision:

- In the entrance storey with unobstructed access.
- Within the principal storey of habitable rooms if this is not at entrance level.
- No higher than the principal storey – stair lift facility to other floors may be considered.
- WC may be located within a bathroom provided that the bath and wash basin are positioned so as not to impede access.
- Access door opens outwards. Inward opening may be considered if there is clear space for door swing and door can be opened outwards in an emergency.
- Compartment to contain clear space as shown in diagrams.

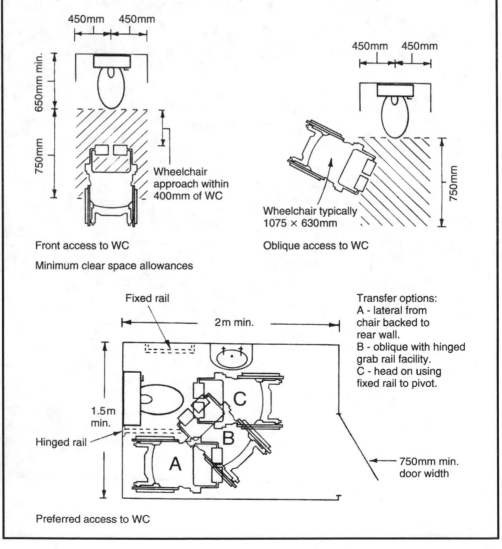

450mm 450mm

650mm min.

750mm

Wheelchair approach within 400mm of WC

Front access to WC

Minimum clear space allowances

450mm 450mm

750mm

Wheelchair typically 1075 × 630mm

Oblique access to WC

Fixed rail

2m min.

Transfer options:
A - lateral from chair backed to rear wall.
B - oblique with hinged grab rail facility.
C - head on using fixed rail to pivot.

1.5m min.

Hinged rail

C

B

A

750mm min. door width

Preferred access to WC

Buildings other than dwellings – at least one unisex WC public lavatory to be provided in cinemas, concert halls, leisure/sports centres, large office buildings, recreational buildings and theatres.

Access dimensions:

Passageway width, min. 1200mm.

- Passageway door opening width, min. 900mm.
- WC compartment size, min. 2200 × 1500mm.
- Door into compartment, min. 1000mm clear width.

Note: Compartment door opens outwards. It should have an emergency release device operated from the outside and a horizontal bar for closing fitted to the inside.

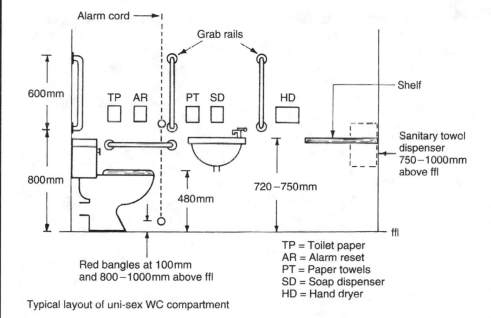

Typical layout of uni-sex WC compartment

TP = Toilet paper
AR = Alarm reset
PT = Paper towels
SD = Soap dispenser
HD = Hand dryer

Facilities for ambulant (not confined to a wheelchair) disabled people should be provided within conventional separate-sex WC and washroom compartments. A suitable compartment will contain some appliances specifically fitted with support rails. There should also be sufficient space to accommodate persons with impaired leg movement and with crutches.

Other provisions and facilities in buildings other than dwellings:

- Support/grab rails of 50mm minimum diameter, each side of a wash basin.
- Hinged or drop-down rail at least 300mm long on the exposed side of a WC.
- WC positioned to allow lateral transfer from a wheelchair.
- WC seat of rigid and robust material, set to a height of 480mm above finished floor level (ffl).
- Means for flushing, maximum 1200mm above ffl.
- Toilet paper dispenser within easy reach and on the side closest to WC seat.
- Wash basin height maximum 750mm and reachable while seated on WC pan.
- Hand dryer preferred to towels. Unit fitted between 800 and 1000mm above ffl. Hot air temperature thermostatically set at a maximum of 35°C.
- Wash basin taps of the quarter-turn lever type, or an electric sensor operated discharge. Water temperature regulated to 35°C maximum.
- Emergency alarm cord suspended from the ceiling, as close as possible to a wall. Cord fitted with two 50mm diameter red bangles set at 100mm and between 800 and 1000mm above ffl.

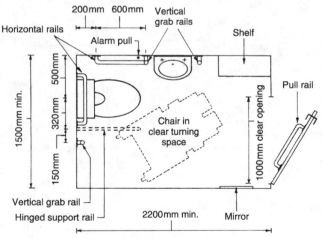

WC compartment for ambulant disabled people

Refs. Building Regulations, Approved Document M: Access to and use of buildings.

BS 8300-1 and 2: Design of an accessible and inclusive built environment.

Disability Discrimination Act.

Foul air from the drain and sewer is prevented from penetrating buildings by applying a water trap to all sanitary appliances. A water seal trap is an integral part of gullies and WCs, being moulded in during manufacture. Smaller fittings (i.e. sinks, basins, etc.) must be fitted with a trap. The format of a traditional tubular trap follows the outline of the letter 'P' or 'S'. The outlet on a 'P' trap is slightly less than horizontal ($2\frac{1}{2}°$) and on an 'S' trap it is vertical. A 'Q' trap has an outlet inclined at an angle of 45°, i.e. halfway between 'P' and 'S'. These are no longer used for sanitation but have an application to gullies.

Depth of water seal:

- WCs and gullies – 50 mm (less than smaller fittings as these are unlikely to lose their seal due to the volume of water retained).
- Sanitary appliances other than WCs with waste pipes of 50 mm nominal diameter or less – 75 mm, where the branch pipe connects directly to a discharge stack. However, because of the slow run-off, seal depth may be reduced to 50 mm for baths and shower trays.
- Sinks, baths and showers – 38 mm, where appliance waste pipes discharge over a trapped gully.

Note: Under working and test conditions, the depth of water seal must be retained at not less than 25 mm.

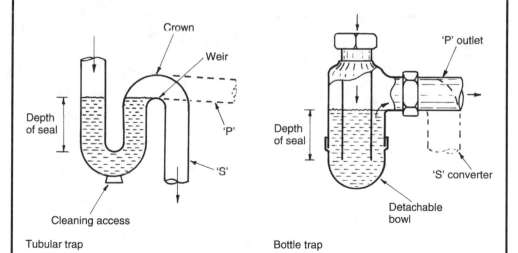

Tubular trap   Bottle trap

Ref. BS EN 274: Waste fitting for sanitary appliances.

457

Loss of Trap Water Seal

The most obvious cause of water seal loss is leakage due to defective fittings or poor workmanship. Otherwise, it may be caused by poor system design and/or installation:

- Self-siphonage – as an appliance discharges, the water fills the waste pipe and creates a vacuum to draw out the seal. Causes are a waste pipe that is too long, too steep or too small in diameter.
- Induced siphonage – the discharge from one appliance draws out the seal in the trap of an adjacent appliance by creating a vacuum in that appliance's branch pipe. Causes are the same as for self-siphonage, but most commonly a shared waste pipe that is undersized. Discharge into inadequately sized stacks can have the same effect on waste branch appliances.
- Back pressure – compression occurs due to resistance to flow at the base of a stack. The positive pressure displaces water in the lowest trap. Causes are a too small radius bottom bend, an undersized stack or the lowest branch fitting too close to the base of the stack.
- Capillary action – a piece of rag, string or hair caught on the trap outlet.
- Wavering out – gusts of wind blowing over the top of the stack can cause a partial vacuum to disturb water seals.

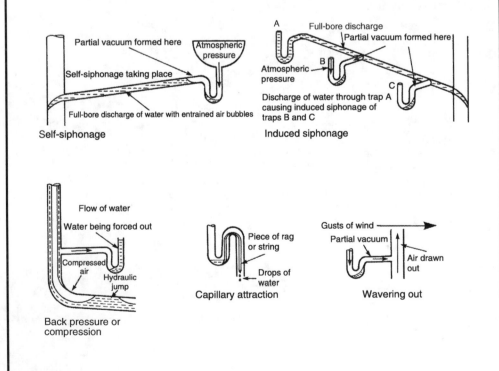

Self-siphonage

Induced siphonage

Back pressure or compression

Capillary attraction

Wavering out

Where trap water seal loss is apparent, the problem may be relieved by fitting either a resealing or an anti-siphon trap. A number of proprietory trap variations are available, some of which include

- McAlpine trap – this has a reserve chamber into which water is retained as siphonage occurs. After siphonage, the retained water descends to reseal the trap.
- Grevak trap – contains an anti-siphonage pipe through which air flows to break any siphonic action.
- Econa trap – contains a cylinder on the outlet into which water flows during siphonic action. After siphonage, the water in the cylinder replenishes the trap.
- Anti-siphon trap – as siphonage commences, a valve on the outlet crown opens, allowing air to enter. This maintains normal pressure during water discharge, preventing loss of water seal.

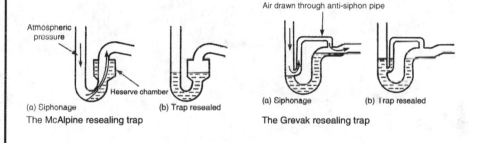

The McAlpine resealing trap

The Grevak resealing trap

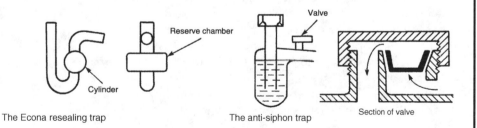

The Econa resealing trap

The anti-siphon trap

Note: Resealing and anti-siphon traps will require regular maintenance to ensure they are functioning correctly. They can be noisy in use.

Self-Sealing Waste Valve

This compact device has been developed by Hepworth Building Products for use on all sanitary appliances with a 32 or 40mm nominal diameter outlet. Unlike conventional water seal traps, it is a straight section of pipe containing a flexible tubular sealed membrane. This opens with the delivery of wastewater and fresh air into the sanitary pipework, resealing or closing after discharge. System design is less constrained, as entry of fresh air into the waste pipework equalises pressures, eliminating the need for traps with air admittance/ anti-siphon valves on long waste pipe lengths.

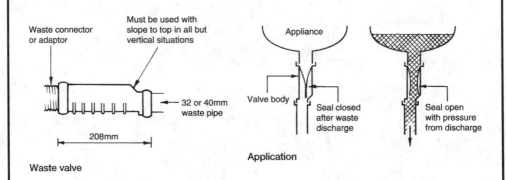

Waste connector or adaptor

Must be used with slope to top in all but vertical situations

32 or 40mm waste pipe

208mm

Waste valve

Appliance

Valve body

Seal closed after waste discharge

Seal open with pressure from discharge

Application

- No siphonage with full-bore discharge.
- Full-bore discharge provides better cleansing of pipework.
- Smaller diameter waste pipes possible as there is no water seal to siphon.
- Anti-siphon and ventilating pipes are not required.
- Ranges of appliances do not need auxiliary venting to stacks.
- No maximum waste pipe lengths or gradients (min. 18mm/m).
- Space saving, i.e. fits unobtrusively within a basin pedestal.
- Tight radius bends will not affect performance.
- In many situations will provide a saving in materials and installation time.

Note: Manufacturers state compliance with British Standard Codes of Practice and Building Regulations, Approved Documents for drainage and waste disposal.

The single-stack system was developed by the Building Research Establishment during the 1960s, as a means of simplifying the extensive pipework previously associated with above-ground drainage. The concept is to group appliances around the stack with a separate branch pipe serving each. Branch pipe lengths and falls are constrained. Initially the system was limited to five storeys, but applications have proved successful in high-rise buildings of over 20 storeys. Branch vent pipes are not required unless the system is modified. Lengths and falls of waste pipes are carefully selected to prevent loss of trap water seals. Water seals on the waste traps must be 75mm (50mm bath and shower).

Branch pipe slope or fall:

Sink and bath – 18 to 90mm/m

Basin and bidet – 20 to 120mm/m

WC – 18 to 90mm/m.

The stack should be vertical below the highest sanitary appliance branch. If an offset is unavoidable, there should be no connection within 750mm of the offset.

The branch bath waste connection must be at least 200mm below the centre of the WC branch to avoid cross-flow. This may require a 50mm nom. dia. parallel pipe to offset the bath waste pipe, or an 'S' trap WC to offset its connection.

The vent part of the stack may reduce to 75mm nom. dia. when it is above the highest branch.

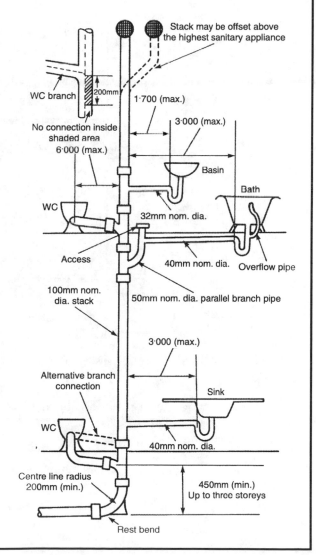

WC branch

200mm

No connection inside shaded area

6·000 (max.)

WC

Access

100mm nom. dia. stack

Alternative branch connection

WC

Centre line radius 200mm (min.)

Rest bend

Stack may be offset above the highest sanitary appliance

1·700 (max.)

3·000 (max.)

Basin

Bath

32mm nom. dia.

40mm nom. dia.

Overflow pipe

50mm nom. dia. parallel branch pipe

3·000 (max.)

Sink

40mm nom. dia.

450mm (min.) Up to three storeys

461

Single-Stack System – Modified

If it is impractical to satisfy all the requirements for waste pipe branches in a standard single-stack system, some modification is permitted in order to maintain an acceptable system performance:

• Appliances may be fitted with resealing or anti-siphon traps (see page 459).
• Branch waste pipes can be ventilated (see pages 464 and 465).
• Larger than standard diameter waste pipes may be fitted.

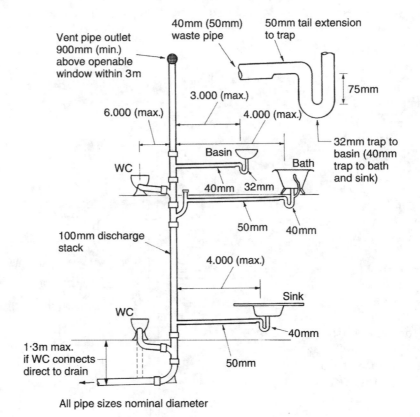

All pipe sizes nominal diameter

Note: Where larger than standard branch pipes are used, the trap size remains as standard. Each trap is fitted with a 50-mm tail extension before connecting to a larger waste pipe.

Refs. Building Regulations, Approved Document H1, Section 1: Sanitary pipework.
BS EN 12056-1: Gravity drainage systems inside buildings.

The collar boss system is another modification to the standard single-stack system. It was developed by the Marley company for use with their uPVC pipe products. The collar is in effect a gallery with purpose-made bosses for connection of waste pipes to the discharge stack without the problem of cross-flow interference. This simplifies the bath waste connection and is less structurally disruptive.

Small-diameter loop vent pipes on (or close to) the basin and sink traps also connect to the collar. These allow the use of 'S' traps and vertical waste pipes without the possibility of siphonage, even when the bath waste discharges and flows into the combined bath and basin waste pipe. Vertical outlets are also likely to be less obtrusive and less exposed than higher level 'P' trap waste pipes.

If the branch waste pipes are kept to minimal lengths, the loop vents may not be required. However, the system must be shown to perform adequately under test without the loss of trap water seals.

All pipe sizes shown are nominal inside diameter. There may be some slight variation between different product manufacturers, particularly those using outside diameter specifications. Note that there is not always compatibility between different manufacturers' components.

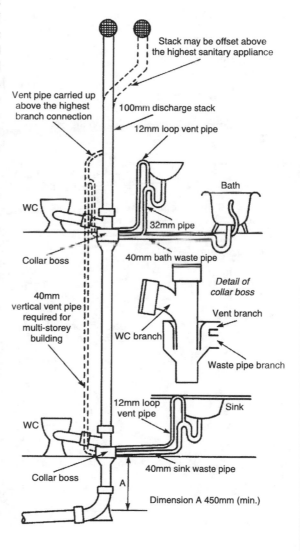

Stack may be offset above the highest sanitary appliance

Vent pipe carried up above the highest branch connection

100mm discharge stack

12mm loop vent pipe

Bath

WC

32mm pipe

Collar boss

40mm bath waste pipe

Detail of collar boss

40mm vertical vent pipe required for multi-storey building

Vent branch

WC branch

Waste pipe branch

12mm loop vent pipe

Sink

WC

40mm sink waste pipe

Collar boss

A

Dimension A 450mm (min.)

463

Modified Single-Stack System

The ventilated stack system is used in buildings where close grouping of sanitary appliances occurs – typical of lavatories in commercial premises. The appliances need to be sufficiently close together and limited in number not to be individually vented.

Requirements –

WCs:

Eight maximum

100mm branch pipe

15m maximum length

Gradient between

9 and 90mm/m

$(\theta = 90\frac{1}{2}° - 95°)$.

Basins:

Four maximum

50mm pipe

4m maximum length

Gradient between

18 and 45mm/m

$(\theta = 91° - 92\frac{1}{2}°)$.

Urinals (bowls):

Five maximum

50mm pipe

Branch pipe as short as possible

Gradient between

18 and 90mm/m.

Urinals (stalls):

Seven maximum

65mm pipe

Branch pipe as for bowls.

All pipe sizes nominal inside diameter.

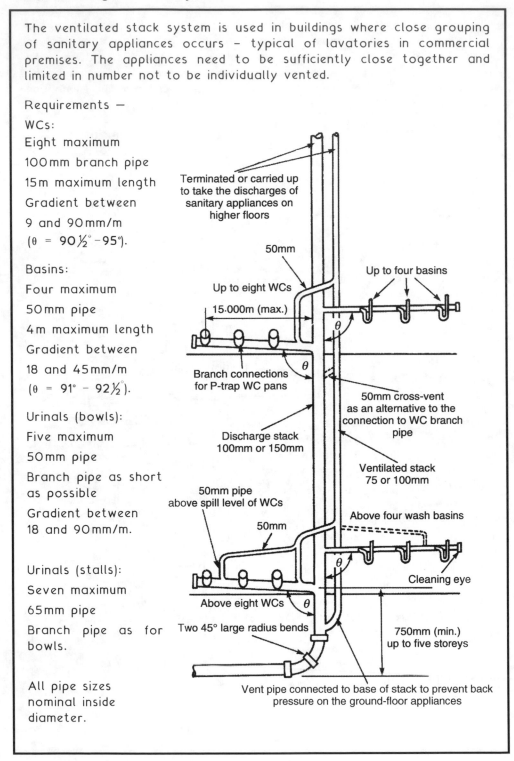

Terminated or carried up to take the discharges of sanitary appliances on higher floors

50mm

Up to eight WCs

Up to four basins

15.000m (max.)

θ

Branch connections for P-trap WC pans

θ

50mm cross-vent as an alternative to the connection to WC branch pipe

Discharge stack 100mm or 150mm

Ventilated stack 75 or 100mm

50mm pipe above spill level of WCs

50mm

Above four wash basins

θ

Above eight WCs

Cleaning eye

θ

Two 45° large radius bends

750mm (min.) up to five storeys

Vent pipe connected to base of stack to prevent back pressure on the ground-floor appliances

The fully vented one-pipe system is used in buildings where there are a large number of sanitary appliances in ranges, e.g. factories, schools, offices and hospitals.

The trap on each appliance is fitted with an anti-siphon or vent pipe. This must be connected within 300mm of the crown of the trap.

Individual vent pipes combine in a common vent for the range, which is inclined until it meets the vertical vent stack. This vent stack may be carried to outside air or it may connect to the discharge stack at a point above the spill-over level of the highest appliance.

The base of the vent stack should be connected to the discharge stack close to the bottom rest bend to relieve any compression at this point.

900mm (min.)

Window opening

If L is less than 3·000m the stack must terminate 900mm above the window opening

Note: The above rule applies to all systems*

Range of wash basins

Range of WCs

40mm

100mm

75mm vent stack

150mm discharge stack

32mm loop vent

50mm loop vent

40mm

Cleaning eye

Easy bend

Rest bend

Size of branch and stack vents:

Discharge pipe or stack (D) (mm)	Vent pipe (mm)
<75	0.67D
75–100	50
>100	0.50D

All pipe sizes nominal inside diameter.

*Building Regulations Approved Document H. Section 1, Sanitary pipework.

465

The Two-Pipe System

This system was devised to comply with the old London County Council requirements for connection of soil (WC and urinal) and waste (basin, bath, bidet, sink) appliances to separate stacks. For modern systems, the terms soil and waste pipes are generally replaced by the preferred terminology, discharge pipes and discharge stacks.

There are many examples of the two-pipe system in use. Although relatively expensive to install, it is still permissible and may be retained in existing buildings that are the subject of refurbishment.

It may also be used where the sanitary appliances are widely spaced or remote and a separate waste stack is the only viable method for connecting these to the drain.

A variation typical of 1930s dwellings has first-floor bath and basin wastes discharging through the wall into a hopper. The waste stack from this and the ground-floor sink waste discharge over a gully.

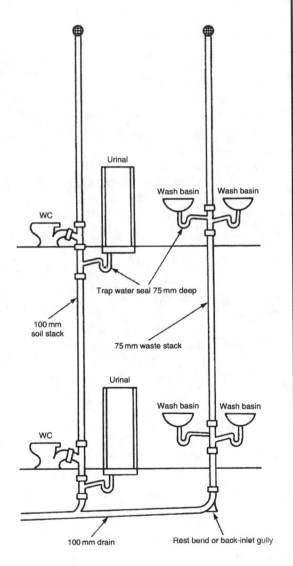

A gully may be used as an alternative to a rest bend before the drain.

These systems are particularly useful where sanitary appliance location is impractical, relative to the existing discharge pipework and stack, e.g. loft conversions and basements. The macerator, pump and small diameter discharge pipe are fairly compact, and unlikely to cause structural disruption on the scale of modifications to a conventional gravity flow system.

There are a variety of proprietary pumping systems, most capable of delivering WC and basin discharge over 20m horizontally and 4m vertically. Only products that have been tested and approved by the European Organisation for Technical Approvals (EOTA) or their recognised members (e.g. British Board of Agrément (BBA)) are acceptable for installation under the Building Regulations.

Installation is at the discretion of the local water and building control authorities. They will not accept the permanent connection of a WC to a macerator and pump, unless there is another WC connected to a gravity discharge system within the same building.

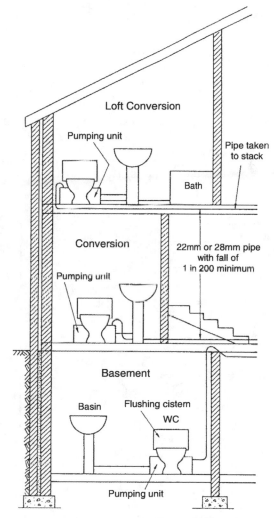

Pipework may be in 22 or 28mm outside diameter copper tube or equivalent in stainless steel or polypropylene. Easy bends, not elbow fittings, must be deployed at changes in direction.

Wash Basins – Waste Arrangements

The arrangement of waste and vent pipes for ranges of basins depends upon the type of building and the number of basins in the range. See BS 6465-1: Sanitary installations. Code of practice for scale of provision, selection and installation of sanitary appliances, to determine exact requirements for different purpose groups of building.

For ranges of up to four basins, branch ventilating pipes are not necessary, providing that the inside diameter of the main waste pipe is at least 50 mm and its slope is between 1° and $2\frac{1}{2}$° (18 mm to 45 mm/m).

For ranges above four basins, the inside diameter and slope is the same, but a 32 mm nominal inside diameter vent pipe is required. Alternatively, resealing or anti-siphon traps may be used.

In schools and factories, a running trap may be used, providing that the length of main waste pipe does not exceed 5 m. Alternatively, the wastes may discharge into a glazed channel with a trapped gully outlet to the stack.

For best-quality installation work, all traps may be provided with a vent or anti-siphon pipework.

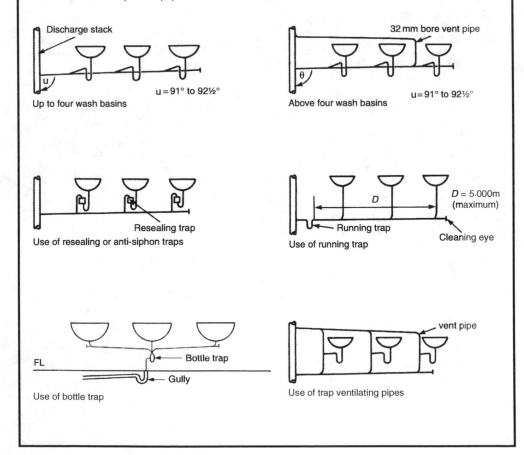

Discharge stack

Up to four wash basins

u = 91° to 92½°

32 mm bore vent pipe

Above four wash basins

u = 91° to 92½°

Resealing trap

Use of resealing or anti-siphon traps

D = 5.000 m (maximum)

Running trap

Cleaning eye

Use of running trap

FL

Bottle trap

Gully

Use of bottle trap

vent pipe

Use of trap ventilating pipes

Waste Pipes from Washing Machines and Dishwashers

The simplest method for discharging the hosepipe from a washing machine or dishwasher is to bend the hosepipe over the rim of the sink. However, this is unattractive and may be inconvenient if the hosepipe creates an obstruction. A more discrete and less obtrusive arrangement is to couple the hose to a tee fitting or purpose-made adaptor located between the trap and waste outlet of the sink. If a horizontal waste pipe is required at low level behind kitchen fitments, it must be independently trapped and some provision must be made for the machine outlet to ventilate to atmosphere (a purpose-made vent must not be connected to a ventilating stack). Alternatively, the machine hosepipe may be inserted loosely into the vertical waste pipe, leaving an air gap between the two pipes.

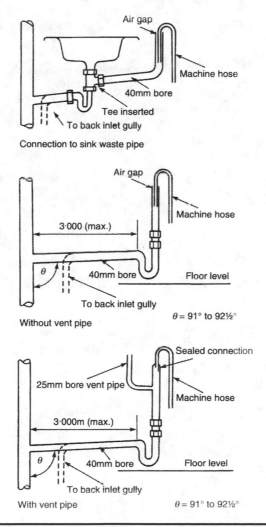

Connection to sink waste pipe

Without vent pipe $\theta = 91°$ to $92\frac{1}{2}°$

With vent pipe $\theta = 91°$ to $92\frac{1}{2}°$

Waste Water Heat Energy Recovery

Application – the transfer of heat energy in wastewater from showers. Applications are not limited to shower wastewater. Energy recovery can also apply to the hot water in waste pipework from other appliances.

Method – the cold water shower supply is indirectly preheated from the warm wastewater discharge by heat energy transfer in a coil attached to the waste pipe.

Concept – up to 90% of the energy in hot water delivered to a shower is lost through discharge down the waste pipe, stack and drain. This simple application provides a worthwhile supplement to other means of energy recovery.

Potential uses – mains fed electric showers, combination boilers and other sealed hot water systems.

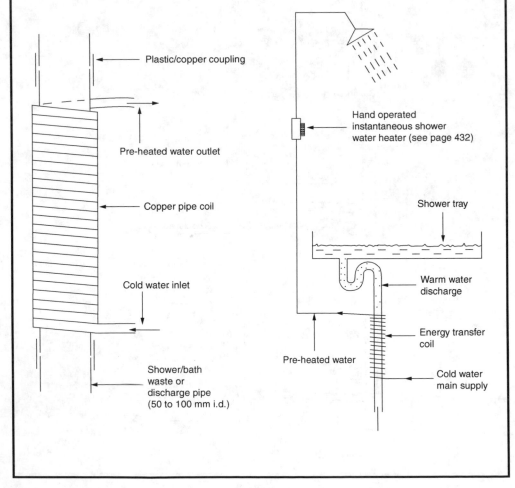

Plastic/copper coupling

Pre-heated water outlet

Copper pipe coil

Cold water inlet

Shower/bath waste or discharge pipe (50 to 100 mm i.d.)

Hand operated instantaneous shower water heater (see page 432)

Shower tray

Warm water discharge

Energy transfer coil

Pre-heated water

Cold water main supply

Approved Document H1 to the Building Regulations provides guidance on an acceptable method for determining airtightness of sanitary pipework systems. Installations must be capable of withstanding an air or smoke test pressure at least equal to a 38mm head of water for a minimum of three minutes. Smoke testing is not recommended for use with uPVC pipework.

Equipment for the air test:

Manometer ('U' gauge), rubber tube, hand bellows and two drain plugs or stoppers.

Procedure:

Stoppers are inserted at the top and bottom of the discharge stack. Each stopper is sealed with water, the lower seal with a flush from a WC. Traps to each appliance are primed to normal depth of seal. The rubber tube connected to the manometer and bellows is passed through the water seal in a WC. Hand bellows are used to pump air into the stack until the manometer shows a 38mm water displacement. After a few minutes for air temperature stabilisation, the water level in the manometer must remain stationary for three minutes. During this time, every trap must maintain at least 25mm of water seal.

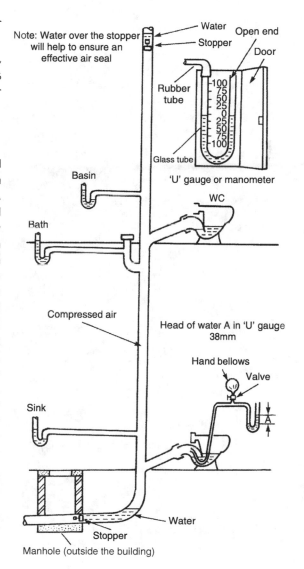

Note: Water over the stopper will help to ensure an effective air seal

Water

Stopper

Open end

Door

Rubber tube

100
75
50
25
0
25
50
75
100

Glass tube

'U' gauge or manometer

Basin

WC

Bath

Compressed air

Head of water A in 'U' gauge 38mm

Hand bellows

Valve

Sink

A

Water

Stopper

Manhole (outside the building)

471

Sanitation – Data (1)

Appliances:

Fitment	Capacity (l)	Discharge flow rate (l/s)
Basin	6	0.6
Basin – spray tap	–	0.06
Bath	80	1.1
Shower	–	0.1
Sink	23	0.9
Urinal	4.5	0.15
Washing machine	180	0.7
Water closet	6	2.3

All appliances in a dwelling are unlikely to be used simultaneously; therefore, the flow rate that stacks and drains have to accommodate is not the summation of their respective discharges. Allowing for normal usage, the anticipated flow rates from dwellings containing one WC, one bath, one or two wash basins and one sink are as follows:

Flow rates per dwelling:

No. of dwellings	Flow rate (l/s)
1	2.5
5	3.5
10	4.1
15	4.6
20	5.1
25	5.4
30	5.8

Discharge stack sizes:

Min. stack size (nom. i.d.)	Max. capacity (l/s)
50	1.2
65	2.1
75	3.4
90	5.3
100	7.2

Stacks serving urinals, not less than 50 mm.

Stack serving one or more washdown WCs, not less than 100 mm.

If one siphonic WC with a 75 mm outlet, stack size also 75 mm.

Discharge pipe and trap sizes:

Fitment	Trap and pipe nom. i.d. (mm)	Trap water seal (mm)
Basin	32	75
Bidet	32	75
Bath	40	75*
Shower	40	75*
Sink	40	75*
Dishwasher	40	75
Washing machine	40	75
Domestic food		
Waste disposal unit	40	75
Commercial food		
Waste disposal unit	50	75
Urinal bowl	40	75
Urinal bowls (2–5)	50	75
Urinal stalls (1–7)	65	50
WC pan – siphonic	75	50‡
WC pan – washdown	100†	50‡
Slop hopper	100†	50‡

*38mm if discharging to a gully.
†Nominally 100mm but approx. 90mm (min. 75mm).
‡Trap integral with fitment.

Bath and shower trays may be fitted with 50mm seal traps. The following materials are acceptable for sanitary pipework:

Application	Material	Standard
Discharge pipes and stacks	Cast iron	BS 416-1 and BS EN 877
	Copper	BS ENs 1254 and 1057
	Galv. Steel	BS 3868
	uPVC	BS EN 1329-1
	Polyethylene	BS EN 1519-1
	Polypropylene	BS EN 1451-1
	MuPVC	BS EN 1329-1
	ABS	BS EN 1455-1
	PVC–C	BS EN 1566-1
	San + PVC	BS EN 1565-1

Offsets

Offsets have two interpretations:

1. Branch waste or discharge pipe connections to the discharge stack. Typically the 200-mm offset required for opposing bath and WC discharge pipes – see page 461. Additional requirements are shown below.

2. Stack offsets – to be avoided, but may be necessary due to the structural outline of the building to which the stack is attached. Large radius bends should be used, and no branch connections are permitted within 750 mm of the offset in buildings up to three storeys. In buildings over three storeys, a separate vent stack may be needed. This is cross-vented to the discharge stack above and below the offset to relieve pressure. Bends and offsets are acceptable above the highest spill-over level of an appliance. They are usually necessary where external stacks avoid eaves projections.

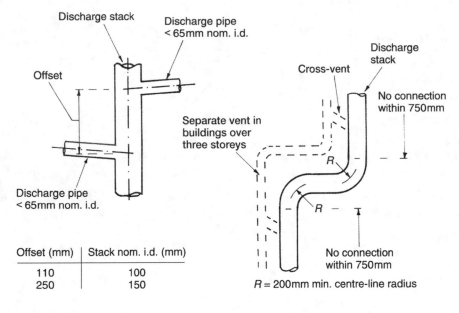

Offset (mm)	Stack nom. i.d. (mm)
110	100
250	150

Discharge pipes offset

R = 200mm min. centre-line radius

Stack offset

Note: Discharge stacks may be located internally or externally to buildings up to three storeys. Above three storeys, stacks should be located internally.

Lowest discharge pipe connection to stack:

Up to three storeys – 450mm min. from stack base (page 461).
Up to five storeys – 750mm min. from stack base (page 464).

Above five storeys, the ground-floor appliances should not connect into the common stack, as pressure fluctuations at the stack base could disturb the lower appliance trap water seals. Above 20 storeys, both ground- and first-floor appliances should not connect into the common stack. Ground- and first-floor appliances so affected can connect directly to a drain or gully or be provided with a stack specifically for lower level use.

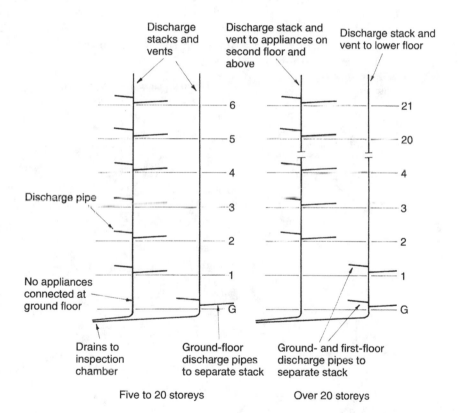

Discharge stacks and vents

Discharge stack and vent to appliances on second floor and above

Discharge stack and vent to lower floor

Discharge pipe

No appliances connected at ground floor

Drains to inspection chamber

Ground-floor discharge pipes to separate stack

Ground- and first-floor discharge pipes to separate stack

Five to 20 storeys

Over 20 storeys

Access – required for clearing blockages. Rodding points should be fitted at the ends of discharge pipes, unless trap removal provides access to the full pipe length. Discharge stacks are accessed from the top and through access plates located midway between floors at a maximum spacing of three storeys apart.

Fire Stops and Seals

For fire protection and containment purposes, the Building Regulations divide parts or units within buildings into compartments. A typical example is division of a building into individual living units, e.g. flats. The dividing compartment walls and floors have fire resistances specified in accordance with the building size and function.

Where pipes penetrate a compartment interface, they must have a means of preventing the spread of fire, smoke and hot gases through the void they occupy. Non-combustible pipe materials may be acceptably sealed with cement and sand mortar, but the most vulnerable are plastic pipes of low heat resistance. The void through which they pass can be sleeved in a non-combustible material for at least 1m each side. One of the most successful methods for plastic pipes is to fit an intumescent collar at the abutment with, or within, the compartment wall or floor. Under heat, these become a carbonaceous char, expand and compress the warm plastic to close the void for up to four hours.

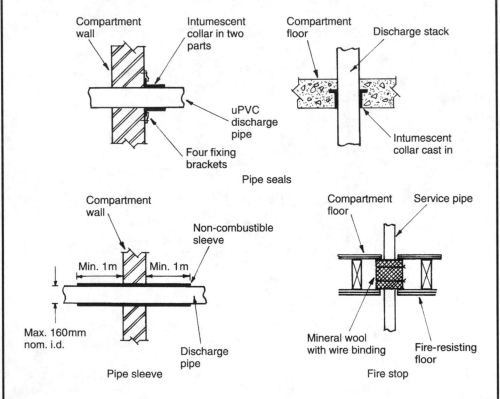

Ref. Building Regulations, Approved Document B3: Internal fire spread (structure).

Note: See also page 725.

Simultaneous demand process – considers the number of appliances likely to be used at any one time, relative to the total number installed on a discharge stack.

Formula:

$$m = np + 1 \cdot 8\sqrt{2np(1 - p)}$$

where m = no of appliances discharging simultaneously

n = no. of appliances installed on the stack

p = appliance discharge time (t) ÷ intervals between use (T).

Average time for an appliance to discharge = 10 seconds (t)

Intervals between use (commercial premises) = 600 seconds (T)

(public premises) = 300 seconds (T)

Commercial premises (e.g. offices, factories, etc.) p = 10 ÷ 600 = 0·017.

Public premises (e.g. cinemas, stadiums, etc.) p = 10 ÷ 300 = 0·033.

E.g. an office building of ten floors with four WCs, four urinals, four basins and one sink on each floor.

Total number of appliances (n) = 13 × 10 floors = 130

Substituting factors for p and n in the formula:

$$m = (130 \times 0 \cdot 017) + 1 \cdot 8\sqrt{2 \times 130 \times 0 \cdot 017\,(1 - 0 \cdot 017)}$$

$$m = 2 \cdot 21 + (1 \cdot 8 \times 2 \cdot 08) = 5 \cdot 96$$

Simultaneous demand factor = m ÷ n

= 5·96 ÷ 130 = 0·045 or 4·5%

Flow rates (see page 472):

Four WCs at 2·3 l/s	= 9·2
Four urinals at 0·15 l/s	= 0·6
Four basins at 0·6 l/s	= 2·4
One sink at 0·9 l/s	= 0·9
Total per floor	= 13·1 l/s
Total for ten floors	= 131 l/s

Allowing 4·5% simultaneous demand = 131 × 4·5% = 5·9 l/s.

Sanitation Flow Rate – Discharge Units

The use of discharge units for drain and sewer design is shown on pages 404 and 405. The same data can be used to ascertain the size of discharge stacks and pipes.

Using the example from the previous page:

Four WCs at 14 DUs	=	56
Four urinals at 0.3 DUs	=	1·2
Four basins at 3 DUs	=	12
One sink at 14 DUs	=	14
Total per floor	=	83·2
Total for ten floors	=	832 discharge units

Discharge units can be converted to flow in litres per second from the chart:

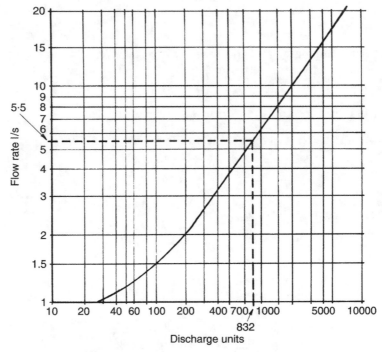

From the chart, a total loading of 832 discharge units can be seen to approximate to 5·5l/s, a fair comparison with the 5·9l/s calculated by formula on the preceding page.

Formula:

$$q = K \sqrt[3]{d^8}$$

where q = discharge or flow rate in l/s
K = constant of 32×10^{-6}
d = diameter of stack in mm.

Transposing the formula to make d the subject:

$$d = \sqrt[8]{(q \div K)^3} \qquad q = 5.5 \text{ l/s (see previous page)}$$

$$d = \sqrt[8]{(5.5 \div 32 \times 10^{-6})^3}$$

$$= 91.9 \text{ mm, i.e. a 100 mm nom. i.d. stack.}$$

Discharge units on stacks:

Discharge stack nom. i.d. (mm)	Max. no. of DUs
50	20
65	80
75	200
90	400
100	850
150	6400

Using the example from the preceding page, 832 discharge units can be adequately served by a 100 mm diameter stack.

Discharge units on discharge branch pipes:

Discharge pipe, nom. i.d. (mm)	Branch gradient		
	1 in 100	1 in 50	1 in 25
32		1	1
40		2	8
50		10	26
65		35	95
75	40	100	230
90	120	230	460
100	230	430	1050
150	2000	3500	7500

Ref. BS EN 12056-2: Gravity drainage systems inside buildings.
Sanitary pipework, layout and calculation.

The discharge unit method of stack and drain design shown on the preceding pages has limitations where a building or group of buildings are of mixed occupancy, e.g. a hotel containing bedrooms, offices, commercial kitchens, etc. In these situations, there are different frequencies of appliance use.

The 'K' factor method is very adaptable. It uses a peak design flow coefficient. This allows for frequency of appliance use, applied to the total possible disposal from all stack or drain connected appliances.

Comparison with discharge units (see page 404):

Application	Discharge unit time interval (min.)	'K' factor coefficient
Domestic	20	0·5
Commercial	10	0·7
Peak/public/congested	5	1·0

Example based on a mixed occupancy application to a single building, containing 60 private apartments and offices:

Each apartment:

Appliances	Disposal based on flow (see page 472)
Two WCs	4·6 (2 × 2·3)
One sink	0·9
Two basins	1·2 (2 × 0·6)
One shower	0·1
One bath	1·1
One washing machine	0·7
One dishwasher	0·2
	8·8 × 60 apartments = 528

Offices:

Appliances		Disposal based on flow (see page 472)
Gents:	Four WCs	9·2 (4 × 2·3)
	Eight urinals	1·2 (8 × 0·15)
	Six basins	3·6 (6 × 0·6)
Ladies:	Ten WCs	23·0 (10 × 2·3)
	Ten basins	6·0 (10 × 0·6)
Kitchen:	Two sinks	1·8 (2 × 0·9)
		44·8

'K' factors: Apartments (domestic) = 0·5   Offices (commercial) = 0·7

To allow for intermittent use of appliances, the following design formula is applied to calculate flow (Q) in litres/second:

$$Q = K \sqrt{\Sigma \text{ disposal}}$$

(continued)

Sanitation and Drainage Design Using 'K' Factors – 2

Before calculating the flow, an adjustment is needed to the lesser figure to represent its proportional disposal. This is achieved by applying a conversion factor from the lesser to the greater flow:

Lesser flow	Greater flow	'K' conversion factor
Domestic	Commercial	0.5 ÷ 0.7 = 0.714
Domestic	Peak/public/congested	0.5 ÷ 1.0 = 0.5
Commercial	Domestic	0.7 ÷ 0.5 = 1.4
Commercial	Peak/public/congested	0.7 ÷ 1.0 = 0.7
Peak/public/congested	Domestic	1.0 ÷ 0.5 = 2.0
Peak/public/congested	Commercial	1.0 ÷ 0.7 = 1.428

In this example, the lesser disposal is from the offices, i.e. 44·8. The commercial – domestic converter is 1·4, therefore $44·8 \times 1·4 = 62·72$. Adding this to the greater domestic disposal of 528 gives a total of 590·72.

Formula application using the 'K' factor for the greater disposal:

$$Q = 0·5 \sqrt{590·72} = 12·15 \text{ l/s}$$

Stack design formula from page 457. Taking Q – q.

$$q = K \sqrt[3]{d^8} \text{ or } d = \sqrt[8]{(q \div K)^3}$$

Note: Do not confuse K in the formula with 'K' factor. K in the formula is a constant of 32×10^{-6}.

Therefore, $d = \sqrt[8]{(12·15 \div 32 \times 10^{-6})} = 124 \text{ mm}$, i.e. 150 mm nom. dia. stack.

Drain design formula from page 405. $Q = V \times A$

where $Q = 0·012 \text{ m}^3/\text{s}$ (12·15 l/s) at a modest velocity (V) of 0·8 m/s.

A = Area of flow in drain (use half-full bore).

$A = Q \div V = 0·012 \div 0·8 = 0·015 \text{ m}^2$ (half bore)

Total area of drainpipe $= 2 \times 0·015 = 0·030 \text{ m}^2$

Pipe area $= \pi r^2$ or $r = \sqrt{\text{Pipe area} \div \pi}$ (r = radius)

$$r = \sqrt{0·030 \div 3·142} = 0·098 \text{m}$$

Pipe diameter $= 2 \times r = 0·196 \text{m}$ or 196 mm

Nearest available standard drainpipe above 196 mm is 225 mm.

Refs. BS EN 12056-2: Gravity drainage systems inside buildings.

BS EN 752: Drain and sewer systems outside buildings.

Gutter and Downpipe Sizing

Examples for estimating gutter and rainwater downpipe sizing are shown on pages 395 and 396. Building Regulations, Approved Document H3: Rainwater Drainage, provide a supplement to these that will also satisfy the requirement for rainwater to be conveyed from roof areas.

The effective roof area to be drained is calculated as shown below:

Roof pitch (degrees)	Multiplier (coefficient)
>70	Roof elevation area × 0.50
60	Plan area × 1.87
45	Plan area × 1.50
30	Plan area × 1.29
Flat	Plan area × 1.00

Interpolate as required. Right-hand column becomes the effective area.

Using the example from page 395. A roof pitch of 45° with a plan area of 40 m^2 to be drained:

$$40 \text{ m}^2 \times 1.50 = 60 \text{ m}^2 \text{ effective roof area.}$$

This figure for effective area can then be used to obtain an acceptable gutter and rainwater outlet pipe size from the guide below:

Roof area max. (m^2)	Gutter size (mm)	Downpipe outlet (mm)	Max. flow (l/s)
18	75	50	0.38
37	100	63	0.78
53	115	63	1.11
65	125	75	1.37
103	150	89	2.16

Note: Gutter dimension is as measured in width for a half-round profile, but allowance should be made for other profiles, e.g. box section.

For 60 m^2 effective area, a gutter size of 125 mm with a downpipe outlet of 75 mm will convey up to 1.37 litres/second of rainwater.

10 GAS INSTALLATION, COMPONENTS AND CONTROLS

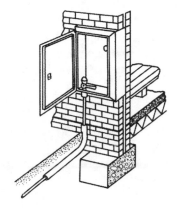

DOI: 10.1201/9781003434894-10

Natural Gas – Combustion

Properties of natural gas are considered on page 242. Some further features include

- Ignition temperature, 700°C.

- Stoichiometric mixture – the quantity of air required to achieve complete combustion of gas. For combustion, the ratio of air volume to natural gas volume is about 10:1. Therefore, about 10% gas to air mixture is required to achieve complete combustion. As air contains about 21% oxygen, the ratio of oxygen to gas is approximately 2:1. Developing this a little further – natural gas is about 90% methane, therefore:

$$CH_4 + 2O_2 = CO_2 + 2H_2O$$

1 part methane + 2 parts oxygen = 1 part carbon dioxide + 2 parts water.

If there is insufficient air supply to a gas burner, incomplete combustion will result. This produces an excess of carbon monoxide in the flue; a toxic and potentially deadly gas.

- Products of complete combustion – water vapour, carbon dioxide and the nitrogen already contained in the air. Correct combustion can be measured by simple tests to determine the percentage of carbon dioxide in flue gases. Analysers shown on pages 529 and 530 are suitable means for this assessment.

- Flues – these are necessary to discharge the products of combustion safely and to enhance the combustion process. The application of flues is considered in more detail later in this chapter. Flue size is normally to the boiler manufacturer's recommendations. The principles for determining the correct flue area and length, with regard to efficient fuel combustion and avoidance of condensation in the flue, are provided on pages 537 to 539. Some gas appliances such as small water heaters and cookers are flueless. Provided they are correctly installed, they will produce no ill-effects to users. The room in which they are installed must be adequately ventilated; otherwise, the room air could become vitiated (oxygen depleted). For a gas cooker, this means an openable window or ventilator. A room of less than 10m³ requires a permanent vent of 5000mm².

Chemical formulae and symbols for constituents of gases –

Butane	C_4H_{10}
Carbon dioxide	CO_2
Carbon monoxide	CO
Ethane	C_2H_6
Hydrogen	H_2
Methane	CH_4
Nitrogen	N_2
Oxygen	O_2
Propane	C_3H_8
Propylene	C_3H_6

Air requirements for combustion – fuel gases contain compounds of carbon and hydrogen. To complete the chemical reaction for combustion to occur, there must be a source of oxygen, and a means for ignition. Combustion produces heat and in the process changes the fuel gases into flue gases, notably carbon dioxide, nitrogen and water vapour. Oxygen is freely available in the atmosphere, occupying about 21% of the air's content. The amount of air required for combustion varies between different types of fuel gases (natural gas, butane and propane), each having a different carbon and hydrogen content.

For every $1m^3$ of gas burned, the volume of air and oxygen required is shown in the table below. The amount of oxygen is calculated on the basis of 21% oxygen content in the air. Fractionally 21/100 or 1/4.76, therefore air volume is divided by 4.76.

Gas ($1m^3$)	Volume of air	Volume of oxygen (air ÷ 4.76)
Natural gas	$10.0m^3$	$2.10m^3$
Butane/air mix*	4.9 ..	1.03 ..
Propane	23.8 ..	5.00 ..

*Note: An LPG (liquid petroleum gas) containing about 81% air (17% oxygen and 64% nitrogen), see page 243.

A more detailed analysis of fuel gas constituents and oxygen requirements for combustion is shown on the next page.

Gas Combustion – Oxygen Requirement

	Percentage by volume	Combustion equation	Volume of oxygen for combustion
Natural gas			
N_2	3.5		
CO_2	0.5		
CH_4	90.0	$CH_4+2O_2 = CO_2+2H_2O$	90/1 x 2 = 180.0
C_2H_6	4.5	$2C_2H_6+7O_2 = 4CO_2+6H_2O$	4.5/2 x 7 = 15.75
C_3H_8	1.0	$C_3H_8+5O_2 = 3CO_2+4H_2O$	1/1 x 5 = 5.0
C_4H_{10}	0.5	$2C_4H_{10}+13O_2 = 8CO_2+10H_2O$	0.5/2 x 13 = 3.25
Total	100		204

Therefore, $1m^3$ of natural gas requires $2.04m^3$ ($204 \div 100$) of oxygen for combustion, approximately $2.10m^3$ (see previous page).

	Percentage by volume	Combustion equation	Volume of oxygen for combustion
Butane air mix			
N_2	64.0		
O_2	17.0		–17.0
C_3H_8	2.5	$C_3H_8+5O_2 = 3CO_2+4H_2O$	2.5/1 x 5 = 12.5
C_4H_{10}	16.5	$2C_4 H_{10}+13O_2 = 8CO_2+10H_2O$	16.5/2 x 13 = 107.25
Total	100		102.75

Therefore, $1m^3$ of butane/air requires $1.0275m^3$ ($102.75 \div 100$) of oxygen for combustion, approximately $1.03m^3$ (see previous page).

	Percentage by volume	Combustion equation	Volume of oxygen for combustion
Propane gas			
C_3H_8	86.0	$C_3H_8+5O_2 = 3CO_2+4H_2O$	86/1 x 5 = 430.0
C_3H_6	12.0	$2C_3H_6+9O_2 = 6CO_2+6H_2O$	12/2 x 9 = 54.0
C_2H_6	1.5	$2C_2H_6+7O_2 = 4CO_2+6H_2O$	1.5/2 x 7 = 5.25
C_4H_{10}	0.5	$2C_4H_{10}+13O_2 = 8CO_2+10H_2O$	0.5/2 x 13 = 3.25
Total	100		492.5

Therefore, $1m^3$ of propane requires $4.925m^3$ ($492.5 \div 100$) of oxygen for combustion, approximately $5.00m^3$ (See previous page).

Gas supply to communities through a network of mains, installed and maintained by National Grid Gas and Cadent Gas. Gas marketing and after-sales services are provided by a number of commercial franchisees for the consumer's choice.

Some of the underground service pipes have been in place for a considerable time. These are manufactured from steel and although protected with bitumen, PVC or grease tape (Denso), they are being progressively replaced with non-corrosive yellow uPVC for mains and polyethylene for the branch supplies to buildings. The colour coding provides for recognition and to avoid confusion with other utilities in future excavation work.

Mains gas pressure is low compared with mains water. It is unlikely to exceed 75mbar (750mm water gauge or 7.5kPa), and this is reduced by a pre-set pressure governor at the consumer's meter to about 20mbar.

A service pipe of 25mm nominal bore is sufficient for normal domestic installations. For multi-installations such as a block of flats, the following can be used as a guide:

Nominal bore (mm)	No. of flats
32	2–3
38	4–6
50*	>6

*Note: Supplies of 50mm nom. bore may be provided with a service valve after the junction with the main. Where commercial premises are supplied and the risk of fire is greater than normal (e.g. a garage), a service pipe valve will be provided regardless of the pipe size and its location will be clearly indicated. Pipes in excess of 50mm nom. bore have a valve fitted as standard.

Gas mains should be protected by at least 375mm ground cover (450mm in public areas).

Refs. The Energy Act.
The Gas Act.
The Gas Safety (Installation and Use) Regulations.

Mains Gas Installation

The details shown below represent two different established installations. Some of these may still be found, but unless there are exceptional circumstances, the meter is no longer located within a building. An exception may be a communal lobby to offices or a block of flats. The preferred meter location for the convenience of meter readers and security of building occupants is on the outside of a building. This can be in a plastic cupboard housing on the external wall or in a plastic box with hinged lid sunken into the ground at the building periphery.

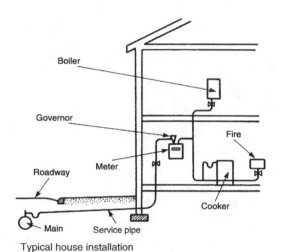

Typical house installation

Prior to conversion to natural gas in the 1960s, a condensate receiver was used to trap moisture from town or coal gas where it was impractical to incline the service pipe back to the main.

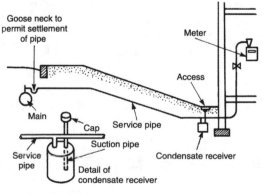

Use of condensate receiver

A service pipe is the term given to the pipe between the gas main and the primary meter control. A polyethylene pipe is used underground and steel or copper pipe where it is exposed. Wherever possible, the service pipe should enter the building on the side facing the gas main. This is to simplify excavations and to avoid the pipe having to pass through parts of the substructure which could be subject to settlement. The service pipe must not:

- pass under the base of a wall or foundations to a building
- be installed within a wall cavity or pass through it except by the shortest possible route
- be installed in an unventilated void space – suspended and raised floors with cross-ventilation may be an exception
- have electrical cables taped to it
- be near any heat source.

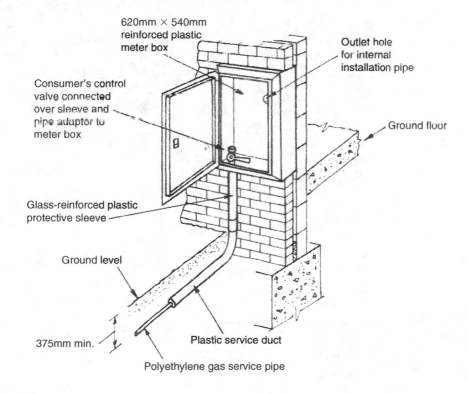

Ref. BS 8499: Specification for domestic gas meter boxes and meter bracket.

489

Gas Service Pipe Intake – 2

Where there is insufficient space or construction difficulties preclude the use of an external meter box or external riser, with certain provisions, the service pipe may be installed under a solid concrete floor or through a suspended floor.

For a solid floor, a sleeve or duct should be provided and built into the wall to extend to a pit of approximately 300 × 300mm plan dimensions. The service pipe is passed through the duct, into the pit and terminated at the meter position with a control valve. The duct should be as short as possible, preferably not more than 2m. The space between the duct and the service pipe is sealed at both ends with mastic and the pit filled with sand. The floor surface is made good to match the floor finish. If the floor is exposed concrete (e.g. a garage), then the duct will have to bend with the service pipe to terminate at floor level and be mastic sealed at this point.

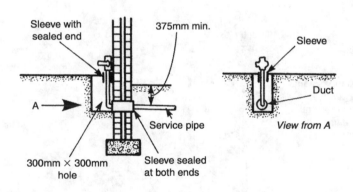

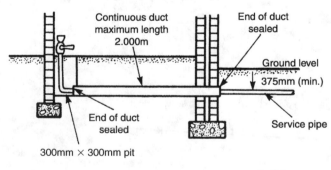

Service pipe entry into solid floor

490

Where a service pipe passes through a wall or a solid concrete floor, it must be enclosed by a sleeve of slightly larger diameter pipe to provide space to accommodate any building settlement or differential movement. The outside of the sleeve should be sealed with cement mortar and the space between the sleeve and service pipe provided with an intumescent (fire-resistant) mastic sealant, such that any gas escaping can only leak to the outside of a building where it will dilute in the external air.

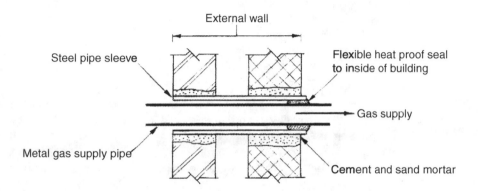

If an internal meter is used, the space or compartment allocated for its installation must be well ventilated. A purpose-made void or air-brick to the outside air is adequate. The surrounding construction should be of at least 30 minutes' fire resistance. In commercial and public buildings, the period of fire resistance will depend on the building size and purpose grouping.

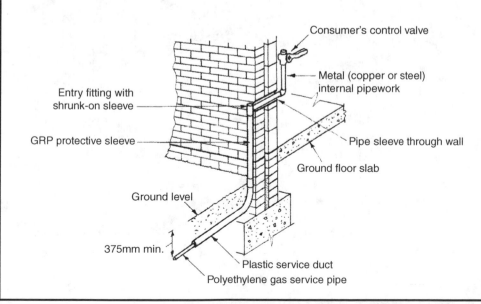

491

Gas Service Pipe in Multi-Storey Buildings

Gas service pipe risers must be installed in fire-protected shafts constructed in accordance with the Building Regulations, Approved Document B: Fire safety. Possible methods for constructing a shaft include:

- A continuous shaft ventilated to the outside at top and bottom. In this situation, a fire-protected sleeve is required where a horizontal pipe passes through the shaft wall.
- A shaft which is fire stopped at each floor level. Ventilation to the outside air is required at both high and low levels in each isolated section.

Shafts are required to have a minimum fire resistance of 60 minutes and the access door or panel a minimum fire resistance of 30 minutes. The gas riser pipe must be of screwed or welded steel and be well supported throughout with a purpose-made plate at its base. Movement joints or flexible pipes and a service valve are provided at each branch.

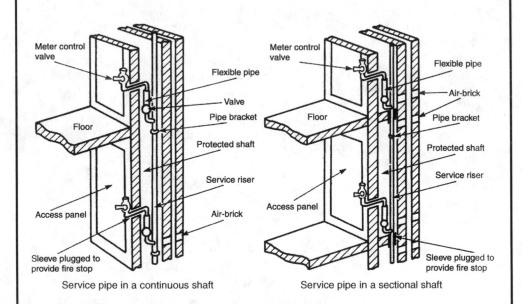

Service pipe in a continuous shaft   Service pipe in a sectional shaft

Refs. Building Regulations, Approved Document B3: Compartmentation.
BS 8313: Code of practice for accommodation of building services in ducts.

Flats and apartments – several meters, each serving individual dwellings can be conveniently grouped together in one compartment. This is usually accessible from a ground floor entrance lobby.

Minimum requirements –

* Accessibility at all times for maintenance and meter reading.
* Compartment to be lockable, or each meter to be contained in a lockable meter box.
* Individual meters to be marked to indicate premises served.
* Enclosed compartment/housing to be fire resisting, ventilated, isolated from heat sources, damp prooofed, detached from electrical installations, not near food storage and with meters positioned so as not to be obstructive.

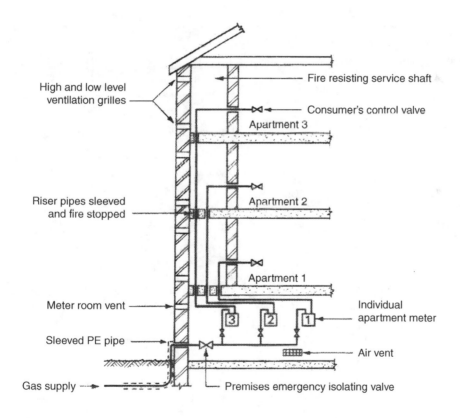

493

Installation of Gas Meters

The gas meter and its associated controls are the property of the gas authority. It should be sited as close as possible to the service pipe entry to the building, ideally in a purpose-made meter cupboard on the external wall. The cupboard should be positioned to provide easy access for meter maintenance, reading and inspection. The immediate area around the meter must be well ventilated, and the meter must be protected from damage, corrosion and heat. A constant pressure governor is fitted to the inlet pipework to regulate the pressure at about 20 mbar (2 kPa or 200 mm w.g.).

Electricity and gas meters should not share the same compartment. If this is unavoidable, a fire-resistant partition must separate them and no electrical conduit or cable should be closer than 50 mm to the gas meter and its installation pipework. One exception is the earth equipotential bond cable. This must be located on the secondary pipework and within 600 mm of the gas meter.

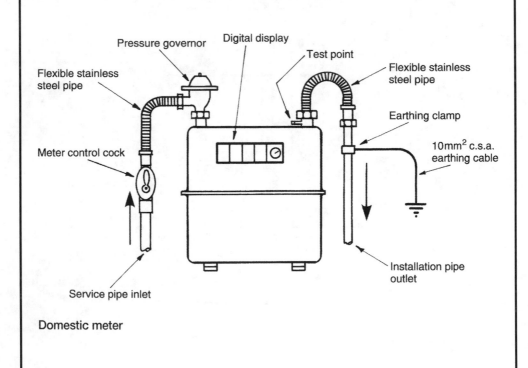

Domestic meter

Gas meters measure the volume of gas in cubic feet or cubic metres consumed within a building. The discharge is converted to kilowatt-hours (kWh): 100 cubic feet or 2.83 cubic metres is approximately 31kWh (see page 535). Some older meters have dials but these have been largely superseded by digital displays which are easier to read.

There are basically four categories of meter:

1. Domestic credit.
2. Domestic prepayment.
3. Industrial credit.
4. Smart (see page 13).

Credit meters measure the fuel consumed, and it is paid for after use at three-monthly billing intervals. Monthly payments can be made based on an estimate, with an annual adjustment made to balance the account.

Prepayment meters require payment for the fuel in advance by means of coins, cards, keys or tokens. Tokens are the preferred method, and these are purchased at energy showrooms, post offices and some newsagents. A variation known as the Quantum meter uses a card to record payment. These cards are purchased at designated outlets and can be recharged with various purchase values.

Industrial meters have flanged connections for steel pipework. Flexible connections are unnecessary due to the pipe strength and a firm support base for the meter. A bypass pipe is installed with a sealed valve. With the supply authority's approval, this may be used during repair or maintenance of the meter.

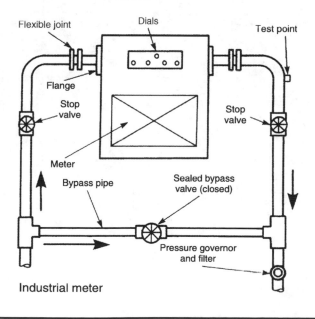

Industrial meter

Gas Controls

A constant pressure governor is fitted at the meter to regulate pressure into the system. It is secured with a lead seal to prevent unqualified tampering. Individual appliances may also have factory-fitted pressure governors, located just before the burners. Gas passes through the valve and also through the bypass to the space between the two diaphragms. The main diaphragm is loaded by a spring, and the upward and downward forces acting upon this diaphragm are balanced. The compensating diaphragm stabilises the valve. Any fluctuation of inlet pressure inflates or deflates the main diaphragm, raising or lowering the valve to maintain a constant outlet pressure.

A meter control cock has a tapered plug which fits into a tapered body. As gas pressures are very low, the valve can operate by a simple 90° turn to align a hole in the plug with the bore of the valve body, and vice versa. The drop-fan safety cock prevents the valve from being turned accidently.

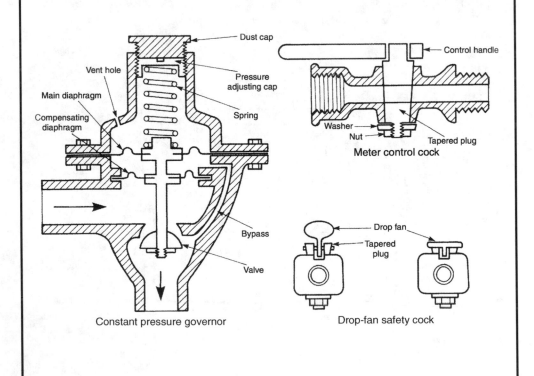

Constant pressure governor

Meter control cock

Drop-fan safety cock

For correct combustion of natural gas, burner design must allow for the velocity of the gas–air mixture to be about the same as the flame velocity. Natural gas has a very slow burning velocity; therefore, there is a tendency for a flame to lift off the burner. This must be prevented as it will allow gas to escape, possibly exploding elsewhere! Correct combustion will occur when the gas pressure and injector bore are correct and sufficient air is drawn in, provided the gas–air velocity is not too high to encourage lift-off. Some control over lift-off can be achieved by a retention flame fitted to the burner. Flame lift-off may also be prevented by increasing the number of burner ports to effect a decrease in the velocity of the gas–air mixture. A box-type burner tray is used for this purpose.

If the gas pressure is too low, or the injector bore too large, insufficient air is drawn into the burner. This can be recognised by a smoky and unstable flame, indicating incomplete combustion and an excess of carbon monoxide. At the extreme, light-back can occur. This is where the flame passes back through the burner to ignite on the injector.

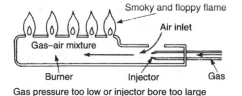

Gas pressure too low or injector bore too large

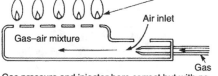

Gas pressure and injector bore correct but with no retention flame

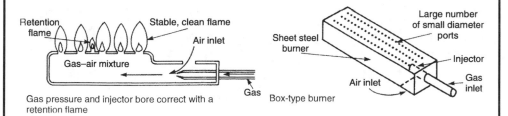

Gas pressure and injector bore correct with a retention flame

Box-type burner

497

Gas Thermostats

A thermostat is a temperature-sensitive device which operates a gas valve in response to a predetermined setting. Hot water heaters and boilers may be fitted with two thermostats:

1. Working thermostat – controls the water flow temperature from the boiler. It has a regulated scale and is set manually to the user's convenience. It engages or disengages the gas valve at a water temperature of about 80°C.

2. High limit thermostat – normally preset by the boiler manufacturer to function at a water temperature of about 90°C. It is a thermal cut-out safety device which will isolate the gas supply if the working thermostat fails.

The rod-type thermostat operates by a difference in thermal response between brass and invar steel. When water surrounding a brass tube becomes hot, the tube expands. This draws the steel rod with it until a valve attached to the rod closes off the fuel supply. The reverse process occurs as the water cools.

The vapour expansion thermostat has a bellows, capillary tube and probe filled with ether. When water surrounding the probe becomes hot, the vapour expands, causing the bellows to respond by closing the fuel valve. Cooling water reverses the process.

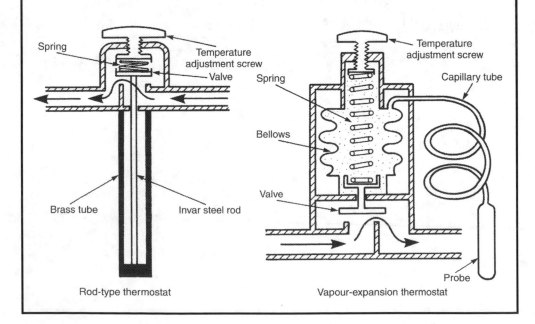

Rod-type thermostat   Vapour-expansion thermostat

A rod-type thermostat is often connected to a relay valve to control gas supply to the burner. When the boiler is operational, gas flows to the burner because valves A and B are open. Gas pressures above and below the diaphragm are equal. When the water reaches the required temperature, the brass casing of the rod thermostat expands and draws the invar steel rod with it to close valve A. This prevents gas from flowing to the underside of the diaphragm. Gas pressure above the diaphragm increases, allowing valve B to fall under its own weight to close the gas supply to the burner. As the boiler water temperature falls, the brass casing of the thermostat contracts to release valve A which reopens the gas supply.

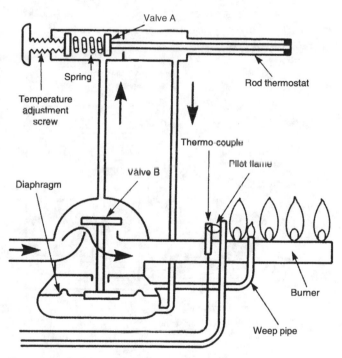

Operating principles of rod thermostat and gas relay valve

Gas Safety Controls

Gas water heaters/boilers and other heat-producing appliances such as air heaters must be fitted with a safety device to prevent gas from flowing in event of the pilot light extinguishing. While functional, the pilot light plays on a thermo-couple suspended in the gas flame. The hot thermo-couple energises an electromagnetic or solenoid valve to open and allow gas to flow. This is otherwise known as a thermo-electric pilot flame failure safety device. The drawing below shows the interrelationship of controls, and the next page illustrates and explains the safety device in greater detail.

To commission the boiler from cold, the thermo-electric valve is operated manually by depressing a push button to allow gas flow to the pilot flame. A spark igniter illuminates the flame while the button is kept depressed for a few seconds, until the thermo-couple is sufficiently warm to automatically activate the valve. Automatic ignition of a non-permanent pilot flame is a feature of new appliances. See comment on page 502.

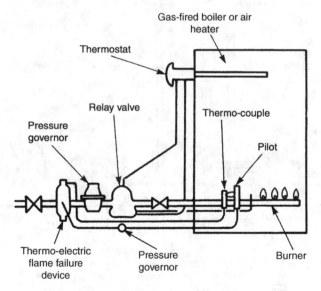

Gas boiler or air heater controls

Ref. BS EN 15502-1 and BS EN 15502-2-1: Gas fired heating boilers.

Thermo-electric – has an ancillary thermo-couple sensing element consisting of two dissimilar metals joined together at each end to form an electrical circuit. When the thermo-couple is heated by the gas pilot flame, a small electric current is generated. This energises an electromagnet in the gas valve which is retained permanently in the open position, allowing gas to pass to the relay valve. If the pilot flame is extinguished, the thermo-couple cools and the electric current is no longer produced to energise the solenoid. In the absence of a magnetic force, a spring closes the gas valve.

Bimetallic strip – has a bonded element of brass and invar steel, each metal having a different rate of expansion and contraction. The strip is bent into a 'U' shape with the brass on the outside. One end is anchored and the other attached to a valve. The valve responds to thermal reaction on the strip. If the pilot flame is extinguished, the bent bimetallic strip contracts, opening to its original position and closing the gas supply and vice versa.

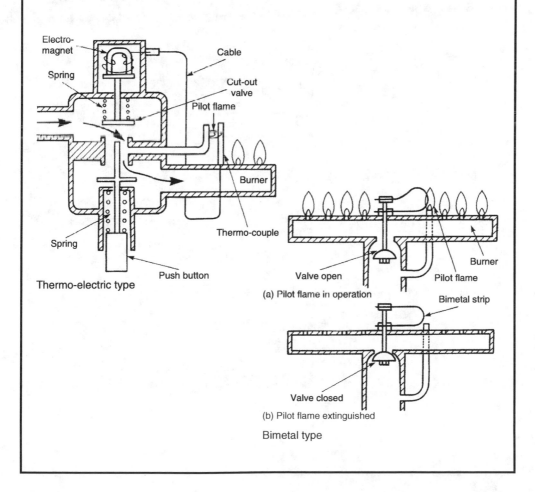

Thermo-electric type

(a) Pilot flame in operation

(b) Pilot flame extinguished

Bimetal type

Gas Ignition Devices

Lighting the pilot flame with matches or tapers is unsatisfactory. It is also difficult to effect while operating the push button control on the gas valve. An integral spark igniter is far more efficient. These are usually operated by mains electricity. An electric charge is compounded in a capacitor, until a trigger mechanism effects its rapid discharge. This electrical energy passes through a step-up transformer to create a voltage of 10 or 15kV to produce a spark. The spark is sufficient to ignite the pilot flame. Spark generation of this type is used in appliances with a non-permanent pilot flame. This is more fuel economic than a permanent flame. The spark operation is effected when the system thermostat engages an automatic switch in place of the manual push switch shown below and a gas supply to the pilot.

A piezoelectric spark igniter contains two crystals. By pressurising them through a cam-and-lever mechanism from a push button, a large electric voltage potential releases a spark to ignite the gas.

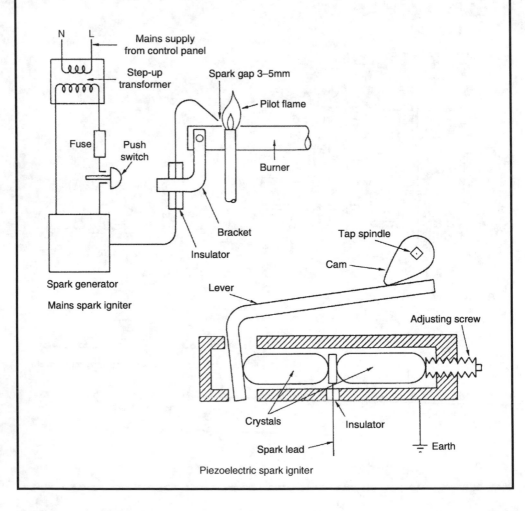

Mains spark igniter

Piezoelectric spark igniter

It is very important that new gas installations are thoroughly purged of air and debris that may remain in the completed pipework. This also applies to existing installations that have been the subject of significant changes. If air is not removed, it is possible that when attempting to ignite the gas, a gas-air mixture will cause a blow-back and an explosion. Before purging, the system should be pressure tested for leakages - see next page.

Procedure:

- Ensure ample ventilation where gas and air will escape from the system.

- Prohibit smoking, use of electrical switches, power tools, etc., in the vicinity of the process.

- Close the main gas control valve at the meter.

- Disconnect the secondary pipework at the furthest fitting.
  Note: if the last appliance has a flame failure safety device, no gas will pass beyond it, therefore remove its test nipple screw.

- Turn on the main gas control valve until the meter is completely purged.

- Purging the meter is achieved by passing through it a volume of gas at least equal to five times its capacity per revolution of the meter mechanism. Most domestic meters show 0.071cu.ft. (0.002m^3) per dial revolution, so: $5 \times 0.071 = 0.355$cu.ft. (0.010m^3) of gas is required.

- Turn off the main gas control valve and reconnect the open end or replace the last appliance test nipple.

- Turn on the main gas control valve and purge any remaining air to branch appliances until gas is smelt.

- Test any previous disconnections by applying soap solution to the joint. Leakage will be apparent by foaming of the solution.

- When all the air in the system has been removed, appliances may be commissioned.

Ref. BS 6891: Specification for the installation and maintenance of low-pressure gas installation pipework of up to 35mm on premises.

Testing Gas Installations for Soundness

Testing a new installation:

- Cap all open pipe ends and turn appliances off.

- Close the main control valve at the meter. If the meter is not fitted, blank off the connecting pipe with a specially prepared cap and test nipple.

- Remove the test nipple screw from the meter or blanking cap and attach the test apparatus by the rubber tubing.

- Level the water in the manometer at zero.

- Pump or blow air through the test cock to displace 300mm water gauge (30mbar) in the manometer. This is approximately one and a half times normal domestic system pressure.

- Wait one minute for air stabilisation, then if there is no further pressure drop at the manometer for a further two minutes, the system is considered sound.

- If leakage is apparent, insecure joints are the most likely source. These are painted with soap solution which foams up in the presence of air seepage.

Testing an existing system:

- Close all appliance valves and the main control valve at the meter.

- Remove the test nipple screw on the meter and attach the test apparatus.

- Open the main control valve at the meter to record a few millimetres' water gauge.

- Close the valve immediately and observe the manometer. If the pressure rises, this indicates a faulty valve.

- If the valve is serviceable, continue the test by opening the valve fully to record a normal pressure of 200 to 250mm w.g. Anything else suggests that the pressure governor is faulty.

- With the correct pressure recorded, turn off the main valve, allow one minute for air stabilisation and for a further two minutes there should be no pressure fluctuation.

- Check for any leakages as previously described.

When used with a flexible tube, hand bellows and control cock, this equipment is suitable for measuring gas installation pressure and testing for leakage. It is also suitable for air testing drains and discharge stacks.

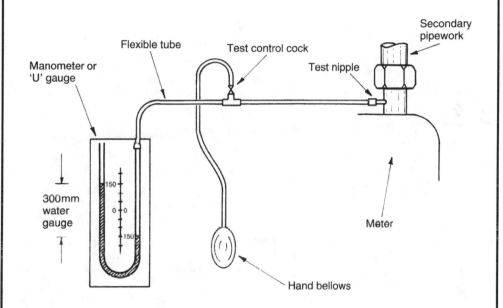

The glass tube is contained in a protective metal or wooden box. It is mounted against a scale graduated in millibars or millimetres. 1mbar is the pressure exerted by a 9.81mm (10mm is close enough) head of water. Water is levelled in the tube to zero on the scale. Care must be taken to note the scale calibration. Some manometers are half scale, which means the measures are in mbar or mm but they are double this to give a direct reading. Others are indirect, as shown. With these, the water displacements either side of the zero must be added.

Fires – these have a relatively low energy rating, usually no more than 3kW net input.* They are set in a fire recess and use the lined flue for extraction of burnt gases. Air from the room in older premises is usually sufficient for gas combustion, as appliances up to 7kW net input do not normally require special provision for ventilation (see page 527). Heat is emitted by convection and radiation.

Decorative fuel-effect fires – these are a popular alternative to the traditional gas fire. They burn gas freely and rely on displacement of heat by the colder air for combustion to encourage burnt gas extraction indirectly into the flue. Sufficient air must be available from a purpose-made air inlet to ensure correct combustion of the gas and extraction of burnt gases. An air-brick with permanent ventilation of at least 10000mm² is sufficient for fires up to 12.7kW net input rating. Log- and coal-effect fires are designed as a visual enhancement to a grate by resembling a real fire, but as a radiant heater, they compare unfavourably with other forms of gas heat emitters.

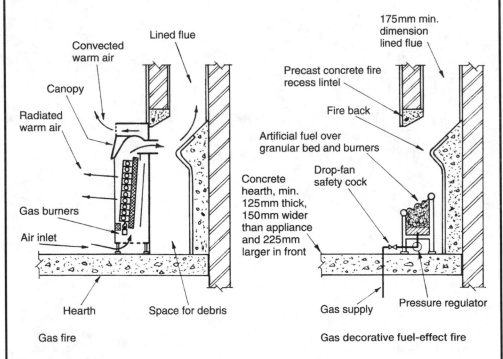

Gas fire

Gas decorative fuel-effect fire

Ref. BS 5871: Specification for the installation and maintenance of gas fires, convector heaters, fire/back boilers and decorative fuel-effect gas appliances (in four parts).

Note: *Gas appliances are rated by maximum heat input rate (kW net). If the rating is given in kW gross, this will include a factor for latent heat of condensation in combustion.

Radiant heaters – in tube format these are simple and effective heat emitters, most suited to high ceiling situations such as industrial units, warehouses and factories. They suspend above the work area and provide a very efficient downward radiation of up to 40kW. Gas is fired into one end of a tube and the combustion gases extracted by fan-assisted flue at the other. The tube may be straight or return in a 'U' shape to increase heat output. A polished stainless steel back plate functions as a heat shield and reflector.

The control box houses an air intake, electronic controls, gas regulators and safety cut-out mechanisms. This includes a gas isolator in the event of fan failure. To moderate burning, the end of the tube has a spiral steel baffle to maintain even temperature along the tube.

Advantages over other forms of heating include a rapid heat response, low capital cost, easy maintenance and high efficiency.

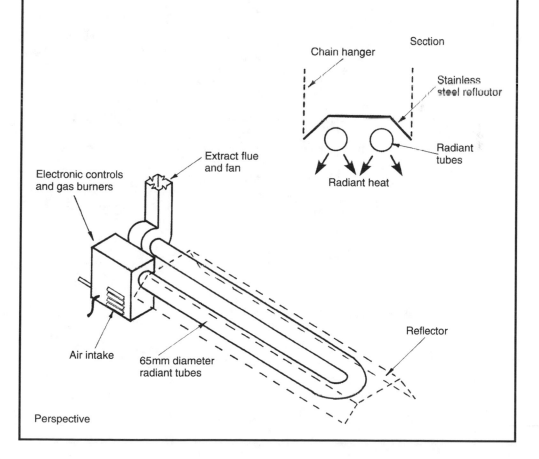

Section

Chain hanger

Stainless steel reflector

Radiant tubes

Radiant heat

Extract flue and fan

Electronic controls and gas burners

Air intake

65mm diameter radiant tubes

Reflector

Perspective

507

Gas Appliances – Convector Heater

Convector – a wall-mounted, balanced flue appliance rated up to about 7kW. They are compact units, room sealed and therefore independent of natural draught from the room in which they are installed. The flue is integral with the appliance and must be installed on an external wall. An exception is when the flue is fan assisted, as this will permit a short length of horizontal flue to the outside wall.

Air for combustion of gas is drawn from outside, through a different pathway in the same terminal as the discharging combusted gases. Correct installation will ensure that the balance of air movement through the terminal is not contaminated by exhaust gases.

About 90% of the heat emitted is by convection, the remainder radiated. Some convectors incorporate a fan, so that virtually all the heat is convected.

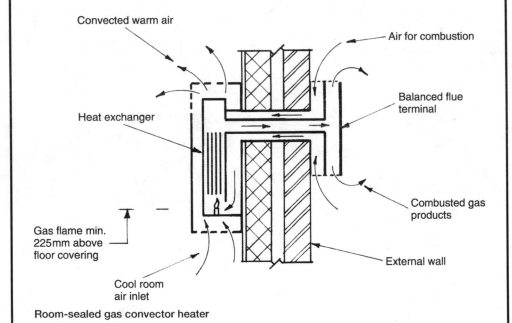

Convected warm air

Air for combustion

Heat exchanger

Balanced flue terminal

Combusted gas products

Gas flame min. 225mm above floor covering

External wall

Cool room air inlet

Room-sealed gas convector heater

Refs. Building Regulations, Approved Document J: Combustion appliances and fuel storage systems. Section 3 – Additional provisions for gas-burning appliances with a rated input up to 70kw (net).

The balanced flue appliance has the air inlet and flue outlet sealed from the room in which it is installed. It is more efficient than a conventional open flue pipe as there is less heat loss in and from the flue. As it is independent of room ventilation, there are no draughts associated with combustion and there is less risk of combustion products entering the room. It is also less disruptive to the structure and relatively inexpensive to install.

A balanced flue is designed to draw in the air required for gas combustion from an area adjacent to where it discharges its combusted gases. These inlets and outlets must be inside a windproof terminal sited outside the room in which the appliance is installed. Gas appliances in a bath or shower room, or in a garage must have balanced flues.*

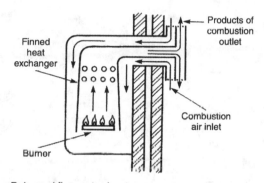

Balanced flue water heater

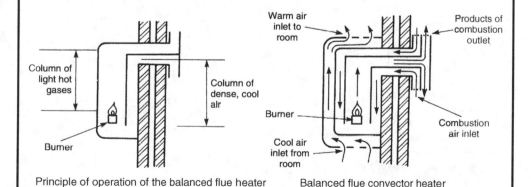

Principle of operation of the balanced flue heater   Balanced flue convector heater

*Ref. Gas Safety (Installation and Use) Regulations.

Balanced Flue Location (Gas) – 1

Balanced flue terminals must be positioned to ensure a free intake of air and safe dispersal of combustion products. Generally, they should be located on a clear expanse of wall, not less than 600mm from internal or external corners and not less than 300mm below openable windows, air vents, grilles, gutters or eaves.

A terminal less than 2m from ground level should be fitted with a wire mesh guard to prevent people contacting with the hot surface. Where a terminal is within 600mm below a plastic gutter, an aluminium shield 1.5m long should be fitted to the underside of the gutter immediately above the terminal. A terminal facing a public access space, e.g. pavement, to be at least 2.1m above ground level.

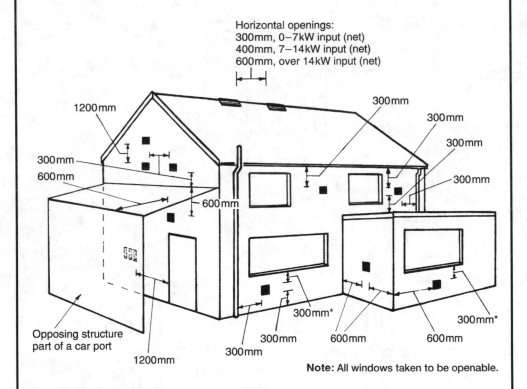

Horizontal openings:
300mm, 0–7kW input (net)
400mm, 7–14kW input (net)
600mm, over 14kW input (net)

1200mm

300mm

300mm

300mm

300mm

300mm

600mm

600mm

Opposing structure part of a car port

1200mm

300mm

300mm

300mm*

600mm

300mm*

600mm

Note: All windows taken to be openable.

*0–7kW input (net) – 300mm, 7–14kW input (net) – 600mm, 14–32kW input (net) – 1500mm, over 32kW input (net) – 2000mm

Balanced flue and ridge natural terminal positions (min. dimensions)

Ref. Building Regulations, Approved Document J: Section 3.

Natural draught flues – appliances discharging flue gases by natural convection are located on an external wall. There must be some regard for the adjacent construction as unsatisfactory location may result in:

- inefficient combustion of fuel
- risk of fire
- combustion products staining the wall
- combustion gases entering the building.

Fan-assisted flues – appliances fitted with these can be located a short distance from an external wall. Smaller terminals are possible due to the more positive extraction of the flue gases. Terminal location is not as critical as for natural draught flues, but due regard must still be given to adjacent construction.

Location of balanced flue terminals (min. distance in mm):

Location of terminal	Natural draught	Fan assisted
Directly under an openable window or a ventilator	300*	300
Under guttering or sanitation pipework	300	75
Under eaves	300	200
Under a balcony or a car port roof	600	200
Horizontally to an opening window	As ridge openings shown previous page	300
Opening in a car port	1200	1200
Horizontally from vertical drain and discharge pipes	300	150 75 < 5kW input (net)
Horizontally from internal or external corners	600	300
Above ground, balcony or flat roof	300	300
From an opposing wall, other surface or boundary	600	600
Opposite another terminal	600	1200
Vertically from a terminal on the same wall	1200	1500
Horizontally from a terminal on the same wall	300	300

*See note on previous page.

Balanced Flue – Condensing Boiler

Installation must be with regard to the intrusive characteristic volume of flue gases that discharge in the form of a plume of moisture droplets. In addition to the flue location guidance given on the previous two pages, a horizontal discharge is not permitted within 2·5m of an opposing wall, a boundary fence or a neighbouring property. A vertical and/or horizontal flue pipe extension may be used to avoid these restrictions. Further, the plume should not intrude

- into a carport
- over a frequently used pedestrian area, such as an access route, a patio or a terrace (see note)
- over a vehicle access route or car-parking area (see note).

Note: An exception is where the flue discharge is at least 2·1m above surface or ground level.

Drainage of the condensation produced by the boiler must also be considered. The condensate can amount to as much as 4 litres in a day, and as it is slightly acid (pH 3–6, see page 21), it must be suitably disposed of. The most convenient means for disposal may be

- to a waste pipe connecting to an internal stack
- into an external gully or rainwater hopper that connects to a combined drainage system
- into a purpose-made soakaway.

Condensate pipes must be fitted with a water seal trap of at least 38mm depth if discharging to an open gully or rainwater hopper. The seal must be 75mm when the condensate pipe connects directly to a sanitation system waste pipe or discharge stack. The principles are as shown below.

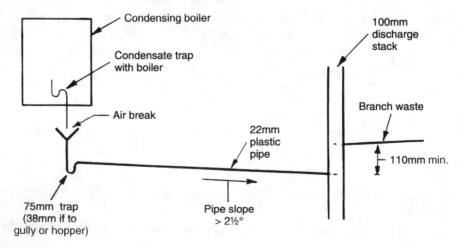

Where gravity discharge is impractical (e.g. from a boiler located in a basement), condensate may be pumped from a sump collector.

A gas appliance may be situated in a fire recess and the chimney structure used for the flue. The chimney should have clay flue linings to BS EN 1457: Chimneys – Clay/ceramic flue liners. A stainless steel flexible flue lining may be installed where the chimney was built before 1 February 1966, provided the lining complies with BS EN 1856-2: Chimneys. Requirements for metal chimneys. Metal flue liners and connecting flue pipes. Prior to installation approval is required from the gas supplier to determine that these standards meet with their specific requirements.

Other suitable flue materials include

Pre-cast hollow concrete flue blocks, pipes made from stainless steel, enamelled steel, cast iron and fibre cement as specified in the Building Regulations (ref. below). Other products may be used that satisfy an acceptable quality standard, such as that awarded by the British Board of Agrément.

Flues must be correctly sized from appliance manufacturer's data (see pages 537 to 539). If a flue is too large or too long, overcooling of the flue gases will produce condensation. This occurs at about 60°C when the gases cool to the dew point of water. The following factors will determine the flue size:

* heat input to the appliance
* resistance to the flow of combustion gases caused by bends and the terminal
* length of the flue.

Spigot and socket flue pipes are installed socket uppermost and joints made with fire cement. For the efficient conveyance of combusted gases, flue pipes should be vertical wherever possible. Where they pass through a floor or other combustible parts of the structure, they should be fitted with a non-combustible sleeve.

A ventilation opening (air-brick) for combustion air is required in the external wall of the room containing the appliance. As a guide, for large boilers in their own plant room a ventilation-free area of at least twice the flue area is required. For domestic appliances, 500mm^2 for each kilowatt of net input rating net is adequate.

Ref. Building Regulations, Approved Document J: Combustion appliances and fuel storage systems. Section 3.

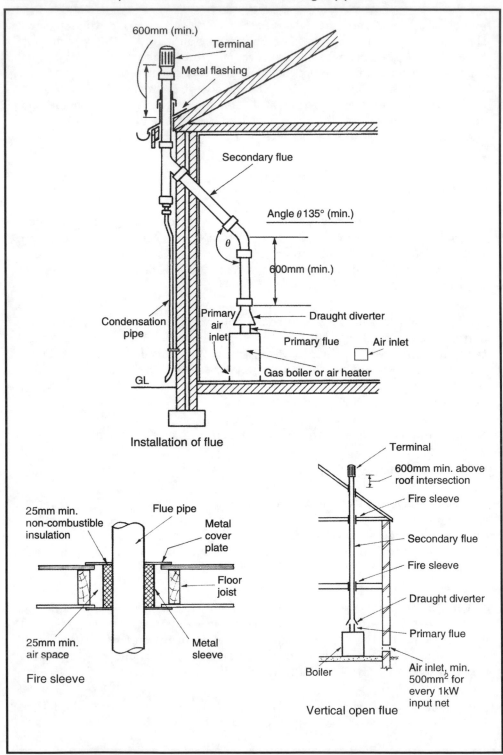

600mm (min.)

Terminal

Metal flashing

Secondary flue

Angle θ 135° (min.)

θ

600mm (min.)

Draught diverter

Condensation pipe

Primary air inlet

Primary flue

Air inlet

Gas boiler or air heater

GL

Installation of flue

25mm min. non-combustible insulation

Flue pipe

Metal cover plate

Floor joist

25mm min. air space

Metal sleeve

Fire sleeve

Terminal

600mm min. above roof intersection

Fire sleeve

Secondary flue

Fire sleeve

Draught diverter

Primary flue

Boiler

Air inlet, min. 500mm^2 for every 1kW input net

Vertical open flue

514

The purpose of a draught diverter is to admit diluting air into the primary flue to reduce the concentration of combustion gases and to reduce their temperature in the flue. The draught diverter, as the name suggests, also prevents flue downdraughts from extinguishing the gas pilot flame by diverting the draughts outside the burners. Draught diverters can be provided in two ways: either as an open lower end to the flue (integral) or as an attachment (separate) to the primary flue.

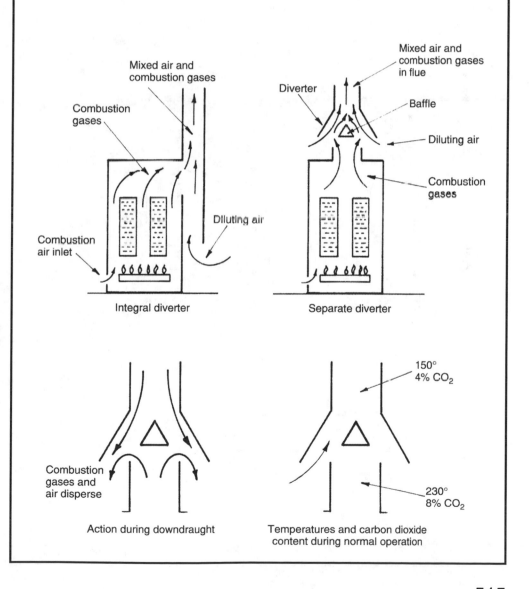

Integral diverter

Separate diverter

Action during downdraught

Temperatures and carbon dioxide content during normal operation

Pre-Cast Concrete Flue Blocks

Pre-cast concrete flue blocks are manufactured from high alumina cement and dense aggregates, to resist the effects of toxic flue gases and condensation. They are jointed with high alumina cement mortar and laid alternately and integrally with the inner leaf of concrete blockwork in a cavity wall. This optimises space and appearance, as there is no chimney structure projecting into the room or unsightly flue pipe. The void in the blocks is continuous until it joins a twin-wall insulated flue pipe in the roof space to terminate at ridge level.

These flue blocks are specifically for gas fires and convectors of relatively low rating. While a conventional circular flue to a gas fire must be at least 12000 mm^2 cross-sectional area, these rectangular flue blocks must have a minimum flue dimension of 90mm and cross-sectional area of 16500mm^2.

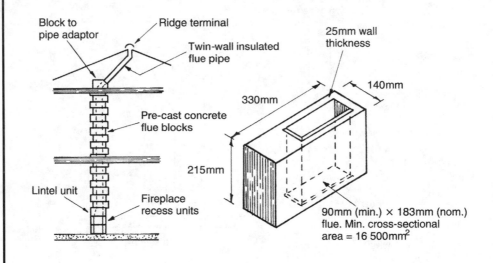

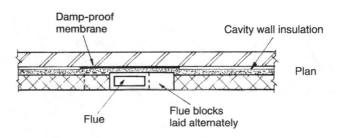

Ref. BS EN 1858: Chimneys. Components. Concrete flue blocks.

516

A flue terminal has several functions:

- to prevent entry of birds, squirrels, etc.
- to prevent entry of rain and snow
- to resist the effects of downdraughts
- to promote flue pull and extraction of combusted gases.

Location – should be with regard to positive and negative wind pressures acting on a roof to permit free wind flow across the terminal and not be too close to windows and other ventilation voids. The preferred location is at or above the ridge of a pitched roof. Elsewhere, the following can be used as guidance:

Location	Min. height (mm) to lowest part of outlet
Within 1·5m horizontally of a vertical surface, e.g. dormer	600 above top of structure
Pitched roof <45°	600 from roof intersection
Pitched roof >45°	1000 " " "
Flat roof	250 " " "
Flat roof with parapet*	600 " " "

*Note: if horizontal distance of flue from parapet is greater than 10 × parapet height, min. flue height – 250mm.

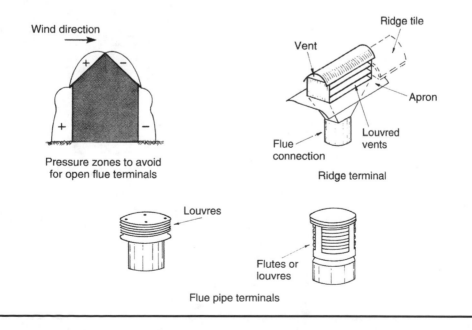

Wind direction

+ −

+ −

Pressure zones to avoid
for open flue terminals

Ridge tile

Vent

Apron

Flue
connection

Louvred
vents

Ridge terminal

Louvres

Flue pipe terminals

Flutes or
louvres

517

Open Flue Terminals – 2

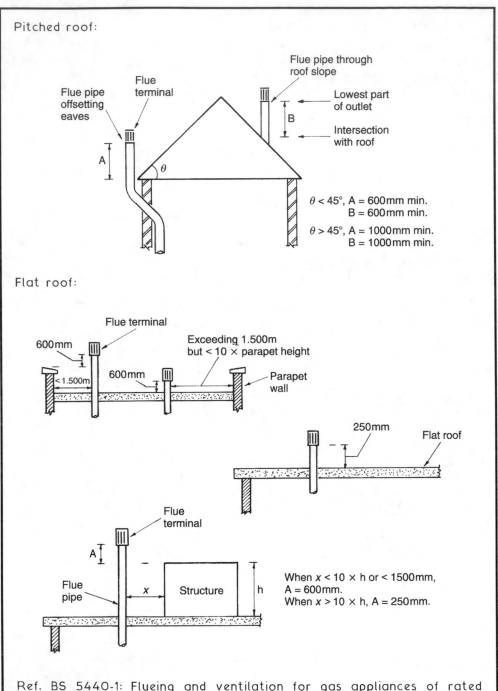

Pitched roof:

Flue pipe through roof slope

Flue terminal

Flue pipe offsetting eaves

Lowest part of outlet

B

Intersection with roof

A

θ

$\theta < 45°$, A = 600mm min.
B = 600mm min.

$\theta > 45°$, A = 1000mm min.
B = 1000mm min.

Flat roof:

Flue terminal

600mm

Exceeding 1.500m but < 10 × parapet height

600mm

< 1.500m

Parapet wall

250mm

Flat roof

Flue terminal

A

Flue pipe

x

Structure

h

When $x < 10 \times h$ or < 1500mm, A = 600mm.
When $x > 10 \times h$, A = 250mm.

Ref. BS 5440-1: Flueing and ventilation for gas appliances of rated input not exceeding 70kW net (first, second and third family gases). Specification for installation of gas appliances to chimneys and for maintenance of chimneys.

Traditional brick chimneys have unnecessarily large flues when used with gas-burning appliances. If an existing unlined chimney is to be used, a flexible stainless steel lining should be installed to prevent the combustion products and condensation from breaking down the old mortar joints. By reducing the flue area with a lining, this will accelerate the discharge of gases (efflux velocity), preventing them from lowering sufficiently in temperature to generate excessive condensation.

Coils of stainless steel lining material are available in 100, 125 and 150 mm diameters to suit various boiler connections. The existing chimney-pot and flaunching are removed to permit the lining to be lowered and then made good with a clamping plate, new flaunching and purpose-made terminal.

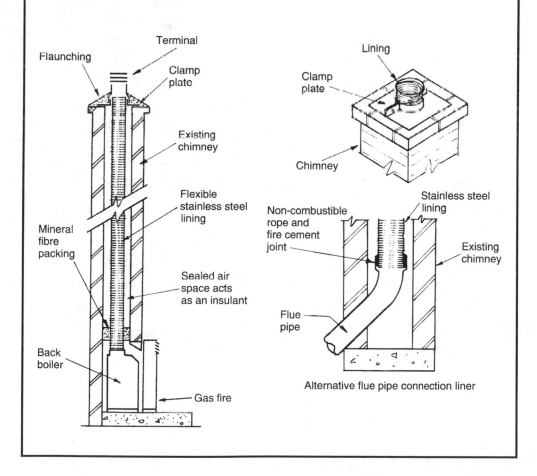

Alternative flue pipe connection liner

Shared Flues – SE-Duct

This is a cost-effective alternative to providing a separate flue for each gas appliance installed in a multi-storey/multi-unit building. It was originally developed by the South-East Gas Board to utilise balanced flues attached to a central ventilated void. Appliances use a central duct for air supply to the gas burners and to discharge their products of combustion. The dilution of burnt gases must be sufficient to prevent the carbon dioxide content from exceeding 1.5% at the uppermost appliance. The size of central void depends on the number of appliances connected. Tables for guidance are provided in BS 5440-1: Flueing and ventilation for gas appliances of rated input not exceeding 70kW net (first, second and third family gases). Specification for installation of gas appliances to chimneys and for maintenance of chimneys.

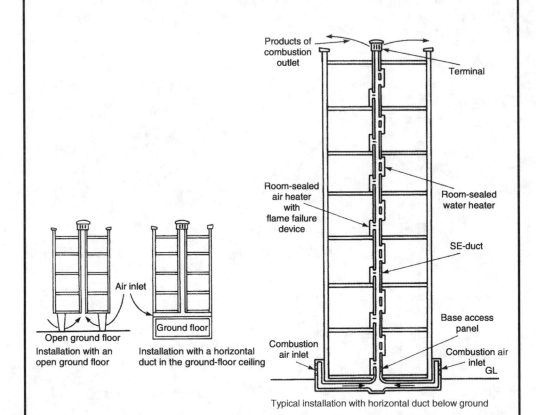

Installation with an open ground floor

Installation with a horizontal duct in the ground-floor ceiling

Typical installation with horizontal duct below ground

Note: A flame failure device is otherwise known as a flame supervision device.

520

The `U´ duct system is similar in concept to the SE-duct, but used where it is impractical to supply air for combustion at low level. The `U´ duct has the benefits of the SE-duct, but it will require two vertical voids which occupy a greater area. The downflow duct provides combustion air from the roof level to appliances. Appliances of the room-sealed type are fitted with a flame failure/supervision device to prevent the build-up of unburnt gases in the duct. They can only connect to the upflow side of the duct. Stable air flow under all wind conditions is achieved by using a balanced flue terminal, designed to provide identical inlet and outlet exposure. As with the SE-duct, the maximum amount of carbon dioxide at the uppermost appliance inlet must be limited to 1.5%.

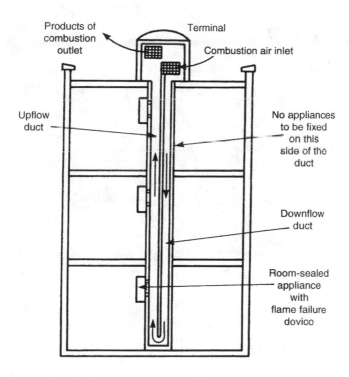

Typical installation of 'U' duct

Shared Flues – Concentric Duct

A metal flue system comprising a flue within a flue. A dual concentric double-wall duct or annular duct flue system uses the inner circular duct as an extract for combusted gases and the outer annular flue as a fresh air supply for fuel combustion. Suitable for use with several individual balanced flue appliances in apartment buildings and similar applications. A space-saving alternative to 'U' duct and SE-duct systems.

Maximum of 20 balanced flue appliances per shared vertical flue riser and no more than two appliances per floor.

Standard flue diameters –

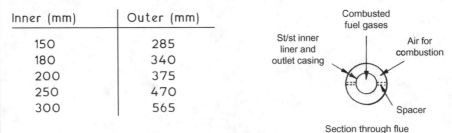

Inner (mm)	Outer (mm)
150	285
180	340
200	375
250	470
300	565

Section through flue

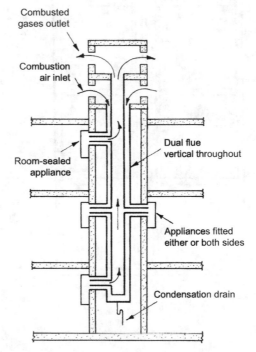

Typical installation of concentric dual duct

Ref. BS EN 1856-1: Chimneys. Requirements for metal chimneys. System chimney products.

Shared Flues – Shunt Duct and Branched Flues

The shunt duct system is applicable to the installation of several conventional appliances with open flues in the same building. It economises in space and installation costs when compared to providing each appliance with an individual flue. It is limited to ten consecutive storeys due to the effects of varying wind pressures and each appliance must be fitted with a draught diverter and a flame failure/ supervision device. Gas fires and water heaters may be connected to this system, provided the subsidiary flue from each is at least 1.2m and 3m long, respectively, to ensure sufficient draught.

Other shared flue situations may be acceptable where conventional open-flued appliances occupy the same room. Consultation with the local gas authority is essential, as there are limitations. An exception is the connection of several gas fires to a common flue. Also, a subsidiary branch flue connection to the main flue must be at least 600mm long measured vertically from its draught diverter.

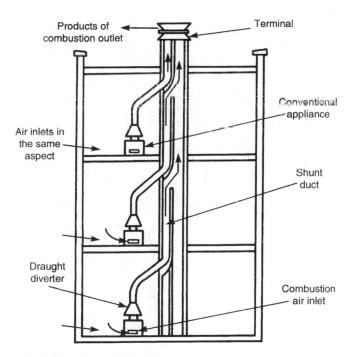

Typical installation of shunt duct

Note: Guidance on sizing of shared flues is provided in BS 5440-1.

523

Fan-Assisted Gas Flues

With high-rise shops, office buildings and flats sharing the same boiler, problems can arise in providing a flue from ground-floor plant rooms. Instead of extending a vertical flue from ground level to the top of a building, it is possible to air dilute the flue gases and discharge them at relatively low level by installing an extract fan in the flue. As the boiler fires, the fan draws fresh air into the flue to mix with the products of gas combustion and to discharge them to the external air. The mixed combustion gases and diluting air outlet terminal must be at least 3m above ground level, and the carbon dioxide content of the gases must not exceed 1%. A draught sensor in the flue functions to detect fan operation. In the event of fan failure, the sensor shuts off the gas supply to the boilers.

The plant room is permanently ventilated with air-bricks or louvred vents to ensure adequate air for combustion. Ventilation voids should be at least equivalent to twice the primary flue area.

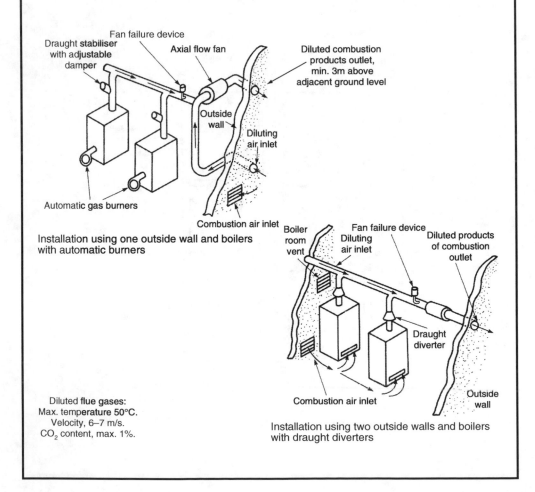

Fan failure device
Draught stabiliser with adjustable damper
Axial flow fan
Diluted combustion products outlet, min. 3m above adjacent ground level
Outside wall
Diluting air inlet
Automatic gas burners
Combustion air inlet

Installation using one outside wall and boilers with automatic burners

Boiler room vent
Fan failure device
Diluting air inlet
Diluted products of combustion outlet
Draught diverter
Combustion air inlet
Outside wall

Diluted flue gases:
Max. temperature 50°C.
Velocity, 6–7 m/s.
CO_2 content, max. 1%.

Installation using two outside walls and boilers with draught diverters

Fan assistance with the dilution and removal of combustion products has progressed from commercial and industrial applications in open flues, to domestic appliance balanced flues. In addition to diluting the CO_2 content at the point of discharge of flue gases, fanned draught-balanced flue systems have the following advantages over standard balanced flues:

- Positive control of flue gas removal without regard for wind conditions.
- Location of flue terminal is less critical – see page 511.
- Flue size (inlet and outlet) may be smaller.
- Flue length may be longer, therefore the boiler need not be mounted on an external wall.
- Heat exchanger may be smaller due to more efficient burning of gas. Overall size of boiler is reduced.

The disadvantages are noise from the fan and the additional features could make the appliance more expensive to purchase and maintain.

If the fan fails, the air becomes vitiated due to lack of oxygen and the flames smother. The flame failure/protection device then closes the gas valve.

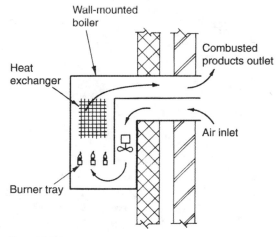

Fan at inlet

Note: Fan may deliver air only to combustion chamber, or an air/gas mixture.

Fan at outlet

Note: Fan will be specifically designed to withstand high flue gas temperatures.

Location of fan

Room-sealed balanced flue appliances do not require a purpose-made air vent for combustion as the air supply is integral with the terminal. Where installed in a compartment or in an enclosure such as a cupboard, an air vent is necessary to remove excess heat. With open or conventional flue appliances, access must be made for combustion air. This equates to at least 500mm^2 of free area per kW (net), e.g. the ventilation area required for an open-flued boiler of 20kW (net) input rating will be at least $20 \times 500 = 10000$mm^2 (see also page 528).

Conventionally flued appliances will also require air for cooling if they are installed in a compartment. This may be by natural air circulation through an air-brick or with fan enhancement.

Flueless appliances such as a cooker or instantaneous water heater require an openable window direct to outside air, plus the following ventilation grille requirements:

Oven, hotplate or grill:

Room volume (m^3)	Ventilation area (mm^2)
<5	10000
5–10	5000 (non-required if a door opens directly to outside air)
>10	non-required

Instantaneous water heater (max. input 11kW (net)):

Room volume (m^3)	Ventilation area (mm^2)
<5	not permitted
5–10	10000
10–20	5000
>20	non-required

Vents should be sited where they cannot be obstructed. At high level, they should be as close as possible to the ceiling, and at low level not more than 450mm above floor level. When installed between internal walls, vents should be as low as possible to reduce the spread of smoke in the event of a fire.

Open-flued gas fires rated below 7kW (net) installed in older premises (built pre-2009) require no permanent ventilation, but decorative fuel-effect fires will require a vent of at least 10000mm^2 free area.

The next page illustrates requirements for room-sealed and open-flued appliances.

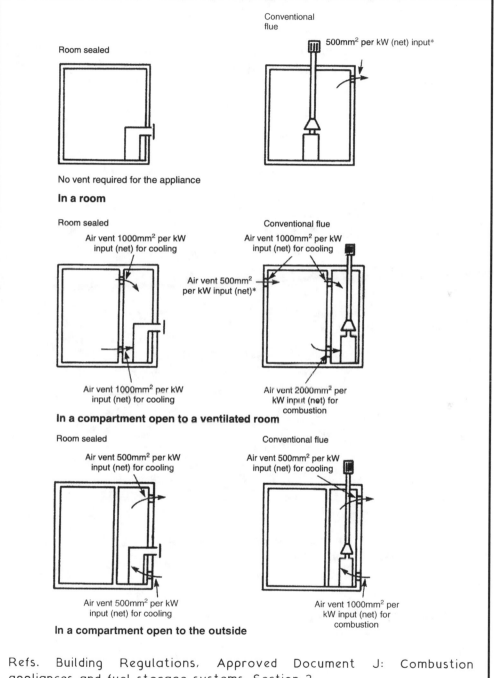

Conventional flue

Room sealed

500mm² per kW (net) input*

No vent required for the appliance

In a room

Room sealed

Air vent 1000mm² per kW input (net) for cooling

Conventional flue

Air vent 1000mm² per kW input (net) for cooling

Air vent 500mm² per kW input (net)*

Air vent 1000mm² per kW input (net) for cooling

Air vent 2000mm² per kW input (net) for combustion

In a compartment open to a ventilated room

Room sealed

Air vent 500mm² per kW input (net) for cooling

Conventional flue

Air vent 500mm² per kW input (net) for cooling

Air vent 500mm² per kW input (net) for cooling

Air vent 1000mm² per kW input (net) for combustion

In a compartment open to the outside

Refs. Building Regulations, Approved Document J: Combustion appliances and fuel storage systems. Section 3.

*Older dwellings (built pre-2009) having air permeability >5 m³/hour per m² at 50 Pa, the first 7kW (net) can be ignored.

Ventilation for Gas Appliances – Calculations

Calculations relate to applications shown on the preceding page.

Example 1: A conventional open flue appliance of 12kW net input rating in an older dwelling (see note on page 506 regarding input and output ratings).

• Installed in a room.

No vent required up to 7kW, but 500mm^2 to be provided per kW thereafter:

12kW − 7kW = 5kW × 500mm^2 = 2500mm^2 air vent area.

• Installed in a cupboard compartment open to a ventilated room.

Air vent area is the same as above.

Vent area for cooling the appliance is 1000mm^2 for every kW rating:

12kW × 1000mm^2 = 12000mm^2

Ventilation, cooling and combustion air area:

12kW × 2000mm^2 = 24000mm^2.

• Installed in a compartment open to the outside.

Air for cooling the appliance is 500mm^2 for every kW rating:

12kW × 500mm^2 = 6000mm^2.

Air for combustion:

12kW × 1000mm^2 = 12000mm^2.

Example 2: A room-sealed balanced flue appliance of 12kW net input rating.

• In a cupboard compartment open to a ventilated room. Air for ventilation and cooling is 1000mm^2 (twice):

12kW × 1000mm^2 = 12000mm^2 (twice).

• In a cupboard compartment open to the outside. Air for ventilation and cooling is 500mm^2 per kW (twice):

12kW × 500mm^2 = 6000mm^2.

Note: Provision for ventilation in walls may be partly by natural infiltration, but where this is insufficient, purpose-made air-bricks are built into the wall. These should not be obscured or covered over.

Simple field tests are available to assess the efficiency of gas combustion with regard to the percentage of carbon monoxide and carbon dioxide in the flue gases.

Draeger analyser – hand bellows, gas sampler tube and a probe. The tube is filled with crystals corresponding to whether carbon monoxide or carbon dioxide is to be measured. The probe is inserted into the flue gases and the bellows pumped to create a vacuum. The crystals absorb different gases and change colour accordingly. Colours correspond with a percentage volume.

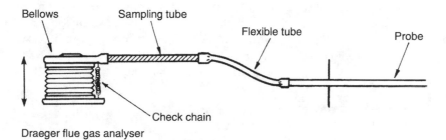

Draeger flue gas analyser

Fyrite analyser – hand bellows, container of liquid reactant and a probe. Flue gases are pumped into the container which is inverted so that the liquid reactant absorbs the gas in solution. The liquid rises to show the percentage carbon dioxide corresponding to a scale on the container. Oxygen content can also be measured using an alternative solution.

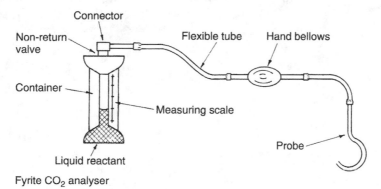

Fyrite CO_2 analyser

Note: Flue gas samples can be taken by inserting the probe below the draught diverter or through the access plate on top of the appliance combustion chamber. Samples can also be taken at the terminal. The above apparatus is retained to illustrate the principles of probe testing. Modern LCD handheld units are now in general use and have the benefit of determining flue gas temperature, O_2, CO and CO_2 content – see next page.

Electronic Analysis of Combusted Gas

Portable electronic battery powered combustion gas analysers have largely superseded the devices shown on the preceding page. Flue gas analysers contain sensors that respond to the various gases emitted from combustion of fuel. Analysers vary in specification depending on whether application is to natural gas, LPG or light oil. Selection of suitable instrument is important, as one type may not fit all situations.

An internal pump draws combusted fuel through a filter from a steel tubular probe inserted into the appliance flue. Basic readings are for oxygen (O_2), carbon monoxide (CO), carbon dioxide (CO_2) and the ratio of CO to CO_2. More sophisticated instruments provide details of flue gas temperature and humidity and have a data storage facility. A gas leak detection function may also be incorporated.

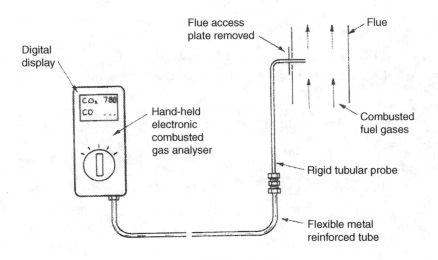

Refs.

BS 7967: Guide for the use of electronic portable combustion gas analysers for the measurement of CO in dwellings and the combustion performance of domestic gas fired appliances.

BS EN 50379 – 1,2 and 3: Specification for portable electrical apparatus designed to measure combustion flue gas parameters of heating appliances.

Calculations relating to the storage, conveyance and combustion of gas include factors for volume, pressure and temperature at constant mass. If not restrained, gas will expand when heated and occupy more than its pre-heated volume. If constrained and the volume of gas is restricted, gas when heated will increase in pressure.

Boyle's law – for a fixed mass of gas at constant temperature, the volume is inversely proportional to its absolute pressure.

$$P = C \div V \quad \text{or} \quad PV = C$$

where

P = pressure (absolute, i.e. gauge pressure + atmospheric pressure)

V = volume   C = constant

By adapting the formula, it is possible to calculate the volume that gas will occupy relative to change in pressure:

$$P_1 V_1 = P_2 V_2$$

where

P_1 = initial pressure (absolute) V_1 – initial volume

P_2 – new pressure (absolute) V_2 = new volume

E.g.

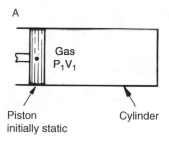

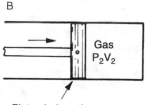

A

Gas
$P_1 V_1$

Piston
initially static

Cylinder

B

Gas
$P_2 V_2$

Piston halves the gas
volume and doubles the
pressure at A

At A, $2V = P \div 2$ At B, $V \div 2 = 2P$

$P \times V$ is always the same. E.g. for $P_1 = 2$ and $V_1 = 20$:

Value of P_2	Value of V_2	Constant sum
4	10	40
8	5	..
10	4	..
20	2	..

Note: At normal operating pressures Boyle's law is reasonably true, but at high pressures there is some variation.

Charles' law – this differs from Boyle's law by considering the effect of temperature on gas. Charles' law states that for a fixed mass of gas at constant pressure, the volume occupied is directly proportional to the absolute or thermodynamic temperature. The proportion is 1/273 of the gas volume at 0°C for every degree rise in temperature. Therefore, if a gas at 0°C is raised to 273°C its volume will double.

Minus 273°C is absolute temperature at zero degrees kelvin (see page 764), the theoretical point at which gas has no volume. Therefore,

$$V \div T = C$$

where

V = volume

T = absolute temperature

C = constant

By adapting the formula, it is possible to calculate the volumes occupied by the same gas at different temperatures at constant pressure:

$$V_1 \div T_1 = V_2 \div T_2$$

where

V_1 = initial volume

T_1 = initial temperature (absolute)

V_2 = new volume

T_2 = new temperature (absolute)

E.g. An underground service pipe containing gas at 5°C supplies a boiler room at 20°C.

$T_1 = 5 + 273 = 278K$ $T_2 = 20 + 273 = 293K$

Transposing Charles' formula to make V_2 the subject:

$V_2 = (V_1 \, T_2) \div T_1$ where V_1 occupies unit volume of gas at $1m^3$

$V_2 = (1 \times 293) \div 278 = 1.054m^3$

This suggests that the consumer would get some free fuel ($0.054m^3$ for every $1m^3$ metered), but gas accounts usually contain a correction factor for the volume conversion.

Changes in the conditions affecting gas will normally include pressure and temperature at the same time. Therefore, if Boyle's and Charles' laws are combined, the three conditions of volume, pressure and temperature can be represented. In this format, the formula is known as the general gas law:

$$PV \div T = C$$

P, V, T and C are as indicated on the previous two pages.

By adapting the general gas law formula, a gas under two different conditions can be compared:

$$(P_1V_1) \div T_1 = (P_2V_2) \div T_2$$

E.g. If a consumer's gas supply is set to 20mbar (millibars) by the meter pressure governor, it will be reduced again at an appliance pressure governor. For this example, say 5mbar.

Note: Atmospheric pressure is taken at 101.3kN/m² or 1013mbar.

For 1m³ initial volume of gas, Boyle's law can be used to show the volume of gas at the reduced pressure of the appliance:

$$P_1V_1 = P_2V_2$$

Transposing:

$$V_2 = (P_1V_1) \div P_2$$
$$= ([1013 + 20] \times [1]) \div (1013 + 5) = 1.015m^3$$

If the gas has a temperature of 10°C at the meter and 16°C at the appliance, the general gas law to determine the new volume (V_2) of gas with regard to pressure and temperature difference can be applied:

$$(P_1V_1) \div T_1 = (P_2V_2) \div T_2$$

where P_1 = 1013 + 20 = 1033mbar
P_2 = 1013 + 5 = 1018mbar
V_1 = 1m³
V_2 = unknown
T_1 = 10 + 273 = 283K
T_2 = 16 + 273 = 289K

Transposing the general gas law to make V_2 the subject:

$$V_2 = (P_1V_1T_2) \div (P_2T_1)$$
$$= (1033 \times 1 \times 289) \div (1018 \times 283) = 1.036m^3$$

Gas Flow Rates in Pipes

The rate of gas flowing in a pipe can be calculated by applying Pole's formula. This is a variation of the D'Arcy fluid flow formula shown on pages 78 and 79.

Pole's formula can be expressed as:

$$q = 0.001978 \times d^2 \times \sqrt{(h \times d) \div (s \times l)} = \text{litres per second (l/s)}$$

$$Q = 0.0071 \times \sqrt{(h \times d^5) \div (s \times l)} = \text{cubic metres per hour (m}^3\text{/h)}$$

where 0.001978 and 0.0071 are constant friction coefficients

h = pressure loss in millibars (mb)

d = pipe diameter (mm)

s = specific gravity of gas (natural gas approx. 0.6)

l = length of pipe conveying gas (m)

The second formula is usually favoured. This provides a figure compatible with gas consumed by an appliance, in m^3/h.

For example, determine the gas flow rate in a 10m length of 15mm o.d. copper tube (13.5mm i.d.) with an acceptable pressure loss of 1mb.

$$Q = 0.0071 \times \sqrt{(1 \times 13.5^5) \div (0.6 \times 10)}$$

$$Q = 0.0071 \times 273.3749 = 1.941 \text{m}^3\text{/h}$$

Pole's formula can be rearranged to make pressure loss (h) the subject:

$$h = (Q^2 \times s \times l) \div (d^5 \times 0.0071^2)$$

It can be seen that the pressure loss (h) is directly proportional to:

• the square of the flow rate (Q)
• the gas specific gravity (s)
• the pipe length (l).

Pressure loss varies inversely with the fifth power of the pipe diameter (d).

If the quantity of gas is doubled, the pressure loss will increase four times, i.e. $(2)^2$.

If the pipe length is doubled, the pressure loss will double.

If the pipe diameter is halved, the pressure loss will increase 32 times, i.e. $(2)^5$.

Note: Pole's formula is limited to normal low-pressure gas installations. Under higher pressure, alternative formulae which incorporate gas compressibility factors are more appropriate.

Typical natural gas consumption figures for domestic appliances:

Boiler   1.6m³/hour
Cooker   1.0 "   "
Fire   0.5 "   "

Exact gas consumption rate (Q) can be calculated from the following formula:

$$Q = \frac{\text{Appliance rating} \times 3600}{\text{Calorific value of gas}}$$

Given that the calorific values for natural gas and propane (LPG) are 38500kJ/m³ and 96000kJ/m³, respectively, the value of Q for a 20kW input boiler is:

$$\text{Nat. gas: } Q = \frac{20 \times 3600}{38500} = 1.87\text{m}^3/\text{h}$$

$$\text{Propane: } Q = \frac{20 \times 3600}{96000} = 0.75\text{m}^3/\text{h}$$

Operating costs – fuel tariffs can be obtained from the various gas suppliers. A typical charge for natural gas is 1.3 pence per kWh. If the 20kW input boiler consumes gas for five hours per day, the operating cost will be

$$1.3 \times 20 \times 5 = \text{£1.30 per day or £9.10 per week}$$

To convert gas metered in units of cubic feet, multiply by 0.0283, i.e. 1 cu. ft. = 0.0283m³.

Gas consumed in kWh:

$$\frac{\text{m}^3 \times \text{volume conversion factor (1.02264)} \times \text{calorific value (MJ/m}^3)}{3.6}$$

where: 1kWh = 3.6MJ (conversion factor)
e.g. 100cu. ft at 2.83m³

$$\frac{2.83 \times 1.02264 \times 38.5}{3.6} = 31\text{kWh}$$

Gas Pipe Sizing

To determine the size of pipework, two factors must be established:

1. The gas consumption (Q).
2. The effective length of pipework.

The effective length of pipework is taken as the actual length plus the following allowances for fittings in installations up to 28mm outside diameter copper tube:

Fitting	Equivalent length (m)
elbow	0.5
tee	0.5
bend (90°)	0.3

The gas discharge in m^3/hour for copper tube for varying effective lengths is as follows:

Tube dia. (mm o.d)	Effective pipe length (m)							
	3	6	9	12	15	20	25	30
8	0.52	0.26	0.17	0.13	0.10	0.07		
10	0.86	0.57	0.50	0.37	0.30	0.22	0.18	0.15
12	1.50	1.00	0.85	0.82	0.69	0.52	0.41	0.34
15	2.90	1.90	1.50	1.30	1.10	0.95	0.92	0.88
22	8.70	5.80	4.60	3.90	3.40	2.90	2.50	2.30
28	18.00	12.00	9.40	8.00	7.00	5.90	5.20	4.70

This table is appropriate for 1mb (10mm w.g.) pressure drop for gas of relative density 0.6.

Example:

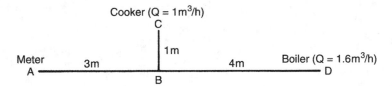

Note: A to B contains 3 elbows and 1 tee

   B to C contains 3 elbows

   B to D contains 4 elbows

Pipe A to B, gas flow = $1m^3$/h + $1.6m^3$/h = $2.6m^3$/h

Actual pipe length = 3m

Effective pipe length = 3 + (3 × 0.5) + (1 × 0.5) = 5m

From the table, a 22mm o.d. copper tube can supply $2.6m^3$/h for up to 23.75 metres (by interpolating between 20 and 25m).

Pressure drop over only 5m will be: 5 ÷ 23.75 = 0.21mb (2.1mm w.g.).

Pipes B to C and B to D can be calculated similarly.

Ref: BS 6891: Specification for the installation and maintenance of low pressure gas installation pipework of up to 35mm on premises.

Open flue, naturally aspirated – a flue pipe equivalent to the size of the appliance outlet is generally adequate. However, some variation may be possible, but care must be taken not to undersize the flue, as this will cause a high efflux velocity and spillage of combustion products. Oversizing is uneconomical, less easy to accommodate and likely to produce condensation in the flue.

Example:

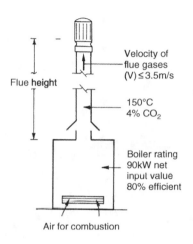

Velocity of flue gases (V) ≤ 3.5m/s

Flue **height**

150°C
4% CO_2

Boiler rating 90kW net input value 80% efficient

Air for combustion

Ambient temperature 20°C

Gross input value of boiler
$= 90kW \times \dfrac{100}{80}$
$= 112.50kW$

Calorific value of natural gas
$= 38500kJ/m^3$

$$\text{Gas consumption rate (Q)} = \frac{\text{Appliance rating} \times 3600}{\text{Calorific value of gas}}$$

$$= \frac{112.50 \times 3600}{38\ 500} = 10.52 m^3/h$$

$$\text{Flue gas volume (v)} = \left| \frac{100}{\% \ CO_2} + 2 \right| \times \text{Gas rate (Q)} \times \frac{^\circ C \text{ absolute} + ^\circ C \text{ flue gas}}{^\circ C \text{ absolute} + ^\circ C \text{ ambient}}$$

$$v = \left| \frac{100}{4} + 2 \right| \times 10.52 \times \frac{(273 + 150)}{(273 + 20)} = 410 m^3/h$$

$$\text{Area of flue pipe (A)} = \frac{\text{Flue gas volume (v)}}{\text{Velocity of flue gas (V)}}$$

where Flue gas volume (v) per second = 410 ÷ 3600 = $0.1139 m^3/s$

$$A = \frac{0.1139}{3.5} = 0.0325 m^2 \qquad \text{From, } A = \pi r^2, \text{ radius(r)} = 0.1018m$$

Therefore, flue diameter = 0.203m, or 203mm (8" standard imperial size)

Gas Appliance Flue Sizing – 2

Induced draught flue – a conventional or open flue with a flue gas extract fan. Extract velocity (V) is between 6 and 7.5m/s.

Using a 112.50kW gross input rated boiler from the example on the previous page, the gas consumption rate (Q) and flue gas volume (v) are $10.52m^3/h$ and $410m^3/h$, respectively.

The flue pipe diameter formula is as shown on the previous page, but with the velocity of fanned flue gases (V) increased to, say, 7m/s.

$$A = \frac{v}{V} = \frac{0.1139}{7} = 0.0163m^2 \qquad \text{From, } A = \pi r^2, \text{ radius (r)} = 0.0719m$$

Therefore, flue diameter = 0.144m, rounded up to 152mm (6" standard imp. size)

$$\text{Velocity check: } \frac{\text{Flue gas volume (v)}}{\text{Flue area (A)}} = \text{Flue gas velocity (V)}$$

$$\frac{0.1139}{\pi \times (0.076)^2} = 6.3m/s \text{ Between 6 and 7.5m/s, therefore}$$
$$\text{152mm flue is satisfactory}$$

Fan air-diluted flue – see page 524 for installation between two side walls and for operating data. Using two of the 112.50kW rated boilers with flue gas extract velocity (V) between 6 and 7m/s, the following formula may be used to obtain the flue gas volume (v):

$$v = 9.7 \times \text{Appliance rating} \times \frac{°C \text{ absolute} + \text{flue gas } °C}{°C \text{ absolute} + \text{amibient } °C}$$

$$v = 9.7 \times 112.50 \times 2 \times \frac{(273 + 50)}{(273 + 20)} = 2406m^3/h \text{ or, } 0.6683m^3/s$$

$$A = \frac{v}{V} = \frac{0.6683}{6} = 0.1114m^2 \text{ flue area}$$

A square flue will be $\sqrt{0.1114} = 334mm \times 334mm$

A circular flue is derived from Area(A) $= \pi r^2$ where, r = radius

$$\text{Therefore, } r = \sqrt{A \div \pi}$$
$$r = \sqrt{0.114 \div 3.1416} = 0.188m \text{ or } 188mm$$
$$\text{diameter of circular duct is } 2 \times r = 376mm$$

The following formula is a guide to the minimum flue height (H) in metres, with regard to the efficient discharge of flue gases from naturally aspirated boilers:

$$H = 6 \times (\text{Boiler rating gross input in MW})^{0.6} + \text{allowance for resistances at bends, etc.}$$

Factors for resistance to flue gas flow can be taken as listed below:

Flue pipe component	Resistance factor
90° bend	0.50
135° bend	0.25
Vertical terminal	0.25
Ridge terminal	1.00
Horizontal flue	0.30/m
Inclined flue (45°)	0.13/m
Vertical flue	Zero

Taking the examples shown in the previous two pages of one 112.50 kW gross input (90 kW net) rated boiler. Assuming that the boiler flue is vertical with the exception of two 135° bends, one metre of 45° inclined flue and a vertical terminal, the formula for minimum flue height can be written:

$$H = 6 \times (0.1125)^{0.6} + (0.25 \times 2) + (0.13) + (0.25) = 2.4975 \text{ i.e. } 2.5\text{m}$$

Condensation within a flue system must be prevented by:

- keeping the flue gas temperature as high as possible
- keeping the dew point of the flue gases low.

In practical terms, this is achieved by:

- correctly sizing the flue to avoid excessive surface areas
- insulating the flue or use of double-walled, insulated flue pipes
- limiting the lengths of flue systems (see graph).

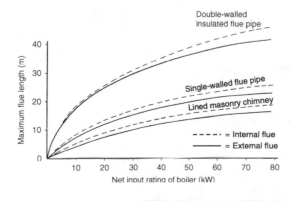

Flue Outlets – Adjacent Roof Lights

Roof lights in line with the roof slope, dormer windows and lantern lights are common features in new build dwellings, extensions to existing buildings and in loft conversions to create habitable accommodation. The position of roof ventilation openings must be with regard to avoiding the discharge of flue gases and other products of combustion. With unvented sealed or closed systems of hot water and heating, the boiler location is not critical. As a space-saving convenience, it is not unusual to install a boiler within the roof space. Where this is the chosen location, consideration must be given to dispersal of combustion gases from a flue outlet and proximity of roof-light ventilation openings.

Specific requirements for positioning solid fuel flue outlets are shown on page 227, page 240 for liquid fuel (oil) and page 518 for gas. The illustration below is supplement to provisions for oil and gas fuels.

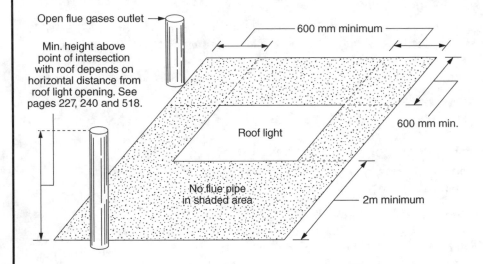

Refs: BS 5440-1: Installation and maintenance of flues.

Building Regulations, Approved Document J: Combustion appliances and fuel storage systems. Clauses: 2.10 (Solid fuel), 3.23 (Gas) and 4.6 (Oil).

11 ELECTRICAL SUPPLY AND INSTALLATIONS

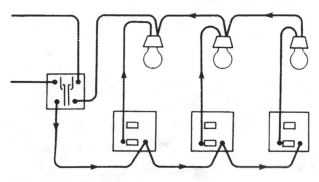

THREE-PHASE GENERATION AND SUPPLY
ELECTRICITY DISTRIBUTION
ELECTRICITY INTAKE TO A BUILDING
EARTHING SYSTEMS AND BONDING
CONSUMER UNIT
POWER AND LIGHTING CIRCUITS
OVERLOAD PROTECTION
INSULATED CABLE
TESTING COMPLETED INSTALLATION
CABLE RATING
DIVERSITY
INDUSTRIAL INSTALLATIONS
ELECTRIC SPACE HEATING
CONTROLS FOR ELECTRIC NIGHT STORAGE
SPACE HEATERS
CONSTRUCTION SITE ELECTRICITY
LIGHT SOURCES, LAMPS AND LUMINAIRES
LIGHTING CONTROLS
EXTRA-LOW-VOLTAGE LIGHTING
LIGHTING DESIGN
DAYLIGHTING
TELECOMMUNICATIONS INSTALLATION
ELECTRIC VEHICLE CHARGER

DOI: 10.1201/9781003434894-11

Three-Phase Generation and Supply

In 1831, Michael Faraday succeeded in producing electricity by plunging a bar magnet into a coil of wire. This is credited as being the elementary process by which we produce electricity today, but the coils of wire are cut by a magnetic field as the magnet rotates. These coils of wire (or stator windings) have an angular spacing of 120°, and the voltages produced are out of phase by this angle for every revolution of the magnets, thus generating a three-phase supply.

A three-phase supply provides 73% more power than a single-phase supply for the addition of a wire. With a three-phase supply, the voltage between two line or phase cables is 1.73 times that between the neutral and any one of the line cables, i.e. 230 volts × 1.73 = 400 volts, where 1.73 is derived from the square root of the three phases.

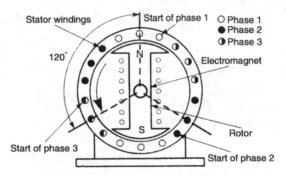

Simplified detail of three-phase generator or alternator

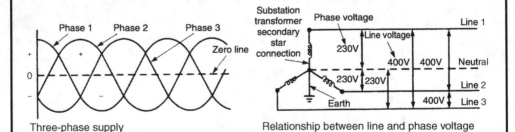

Three-phase supply   Relationship between line and phase voltage

Note: The following section on electrical systems should be read with regard to Building Regulations, Approved Document P: Electrical safety, and BS 7671: Requirements for Electrical Installations, IET Wiring Regulations 18th edition.

Voltage optimisation – controlled variation in voltage received by a consumer for the purpose of regulating energy use.

European harmonisation – before 1995, the standardised three-phase and single-phase supply of electricity in the UK was at 415 volts and 240 volts, respectively. In continental Europe, it was 380 volts and 220 volts. Since, 1995 the nominal voltage across Europe has been harmonised at 400 volts and 230 volts with a single-phase tolerance of between –6% and +10% in the UK, i.e. 216 volts to 253 volts. In 2008, the 230 volt single-phase power supply was revised to have tolerances of ±10%, i.e. 207 volts to 253 volts.

Voltage variation – older electrical appliances and equipment manufactured for UK operation at 240 volts should function perfectly well with European harmonised voltage power supply. Most electrical equipment is now produced to operate at 220 volts, optimising power supply towards the lower end of the statutory voltage range. The gap of 20 volts between supply and functional voltage for older 240 volt over-voltage equipment, will incur significantly more use of energy. This energy inefficiecy can in theory be corrected by the fitting of voltage optimisation to individual pieces of equipment, but in reality it is not economically viable. For industrial and commercial developments, and for community power-stations, particularly where electricity is produced by private power sources, over-voltage (and under-voltage) supply to buildings can be regulated by large scale voltage optimisation plant. This is effectively use of a transformer between the mains distribution or source and mains low voltage single-phase distribution.

Selected other world-wide domestic nominal voltages –

African states	220–250	Australia	230
Canada	120	Caribbean	115 &/or 230
China	220	India	220–250
Japan	110	Middle East	220/230
S. America	120 &/or 220	USA	120

Electricity Distribution

In the UK, electricity is produced at power-generating stations at 25 kilovolt (kV) potential, in three-phase supply at 50 cycles per second or hertz (Hz). Thereafter, it is processed by step-up transformers to 132, 275 or 400kV before connecting to the national grid. Power to large towns and cities is by overhead lines at 132kV or 33kV where it is transformed to an 11kV underground supply to substations. From these substations, the supply is again transformed to the lower potential of 400 volts, three-phase supply and 230 volts, single-phase supply for general distribution.

The supply to houses and other small buildings is by an underground ring circuit from local substations. Supplies to factories and other large buildings or complexes are taken from the 132 or 33kV main supply. Larger buildings and developments will require their own transformer, which normally features a delta-star connection to provide a four-wire, three-phase supply to the building.

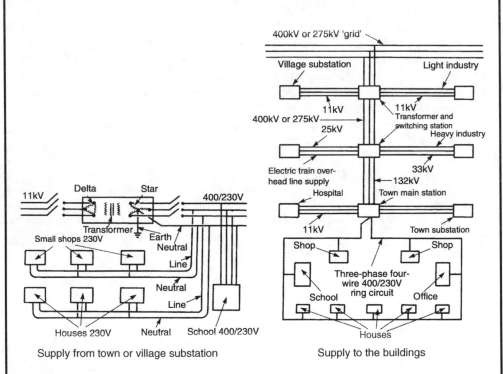

Supply from town or village substation

Supply to the buildings

Note: For easy identification, each phase cable has colour-coded plastic insulation of brown (red), black (yellow) or grey (blue). The neutral is colour coded blue (black). An outer sheathing of red or black provides for future identification. Older installations will have colour codes as indicated in brackets.

A substation is required for the conversion, transformation and control of electrical power. It is used where large buildings or complexes of buildings require greater power than the standard low or medium potential of 230 and 400 volts. A substation must be constructed on the customer's premises. It is supplied by high-voltage cables from the electricity authority's nearest switching station. The requirements for a substation depend upon the number and size of transformers and switchgear.

A transformer is basically two electric windings, magnetically interlinked by an iron core. An alternating electromotive force applied to one of the windings produces an electromagnetic induction corresponding to an electromotive force in the other winding. If the number of turns in the secondary coil is reduced, the voltage is reduced and the current increased, and vice versa. See next page for more detail.

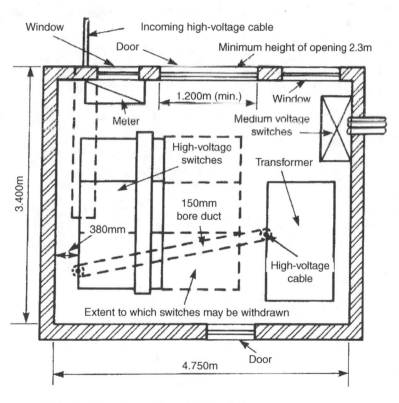

Typical construction and layout of sub-station

Power Transformers

High-voltage electricity transmitted through the national grid has to be reduced in potential before serving end users, the amount of reduction varying depending on whether the supply is for industrial scale premises, commercial or domestic use. In alternating current supply systems, transformers are used to convert voltage and/or current into some other voltage and/or current. Transformers are in principle fairly simple devices, having no moving or rotating parts. They comprise three main components:

- Primary coil – several turns of wire carrying alternating current (AC) from a mains supply.
- Core of magnetic material – traditionally laminations of soft iron, but variations using silicon steel are now preferred.
- Secondary coil – several turns of wire in which an electro-magnetic force is produced by the change in magnetism in the metal core it surrounds.

Transformer principle –

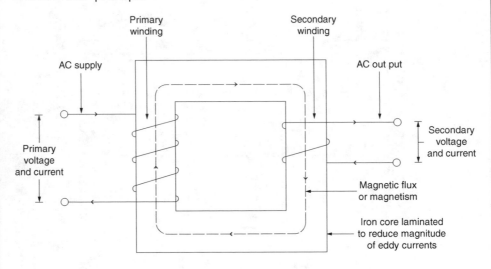

The above principle is based on a fluctuating or alternating electric current passing through a wire to generate a magnetic field in the metal core. This electro-magnetic induction is activated through the core from the primary coil to a secondary coil. The number of turns or windings in each coil determines the change in voltage potential and current.

Step-down transformers – as shown on the preceding page, the primary coil has more turns than the secondary coil. Therefore, the secondary coil voltage is less than that of the primary coil.

$$\frac{\text{Secondary voltage}}{\text{Primary voltage}} = \frac{\text{No. of secondary turns}}{\text{No. of primary turns}}$$

If the secondary coil has half as many turns than the primary coil, e.g. 4 and 8, respectively, then the secondary voltage will be four-eighths or half that of the primary voltage, and so on in proportion.

E.g. 1: As above with a primary voltage of 230 V.

$$\frac{\text{Secondary voltage}}{230} = \frac{4}{8} = 115 \text{ secondary volts}$$

E.g. 2: Primary voltage of 230 V, with a secondary coil having one-tenth as many turns as the primary coil.

$$\frac{\text{Secondary voltage}}{230} = \frac{1}{10} = 23 \text{ secondary volts}$$

Current is transformed in the opposite way:

$$\frac{\text{Secondary current}}{\text{Primary current}} = \frac{\text{No. of primary turns}}{\text{No. of secondary turns}}$$

E.g. Ten times more turns in the primary coil than the secondary coil. As shown above, the voltage will reduce to one-tenth, but the current increases by a factor of ten.

$$\frac{\text{Secondary current}}{\text{Primary current}} = \frac{10}{1}$$

Theoretically the power expressed in watts remains the same in primary and secondary coils, i.e. watts = volts x amps. In reality, some power is lost through magnetic flux and heat from the core.

Step-up transformers – this reversal transforms low voltage into a higher voltage by having more turns on the secondary coil than the primary coil. With a higher secondary voltage, there will be a smaller current. The voltage and current formulae shown on the preceding page for step-down transformers are applicable:

E.g. Primary voltage of 230 volts with a ratio of 10:1 secondary to primary turns.

$$\frac{Secondary\ voltage}{230} = \frac{10}{1} = 2300\ secondary\ volts$$

$$\frac{Secondary\ current}{Primary\ current} = \frac{1}{10}$$

Transformer cooling – the energy transfer between coils will generate heat in the wires and the metal core. This waste of energy reduces transformer efficiency and power output. Heat energy occurs with the resistance to current in both primary and secondary windings and eddy currents in the metal core. Surplus heat energy must be dissipated to surrounding air or be transferred to a heat recovery unit for re-use. Cooling is essential to prevent damage and deterioration of coil wire insulation. Depending on the size of transformer and operating voltage, cooling processes include

- Air cooled by exposure.
- Fan air cooled.
- Liquid water cooled.
- Liquid oil cooled.

Note: Liquid oil cooling has the additional benefit of insulating the coils.

Refs. Estimating energy saving potential from transformers and evaluating their impact on the feasibility of renewable energy systems – DEFRA.

EU Commission Regulation 548/2014. Directive 2009/125/EC: Small, medium and large power transformers.

Alternating current (AC) to direct current (DC) transformers also known as adapters, inverters and converters depending on the application and use. For construction site purposes, the two main advantages are the safety feature of a reduced mains voltage in hand held tools and where tools feature a charged battery/power pack, they can be used in isolation without trailing leads. Electric arc welding units and other fairly heavy equipment require a 230 volt AC mains connection. Transformed output is usually about 50 volts DC but can be AC depending on manufacturer's specification. Arcing current is variable, usually between 100 and 200 amps. Inverter type arc welders have a comparatively small transformer. In relative terms, this reduces unit weight, size and heat energy losses. Inverters use switched mode power supplies (SMPSs), a process of rectifying voltage to a switching circuit that contains a high frequency transformer outputting DC. For practical purposes, this is usually preferred to the fluctuating characteristic of AC output.

Typical circuit diagram for AC to DC transformer

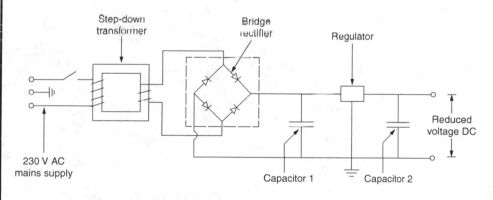

Rectifier   – converts AC to DC.
Capacitor 1 – smooths rectifier output ripples.
Regulator   – maintains constant voltage with varying load.
Capacitor 2 – filters noise or interference.

Electricity Intake to a Building

The termination and metering of services cables to buildings is determined by the electricity authority's supply arrangements. Most domestic supplies are underground with the service cable terminating at the meter cupboard, as shown. Depth of cover to underground cables should be at least 750mm below roads and 450mm below open ground. In remote areas, the supply may be overhead. Whatever method is used, it is essential that a safety electrical earthing facility is provided and these are considered on the next page. All equipment up to and including the meter is the property and responsibility of the supplier. This also includes a fusible cut-out, neutral link and in some situations a transformer. Meters are preferably sited in a purpose-made reinforced plastic compartment set in or on the external wall of a building.

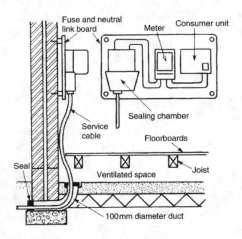

Underground service entry

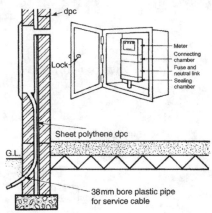

Alternative underground service entry using external meter cabinet

Note: All domestic internal distribution systems must be undertaken by a 'competent person', i.e. a qualified electrician. Electrical contractors certified as competent can 'self-certificate' their work. Work completed by lesser qualified people must be referred to the Local Authority Building Control Department and a fee paid for inspection by their appointed qualified inspector. Minor work, such as replacing socket outlets, control switches and ceiling fittings can be undertaken without contravention.

Ref. Building Regulations, Approved Document P: Electrical Safety.

Supply systems require a safety electrical earthing facility. The manner in which this is effected will depend on whether the supply is overhead or underground and the conductive property of the ground surrounding the installation. Systems are classified in accordance with a letter coding:

First letter – type of earthing:

T – at least one point of the supply is directly earthed.
I – the supply is not directly earthed, but connected to earth through a current limiting impedance. Not acceptable for public supplies in the UK.

Second letter – installation earthing arrangement:

T – all exposed conductive metalwork is directly earthed.
N – all exposed conductive metalwork is connected to an earth provided by the supply company.

Third and fourth letters – earth conductor arrangement:

S – earth and neutral conductors separate.
C – earth and neutral conductors combined.

Common supply and earthing arrangements are
T I (shown below).
TN-S and TN C S (shown on next page).

TT system:

Most used in rural areas where the supply is overhead. An earth terminal and electrode is provided on site by the consumer. As an extra safety feature, a residual current device (RCD), generally known as a trip switch, is located between the meter and consumer unit. The RCD in this situation should be of the time-delayed type – see page 569.

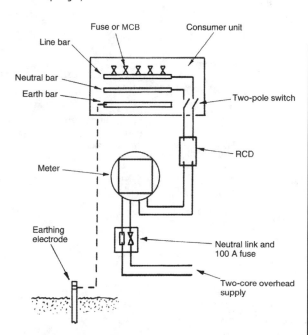

551

TN-S system – this is widely used in the UK, with the electricity supply company providing an earth terminal with the intake cable. This is usually the metal sheathing around the cable, otherwise known as the supply protective conductor. It connects back to the star point at the area transformer, where it is effectively earthed.

TN-C-S system – this is as the TN-S system, but a common conductor is used for neutral and earth supply. The supply is therefore TN-C, but with a separated neutral and earth in the consumer's installation it becomes TN-C-S. This system is also known as protective multiple earth (PME). The advantage is that a fault to earth is also a fault to neutral, which creates a high fault current. This will operate the overload protection (fuse or circuit-breaker) rapidly.

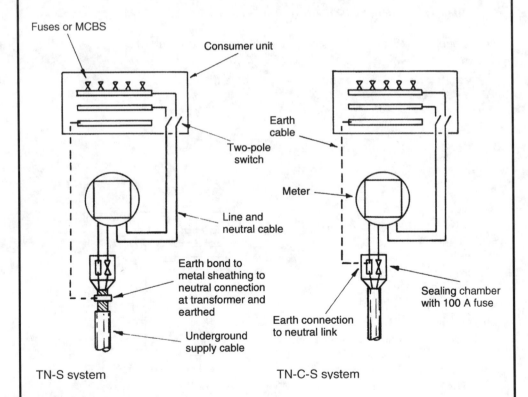

TN-S system

TN-C-S system

Note: Specification of installation cable between supply company's sealing chamber and consumer unit – phase/line and neutral 25mm², earth 16mm² cross-sectional area.

Pages 542, 544 and 552 show that the consumer's earth conductor is connected to the neutral and earthed at the local transformer. For below-ground supplies, this arrangement provides a path of low resistance for an electrical fault. With an overhead supply typical of rural areas, individual consumers must provide a suitable earth terminal or electrode as shown on page 551.

Unless wet, the ground surface is not usually a very good conductor, therefore ground contact is made at about 1.5 to 2m below the surface. In the past, this was achieved by earth bonding to metal water and gas mains. Since the introduction of plastic pipe materials, this is of course no longer acceptable. Current practices include burying a metal plate or a metal tape mesh arranged over several square metres, or driving a metal rod electrode into the ground. The latter is normally adequate for domestic and other small-scale installations. In some instances, the electrode is housed as shown below. Whatever earth method is used, a low resistance to an electrical fault is essential. The IET Wiring Regulations recommend that the earth electrode resistance should not exceed 200ohms.

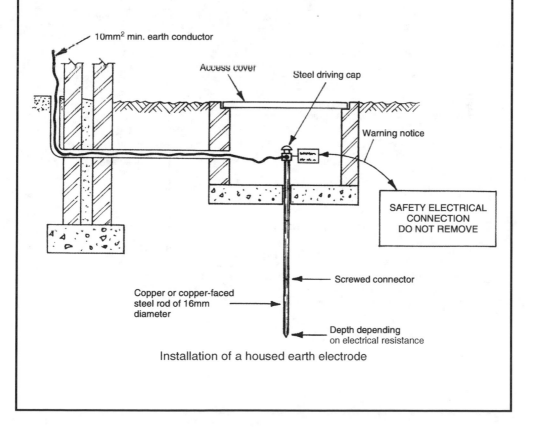

Installation of a housed earth electrode

Equipotential Bonding of Services and Extraneous Metalwork

Metal sheathed and armoured cables, earthed metal trunking, metal service pipework associated with electrical equipment and fixed structural steelwork liable to introduce a potential to be protectively earthed by bonding together and connecting to earth. This ensures that no dangerous potential differences can occur as there will be a low-resistance return path to earth to promptly operate the overload protection device.

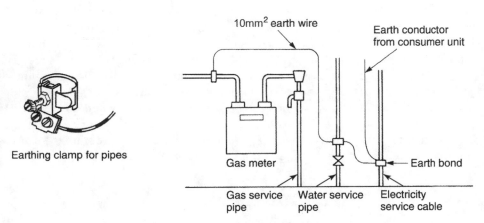

Earthing clamp for pipes

Main equipotential bonding of services at intake

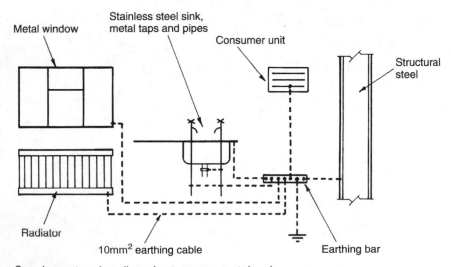

Supplementary bonding of extraneous metalwork

As indicated on the previous four pages, every part of an electrical installation must be earthed. This is achieved by connecting all exposed conductive parts with a circuit protective conductor (cpc) and joining this to the main earthing terminal. The cpc is usually a single core cable with distinct green and yellow insulation, although metal trunking and conduit used for cable conveyance may also function as the cpc.

Earthing provision of exposed and extraneous metal parts is shown on the preceding page. This ensures that no dangerous potential difference can occur between possible conductive parts.

- Main equipotential bonding – of at least 10 mm² cross-sectional area (csa) is attached to the gas and water supplies with an earth clamp as shown on the preceding page. Connection to the gas pipe is within 600 mm of the meter on the consumer's side (see page 494) and above the water supply stop valve if the supply to the valve is in plastic. If the water supply pipe is metal, connection is before the valve.

- Supplementary bonding – provided for fixed metalwork or extraneous conductive parts, i.e. metalwork that is not directly associated with the electrical installation but could accidentally come into contact with it and become live. This will include taps (electric immersion heater), radiator (central heating pump), window (cable through to garden), etc. All extraneous metalwork in a bathroom must be bonded.

A minimum of 4 mm² csa supplementary bonding conductor satisfies most domestic situations, but if the cpc exceeds 10 mm² csa the supplementary bonding conductor must have at least half this csa. E.g. a 16 mm² csa cpc will require 10 mm² csa supplementary bonding (6 mm² is too small and 8 mm² csa is not a standard commercially available specification). Supplementary bonding conductors of less than 16 mm² must not be of aluminium.

Ref. BS 951: Electrical earthing. Clamps for earthing and bonding. Specification.

Consumer Unit

Historically, electrical installations required a separate fuse and isolator for each circuit. Modern practice is to rationalise this into one 'fuse box', known as a consumer's power supply control unit or consumer unit for short. This unit contains a two-pole switch isolator for the phase/line and neutral supply cables and three bars for the line, neutral and cpc to earth terminals. The line bar is provided with several fuseways or miniature circuit-breakers (up to 16 in number for domestic use) to protect individual circuits from overload. Each fuse or MCB is selected with a rating in accordance with its circuit function. Traditional fuses are rated at 5, 15, 20, 30 and 45 amps while the more modern MCBS are rated in accordance with BS EN 60898 Electrical accessories: Circuit-breakers for over-current protection for household and similar installations.

Circuit	MCB rating (amps)
Lighting	6
Immersion heater	16 or 20*
Socket ring	32
Cooker	40 or 45*
Shower	40 or 45*

*Depends on the power rating of appliance. A suitable MCB can be calculated from: amps = watts ÷ voltage.

E.g. A 3kW immersion heater: amps = 3000 ÷ 230 = 13.

Therefore, a 16 amp rated MCB is adequate.

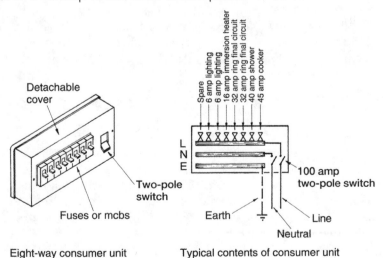

Eight-way consumer unit · Typical contents of consumer unit

Ref. BS EN 61439-3: Low-voltage switchgear and controlgear assemblies.

556

A split load consumer unit provides for additional and specific protection to outgoing circuits that may supply electricity to portable equipment for use outdoors. This is particularly appropriate for ground-floor sockets that could have an extension lead attached, e.g. cooker control panel, kitchen ring final circuit and ground-floor ring final circuit.

These ground-floor circuits have a dedicated line and neutral bar within the consumer unit and an RCD (RCCB) protection device in addition to miniature circuit-breakers for each individual circuit. A typical disposition of components within a split load consumer unit is as shown.

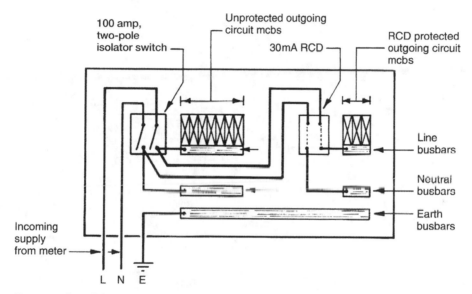

Contents of a split load consumer unit

Types of protection against residual current by residual current devices (RCDs):

- RCCB – Residual current circuit-breaker. An incoming switch disconnecting device activated by an earth leakage fault.
- RCBO – Residual current circuit breaker with integral overload protection. An alternative to a miniature circuit-breaker (MCB) as an outgoing individual circuit protection device. It has a dual function, combining earth leakage protection with the current overload protection provided by an MCB.

Supplementary Consumer Unit

Where an existing consumer unit is in good order but of insufficient capacity to accept additional fuseways/MCBs, replacement with a larger unit is not always necessary. It is not acceptable to connect more than one circuit to a fuseway. If there is adequate space, an additional consumer unit can be added in parallel to the existing unit. When upgrading any of the intake fitments, the electricity supply authority should be consulted to determine that their supply equipment and facility would not be overloaded, particularly where it is proposed to provide for high-powered appliances and fittings such as a cooker and/or a shower. Line (phase) and neutral cable connections between meter and consumer unit(s) may need to be upgraded to 25mm² csa and the earth conductor to 16mm² csa.

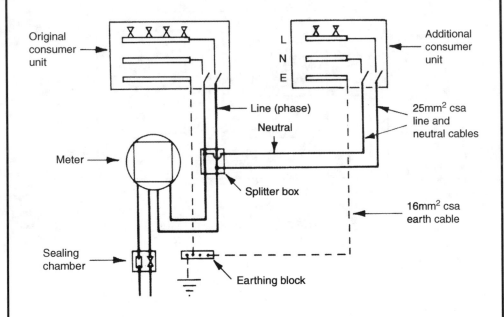

With the intake isolated, a service connector box or splitter box is fitted to the live and neutral supply cables between the meter and the existing consumer unit. From this connection, supply cables extend to the additional consumer unit. See note on page 550 regarding competence of installer.

A ring final circuit is used for single-phase power supply to three-pin sockets. It consists of PVC sheathed cable containing line and neutral conductors in PVC insulation and an exposed circuit protective conductor to earth looped into each socket outlet. In a domestic building, a ring final circuit may serve an unlimited number of sockets up to a maximum floor area of $100\,m^2$. A separate circuit is also provided solely for the kitchen, as this contains relatively high rated appliances. Plug connections to the ring have small cartridge fuses up to 13 amp rating to suit the appliance wired to the plug. The number of socket outlets from a spur should not exceed the number of socket outlets and fixed appliances on the ring. Maximum one socket (single or double) to each non-fused spur, or maximum one fixed appliance.

Cable rating:

$2.5\,mm^2$ csa

Consumer unit:

BS EN 61439-1, 2 and 3.

Three-pin plugs and sockets:
BS 1363-1 and 2.

Plug cartridge fuses:
BS 1362.

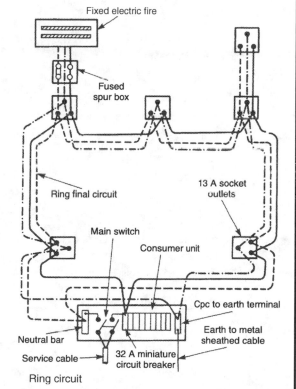

Fixed electric fire

Fused spur box

Ring final circuit

13 A socket outlets

Main switch

Consumer unit

Cpc to earth terminal

Earth to metal sheathed cable

Neutral bar

Service cable

32 A miniature circuit breaker

Ring circuit

Note: Fixed appliances such as fires, heating controls and low-powered water heaters can be connected to a fused spur from a ring socket. Appliances and installations with a load factor above 3kW (e.g. immersion heater, cooker, extension to an outbuilding, etc.) must not be connected to any part of a ring final circuit. These are supplied from a separate radial circuit from the consumer unit.

Power Sockets

Power sockets should be positioned between 150mm and 250mm above work surfaces and between 450mm and 1200mm above floor levels. An exception is in buildings designed for the elderly or infirm, where socket heights should be between 750 and 900mm above the floor. Every socket terminal should be fitted with a double outlet to reduce the need for adaptors. Disposition of sockets would limit the need for lead lengths to no more than 2m.

The following provides guidance on the minimum provision for power sockets in domestic accommodation:

Location	Minimum quantity of sockets
Living rooms	8
Kitchen	6
Master bedroom	6
Dining room	4
Study bedroom	4
Utility room	4
Single bedrooms	4
Hall and landing	2
Garage/workshop	2
Bathroom	1 – double-insulated shaver socket

Maximum appliance load (watts) and plug cartridge fuse (BS 1362) selection for 230 volt supply:

Maximum load (W)	Plug fuse rating (amp)
230	1
460	2
690	3
1150	5
1610	7
2300	10
2900	13

Calculated from: watts = amps × voltage.

A radial circuit may be used as an alternative to a ring final circuit to supply any number of power sockets, provided the following limitations are effected:

Cable csa (mm^2)	Minimum overload protection (amps)	Remarks
2.5	20	Max. 20m^2 floor area, 17m cable
4.0	30	Max. 50m^2 floor area, 21m cable

With 2.5mm^2 cable length limitation of 17m over 20m^2 floor area for a radial supply to sockets, a ring main with a maximum cable length of 54m over 100m^2 will usually prove to be more effective. Therefore, radial circuits are more suited to the following:

Application	Cable csa (mm^2)	Minimum overload protection (amps)	Remarks
Lighting	1.5	5	Max. 10 light fittings
Immersion heater	2.5	15	Butyl rubber flex from two-pole control switch
Cooker	6	30	Cable and fuse
	10	45	ratings to suit cooker rating
Shower	4, 6 or 10	30 to 45	See page 431
Storage radiator	2.5	20	See page 583
Outside extension	2.5	20	Nominal light and power
	4	30	Max. five sockets and 3 amp light circuit (next page)

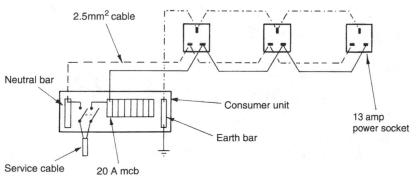

Radial circuit to power sockets

Radial Extension to an Outbuilding

An electricity supply to an outside building may be overhead at a height not less than 3.5m. It may be supported in a conduit or from a catenary suspension wire. An underground supply is less obtrusive and should be at least 500mm below the surface. The cable should be armoured PVC sheathed or copper sheathed mineral insulated (MICC). Standard PVC insulated cable may be used, provided it is enclosed in a protective conduit. Fused isolators are required in the supply building and the outside building, and a residual current device (RCD) 'trip switch' should also be installed after the fused switch control from the consumer unit. 2.5mm^2 csa cable is adequate for limited installations containing no more than a power socket and lighting. In excess of this, a 4mm^2 csa cable is preferred, particularly if the outbuilding is some distance to overcome the voltage drop.

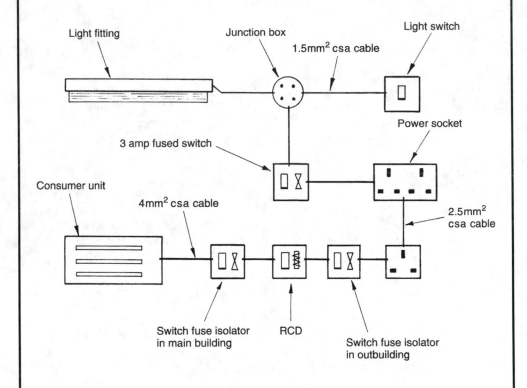

Schematic diagram of electricity supply to an outbuilding

Lighting circuits can incorporate various switching arrangements. In a one-way switch circuit, the single-pole switch must be connected to the line conductor. To ensure that both line and neutral conductors are isolated from the supply a double-pole switch may be used, although these are generally limited to installations in larger buildings where the number and type of light fittings demand a relatively high current flow. Provided the voltage drop (4% max., see page 577) is not exceeded, two or more lamps may be controlled by a one-way single-pole switch.

In principle, the two-way switch is a single-pole change-over switch interconnected in pairs. Two switches provide control of one or more lamps from two positions, such as that found in stair/landing, bedroom and corridor situations. In large buildings, every access point should have its own lighting control switch. Any number of these may be incorporated into a two-way switch circuit. These additional controls are known as intermediate switches. See lower details below and page 565.

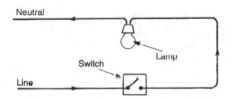

One-way single-pole switch circuit controlling one lamp.

One-way single-pole switch circuit controlling two or more lamps

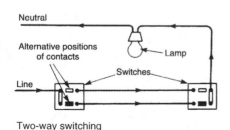

Two-way switching

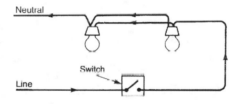

Two-way switching with one intermediate switch

Note: Cpc to earth also required between switch and lamp fitments, but omitted here for clarity.

The purpose of a `master´ switch is to limit or vary the scope of control afforded by other switches in the same circuit. If a `master´ switch (possibly one with a detachable key option) is fixed near the main door of a house or flat, the householder is provided with a means of controlling all the lights from one position.

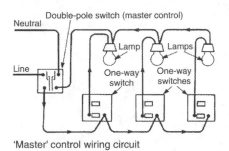

'Master' control wiring circuit

Note: Cpc omitted as indicated on preceding page.

A sub-circuit for lighting is generally limited to a total load of ten, 100 watt light fittings. It requires a 5 amp fuse or 6 amp MCB overload protection at the consumer unit. The importance of not exceeding these ratings can be seen from the simple relationship between current (amps), power (watts) and potential (voltage), i.e. amps = watts ÷ volts. To avoid overloading the fuse or MCB, the limit of 10 lamps @ 100 watts becomes

amps = (10 × 100) ÷ 230 = 4.3

i.e. < 5 amps fuse protection.

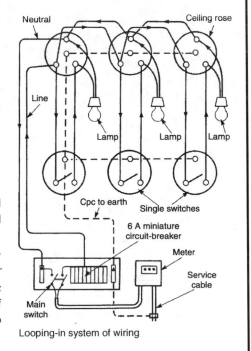

Looping-in system of wiring

In large buildings, higher rated overload protection is often used due to the greater load.

Wiring for lighting is usually undertaken using the `looping-in´ system, although it is possible to use junction boxes instead of ceiling roses for connections to switches and light fittings.

Two-way switching is convenient for hall/landing lighting control and for bedroom door/bedside control. Intermediate switching has application to long corridors and multi-flight stairways.

Two-way switching

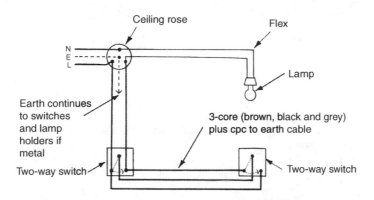

Intermediate switching

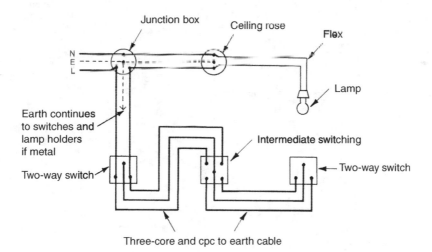

Sleeving – in addition to using green and yellow striped sleeving to all exposed earth conductors (see page 571), brown over-sleeving is used specifically in lighting circuits to part cover the blue, black and grey insulated conductors at switches and other terminals to identify where they provide continuity to the brown insulated line conductor.

Table and standard lamps, and up-lighters can plug into 13 amp plug power circuit sockets provided the plug is fitted with a low-amperage (3 amp) fuse. This may occupy power sockets that might be better used for appliances. Also, these sockets are considerably overrated for most supplementary light fittings. Therefore, a dedicated sub-circuit can be provided for light fittings from a socket spur as shown below:

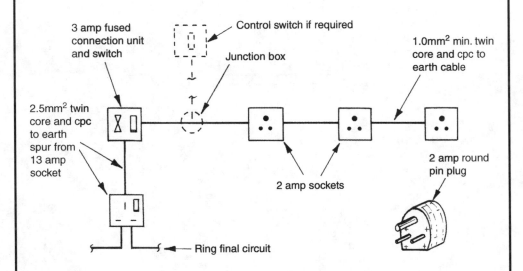

Features:

- Light fitting flex attached to small round pin plugs (historically used for old-style 2 amp power circuits – now obsolete practice).
- Unswitched 2 amp rated socket face plates purpose made for small round pin plugs fitted to single back boxes.
- Switched and fused (3 amp) connection unit spurred off an existing 13 amp power socket with 2.5mm² csa cable.
- 1.0mm² min. csa cable from fused connection unit to each 2 amp socket.
- Sub-circuit max. power output of 690 watts, derived from 3 amp circuit protection × 230 volt supply.
- Individual lamps controlled with their own fitment switch.

The Building Regulations require reasonable provision for people, whether ambulant or confined to a wheelchair, to be able to use a building and its facilities. Facilities include wall-mounted switches and sockets located within easy reach, to be easily operated, visible and free of obstruction.

Dwellings – switches and sockets between 450 and 1200mm from finished floor level (ffl).

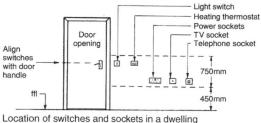

Location of switches and sockets in a dwelling

Non-domestic buildings – basic requirements for switches, outlets and controls:

- Conventional and familiar.
- Contrasting in colour to their surroundings.
- Large push pad preferred or extra-wide rocker switches.
- Pictogram to clarify use and purpose where multiple switches occur.
- Separation or gap between individual switches where multiples exist.

Recommendations for location of wall-mounted switches and sockets in non-domestic buildings:

- Sockets for TV, power and telephone: 400 to 1000mm above ffl and ≥350mm from corners. Power socket switches to indicate whether they are 'ON'.
- Switches to permanently wired appliances: 400 to 1200mm above ffl.
- Controls requiring precise hand movement: 750 to 1200mm above ffl.
- Push buttons, e.g. lift controls: ≤ 1200mm above ffl.
- Pull cords for emergencies, coloured red and located close to a wall and to have two, 50mm diameter bangles set 100mm and 800–900mm above ffl.
- Controls that require close visual perception, e.g. thermostat, located 1200–1400mm above ffl for convenience of people sitting or standing.
- Light switches for general use of the push pad type and located at 900–1100mm height. Alternatively, a pull cord with 50mm diameter bangle set at the same height. The pull cord should be distinguishable from any emergency pull.
- Main and circuit isolators to clearly indicate that they are 'ON' or 'OFF'.
- Pattress or front plate to visually contrast with background.
- Operation of switches and controls to be from one hand, unless both hands are required for safety reasons.

Note: Exceptions to the above may occur in unavoidable design situations such as open plan offices with fitted floor sockets.

Refs. Building Regulations, Approved Document M: Access to and use of buildings. Disability Discrimination Act.

BS 8300: Design of an accessible and inclusive built environment – Buildings.

Overload Protection

Electrical installations must be protected from current overload, otherwise appliances, cables and people using the equipment could be damaged. Historically, protection devices were considered in two categories:

1. Semi-enclosed (rewirable) fuses.
2. High breaking or rupturing capacity (HBC or HRC) cartridge fuses.

Neither of these devices necessarily operate instantly. Their efficiency depends on the degree of overload. Rewirable fuses can have a fusing factor of up to twice their current rating and cartridge fuses up to about 1.6. Contemporary installations use miniature circuit breakers (MCBs). These can carry some overload, but will be instantaneous (0.01 seconds) at very high currents.

Characteristics:

- Semi-enclosed rewirable fuse:

  Obsolete.

  Inexpensive.

  Simple, i.e. no moving parts.

  Prone to abuse (wrong wire could be used).

  Age deterioration.

  Unreliable with temperature variations.

  Cannot be tested.

- Cartridge fuse:

  Obsolete.

  Compact.

  Fairly inexpensive, but costs more than rewirable.

  No moving parts.

  Not repairable.

  Could be abused.

- Miniature circuit-breaker:
  Relatively expensive.

  Factory tested.

  Instantaneous in high current flow.

  Unlikely to be misused.

  MCBs are generally used for low energy domestic applications of < 100 amps.

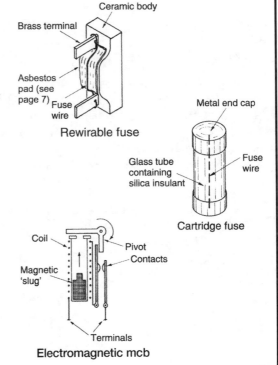

Rewirable fuse

Cartridge fuse

Electromagnetic mcb

BS EN 60898-1 and 2: Electrical accessories. Circuit-breakers for over-current protection for household and similar installations.

Residual Current Devices (RCD) are required where a fault to earth may not produce sufficient current to operate an overload protection device, e.g. an overhead supply. If the impedance of the earth fault is too high to enable enough current to effect the overload protection, it is possible that current flowing to earth may generate enough heat to start a fire. Also, the metalwork affected may have a high potential relative to earth and if touched could produce a severe shock.

An RCD has the load current supplied through two equal and opposing coils, wound on a common transformer core. When the line and neutral currents are balanced (as they should be in a normal circuit), they produce equal and opposing fluxes in the transformer or magnetic coil. This means that no electromotive force is generated in the fault detector coil. If an earth fault occurs, more current flows in the line coil than the neutral and an alternating magnetic flux is produced to induce an electromotive force in the fault detector coil. The current generated in this coil activates a circuit-breaker. While a complete system can be protected by a 100mA (milliamp) RCD, it is possible to fit specially equipped sockets with a 30mA RCD where these are intended for use with outside equipment. Plug-in RCDs are also available for this purpose. Where both are installed, it is important that discrimination comes into effect. Lack of discrimination could effect both circuit-breakers simultaneously, isolating the whole system unnecessarily. Therefore, the device with the larger operating current should be specified with a time delay mechanism. The test resistor provides extra current to effect the circuit-breaker. This should be operated periodically to ensure that the mechanics of the circuit-breaker have not become ineffective due to dirt or age deterioration. A notice to this effect is attached to the RCD.

RCCB – Residual Current Circuit Breaker. An electrical wiring device that will disconnect a circuit immediately there is a current leak to the earth conductor.

RCBO – Residual Current Breaker with Over-current. Combines the function of an MCB and RCCB. Where a current leakage occurs, a RCBO disconnects the whole circuit.

Refs. BS EN 61008-1 (RCCB) and BS EN 61009-1 (RCBO): Residual current operated circuit breakers.

An RCD is not appropriate for use with a TN-C system (i.e. combined neutral and earth used for the supply), as there will be no residual current when an earth fault occurs as there is no separate earth pathway.

They are used primarily in the following situations:

- Where the electricity supply company does not provide an earth terminal, e.g. a TT overhead supply system.
- In bedrooms containing a shower cubicle.
- For socket outlets supplying outdoor portable equipment.

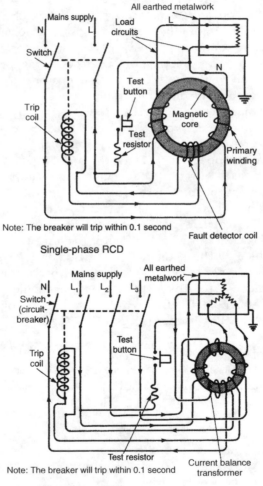

Note: The breaker will trip within 0.1 second

Single-phase RCD

Note: The breaker will trip within 0.1 second

Three-phase RCD

A three-phase device operates on the same principle as a single-phase RCD, but with three equal and opposing coils.

PVC insulated and sheathed cable is now used for most electrical installations. The exceptions are for specialised applications using armoured or mineral insulated cable, and heat-resisting butyl rubber-insulated flex for connecting to heat-producing fittings such as immersion heater and cooker. Historically, lead outer sheathed rubber-insulated cable was used and may still be found in dated and unmodernised buildings.

PVC insulated cable has a temperature limitation of between 0°C and 70°C. Below zero it becomes brittle and can be damaged. At the higher temperature, it softens which could encourage the conductor to migrate through the sheathing. Outside these temperatures, the cable must be protected or an appropriate rubber insulant specified.

Cables are manufactured with one, two or three insulated conductors, with or without a circuit protective conductor (cpc). The cpc is bare (i.e. uninsulated) and must be protected and easily identified with green and yellow sleeving where exposed at junction boxes, sockets, etc.

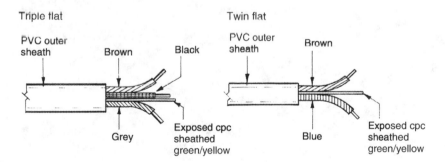

Note: Pre-2006 installations may have the line (phase) conductor colour coded red and the neutral black. Under a European harmonisation requirement, these are now identified brown and blue. For two-way lighting, the old colours were red, yellow and blue, but are now identified as brown, black and grey.

Ref. BS EN 50525 series: Electric cables. Low-voltage energy cables of rated voltage up to and including 450/750V.

Conduit and Trunking

Conduit tubing is used for containment of electrical cables, continuity of support and protection against physical damage and heat. Conduit is available in galvanised or black-painted steel with screw-threaded connections and couplings. Lightweight plastic conduit has push-fit connections. Standard outside diameters are 20, 25, 32, 40, 50 and 65mm in lengths of 3.750m. Shorter length requirements can be cut and threaded. Box section trunking is usually specified where a large capacity is required for multi-cable installations. With smaller diameter conduit, bends and offsets are formed with purpose-made portable tube benders. These are very similar to those used in plumbing for copper tube.

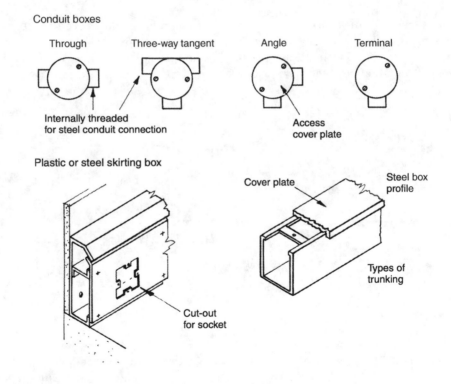

Conduit boxes

Through   Three-way tangent   Angle   Terminal

Internally threaded
for steel conduit connection

Access
cover plate

Plastic or steel skirting box

Cover plate   Steel box
profile

Cut-out
for socket

Types of
trunking

Ref. BS EN 61386-1: Conduit systems for cable management. General requirements.

572

Mineral insulated copper-covered (MICC) cable, otherwise known as mineral insulated metal sheathed (MIMS), has copper conductors insulated with highly compressed magnesium oxide powder inside a copper tube. It is resistant to most corrosive atmospheres and temperatures up to 150°C, hence its suitability with fire alarms. Cutting and jointing the cable requires particular procedures that seal the insulant from penetration of dampness. The outer copper cover provides for circuit-protective continuity to earth.

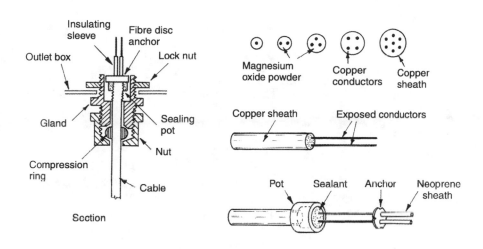

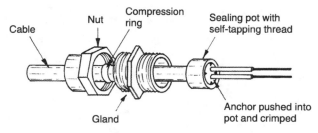

Termination components

Ref. BS EN 60702 (Parts 1 and 2): Mineral insulated cables and their terminations with a rated voltage not exceeding 750V [Cables – Part 1; Terminations – Part 2].

Armoured Cable

Armoured cable is used for underground main and sub-main supplies. It terminates at electricity substations or transformers and at the intake to high-voltage supplies for large buildings. Steel wire armour provides protection against physical damage.

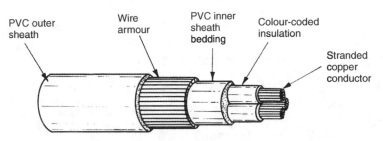

Armoured three-phase cable for underground mains supply

It is rarely used for domestic and other relatively low-voltage applications, but is available if required for underground supplies to outbuildings such as a garage or workshop and for garden lighting.

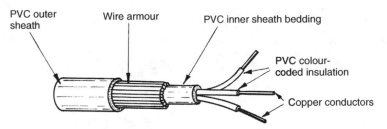

Armoured cable for low-voltage single-phase use

Refs. BS 6724 and BS 7846: Electric cables. Thermosetting insulated, armoured cables.

Electrical installations must be tested on completion to verify that the system will operate efficiently and safely. The tests are extensive, as defined in the Institution of Engineering and Technology Regulations. They can only be carried out by a competent person, i.e. a qualified electrician or electrical engineer. The following tests are an essential part of the proceedings:

- Continuity.
- Insulation.
- Polarity.

Testing is undertaken by visual inspection and the use of a multipurpose meter (multimeter) or an instrument specifically for recording resistance, i.e. an ohmmeter.

Continuity – there are several types of continuity test for ring final circuits. Each is to ensure integrity of the line, neutral and circuit protective (earth) conductors without bridging (shorting out) of connections. The following is one established test to be applied to each conductor:

- Record the resistance between the ends of the ring circuit (A).
- Record the resistance between closed ends of the circuit and a point midway in the circuit (B).
- Check the resistance of the test lead (C).
- Circuit integrity is indicated by: A ÷ 4 approx. = B – C

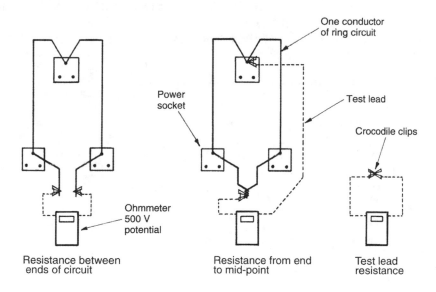

One conductor of ring circuit

Power socket

Test lead

Crocodile clips

Ohmmeter 500 V potential

Resistance between ends of circuit

Resistance from end to mid-point

Test lead resistance

Note: Resistances A, B and C are also referred to as R1, R2 and R3.

Insulation – this test is to ensure that there is a high resistance between line and neutral conductors and these conductors and earth. A low resistance will result in current leakage and energy waste which could deteriorate the insulation and be a potential fire hazard. The test to earth requires all lamps and other equipment to be disconnected, all switches and circuit-breakers closed and fuses left in. Ohmmeter readings should be at least 1MΩ.

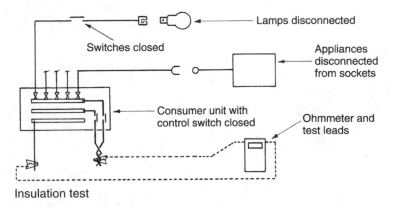

Insulation test

Polarity – this is to ensure that all switches and circuit-breakers are connected in the phase or line conductor. An inadvertent connection of switchgear to a neutral conductor would lead to a very dangerous situation where apparent isolation of equipment would still leave it live! The test leads connect the line bar in the disconnected consumer unit to line terminals at switches. A very low resistance reading indicates the polarity is correct and operation of the switches will give a fluctuation on the ohmmeter.

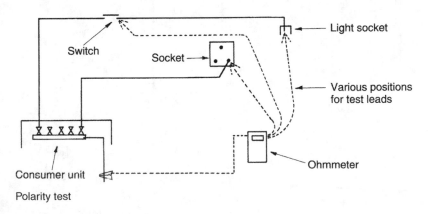

Polarity test

Ref. BS EN 61010-1: Safety requirements for electrical equipment for measurement, control and laboratory use.

Standard applications	Cable specification (mm² csa.)
Lighting	1 or 1·5
Immersion heater	1·5 or 2·5
Sockets (ring)	2·5
Sockets (radial)	2·5 or 4 (see page 561)
Cooker	6 or 10
Shower	4, 6 or 10 (see page 431)

Some variations occur as the specification will depend on the appliance or circuit loading – see calculation below. Where non-standard circuits or special installations are necessary, the cable specification must be calculated in the following stages:

- Determine the current flowing.
- Select an appropriate cable (see table below).
- Check that the voltage drop is not greater than 4%.

Current ratings and voltage reduction for sheathed multi-core PVC insulated cables:

csa (mm²)	Current carrying capacity (amps)		Voltage drop (mV/amp/m)
	In conduit	Clipped	
1	13	15	44
1·5	16·5	19.5	29
2·5	23	27	18
4	30	36	11
6	38	46	7·3
10	52	63	4·4

E.g. a 7.2kW shower with a clipped cable length of 10m:

amps = watts ÷ volts = 7200 ÷ 230 = 31·3

From table, select 4mm² csa (36 amps)

Voltage drop = (mV × current flowing × cable length) ÷ 1000
= (11 × 31.3 × 10) ÷ 1000 = 3.44 volts

Maximum voltage drop = 230 × 4% = 9.2 volts.

Therefore, 4mm² csa cable is satisfactory.

Note: Correction factors may need to be applied, e.g. when cables are grouped, insulated or in an unusual temperature. The IET Wiring Regulations should be consulted to determine where corrections are necessary.

Diversity

Diversity in electrical installations permits specification of cables and overload protection devices with regard to a sensible assessment of the maximum likely demand on a circuit. For instance, a ring circuit is protected by a 30 amp fuse or 32 amp MCB, although every socket is rated at 13 amps. Therefore, if only three sockets were used at full rating, the fuse/MCB would be overloaded. In practice, this does not occur, so some diversity can be incorporated into calculations.

Guidance for diversity in domestic installations:

Circuit	Diversity factor
Lighting	66% of the total current demand.
Power sockets	100% of the largest circuit full load current + 40% of the remainder.
Cooker	10 amps + 30% full load + 5 amps if a socket outlet is provided.
Immersion heater	100%.
Shower	100% of highest rated + 100% of second highest + 25% of any remaining.
Storage radiators	100%.

E.g. a house with 7.2kW shower, 3kW immersion heater, three ring circuits and three lighting circuits of 800W each:

Appliance/circuit	Current demand (amps)	Diversity allowance (amps)
Shower	$\frac{7200}{230} = 31.3$	$31.3 \times 100\% = 31.3$
Ring circuit-1	30	$30 \times 100\% = 30$
Ring circuit-2	30	$30 \times 40\% = 12$
Ring circuit-3	30	$30 \times 40\% = 12$
Lighting	$3 \times 800 = \frac{2400}{230} = 10.4$	$10.4 \times 66\% = 6.9$
		Total = 92.2 amps

For a factory of modest size where the electrical load is not too high, a three-phase, four-wire, 400 volts supply will be sufficient. The distribution to three-phase motors is through exposed copper busbars in steel trunking running around the periphery of the building. Supply to individual motors is through steel conduit via push button switchgear. In addition to providing protection and support, the trunking and conduit can be used as earth continuity.

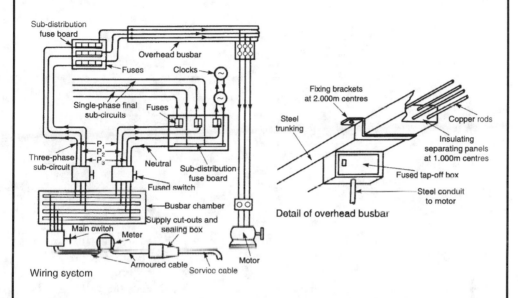

Switches must be within easy reach of machinery operators and contain a device to prevent restarting of the motor after a power failure stoppage.

Overhead busbars provide an easily accessible means of connecting supplies to machinery by bolting the cable to the busbars.

Lighting and other single-phase circuits are supplied through separate distribution fuse boards.

Refs. BS ENs 61439-1 and 2: Low-voltage switchgear and controlgear assemblies.

Electricity Supply to Groups of Large Buildings

For large developments containing several buildings, either radial or ring distribution systems may be used.

Radial system – separate underground cables are laid from the substation to each building. The system uses more cable than the ring system, but only one fused switch is required below the distribution boards in each building.

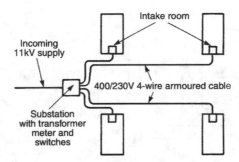

Radial distribution (block plan)

Ring circuit system – an underground cable is laid from the substation to loop into each building. To isolate the supply, two fused switches are required below the distribution boards in each building. Current flows in both directions from the intake, to provide a better balance than the radial system. If the cable on the ring is damaged at any point, it can be isolated for repair without loss of supply to any of the buildings.

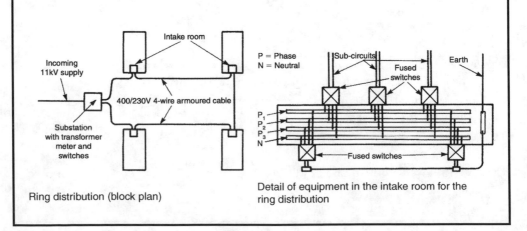

Ring distribution (block plan)

Detail of equipment in the intake room for the ring distribution

The rising main supply system is used in high-rise offices and flats. Copper busbars run vertically inside trunking and are given support by insulated bars across the trunking chamber. The supply to each floor is connected to the rising main by means of tap-off units. To balance electrical distribution across the phases, connections at each floor should be spread between the phase bars. If a six-storey building has the same loading on each floor, two floors would be supplied from separate phases. Flats and apartments will require a meter at each tap-off unit.

To prevent the spread of fire and smoke, fire barriers are incorporated with the busbar chamber at each compartment floor level. The chamber must also be fire stopped to the full depth of the floor.

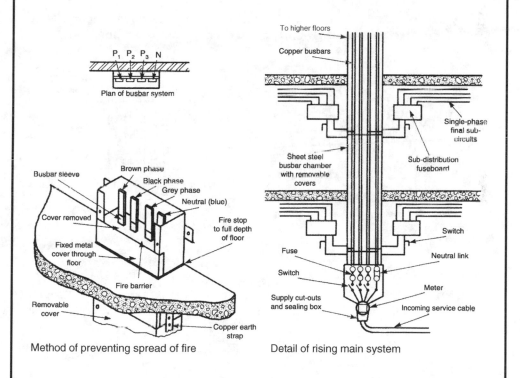

Method of preventing spread of fire

Detail of rising main system

Ref. Building Regulations, Approved Document B3: Internal fire spread (structure).

Electric Space Heating – 1

It is uneconomical to shut down electricity-generating plant overnight, even though there is considerably less demand. To encourage the use of off-peak energy, the electricity supply companies offer it at an inexpensive tariff. A timer and white meter or Economy 7 (midnight to 0700) meter controls the supply to an energy storage facility. See also Economy 10, page 584.

Underfloor – makes use of the thermal storage properties of a concrete floor. High-resisting insulated conductors are embedded in the floor screed at 100 to 200mm spacing, depending on the desired output. This is about 10 to 20W/m of cable. To be fully effective, the underside of the screed should be completely insulated and thermostatic regulators set in the floor and the room.

Block heaters – these are rated between 1kW and 6kW and incorporate concrete blocks to absorb the off-peak energy (see next page).

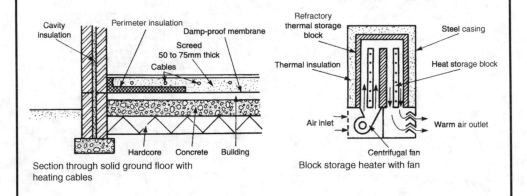

Section through solid ground floor with heating cables

Block storage heater with fan

Electrically heated ceilings use standard tariff electricity supply. The heating element is flexible glasscloth with a conducting silicone elastomer.

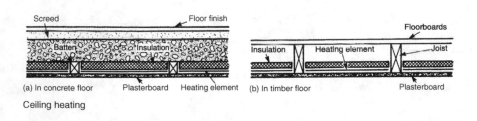

(a) In concrete floor

(b) In timber floor

Ceiling heating

582

Night storage heaters – these have developed from very bulky cabinets containing concrete blocks which effectively absorb the overnight electrical energy and dissipate it gradually during the next day. Improvements in storage block material have considerably reduced the size of these units to compare favourably with conventional hot water radiators. They contain a number of controls, including a manually set input thermostat on each heater, an internal thermostat to prevent overheating and a time-programmed fan. Manufacturers provide design tables to establish unit size. As a rough guide, a modern house will require about 200W output per square metre of floor area. Storage heaters are individually wired on radial circuits from the off-peak time-controlled consumer unit.

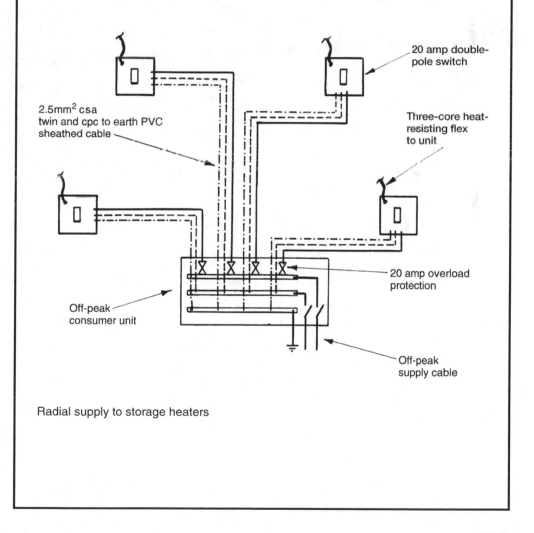

20 amp double-pole switch

2.5mm² csa twin and cpc to earth PVC sheathed cable

Three-core heat-resisting flex to unit

20 amp overload protection

Off-peak consumer unit

Off-peak supply cable

Radial supply to storage heaters

Economy 10 – a variation on Economy 7 (see pages 112 and 582) that is sometimes referred to as 'Warmwise'. It is suitable where discounted electricity is used as the energy source for both space heating and hot water supply. Often installed as an upgrade to an Economy 7 system in homes where the hot water and storage heaters are losing heat by the evening, typical of households where occupancy has increased.

All electric heating system emitters have improved from their origins as bulky storage radiators that contain heavy concrete blocks as the off-peak electrical energy-absorbing material. Contemporary electric combination radiators are much slimmer, with overall dimensions that compare favourably with hot water convector radiators. Also, they do not require separate provision for a centralised hot water boiler, extensive pipework, cisterns and associated controls, effecting a considerable saving in space and installation time. Electric emitters can plug into a standard socket, but for maximum economy, a dedicated spur is provided from an Economy 10 meter as indicated on the previous page. The ten off-peak reduced tariff hours can vary depending on supplier, but are usually either:

Midnight to 7 a.m. and 1 p.m. to 4 p.m., or
Midnight to 5 a.m., 1 p.m. to 4 p.m. and 8 p.m. to 10 p.m.

Radiator output – individual units ranging from 500 to 2500 watts.

Function – comprise heat-retaining ceramic tablets or cells that contain an embedded heating element.

Control – each emitter has a built-in thermostat for modulating heat output. This can be set manually for independent control or be controlled centrally from a remote room thermostat that also regulates all other emitters to effectively centrally heat a whole building.

Application – most suited for apartments and dwellings of limited size. Also as a retro-fit in existing dwellings, as cable installation is considerably less disruptive to the structure than a hot water system.

Electrically heated warm air systems are a development of the storage heater concept – see previous three pages. A central unit rated from 6kW to 12kW absorbs electrical energy off-peak and during the day delivers this by fan to various rooms through a system of insulated ducting. A room thermostat controls the fan to maintain the air temperature at the desired level. Air volume to individual rooms is controlled through an outlet register or diffuser.

Stub duct system – the unit is located centrally and warm air is conveyed to rooms through short ducts with attached outlets.

Radial duct system – warm air from the unit is supplied through several radial ducts designated to specific rooms. Outlet registers are located at the periphery of rooms to create a balanced heat distribution.

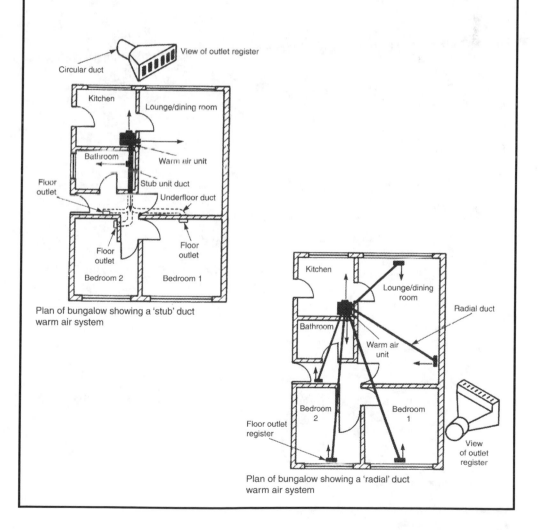

Plan of bungalow showing a 'stub' duct warm air system

Plan of bungalow showing a 'radial' duct warm air system

There are numerous types of independent heat emitters for use with 13 amp power sockets or fused spur sockets.

Panel heater – the heat output is mainly radiant from a surface operating temperature of between 204°C and 240°C. For safety reasons, it is mounted at high level and may be guarded with a mesh screen.

Infra-red heater – contains an iconel-sheathed element or nickel chrome spiral element in a glass tube, backed by a curved reflector. May be used at high level in a bathroom and controlled with a string pull.

Oil-filled heater – similar in appearance to steel hot water radiators, they use oil as a heat-absorbing medium from one or two electrical elements. Heat is emitted by radiant and convected energy. An integral thermostat allows for manual adjustment of output.

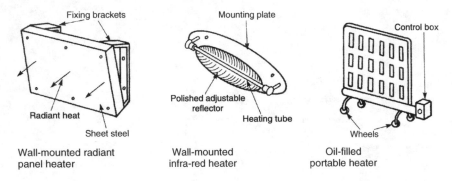

Wall-mounted radiant panel heater

Wall-mounted infra-red heater

Oil-filled portable heater

Convector heater – usually has two electrical elements with independent control to vary the output. May be used where a constant level of background warmth is required.

Parabolic reflector fire – has the heating element in the focal point to create efficient radiant heat output.

Wall-mounted fan heaters – usually provided with a two-speed fan to deliver air through a bank of electrical elements at varying velocities. Direction is determined by adjustable louvres.

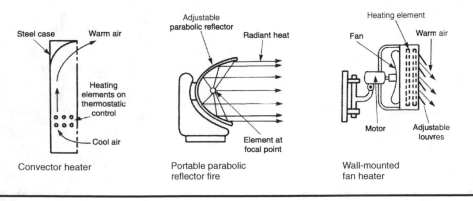

Convector heater

Portable parabolic reflector fire

Wall-mounted fan heater

Controls for Electric Night Storage Space Heaters

Controls vary from simple switches and sensors integrated with appliances, to overall system management programmed through time switches and optimisers:

- Manual charge control – set by the user to regulate energy input and output. The effect can be variable and unreliable as it does not take into account inconsistencies such as daily variations in temperature.

- Automatic charge control – sensors within the heater and room are pre-set to regulate the electrical input charge. When room temperature is high, the sensor in the heater reduces the energy input. Conversely, the energy input is increased when the room temperature is low.

- Heat output control – this is a damper within the heater casing. It can be adjusted manually to regulate heat emission and prevent a room from overheating. A variable speed fan can be used to similar effect or to vary the amount of heat emission and its distribution.

- Time switch/programmer and room thermostat – the simplest type of programmed automatic control applied individually to each heater or as a means of system or group control. Where applied to a system of several emitters, individual heaters should still have some means of manual or preferably automatic regulation. This type of programmed timing is also appropriate for use with direct acting thermostatically switched panel-type heaters.

- 'CELECT-type' controls – this is a type of optimiser control which responds to pre-programmed times and settings, in addition to unknown external influences such as variations in the weather. Zones or rooms have sensors which relate room information to the controller or system manager, which in turn automatically adjusts individual storage heater charge periods and amount of energy input to suit the room criteria. This type of control can also be used for switching of panel heaters.

Construction Site Electricity – 1

A temporary supply of electricity for construction work may be obtained from portable generators. This may be adequate for small sites but most developments will require a mains supply, possibly up to 400 volts in three phases for operating hoists and cranes. Application must be made in good time to the local electricity authority to ascertain the type of supply and the total load. The incoming metered supply provided by the electricity company will be housed in a temporary structure constructed to the authority's approval. Thereafter, site distribution and installation of reduced voltage transformers is undertaken by the developer's electrical contractor subject to the supply company's inspection and testing.

Typical site distribution –

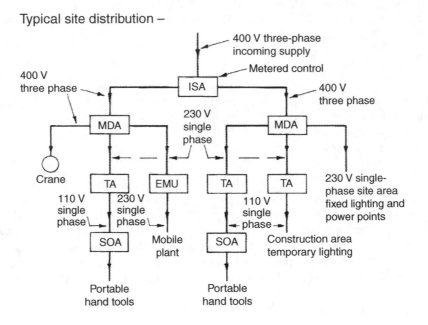

See next page for key to abbreviations.

Site electricity installations are temporary and will be subjected to ongoing alterations as site demands change. Therefore, they must be inspected, tested and documented as they occur and at frequent intervals.

Equipment:

Incoming site assembly (ISA) – provided by the local electricity supply company. It contains their switchgear, overload protection, transformers and meters for a 400 volt, three-phase supply at 300, 200 and 100 amps.

Main distribution assembly (MDA) – contains three-phase and single-phase distribution boards, overload protection and lockable switchgear. May be combined with the ISA to become an ISDA.

Transformer assembly (TA) – supplied from the MDA to transform voltage down to 110V, 50V and possibly 25V for use in very damp situations.

Earth monitor unit (EMU) – used where mobile plant requires flexible cables at mains voltage. A very low-voltage current is conducted between plant and EMU and earth conductor, so that if this is interrupted by a fault, a monitoring unit disconnects the supply.

Socket outlet assembly (SOA) – a 110 volt supply source at 32 amps with switchgear and miniature circuit-breakers for up to eight 16 amp double-pole sockets for portable tools.

Cable colour codes and corresponding operating voltage.

Colour	Voltage
Violet	25
White	50
Yellow	110
Blue	230
Red	400
Black	500/650

Refs. BS 4363: Specification for distribution assemblies for reduced low-voltage electricity supplies for construction and building sites.

BS 7375: Distribution of electricity on construction and demolition sites. Code of practice.

BS EN 61439-4: Low-voltage switchgear and controlgear assemblies. Particular requirements for assemblies for construction sites.

Electrical transformers, distribution units, tools and equipment – must be appropriate for the various applications, situations and supply voltages to which they are connected and for the purpose for which they are designed. Plugs and socket profiles differ relative to the supply voltage to prevent misuse and interchangeability. Voltage potential commonly used on building sites includes

- 25 volt single phase safety extra low voltage (SELV). Portable hand held lamps in damp and confined locations.
- 50 volt single phase centre point earthed (25 volts to earth). May also be used for hand held lamps in damp and confined locations.
- 110 volt single phase (usually centre point earthed, i.e. 55 volts) for portable hand tools and equipment. Maximum load is 2kW.
  Note: A 110 volt three-phase supply may be used for equipment with a load up to 3.75kW.
- 230 volt single phase to site accommodation and other fixed situations.
- 400 volt three phase for fixed and transportable plant with a rated load exceeding 3.75kW.

Overhead power lines – temporary installations often require overhead distribution of cables. Also, there may be existing overhead supplies on new development sites. Provisions must be made to protect these power lines and site personnel from contact and damage by contractor's plant and equipment.

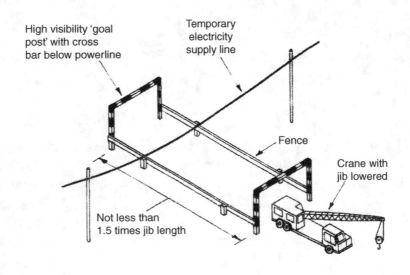

High visibility 'goal post' with cross bar below powerline

Temporary electricity supply line

Fence

Crane with jib lowered

Not less than 1.5 times jib length

Light is a form of electromagnetic radiation. It is similar in nature and behaviour to radio waves at one end of the frequency spectrum and X-rays at the other. Light is reflected from a polished (specular) surface at the same angle that it strikes it. A matt surface reflects in a number of directions and a semi-matt surface responds somewhere between a polished and a matt surface.

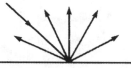

Angle of incidence θ_1 =
Angle of reflection θ_2

Light reflected in all directions

Some light is scattered and some light is reflected directionally

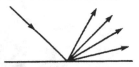

Light reflected from a polished surface

Light reflected from a matt surface

Light scattered and reflected from a semi-matt surface

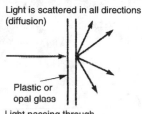

Light is scattered in all directions (diffusion)

Plastic or opal glass

Light passing through a diffusing screen

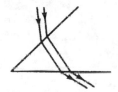

Light is bent or refracted when passing through a surface between two media

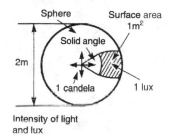

Sphere

Surface area $1m^2$

Solid angle

2m

1 candela

1 lux

Intensity of light and lux

Illumination produced from a light source perpendicular to the surface:

$$E = I \div d^2$$

E = illumination on surface (lux)

I = Illumination intensity from source (candela or cd)

d = distance from light source to surface (metre or m).

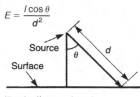

$$E = \frac{I \cos \theta}{d^2}$$

Source

Surface

θ

d

Illumination produced from a light source not perpendicular to the surface

The inverse square law – intensity of illumination from a point source of light decreases inversely with the square of the distance from the source. The illustration below represents this principle.

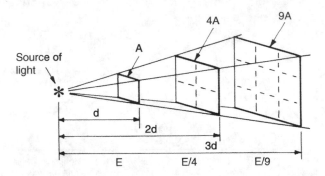

E.g. 1 – A spotlight of luminous intensity 20000 candelas directed perpendicularly onto a flat surface at 6m distance. Using the formula shown on the previous page, applying the inverse square law principle:

$$E = I \div d^2$$
$$E = 20{,}000 \div 6^2$$
$$E = 556 \text{ lux or lumens/m}^2$$

Cosine illumination law – this provides a correction to the inverse square law formula to allow for the subject area being at an angle from the light source. This is appropriate for most lighting applications as large parts of a surface will not receive light directly in the perpendicular. The modified formula and concept is shown on the previous page.

E.g. 2 – The light from the same spotlight in example 1 is directed at 30° (angle θ) onto a subject 6m away. The illumination will be:

$$E = (I \cos \theta) \div d^2$$
$$E = (20{,}000 \times 0.866) \div 6^2$$
$$E = 481 \text{ lux or lumens/m}^2$$

Definitions and units of measurement:

- Luminous intensity – candela (cd), a measurement of the magnitude of luminance or light reflected from a surface, i.e. cd/m^2.
- Luminous flux – lumen (lm), a measurement of the visible light energy emitted.
- Illuminance – lumens per square metre (lm/m^2) or lux (lx), a measure of the light falling on a surface.
- Efficacy – efficiency of lamps in lumens per watt (lm/W). Luminous efficacy = luminous flux output ÷ electrical power input.
- Glare index – a numerical comparison ranging from about 10 for shaded light to about 30 for an exposed lamp. Calculated by considering the light source size, location, luminances and effect of its surroundings.

Examples of acceptable illumination levels and limiting glare indices for different activities:

Activity/location	Illuminance (lux)	Limiting glare index
Assembly work: (general)	250	25
(fine)	1000	22
Computer room	300	16
House	50 to 300*	n/a
Laboratory	500	16
Lecture/classroom	300	16
Offices: (general)	500	19
(drawing)	750	16
Public house bar	150	22
Shops/supermarkets	500	22
Restaurant	100	22

*Varies from 50 in bedrooms to 300 in kitchen and study.

The Building Regulations, Approved Document L – Vol. 2 requires non-domestic buildings to have lighting systems appropriate to the activity, but not over-illuminated. The CIBSEs Lighting Handbook (SLL) provides an acceptable design guide.

Electric Lamps – 1

Filament lamps – the tungsten iodine lamp is used for floodlighting. Evaporation from the filament is controlled by the presence of iodine vapour. The gas-filled, general-purpose filament lamp has a fine tungsten wire sealed within a glass bulb. The wire is heated to incandescence (white heat) by the passage of an electric current.

Discharge lamps – these do not have a filament, but produce light by excitation of a gaseous mixture of vapourised mercury, metal halides and argon. When voltage is applied to the two electrodes, within an inner electric arc tube, ionisation occurs until a critical value is reached when current flows between them. As the temperature rises, the mercury vaporises and electrical discharge between the main electrodes causes light to be emitted. The inner tube contains argon gas to facilitate initial starting of the arc. Operating pressure varies, typically 400 to 2000 kPa depending on application.

Fluorescent tube – this is a low-pressure variation of the mercury discharge lamp. Energised mercury atoms emit ultraviolet radiation and a blue/green light. The tube is coated internally with a fluorescent powder which absorbs the ultraviolet light and re-radiates it as visible light.

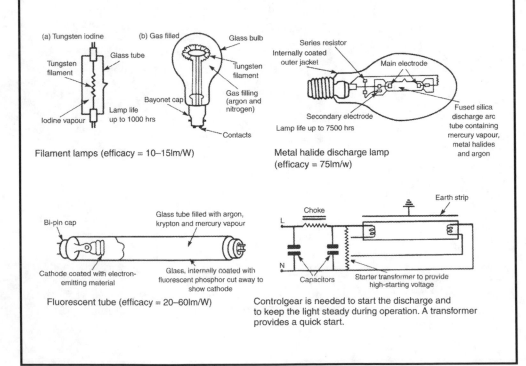

Filament lamps (efficacy = 10–15lm/W)

Metal halide discharge lamp (efficacy = 75lm/w)

Fluorescent tube (efficacy = 20–60lm/W)

Controlgear is needed to start the discharge and to keep the light steady during operation. A transformer provides a quick start.

Fluorescent strip lamps have many applications. The fittings and reflectors shown are appropriate for use in industrial locations, with a variation which creates an illuminated ceiling more suited to shops and offices. A false ceiling of thermaluscent panels provides well-diffused illumination without glare and contributes to the insulation of the ceiling. Other services should not be installed in the void as they will cast shadows onto the ceiling. Tubes are mounted on batten fittings and the inside of the void should be painted white to maximise effect.

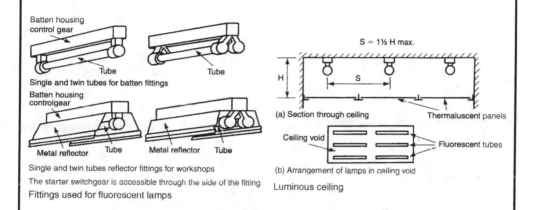

Batten housing control gear

Tube

Tube

Single and twin tubes for batten fittings

Batten housing controlgear

Metal reflector   Tube   Metal reflector   Tube

Single and twin tubes reflector fittings for workshops

The starter switchgear is accessible through the side of the fitting

Fittings used for fluorescent lamps

S = 1½ H max.

H   S

(a) Section through ceiling   Thermaluscent panels

Ceiling void   Fluorescent tubes

(b) Arrangement of lamps in ceiling void

Luminous ceiling

High-pressure sodium discharge lamps produce a consistent golden white light in which it is possible to distinguish colours. They are suitable for floodlighting, commercial and industrial lighting and illumination of highways. The low-pressure variant produces light that is virtually monochromatic. The colour rendering is poor when compared to the high-pressure lamp limiting its yellowish light to outdoor street and tunnel illumination. Sodium vapour pressure varies between 0.5Pa and 33kPa. Typical efficacy is between 125 and 180lm/W.

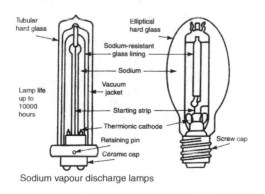

Tubular hard glass

Elliptical hard glass

Sodium-resistant glass lining

Sodium

Lamp life up to 10000 hours

Vacuum jacket

Starting strip

Thermionic cathode

Retaining pin

Screw cap

Ceramic cap

Sodium vapour discharge lamps

595

Ceramic discharge metal halide (CDM) lamps – similar in principle to the mercury vapour metal halide discharge lamp (page 594), but with a sintered alumina discharge arc tube instead of a fused silica (quartz) arc tube. This improvement reduces the tendency for ion creep and associated loss of light emitting qualities, a characteristic of the earlier type of discharge lamps as they age. Compared to a standard incandescent filament lamp, CDM lamps consume only about one-fifth of the power for a similar output of 75 to 100 lm/W. Light output is more consistent with closer to daylight white light emission than other discharge lamps. Applications vary considerably and include lighting to shops and architectural features. Many local authorities are now using this type of lamp as a more effective and more economic replacement for high-pressure sodium street lighting. Colour rendition is generally about 80 (see below) and can be in the upper nineties depending on the metal halide content.

Colour rendering index (CRI) – otherwise known as the colour rendition index. A numerical scale not exceeding 100. Used for measuring the effectiveness of an artificial light source to reproduce colour when compared to natural light. The most obvious application is to photography and cinematography, but the index is also useful for categorising artificial light applications to building interiors. Low-pressure sodium discharge lamps have a poor CRI rated below zero. A guide to more positive lamp ratings is listed below:

Candle	100
Cool white fluorescent	60 – 65
Daylight bulb	80
Fluorescent containing tri-phosphors	80 – 85
High-pressure sodium	25
Incandescent	95
LEDs	70 – 90
Sunlight bulb	80 – 85
Tungsten-halogen	95 – 100

Fittings for lighting may be considered in three categories:

1. General utility – designed to be effective, functional and economic.
2. Special – usually provided with optical arrangements such as lenses or reflectors to give directional lighting.
3. Decorative – designed to be aesthetically pleasing or to provide a feature, rather than to be functional.

From an optical perspective, the fitting should obscure the lamp from the discomfort of direct vision to reduce the impact of glare.

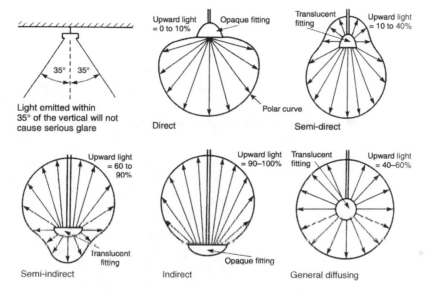

Upward light percentage is the proportion of upward light in luminous flux (visible light energy emitted), usually referred to as the Upward Light Output Ratio (ULOR). See next page.

Ventilated fittings can be specified where it is praticable to recirculate the heat produced by the lamps through a ceiling void as a heat energy recovery facility that can supplement a warm air ventilation or air-conditioning system. The cooling effect on the lamp will also improve its efficiency. See pages 307, 308 and 310.

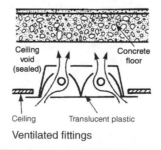

Ventilated fittings

597

Light Output Ratio (LOR)

LOR – a figure used to indicate how much light is lost within a luminaire. It is defined in BS EN 13032-2: *Light and lighting*, as the '... ratio of the luminous flux of the luminaire to the lumens of the lamp used'. For example, apart from the area obscured by a bayonet or screw fixing, a standard exposed light bulb will emit unrestricted light in all directions. Within a lampshade or diffuser, some light from the bulb will be lost or absorbed.

$$\text{LOR} = \frac{\text{Total light output from a luminaire}}{\text{Total lamp output}} \, \%$$

Note: Output is measured in lumens (luminous flux).

ULOR and DLOR – Upward and Downward Light Output Ratios are the amounts of light from the upper and lower halves of a luminaire, respectively, combining to produce the following summary:

LOR = ULOR + DLOR

where

$$\text{ULOR} = \frac{\text{Light output from a luminaire upwards}}{\text{Total lamp output}} \, \%$$

$$\text{DLOR} = \frac{\text{Light output from a luminaire downwards}}{\text{Total lamp output}} \, \%$$

E.g. LOR = 1200 lumens
    ULOR = 400 ..
    DLOR = 600 ..

ULOR = 400 ÷ 1200 = 0.33 or 33%
DLOR = 600 ÷ 1200 = 0.50 or 50%
LOR = 0.83 or 83%

Therefore, the amount of light energy absorbed by the luminaire is 100% – 83% = 17%

The five types or classifications of luminaires shown on page 599 are defined by the proportion of light emitted upwards. The remainder is apportioned downwards with a deduction in both directions for light absorbed or lost to the shade or diffuser.

Classification	LORs		Application examples
Direct	ULOR	0 – 10%	Luminous ceilings
	DLOR	90 – 100%	and downlighters
Semi-direct	ULOR	10 – 40%	Bare tubes and
	DLOR	60 – 90%	fluorescent luminaires
General diffusing	ULOR	40 – 60%	Spherical
	DLOR	40 – 60%	luminaires
Semi-indirect	ULOR	60 – 90%	Uplighters and
	DLOR	10 – 40%	portable luminaires
Indirect	ULOR	90 – 100%	Uplighters and
	DLOR	0 – 10%	cornice lighting

Polar curves, Light Output Ratios (LORs) and British Zonal (BZ) classifications for a sample of luminaires –

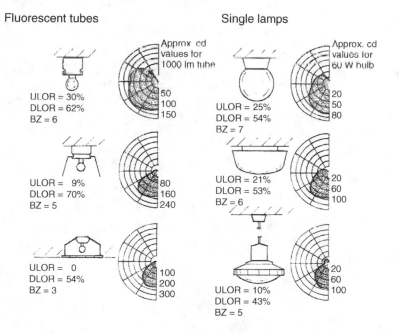

Fluorescent tubes   Single lamps

Approx cd values for 1000 lm tube

ULOR = 30%
DLOR = 62%
BZ = 6
50 / 100 / 150

ULOR = 9%
DLOR = 70%
BZ = 5
80 / 160 / 240

ULOR = 0
DLOR = 54%
BZ = 3
100 / 200 / 300

Approx. cd values for 60 W bulb

ULOR = 25%
DLOR = 54%
BZ = 7
20 / 50 / 80

ULOR = 21%
DLOR = 53%
BZ = 6
20 / 60 / 100

ULOR = 10%
DLOR = 43%
BZ = 5
20 / 60 / 100

For further examples, consult lighting manufacturers' catalogues and lighting design guides.

British Zonal (BZ) Classification

BZ classification of luminaires – the established UK numerical referencing system for comparing downward light distribution. Ten different standards of polar curves plot distribution of light in the lower hemisphere only. The curves derived are classified from BZ1 to BZ10. BZ1 has a light distribution that is mostly downward. Thereafter, as the number increases, the light distribution becomes more outward up to BZ10, as shown below.

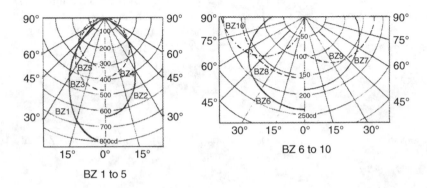

BZ 1 to 5   BZ 6 to 10

Purpose – polar curves for particular light fixtures can be matched as close as possible to one of the BZ numbers, thereby providing catalogue reference data for manufacturers and a means of lighting output assessment for lighting designers. Some examples are shown on the preceding page.

Guide to spacing ratios of light fittings to mounting height –

BZ class	Max. spacing to mounting height ratio
1 and 2	1:1
3 and 4	1.25:1
5 to 10	1.5:1

Luminaire – a word to describe the complete lighting unit including the lamp. When selecting a lamp type, it is important to select a luminaire to complement the lamp both functionally and aesthetically. A luminaire has several functions: it defines the lamp position, protects the lamp and may contain the lamp control mechanism. In the interests of safety, it must be well insulated, in some circumstances resistant to moisture, have adequate appearance for purpose and be durable.

Polar curve – shows the directional qualities of light from a lamp and luminaire by graphical representation, as shown in outline on pages 597 and 599. A detailed plot can be produced on polar coordinated paper from data obtained by photometer readings at various angles from the lamp. The coordinates are joined to produce a curve.

Typical representation:

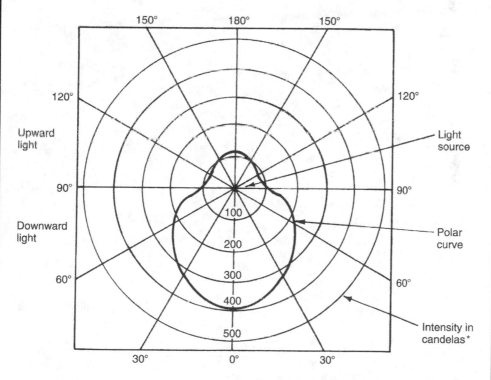

*A candela (cd) is a measure of luminous intensity based on the power of one candle having an intensity of one watt. Therefore, a 60W bulb has approximately 60cd spherical intensity.

Compact Fluorescent Lamps

Compact fluorescent lamps are a smaller variation and development of the standard fluorescent tube fitting. They are manufactured with conventional bayonet or screw fittings. Unit cost is higher than tungsten filament bulbs but will last for over 8000 hours, consuming only about 25% of the energy of a conventional bulb. Tungsten filament bulbs have a life expectancy of about 1000 hours.

The comfort type produces gentle diffused light and is suitable where continuous illumination is required. The prismatic types are more robust and are suitable for application to workshops and commercial premises. Electronic types are the most efficient, consuming only 20% of the energy that would be used in a tungsten filament bulb. Compact fluorescent lamps are not appropriate for use with dimmer switches.

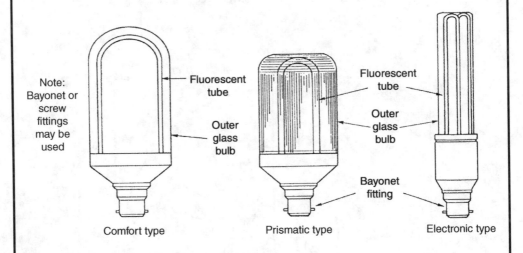

Comfort type   Prismatic type   Electronic type

The Buildings Regulations, Approved Document L, lists compact fluorescent lamps as an acceptable means for lighting buildings.

Energy Saving Chart

Energy saver	Ordinary light bulb	Energy saving	Over 8000 hours save up to (£)
25 W	100 W	80%	72.00
18 W	75 W	73%	54.72
11 W	60 W	80%	47.04
9 W	40 W	72%	29.76

Domestic energy costed at 12p/kWh

Only a small proportion of the energy in a light fitting is converted into light. All the energy dissipated is a measure of heat.

Tungsten filament lamp – heat contribution is the power rating quoted on the bulb.

Fluorescent tube – heat contribution is the power rating plus about 25% attributed to heat energy from the control gear.

High levels of artificial lighting can make a significant contribution to the heating load of a building. This may be useful in winter, but at other times, it can cause an overheating problem. A possible solution is combination duct extract/luminaires as shown on pages 307, 308 and 310. Some 40–50% of the lighting heat energy can be directed through a controlled extract or preferably recycled through a heat exchanger. Also, the cooling effect on the light fitting should contribute to its life expectancy. Polyphosphor tubes should not be used in extract luminaries, as the illuminance effect will be reduced.

The following table indicates the approximate heat dissipation, relative to the type of light fitting and level of illuminance:

Illuminance (lux)	Heat dissipation (W/m² floor area)				
	Tungsten lamp		Fluorescent tube		
	Open reflector	Diffuser	Open trough	Enclosed diffuser	Louvred ceiling
100	19–25	28–32	4–5	6–8	≅6
200	26–36	33–45	<8	<11	<11
300	37–50	46–69	<11	<16	<17
400	51–65		<15	<22	<23
500	66–88		<25	<27	<30
1000			<38	<54	<60

Proportionate distribution of energy from lamps and tubes:

Energy transfer	Energy dissipated by fitting type (%)		
	Fluorescent	Tungsten	Discharge
Conduction and convection	55	15	40
Radiation	45	85	60

Lighting Controls – Dwellings

Interior lighting – the energy consumed by lighting in dwellings depends on the overall performance and efficiency of luminaires, lamps and control gear. The Building Regulations require that lighting circuits, whether in new-build or to existing dwellings are fitted with low energy efficient lamps. The lighting level determined appropriate to the use and purpose of the occupied space with regard to not over illuminating.

Installations are required to have every light fitting with efficient lamps of minimum luminous efficacy of 75 light source lumens per circuit-watt.

- Lumen is a measure of light output.
  See page 593.
- The term circuit-watt is used instead of the usual term for electrical power in watts, as it includes the power used by the lamp in addition to the installation and control system.

Each separate space or room to be provided with a dedicated control facility. This may be manual (light switch) or by automatic programmer, or a combination of both.

Hall, stairs and landing are regarded as one room.

An integral (attached to the building) conservatory is considered a room.

Fixed exterior lighting – reasonable provisions are required for economical use. This should include both of the following:

- automatic timed switching control where luminous efficacy is equal to or less than 75 light source lumens per circuit-watt that has non-occupancy detection – possibly infra-red unit – after the space has been lit. Manual over-ride control is acceptable where luminous efficacy exceeds 75 light source lumens per circuit-watt.
- photo-electric switching control that operates in response to daylight, i.e. switches luminaires off at predetermined natural light levels.

Note: Lamps that satisfy the criteria of efficiency include fluorescent tubes and compact fluorescent lamps. Special socket fittings are made to prevent interchange with unsuitable standard tungsten lamps. Notably of a square fit profile.

Refs. Building Regulations, Approved Document L: Conservation of fuel and power. Vol. 1 Dwellings.

BS EN 15193: Energy Performance of Buildings – Energy Requirements for Lighting.

Lighting control objectives:

- to maximise daylight.
- to avoid unnecessary use of artificial lighting when spaces are unoccupied.
- to achieve an appropriate level of illumination.

Lighting should be designed accordingly. Exceptions may include situations where lighting is low level and used occasionally, e.g. toilets. Specialised applications such as spotlighting are not included in Building Regulations.

General lighting requirements

Satisfy either:

- Average luminaire efficacy of 95 luminaire lumens per circuit-watt.
- Lighting Energy Numeric Indicator (LENI) calculation.

LENI: Formula to determine total energy as a summation of day and night time use. Tabulated data (AD L – Table B1) for new and existing buildings provides max. values in kWh per sq. m per year.

Display lighting requirements

Satisfy any of:

- Ave. light source efficacy of 80 light source lumens per circuit-watt.
- Power use rate $\leq$ 0.3 W/m^2 per room.
- LENI methodology.

High excitation purity of light sources. See Illuminating Engineering Society definitions.

- Ave. light source efficacy 65 light source lumens per circuit-watt.

General and display lighting to have either:

- Meter (kWh) dedicated for each circuit.
- Meter integrated with lighting control of a dedicated management system. To calculate energy consumed and to avail it to a building management system.

Control facilities:

- Local, easily accessible manual switches or remote devices including infra-red transmitters, sonic, ultrasonic and telecommunication controls.
- Plan distance from switch to luminaire, maximum 6 metres or two times fitting height above floor (take greater).
- Time switches as appropriate to occupancy.
- Photoelectric light metering switches.
- Automatic infra-red sensor switches which detect the absence or presence of occupants.

Controls specific to display lighting include dedicated circuits that can be manually switched off when exhibits or merchandise presentations are not required. Timed switching that automatically switches off when premises are closed.

Refs. Building Regulations, AD L, Vol. 2.

BRE Information Paper 2/99, Photoelectric control of lighting.

BRE Digest 498: Selecting Lighting Controls.

BS EN 12464-1: Light and lighting – Lighting of work places.

CIBSE: Lighting Handbook – SLL.

Extra-low-voltage lighting has application to display lighting for shops and exhibitions. It is also used as feature lighting in domestic premises where set in the ceiling in kitchens and bathrooms. These situations benefit from the low heat emission, good colour rendering and very low running costs of this form of lighting. System potential is only 12 volts AC, through a transformed 230 volt mains supply. High-performance 50 watt tungsten halogen dichroic lamps are compact and fit flush with the mounting surface.

Electricity is supplied from the transformer through a fused splitter to provide a fairly uniform short length of cable to each lamp. Similarity in cable lengths is important to maintain equivalent voltage drop and a short length of cable will minimise voltage drop. Lamps are very sensitive to change in voltage; therefore, the correct selection of transformer is essential. A voltage drop of 6% (approx. 0.7 volts) will reduce the illuminating effect by about 30%. Cable sizing is also critical with regard to voltage drop. The low voltage creates a high current, i.e. just one 50 watt bulb at 12 volts = 4.17 amps (see page 577 for cable sizing).

Schematic ELV lighting:

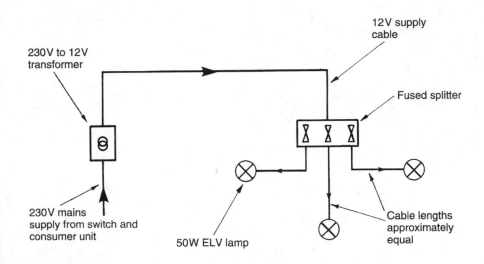

12V supply cable

230V to 12V transformer

Fused splitter

230V mains supply from switch and consumer unit

Cable lengths approximately equal

50W ELV lamp

Note: A variation is the use of individual low-voltage lamps which contain their own transformer. However, these are relatively expensive items and are attached to special fittings.

Emission from a tungsten-halogen bulb is up to three times that of a filament bulb, e.g. a 50 watt halogen bulb has comparable light output to one 150 watt filament bulb.

A guide or 'rule of thumb' that can be used to estimate the number of halogen bulbs required is: one 20W lamp per square metre of floor or one 50W lamp per one-and-a-half square metres of floor.

Alternative applications to that shown on the previous page:

New circuit

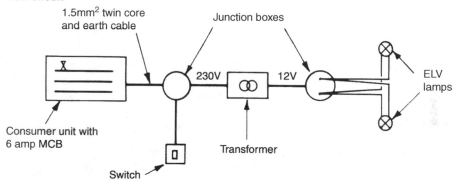

Existing ceiling rose

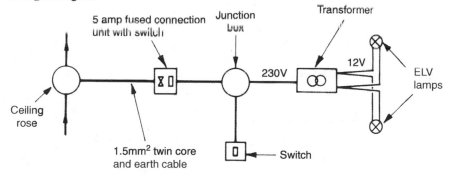

Note: neither the 12V light fittings nor the transformer are earthed.

Definitions:

Low voltage – < 1000 volts AC between conductors

< 600 volts AC between conductors and earth

Extra-low voltage – < 50 volts AC between conductors and earth

Reduced voltage – < 110 volts AC between conductors

< 55 volts AC to earth (single phase)

< 65 volts AC to earth (three phase)

(see Construction Site Electricity, pages 588–590)

Lumen Method of Lighting Design

The lumen method of lighting design is used to determine a lighting layout that will provide a design-maintained illuminance. It is valid if the luminaires are mounted above the working plane in a regular pattern. The method uses the formula: $N = (E \times A) \div (F \times U \times M)$.

N = number of lamps

E = average illuminance on the working plane (lux)

A = area of the working plane (m²)

F = flux from one lamp (lumens)

U = utilisation factor

M = maintenance factor.

The utilisation factor (U) is the ratio of the lumens received on the working plane to the total flux output of lamps in the scheme. The maintenance factor (M) is a ratio which takes into account the light lost due to an average expectation of dirtiness of light fittings and surfaces.

Both can be calculated from the many variable situations affecting room layout, usage and requirements. Tabulated data for various applications are provided in the Lighting Industry Federation's guide – Interior Lighting Design. Product manufacturers' catalogues and design guides may also contain this information.

Spacing-to-height ratio (SHR) is the centre-to-centre (S) distance between adjacent luminaires to their mounting height (H) above the working plane. Manufacturers' catalogues can be consulted to determine maximum SHRs, e.g. a luminaire with trough reflector is about 1·65 and an enclosed diffuser about 1·4.

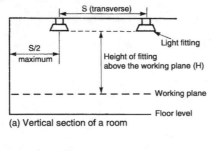

(a) Vertical section of a room

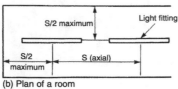

(b) Plan of a room

Method of spacing fluorescent tubes

Example. An office 8m long by 7m wide requires an illumination level of 400 lux on the working plane. It is proposed to use 80W fluorescent fittings having a rated output of 7375 lumens each. Assuming a utilisation factor of 0.5 and a maintenance factor of 0.8 design the lighting scheme.

$$N = \frac{E \times A}{F \times U \times M} \quad \therefore \quad N = \frac{400 \times 8 \times 7}{7375 \times 0.5 \times 0.8} \quad N = 7.59, \text{ use 8 fittings}$$

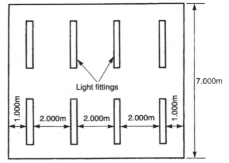

Layout of fluorescent tubes for the office

Permanent Supplementary Lighting of Interiors

Illumination of building interiors is a very important factor for designers. This will relate to user convenience and visual impact of the building. Overall considerations fall into three categories:

A – daylighting alone, in which the window area occupies about 80% of the facades

B – permanent supplementary artificial lighting of interiors, in which the window area is about 20% of the facades

C – permanent artificial lighting of interiors in which there are no windows.

Occupants of buildings usually prefer a view to the outside. Therefore, the choice of lighting for most buildings is from type A or B. With type B, the building may be wider, because artificial lighting is used to supplement daylighting. Although the volume is the same as type A, the building perimeter is less, thus saving in wall construction. Type B building also has lower heat gains and energy losses through the glazing, less noise from outside and less maintenance of windows.

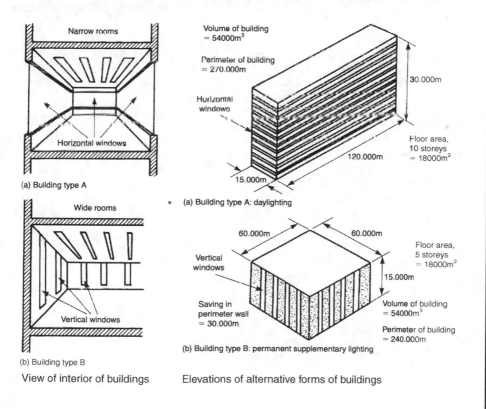

(a) Building type A

(b) Building type B

View of interior of buildings

Volume of building = 54000m³

Perimeter of building = 270.000m

Horizontal windows

30.000m

Floor area, 10 storeys = 18000m²

120.000m

15.000m

(a) Building type A: daylighting

60.000m   60.000m

Vertical windows

Floor area, 5 storeys = 18000m²

15.000m

Saving in perimeter wall = 30.000m

Volume of building = 54000m³

Perimeter of building = 240.000m

(b) Building type B: permanent supplementary lighting

Elevations of alternative forms of buildings

Ref. BS EN 12464-1: Light and lighting.
Lighting of workplaces. Indoor workplaces.

The daylight received inside a building can be expressed as 'the ratio of the illumination at the working point indoors, to the total light available simultaneously outdoors'. This can also be expressed as a percentage and it is known as the 'daylight factor'.

The daylight factor includes light from:

- Sky component – light received directly from the sky; excluding direct sunlight.
- External reflected component – light received from exterior reflecting surfaces.
- Internal reflected component – light received from internal reflecting surfaces.

If equal daylight factor contours are drawn for a room, they will indicate how daylighting falls as distance increases from a window.

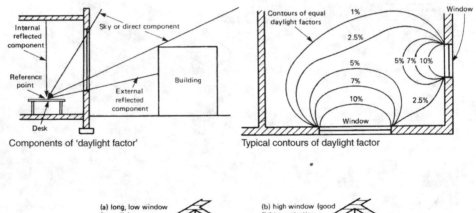

Components of 'daylight factor'

Typical contours of daylight factor

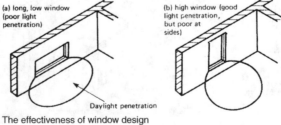

The effectiveness of window design

Refs. BRE Digests 309 and 310: Estimating daylight in buildings.
BS EN 17037: Daylight in Buildings.

The effect of daylight in a room can be studied by using scaled models. Providing that textures and colours of a room surface are the same, an approximate result may be obtained.

An estimate of the effect of daylight in a room may also be made from daylight factor protractors and associated tables of data. These were developed by the Building Research Establishment for use with scaled drawings to determine the sky component from a sky of uniform luminance.

There are pairs of protractors to suit different window types. Protractor No. 1 is placed on the cross section as shown. Readings are taken where the sight lines intersect the protractor scale.

In the diagram, the sky component = 8·5 – 4 = 4·5% and an altitude angle of 30°. The sky component of 4·5% must be corrected by using protractor No. 2. This is placed on the plan as shown. Readings from protractor No. 2 are 0·25 and 0·1, giving a total correction factor of 0·35. Therefore, 4·5 × 0·35 = 1·6%.

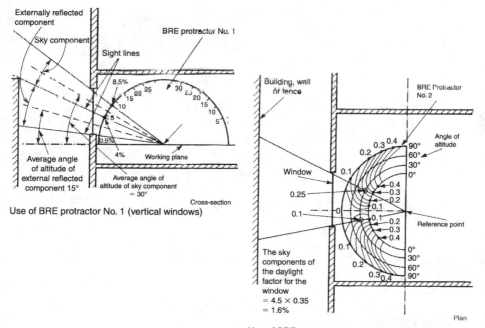

Use of BRE protractor No. 1 (vertical windows)

Use of BRE protractor No. 2 (vertical windows)

Note: Daylight protractors numbers 1 to 10. They are available with a guide from the Building Research Establishment, ref. Publication code AP 68.

The external reflected component of the daylight factor for a uniform sky may be taken as approximately $0.1 \times$ the equivalent sky component. Using the diagrams shown in Daylighting – 2, the value may be found as follows:

- Readings from protractor No. 1 are 4% and 0.5%.

- Equivalent sky component = 4% − 0.5% = 3.5%.

- Average angle of altitude = 15°.

- Readings on protractor No. 2 are 0.27 and 0.09 (for 15°).

- Correction factor = 0.27 + 0.09 = 0.36.

- Equivalent uniform sky component = 3.5% × 0.36 = 1.26%.

- Externally reflected component = 0.1 × 1.26% = 0.126%.

To establish the daylight factor, the internal reflected component is calculated and added to both the sky and externally reflected components – see example.

Example: Find the minimum internally reflected component of the daylight factor for a room measuring 10m × 8m × 2.5m high, having a window in one wall with an area of 20m². The floor has an average reflection factor of 20% and the walls and ceiling average reflection factors of 60% and 70%, respectively.

$$\text{Window area as a percentage of floor area} = \frac{20}{80} \times \frac{100}{1} = 25\%$$

Referring to Table 2 (page 613), the minimum internally reflected component = 1.3%.

Allowing a maintenance factor of 0.9 for dirt on the windows, the value will be modified to 1.3 × 0.9 = 1.17%.

For the example given in daylighting 2 and 3, the daylight factor will be the addition of the three components = 1.6 + 0.126 + 1.17 = 2.9%.

Table 1 Reflection factors

Reflection factors (%)		Reflection factors (%)	
White	75–88	Golden yellow	62
Light stone	53	Orange	36
Middle stone	37	Eau-de-nil	48
Light buff	60	Sky blue	47
Middle buff	43	Turquoise	27
Light grey	44	Light brown	30
Dark grey	26	Middle brown	20
Pale cream	73	Salmon pink	42

Table 2 Minimum internally reflected component of the daylight factor (%)

Ratio of window area to floor area	Window area as a per-centage of floor area	Floor reflection factor (%)											
		10				20				40			
		Wall reflection factor (%)											
		20	40	60	80	20	40	60	80	20	40	60	80
		%	%	%	%	%	%	%	%	%	%	%	%
1:50	2			0·1	0·2		0·1	0·1	0·2		0·1	0·2	0·2
1:20	5	0·1	0·1	0·2	0·4	0·1	0·2	0·3	0·5	0·1	0·2	0·4	0·6
1:14	7	0·1	0·2	0·3	0·5	0·1	0·2	0·4	0·6	0·2	0·3	0·6	0·8
1:10	10	0·1	0·2	0·4	0·7	0·2	0·3	0·6	0·9	0·3	0·5	0·8	1·2
1:6·7	15	0·2	0·4	0·6	1·0	0·2	0·5	0·8	1·3	0·4	0·7	1·1	1·7
1:5	20	0·2	0·5	0·8	1·4	0·3	0·6	1·1	1·7	0·5	0·9	1·5	2·3
1:4	25	0·3	0·6	1·0	1·7	0·4	0·8	1·3	2·0	0·6	1·1	1·8	2·8
1:3·3	30	0·3	0·7	1·2	2·0	0·5	0·9	1·5	2·4	0·8	1·3	2·1	3·3
1:2·9	35	0·4	0·8	1·4	2·3	0·5	1·0	1·8	2·8	0·9	1·5	2·4	3·8
1:2·5	40	0·5	0·9	1·6	2·6	0·6	1·2	2·0	3·1	1·0	1·7	2·7	4·2
1:2·2	45	0·5	1·0	1·8	2·9	0·7	1·3	2·2	3·4	1·2	1·9	3·0	4·6
1:2	50	0·6	1·1	1·9	3·1	0·8	1·4	2·3	3·7	1·3	2·1	3·2	4·9

Note: The ceiling reflection factor is assumed to be 70%.

There are other methods for determining daylight factor. Some are simple rules of thumb and others more detailed formulae. An example of each is shown below.

- Rule of thumb – $D = 0.1 \times P$

where D = daylight factor

P = percentage of glazing relative to floor area.

E.g. a room 80m² floor area with 15m² of glazing.

$$D = 0.1 \times 15/80 \times 100/1 = 1.875\%$$

- Formula –

$$D = \frac{T \times G \times \theta \times M}{A(1 - R^2)}$$

where D = average daylight factor

T = transmittance of light through glass (clear single glazing = 0.85, clear double glazing = 0.75)

G = glazed area (m²)

θ = angle of sky component

M = maintenance factor (see page 608)

A = total area of interior surfaces, inc. windows (m²)

R = reflection factors (see page 613).

E.g. using the data from the example on page 612 and assuming a 50% reflection factor, double glazing and a sky component angle of 35°:

$$D = \frac{0.75 \times 20 \times 35 \times 0.9}{250 \,(1 \,= \,[50/100]^2)} = 2.52\%$$

All calculations and estimates of daylight factor and glazing area must conform with the energy-saving requirements defined in the Building Regulations, Approved Document L – Conservation of Fuel and Power.

Previously this has included a maximum allowance for glazed areas relative to floor and external wall areas, but with the availability of quality double-glazed units, these limitations are now relaxed. See also page 197 and associated references.

Cabling systems that were originally used solely for telephone communications now have many other applications. These include fire alarms, security/intruder alarms, computer networking, teleprinters, facsimile machines, etc. The voltage and current are very low and have no direct connection to the mains electricity in a building. Therefore, telecommunications and mains cabling should be distinctly separated in independent conduits and trunking for reasons of safety and to prevent interference.

External telecommunications cables may supply a building from overhead or underground, the latter being standard for new building work. The intake is below surface level at a point agreed with the cable supplier. In large buildings, the incoming cable supplies a main distribution unit which has connections for the various parts of the building. Cables supply both switchboards and individual telephones from vertical risers. There may be limitations on the number of cables supplied from risers, and early consultation with the cable supplier is essential to determine this and any other restrictions.

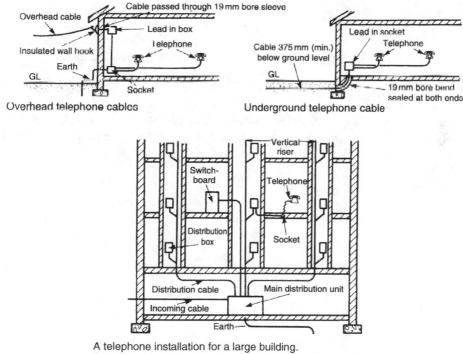

Overhead telephone cables

Underground telephone cable

A telephone installation for a large building.
Cables inside the building (not the flexible cord)
must be concealed in ducts and the system earthed.

Installation – since 2017 all new buildings and existing buildings that are subjected to extensive renovation, refurbishment and/or conversion must be provided with a facility to accommodate the installation of high-speed electronic communications. Some categories of building may be exempt. Examples include sheds, garages and monuments; a full listing exists under Section 1.5 of the Building Regulations, Approved Document R. The requirement is not for cabling and equipment, but the physical infrastructure or means for this to be installed. An access terminal is located on the outside of a building or within a common entrance area. A network terminal is provided inside each dwelling or unit with a duct or conduit to convey copper or fibre-optic cables from the access terminal. Wireless devices may also be used to link the two. See next page.

Bandwidth or data transfer rate – expressed in units averaging the number of megabits per second (Mbps). Sometimes expressed in millions of bits per second. A bit is a binary digit with a value of zero (0) or one (1), the smallest unit of data in a computer. The minimum delivery broadband speed is 30 Mbps. Most service providers deliver at up to 70 Mbps.

External and internal installation work – the site-wide cabling network installation is not a requirement under Building Regulations, nor is the internal cabling distribution. These facilities are by arrangement between the site developer and their (or their client's) appointed broadband internet service provider. However, the NHBC recommends that new-build homes are future-proofed by incorporating hard-wired communications cabling during construction.

References and further reading –

Building Regulations, Approved Document R: Physical infrastructure for high speed electronic communications networks.

BS PAS 2016: 2010: Next generation access for new build homes – Guide.

NHBC – The connected home. Designing and building technology into today's new homes – (NF67).

BT Openreach – Developers' Handbook. Guides and Handbooks.

Schematic installation to a dwelling house –

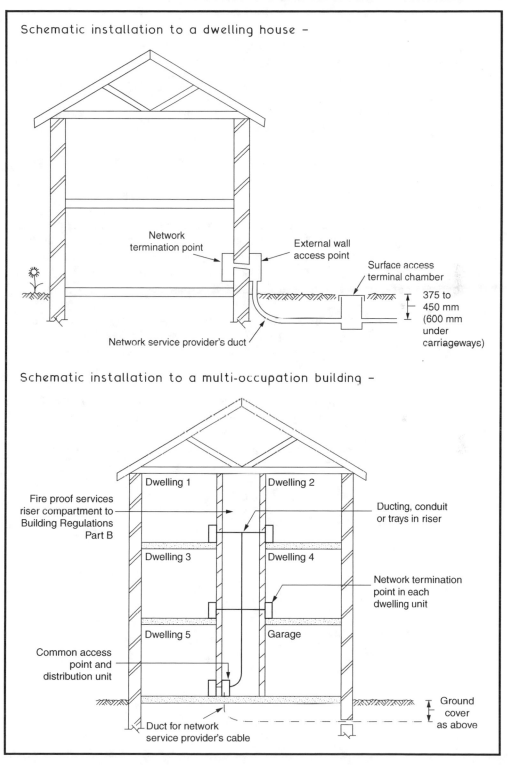

Network termination point

External wall access point

Surface access terminal chamber

375 to 450 mm (600 mm under carriageways)

Network service provider's duct

Schematic installation to a multi-occupation building –

Dwelling 1

Dwelling 2

Fire proof services riser compartment to Building Regulations Part B

Ducting, conduit or trays in riser

Dwelling 3

Dwelling 4

Network termination point in each dwelling unit

Dwelling 5

Garage

Common access point and distribution unit

Ground cover as above

Duct for network service provider's cable

Electric Vehicle Charger Point

As part of the UK government's commitment to a net zero carbon environment by 2050, there is an expectation that there will be no petrol and diesel car sales after 2040.

Building Regulations, A.D. S establish a requirement for all new buildings and existing buildings subject to refurbishment, alteration and material changes to be provided with sufficient charger points.

As a significant electrical load is needed for a considerable time, a dedicated circuit is usually required from the intake. Metered economies may be achieved with a "time-of-use" (off-peak) tariff.

A charger point should be located in a ventilated area, out of direct sunlight. Installation should be undertaken by a 'competent person' (pg. 550). Configurations vary depending on individual supply systems, see pages 551–552. For example, a TN-C supply is not compatible as it has a protective earth neutral (PEN) conductor, i.e. a combined function of earthing and neutral which may also affect TN-C-S (PME) installations unless provided with a dedicated EV earthing electrode. Without that, there is a possibility of over-heating as a diverted neutral fault current may return to the supply which normal overload protection may not react to.

Numerous wiring options exist, some with complimentary protection based on voltage monitoring and some with supplementary earthing. Local supply authority to be consulted. Schematic diagram below shows a possible after meter connection from the intake to a charger point.

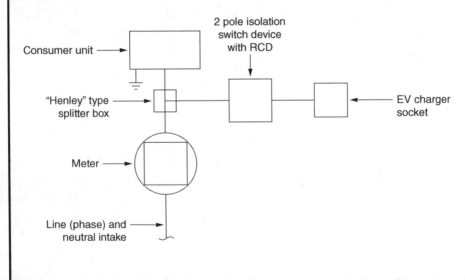

12 MECHANICAL CONVEYORS – LIFTS, ESCALATORS AND TRAVELATORS

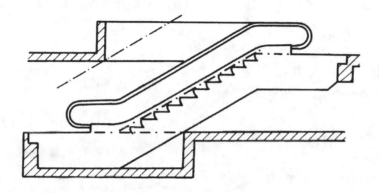

DOI: 10.1201/9781003434894-12

Planning Lift Installations

To function efficiently and to provide access for the elderly and disabled, modern offices and public buildings are provided with suitably designed lift installations. Planning (as with all services) should commence early in the design programme. Priority must be given to locating lifts centrally within a building to minimise horizontal travel distance. Consideration must also be given to position, relative to entrances and stairs. Where the building size justifies several passenger lifts, they should be grouped together. In large buildings, it is usual to provide a group of lifts near the main entrance and single lifts at the ends of the building. The lift lobby must be wide enough to allow pedestrian traffic to circulate and pass through the lift area without causing congestion. For tall buildings in excess of 15 storeys, high-speed express lifts may be used which bypass the lower floors.

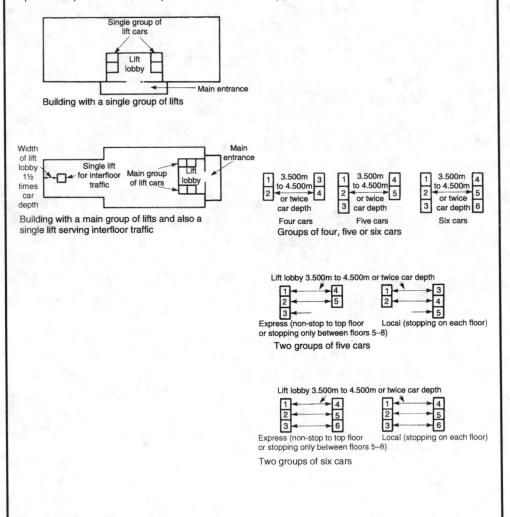

Building with a single group of lifts

Building with a main group of lifts and also a single lift serving interfloor traffic

Groups of four, five or six cars

Two groups of five cars

Two groups of six cars

Requirements:

- Necessary in all buildings over three storeys high.
- Essential in all buildings over a single storey if they are accessed by the elderly or disabled.
- Minimum standard – one lift per four storeys.
- Maximum walking distance to access a lift – 45m.
- Floor space and lift car capacity can be estimated at $0.2m^2$ per person.

Lift speed:

Type	Car speed (m/s)
Goods (electric or hydraulic)	0·2–1
Electric passenger <4 floors	0·3–0·8
4–6 floors	0·8–1·2
6–9 floors	1·2–1·5
9–15 floors*	5–7
Paternoster	<0·4
Hydraulic passenger†	0·1–1·0

*Express lift that does not stop at the lower floor levels. The upper speed limit is 7m/s because of the inability of the human ear to adapt to rapidly changing atmospheric conditions.
†Overall theoretical maximum travel distance is 21m vertically, therefore limited to four or five storeys.

Electric motor – low-speed lifts operate quite comfortably with an AC motor to drive the traction sheave through a worm gear (see page 628). For faster speed applications, a DC motor is preferable. This is supplied via a mains generator for each lift motor. DC motors have historically provided better variable voltage controls, more rapid and smoother acceleration, quieter operation, better floor levelling and greater durability in resisting variable demands. Recent developments with AC motors have made them more acceptable, and these are now becoming more widely used.

Refs. BS 5655: Lifts and service lifts (several parts).

BS EN 81: Safety rules for the construction and installation of lifts (several parts).

BS 8486: Examination and test of new lifts before putting into service (Several Parts).

Roping Systems for Electric Lifts – 1

High-tensile steel ropes are used to suspend lift cars. They have a design factor of safety of ten and are usually at least four in number. Ropes travel over grooved driving or traction sheaves and pulleys. A counterweight balances the load on the electric motor and traction gear.

Methods for roping vary:

Single wrap 1:1 – the most economical and efficient of roping systems but is limited in use to small-capacity cars.

Single wrap 1:1 with diverter pulley – required for larger capacity cars. It diverts the counterweight away from the car. To prevent rope slip, the sheave and pulley may be double wrapped.

Single wrap 2:1 – an alternative for use with larger cars. This system doubles the load-carrying capacity of the machinery but requires more rope and also reduces the car speed by 50%.

Double wrap – used to improve traction between the counterweight, driving sheave and steel ropes.

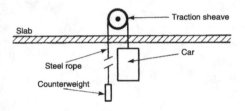

Single wrap 1 : 1 roped

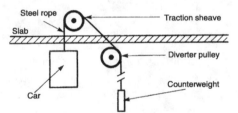

Single wrap 1 : 1 roped with diverter pulley

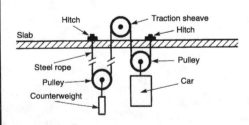

Single wrap 2 : 1 roped

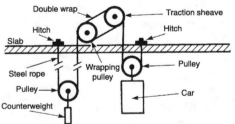

Double wrap 2 : 1 roped (for high speed and medium to heavy duty loads)

Single wrap 3:1 – used for heavy goods lifts where it is necessary to reduce the force acting upon the machinery bearings and counterweight. The load-carrying capacity is increased by up to three times that of uniform ratio, but the capital costs are higher with increased pulleys and greater length of rope. By comparison, the car speed is also reduced to one-third.

Drum drive – a system with one set of ropes wound clockwise around the drum and another set anticlockwise. It is equally balanced, as one set unwinds the other winds. The disadvantage of the drum drive is that as height increases the drum becomes less controllable, limiting its application to rises of about 30m.

Compensating rope and pulley – used in tall buildings where the weight of the ropes in suspension will cause an imbalance on the driving gear and also a possible bouncing effect on the car. The compensating ropes attach to the underside of the car and counterweight to pass around a large compensating pulley at low level.

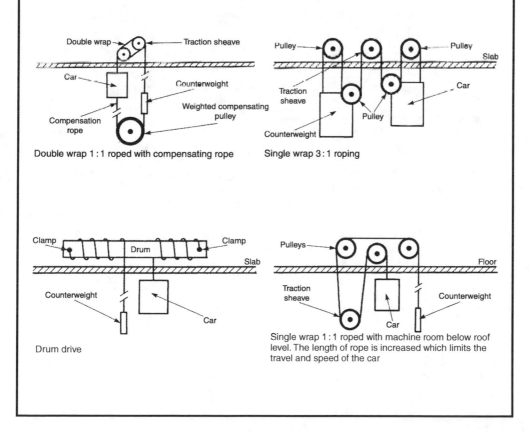

Double wrap 1:1 roped with compensating rope

Single wrap 3:1 roping

Drum drive

Single wrap 1:1 roped with machine room below roof level. The length of rope is increased which limits the travel and speed of the car

623

Single Automatic Lift Control

The single automatic push button system is the simplest and least sophisticated of controls. The lift car can be called and used by only one person or group of people at a time. When the lift car is called to a floor, the signal lights engraved 'in use' are illuminated on every floor. The car will not respond to any subsequent landing calls, nor will these calls be recorded and stored. The car is under complete control of the occupants until they reach the required floor and have departed the lift. The 'in use' indicator is now switched off, and the car is available to respond to the next landing call. Although the control system is simple and inexpensive by comparison with other systems, it has its limitations for user convenience. It is most suited to light traffic conditions in low-rise buildings such as nursing homes, small hospitals and flats.

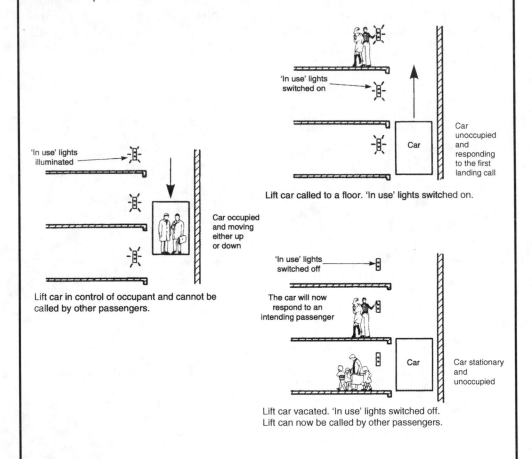

'In use' lights switched on

Car unoccupied and responding to the first landing call

Lift car called to a floor. 'In use' lights switched on.

'In use' lights illuminated

Car occupied and moving either up or down

Lift car in control of occupant and cannot be called by other passengers.

'In use' lights switched off

The car will now respond to an intending passenger

Car stationary and unoccupied

Lift car vacated. 'In use' lights switched off. Lift can now be called by other passengers.

Ref. BS ISO 4190-5: Lift (elevator) installation. Control devices, signals and additional fittings.

Down collective – store calls made by passengers in the car and those made from the landings. As the car descends, landing calls are answered in floor sequence to optimise car movement. If the car is moving upwards, the lift responds to calls made inside the car in floor sequence. After satisfying the highest registered call, the car automatically descends to answer all the landing calls in floor sequence. Ony one call button is provided at landings. This system is most suited to flats and small hotels, where the traffic is mainly between the entrance lobby and specific floors.

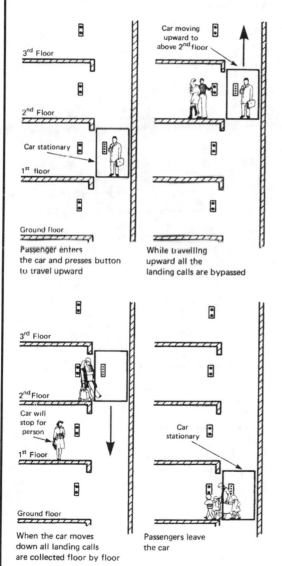

3rd Floor

2nd Floor

Car stationary

1st floor

Ground floor

Passenger enters
the car and presses button
to travel upward

Car moving
upward to
above 2nd floor

While travelling
upward all the
landing calls are bypassed

3rd Floor

2nd Floor

Car will
stop for
person

1st Floor

Ground floor

When the car moves
down all landing calls
are collected floor by floor

Car
stationary

Passengers leave
the car

Full or directional collective – a variation in which car and landing calls are immediately stored in any number. Upward and downward intermediate landing calls are registered from one of two directional buttons. The uppermost and lowest floors only require one button. The lift responds to calls in floor order independent of call sequence, first in one direction and then in the other. It has greater flexibility than the down collective system and is appropriate for offices and departmental stores where there is more movement between intermediate floors.

Controls for Two or More Cars

Two cars may be coordinated by a central processor to optimise efficiency of the lifts. Each car operates individually on a full or down collective control system. When the cars are at rest, one is stationed at the main entrance lobby and the other, which has call priority, at a mid-point within the building or at another convenient floor level. The priority car will answer landing calls from any floor except the entrance lobby. If the priority car is unable to answer all call demands within a specific time, the other car if available will respond. A similar system may also apply to three cars, with two stationary at the entrance lobby and one available at mid-point or the top floor.

With the supervisory control system, each car operates on full collective control and will respond to calls within a dedicated zone. A micro-processor determines traffic demand and locates cars accordingly to each operating zone.

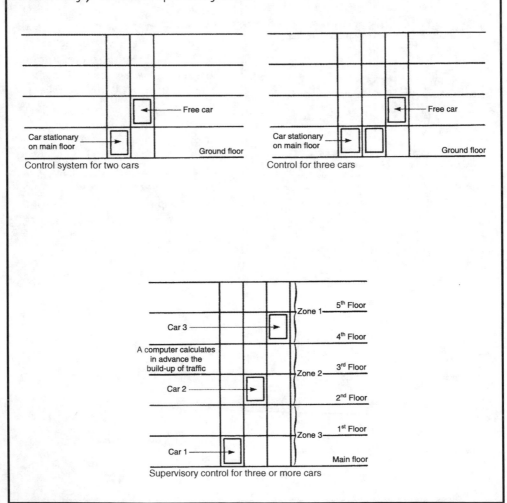

Control system for two cars

Control for three cars

Supervisory control for three or more cars

Door operation is by an electric motor through a speed reduction unit, clutch drive and connecting mechanism. The types of entrance and doors form a vital part of the lift installation. The average lift car will spend more time at a floor during passenger transfer time than it will during travel. For general passenger service, either side opening, two-speed or even triple-speed side-opening doors are preferred. The most efficient in terms of passenger handling is the two-speed centre opening. The clear opening may be greater and usable clear space becomes more rapidly available to the passengers. Vertical centre-bi-parting doors are suitable for very wide openings, typical of industrial applications.

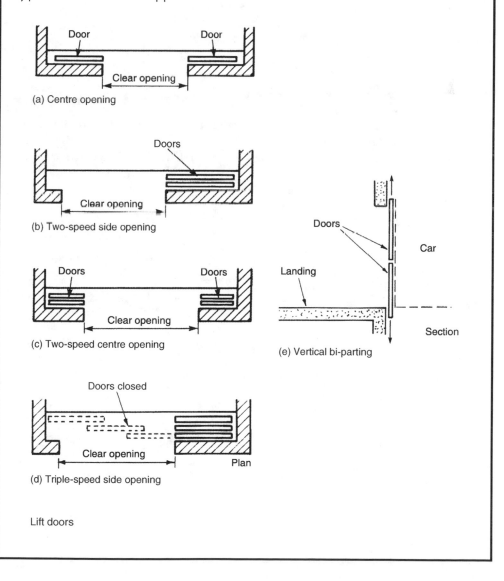

(a) Centre opening

(b) Two-speed side opening

(c) Two-speed centre opening

(d) Triple-speed side opening

(e) Vertical bi-parting

Lift doors

Lift Machine Room and Equipment

Wherever possible, the machine room should be sited above the lift shaft. This location minimises the length of ropes and optimises efficiency. The room should be ventilated, but the vent opening must not be over the equipment. Machinery must be well secured to a concrete base. To reduce sound transmission and vibration, compressed cork, dense rubber or a composite layer is used as an intermediate mounting.

A steel lifting beam is built into the structure above the machinery for positioning or removing equipment for maintenance and repair. Sufficient floor space is necessary for the inspection and repair of equipment. Recommended machine room dimensions are given in BS 5655-6: Lifts and service lifts. Code of practice for the selection, installation and location of new lifts.

To prevent condensation, the room must be well insulated and heated to provide a design air temperature between 10°C and 40°C. Walls, ceiling and floor should be smooth finished and painted to reduce dust formation. A regular pattern of room cleaning and machinery maintenance should be scheduled.

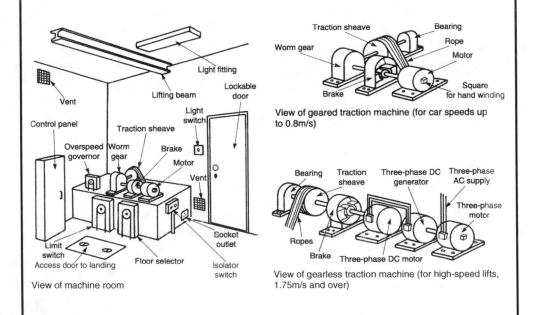

View of machine room

View of geared traction machine (for car speeds up to 0.8m/s)

View of gearless traction machine (for high-speed lifts, 1.75m/s and over)

628

Buffers – located at the base of the shaft. They are usually oil loaded for lift speeds >1·5 m/s and otherwise spring loaded. Some variations use compressible plastics.

Overspeed governor – a steel rope passes round a tension pulley in the pit and a governor pulley in the machine room. It also attaches to the lift car's emergency braking system. Overspeeding locks the governor as it responds to spring-loaded flyweight inertia from the centrifugal force in its accelerating pulley. This also switches off power to the lift. The tightening governor rope actuates the safety braking gear.

Safety gear – hardened steel wedges are arranged in pairs each side of the lift car to slow down and stop the car by frictional contact with the car guide rail. Slow- and medium-speed lifts have pairs of hardened steel cams which instantaneously contact a steel channel secured to the lift wall.

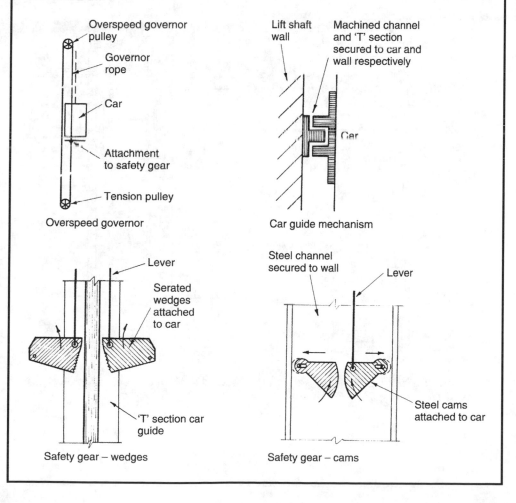

Overspeed governor

Car guide mechanism

Safety gear – wedges

Safety gear – cams

Details of an Electric Lift Installation

To satisfy the need for economies in lift-manufacturing processes, dimensions are limited. For this purpose, guidance in BS 5655-6 refers to the internationally agreed standards, BS ISO 8100. Therefore, architects will have to establish passenger transport requirements as a preliminary design priority. The size of lift shaft will depend upon the car capacity and the space required for the counterweight, guides and landing door. The shaft extends below the lowest level served to provide a pit. This permits a margin for car overtravel and a location for car and counterweight buffers. The pit must be watertight and have drainage facilities. Shaft and pit must be plumb and the internal surfaces finished smooth and painted to minimise dust collection. A smoke vent with an unobstructed area of 0·1m² is located at the top of the shaft. The shaft is of fire-resistant construction as defined for 'protected shafts' in the Building Regulations. This will be at least 30 minutes and is determined by building function and size. No pipes, ventilating ducts or cables (other than those specifically for the lift) must be fitted within the shaft. A clearance is required at the top of the lift for car overtravel.   Counterweight location is at the back or side of the car.

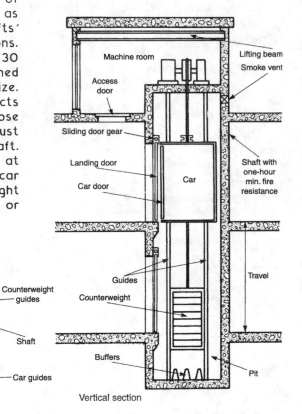

Plan of lift

Vertical section

Refs. BS 5655-6: Lifts and service lifts. Code of practice for selection, installation and location of new lifts.

BS EN 81: Safety rules for the construction and installation of lifts.

BS ISO 8100: Lifts for transportation of persons and goods.

Building Regulations, Approved Document B3: Internal fire spread (structure).

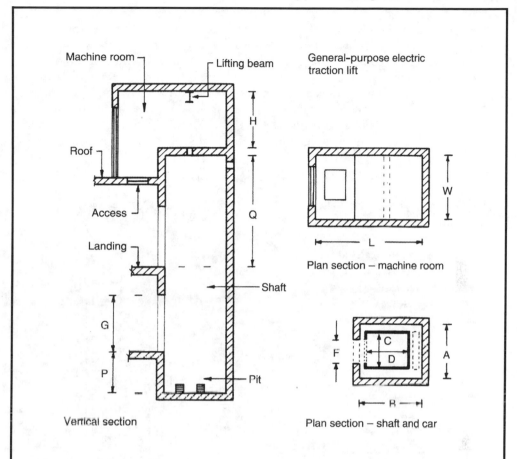

Vertical section

Plan section – machine room

Plan section – shaft and car

All dimensions in metres:

Shaft size		Car size			Door size		Pit	Machine room			
A	B	C	D	E	F	G	P	Q	H	L	W
1·8	2·1	1·1	1·4	2·2	0·8	2·0	1·7	4/4·2	2·6	3·7	2·5
1·9	2·3	1·35	1·4	2·2	0·8	2·0	1·7	4/4·2	2·6	3·7	2·5
2·4	2·3	1·6	1·4	2·3	1·1	2·1	1·8	4·2	2·7	4·9	3·2
2·6	2·3	1·95	1·4	2·3	1·1	2·1	1·9	4·4	2·7	4·9	3·2
2·6	2·6	1·95	1·75	2·3	1·1	2·1	1·9	4·4	2·8	5·5	3·2

Note: Dimension E refers to the car door height.

Paternoster Lifts

A paternoster consists of a series of open-fronted two-person cars suspended from hoisting chains. Chains run over sprocket wheels at the top and bottom of the lift shaft. The lift is continuously moving and provides for both upward and downward transportation of people in one shaft. Passengers enter or leave the car while it is moving; therefore, waiting time is minimal. Passengers will have to be fairly agile, which limits this type of installation to factories, offices, universities, etc. It is not suitable in buildings that accommodate the infirm or elderly! When a car reaches its limit of travel in one direction, it moves across to the adjacent set of hoisting chains to engage with car guides and travel in the other direction. In the interests of safety, car speed must not exceed 0.4m/s.

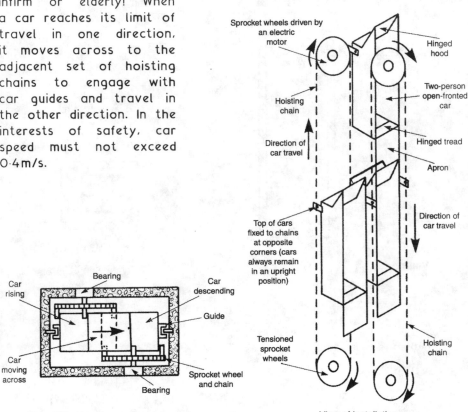

Plan of lift at top change-over

View of installation

Paternosters convey about 600 persons per hour. This type of lift has the advantage of allowing passengers to begin their journeys undelayed, regardless of travel direction. Simplicity of control gear adds to the advantages, resulting in fewer breakdowns by eliminating normal processes of stopping, starting, accelerating and decelerating. They are most suited to medium-rise buildings.

Direct acting – the simplest and most effective method, but it requires a borehole below the pit to accommodate the hydraulic ram. The ram may be one piece or telescopic. In the absence of a counterweight, the shaft width is minimised. This will save considerably on construction costs and leave more space for general use.

Side acting – the ram is connected to the side of the car. For large-capacity cars and heavy goods lifts, two rams may be required, one each side of the car. A borehole is not necessary, but due to the cantilever design and eccentric loading of a single ram arrangement, there are limitations on car size and load capacity.

Direct side acting – the car is cantilevered and suspended by a steel rope. As with side acting, limitations of cantilever designs restrict car size and payload. Car speed may be increased.

Indirect side acting – the car is centrally suspended by steel rope and the hydraulic system is inverted.

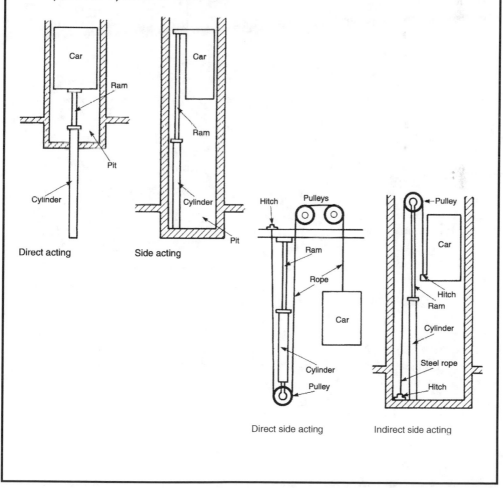

Direct acting

Side acting

Direct side acting

Indirect side acting

Details of Oil-Hydraulic Lift Installation

Originally, hydraulic lifts used mains water supply as the operating medium. The main was pressurised from a central pumping station to service lift installations in several buildings. The oil-hydraulic system has oil pressure fed by a pump into a cylinder to raise the ram and lift the car. Each lift has its own pumping unit and controller. These units are usually sited at or near to the lowest level served, no more than 10m from the shaft. The lift is ideal in lower rise buildings where moderate speed and smooth acceleration is preferred.

Car speed ranges from 0·1 to 1m/s, and the maximum travel is limited to about 21m. The lift is particularly suitable for goods lifts and for hospitals and old people's homes. Most hydraulic lifts carry the load directly to the ground, therefore as the shaft does not bear the loads, construction is less expensive than for a comparable electric lift installation.

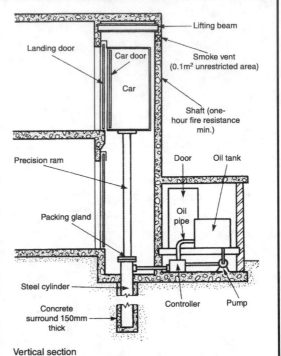

Vertical section

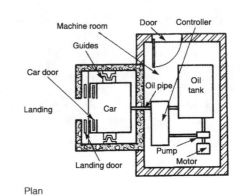

Plan

Refs. BS 5655-10.2.1: Lifts and service lifts. Specification for the testing and examination of lifts and service lifts. Hydraulic lifts. Commissioning tests for new lifts.

BS EN 81-20: Safety rules for the construction and installation of lifts. Hydraulic lifts.

Upward movement – the oil pressure must be gradually increased. The up solenoid valve is energised by an electric current and opens to allow oil to enter above piston D. As the area of piston D is greater than valve C, the oil pressure closes the valve and allows high-pressure oil to flow to the cylinder and lift the ram and the car.

Downward movement – the oil pressure must be gradually decreased. The lowering solenoid valve is energised by an electric current and opens, allowing oil to flow back to the tank through the bypass. As the area of piston A is greater than valve B, the reduced oil pressure behind the piston allows valve B to open. Oil flows into the tank and the car moves downward.

A special packing gland with several seals is required between the cylinder and ram.

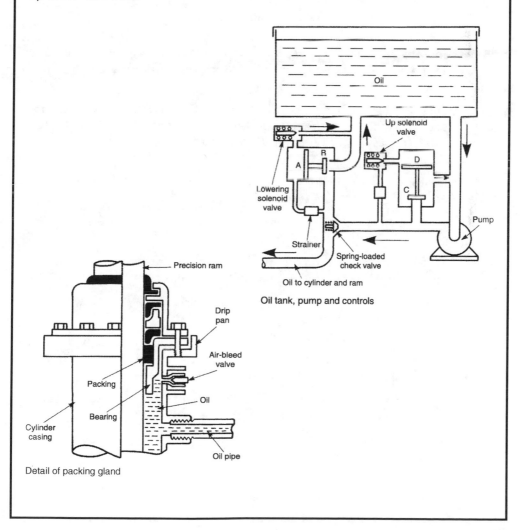

Oil tank, pump and controls

Detail of packing gland

Lift Performance

Lift performance depends on:

- acceleration;
- retardation;
- car speed;
- speed of door operation; and
- stability of speed and performance with variations of car load.

The assessment of population density may be found by allowing between one person per $9.5m^2$ of floor area to $11.25m^2$ of floor area. For unified starting and finishing times, 17% of the population per five minutes may be used. For staggered starting and finishing times, 12% of the population may be used.

The number of lifts will have an effect on the quality of service. Four 18-person lifts provide the same capacity as three 24-person lifts but the waiting time will be about twice as long with the three-car group.

The quality of service may be found from the interval of the group: 25–35 seconds interval is excellent, 35–45 seconds is acceptable for offices, 60 seconds for hotels and 90 seconds for flats.

Further criteria for the comfort and convenience of lift users:

- Directional indication of location of the lift lobby for people unfamiliar with the building.

- Call buttons at landings and in the car positioned for ease of use with unambiguous definition for up and down directions.

- Call buttons to be at a level appropriate for use by people with disabilities and small children.

- Call display/car location display at landings to be favourably positioned for a group of people to watch the position of all cars and for them to move efficiently to the first car arriving.

- Call lights and indicators with an audible facility to show which car is first available and in which direction it is travelling.

- Lobby space of sufficient area to avoid congestion by lift users and general pedestrian traffic in the vicinity.

Example: An office block with 20 storeys above ground floor having unified starting and stopping times is to have a floor area above the ground floor of 8000m² and floor pitch of 3m. A group of four lifts, each car having a capacity of 20 persons and a car speed of 2·5m/s, are specified. The clear door width is to be 1·1m, and the doors are to open at a speed of 0·4m/s. Estimate the interval and quality of service that is to be provided.

1. Peak demand for a five-minute period $= \dfrac{8000m^2 \times 17\%}{11m^2/person \times 100}$

$$= 124 \text{ persons}$$

2. Car travel $= 20 \times 3m = 60m$

3. Probable number of stops $= S - S\left(\dfrac{S-1}{S}\right)^n$

(where S = maximum number of stops)

∴ Probable number of stops $= 20 - 20\left(\dfrac{20-1}{20}\right)^{16}$

$$= 11$$

(where n = number of passengers usually approximately 80% of capacity)

4. Upward journey time $= S_1\left(\dfrac{L}{S_1 V} + 2V\right)$

where S_1 = probable number of stops L = travel V = speed

∴ Upward journey time $= 11\left(\dfrac{60}{11 \times 2·5} + 2 \times 2·5\right)$

$$= 79 \text{ seconds}$$

5. Downward journey time $= \left(\dfrac{L}{V} + 2V\right)$

$$= \dfrac{60}{2·5} + 2 \times 2·5$$

$$= 29 \text{ seconds}$$

6. Door operating time $= 2(S_1 + 1)\dfrac{W}{Vd}$

where W = width of door opening; Vd = opening speed

∴ Door operating time $= 2(11 + 1)\dfrac{1·1}{0·4} = 66 \text{ seconds}$

7. The average time taken for each person to get into and out of a lift car may be taken as two seconds

∴ Transfer time $= 2n = 2 \times 16 = 32$ seconds

8. Round-trip time $= 79 + 29 + 66 + 32 = 206$ seconds

9. Capacity of group $= \dfrac{5 \text{ mins} \times 60 \times 4 \times 20 \times 0·8}{206}$

$$= 93 \text{ persons per 5 minutes}$$

10. Interval for the group $= \dfrac{206}{4} = 51.5$ seconds

The capacity of the group of lifts and the interval for the group are satisfactory. (Note: Cars less than 12 capacity are not satisfactory.)

Firefighting Lifts – 1

During the early part of the twentieth century, it became apparent that the growing number of high-rise buildings would require special provisions for fire control. The firefighting lift was conceived as a means of rapidly accessing the upper floors. Early innovations prioritised the passenger lift by means of a 'break-glass' key switch which brought the lift to the ground floor immediately. This is now unlikely to be acceptable to building insurers and the fire authorities. It is also contrary to current building standards which specify a separate lift installation specifically for firefighting purposes.

Special provisions for firefighting lifts:

- Minimum duty load of 630 kg.
- Minimum internal dimensions of 1·1 m wide × 1·4 m deep × 2·0 m high.
- Provision of an emergency escape hatch in the car roof.
- Top-floor access time – maximum 60 seconds.
- Manufactured from non-combustible material.
- A two-way intercommunications system installed.
- Door dimensions at least 0·8 m wide × 2·0 m high of fire-resisting construction.
- Two power supplies – mains and emergency generator.

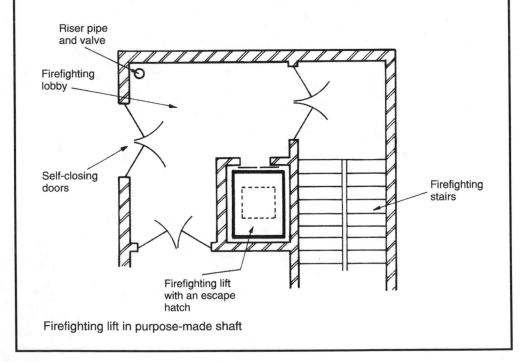

Riser pipe and valve

Firefighting lobby

Self-closing doors

Firefighting stairs

Firefighting lift with an escape hatch

Firefighting lift in purpose-made shaft

Building Regulations – structures with floors at a height greater than 18m above fire service vehicle access (usually ground level), or with a basement greater than 10m below fire service vehicle access, should have accessibility from a purpose-made firefighting lift. All intermediate floors should be served by the lift. Firefighting lifts for shop and commercial premises will require two firefighting shafts if they have a storey area over 900m² and a storey height greater than 7.5m above the fire service vehicle access.

Maximum distance of firefighting lift shaft to any part of a floor is 60m with sprinklers, 45m without. Hydrant outlets should be located in the firefighting lobby.

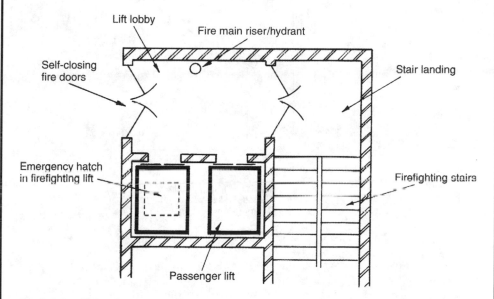

Firefighting lift in shared shaft

Refs. Building Regulations, Approved Document B: Fire safety, Volume 2, Part B5, Section 17: Access to buildings for firefighting personnel for specific applications.

BS 9999: Code of practice for fire safety in the design, management and use of buildings.

BS 8899: Improvement of firefighting and evacuation provisions in existing lifts. Code of practice.

Vertical Transportation for the Disabled

A passenger lift is the most suitable means for conveying wheelchair occupants between floor levels. However, a platform lift (see next page) or a platform stair lift may be used if access is only between two levels. Platform lifts must not be used where they would obstruct a designated means of fire escape on a stairway.

Lift provisions:

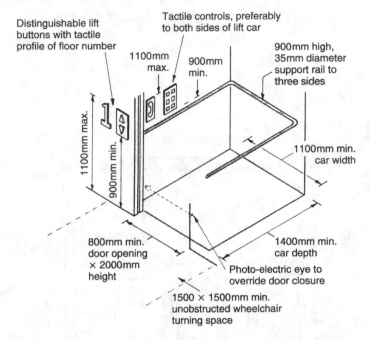

Distinguishable lift buttons with tactile profile of floor number

Tactile controls, preferably to both sides of lift car

900mm high, 35mm diameter support rail to three sides

1100mm max.

900mm min.

1100mm min. car width

1100mm max.

900mm min.

800mm min. door opening × 2000mm height

Photo-electric eye to override door closure

1400mm min. car depth

1500 × 1500mm min. unobstructed wheelchair turning space

- Landing space in front of lift doors should be sufficient to allow a wheelchair to turn and reverse into a lift car.
- Control/call panel should be prominent and easily distinguishable from its background.
- Time delay on door opening to be sufficient to allow wheelchair access. Doors fitted with a reactivation device to prevent people and/or wheelchair from being trapped in closing doors.
- Control panel in lift car positioned on a side wall, at least 400mm from a corner at a height accessible while seated.
- Control panel floor numbers to be raised on buttons to assist the visually impaired.
- Audible announcement of the floor levels served to help people with visual difficulties.
- Visual display of floor levels served to assist people with hearing impairments.
- Emergency telephones to be provided with inductive couplers for the benefit of hearing aid users. Location at an accessible height from a wheelchair.
- Alarm controls provided at an accessible height with a visual display to confirm the bell has responded for the benefit of lift users with hearing difficulties.

Passenger lifts can accommodate wheelchairs. Purpose-designed platform lifts can also be provided for wheelchair users in domestic and public buildings.

Power – Electricity, three phase 400V motor and 230V single phase for controls. For domestic use, hydraulic power with 230V single-phase control is more common. Lift speed – 0.15m/s maximum.

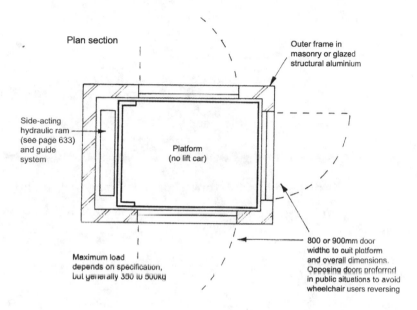

Plan section

Outer frame in masonry or glazed structural aluminium

Side-acting hydraulic ram (see page 633) and guide system

Platform (no lift car)

Maximum load depends on specification, but generally 350 to 500kg

800 or 900mm door widths to suit platform and overall dimensions. Opposing doors preferred in public situations to avoid wheelchair users reversing

Typical dimensions (mm) –

Platform depth	Platform width	Application
1250	900	Domestic
1400	1100	Public

Controls – can be automatic, but usually maintained command, otherwise known as push or hold to run, i.e. continuous hand pressure.

Refs. Disability Discrimination Act.

European Machinery Directive, 2006/42/EC (2nd Edition 2010).

Building Regulations Approved Document M: Access to and use of buildings.

BS 6440: Powered vertical lifting platforms having non-enclosed or partially enclosed liftways intended for use by persons with impaired mobility. Specification.

BS 8300: Design of an accessible and inclusive built environment. Buildings. Code of practice.

Builder's work – machine room:

- Door and window openings sealed against the weather.
- Lockable and safe access for lift engineers and building facilities manager.
- Provide and secure a trapdoor to raise and lower machinery.
- Secure all non-structural floors, decking and temporary scaffolding in position.
- Temporary guards and ladders to be secured in position.
- Dimensions to the requirements of BS 5655 or lift manufacturer's specification.
- Provide reinforced concrete floor and plinths to include at least nine rope holes.
- Treat floor to prevent dust.
- Provide lifting beam(s) and pad stone support in adjacent walls.
- Heating and ventilation to ensure a controlled temperature between 4°C and 40°C.

Electrical work:

- Reduced voltage temporary lighting and power supplies for portable tools during construction.
- Main switch fuse for each lift at the supply company's intake.
- Run power mains from intake to the motor room and terminate with isolating switches.
- Lighting and 13 amp power supply in the machine room.
- Independent light supply from the intake to the lift car with control switchgear in the machine room or halfway down the well.
- Lighting to the pit with a switch control in the lowest floor entrance.
- Permanent lighting in the well to consist of one lamp situated 500mm maximum from the highest and lowest points with intermediate lamps at 7m maximum spacing.

Builder's work – lift well:

- Calculations with regard to the architect's plans and structural loadings.

- Form a plumb lift well and pit according to the architect's drawings and to tolerances acceptable to the lift manufacturer (known as Nominal Minimum Plumb – the basic figures in which the lift equipment can be accommodated).

- Minimum thickness of enclosing walls – 230 mm brickwork or 130 mm reinforced concrete.

- Applying waterproofing or tanking to the pit and well as required.

- Paint surfaces to provide a dust-free finish.

- Provide dividing beams for multiple wells and inter-well pit screens. In a common well, a rigid screen extending at least 2.5 m above the lowest landing served and a full depth of the well between adjacent lifts.

- Secure lift manufacturer's car guides to lift well walls.

- Make door opening surrounds as specified and secure one above the other.

- Build or cast in inserts to secure lift manufacturer's door sills.

- Perform all necessary cutting away and making good for landing call buttons, door and gate locks, etc.

- Provide smoke vents of at least 0.1m^2 free area per lift at the top of the shaft.

- Apply finishing coat of paintwork to all exposed steelwork.

- Provide temporary guards for openings in the well.

- Supply and install temporary scaffolding and ladders to lift manufacturer's requirements.

- Offload and store materials, accessories, tools and clothing in a secure, dry and illuminated place protected from damage and theft.

- Provide mess rooms, sanitary accommodation and other welfare facilities in accordance with the Construction (Health, Safety and Welfare) Regulations. Note: These Regulations are now integrated with the CDM Regs. (see page 6).

- Provide access, trucking and cranage for equipment deliveries.

Escalators

Escalators are moving stairs used to convey people between floor levels. They are usually arranged in pairs for opposing directional travel to transport up to 12000 persons per hour between them.

The maximum carrying capacity depends on the step width and conveyor speed. Standard steps widths are 600, 800 and 1000mm, with speeds of 0.5 and 0.65m/s. Control gear is less complex than that required for lifts as the motor runs continuously with less load variations. In high-rise buildings, space for an escalator is unjustified, since the full height and the high speed of modern lifts provides for a better service.

To prevent the exposed openings from facilitating fire spread, a water sprinkler installation (see Part 13) can be used to automatically produce a curtain of water over the well. An alternative is a fireproof shutter actuated from a smoke detector or fusible links.

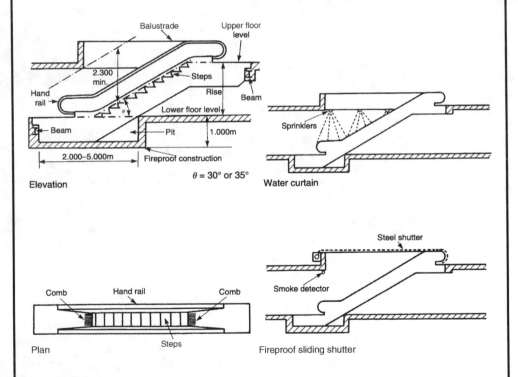

Refs. BS 5656-1 and 2: Safety rules for the construction and installation of escalators and moving walks.

BS EN 115-1 and 2: Safety of escalators and moving walks.

Escalator configurations vary depending on the required level of service. The one-directional single bank avoids interruption of traffic, but occupies more floor space than other arrangements.

A criss-cross or cross-over arrangement is used for moving traffic in both directions.

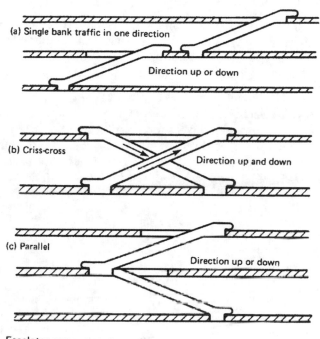

(a) Single bank traffic in one direction

Direction up or down

(b) Criss-cross

Direction up and down

(c) Parallel

Direction up or down

Escalator arrangements

Escalator capacity formula to estimate the number of persons (N) moved per hour:

$$N = \frac{3600 \times P \times V \times \text{cosine } \theta}{L}$$

where P = number of persons per step
V = speed of travel (m/s)
θ = angle of incline
L = length of each step (m).

E.g. an escalator inclined at 35°, operating with one person per 400mm step at 0·65m/s.

$$N = \frac{3600 \times 1 \times 0·65 \times 0·8192}{0·4} = 4792 \text{ persons per hour}$$

Travelators

Travelators – also known as autowalks, passenger conveyors and moving pavements. They provide horizontal conveyance for people, prams, luggage trolleys, wheelchairs and small vehicles for distances up to about 300 metres. Slight inclines of up to 12° are also possible, with some as great as 18°, but these steeper pitches are not recommended for use with wheeled transport.

Applications range from retail, commercial and store environments to exhibition centres, railway and airport terminals. Speeds range between 0·6 and 1·3m/s; any faster would prove difficult for entry and exit. When added to walking pace, the overall speed is about 2·5m/s.

There have been a number of experiments with different materials for the conveyor surface. These have ranged from elastics, rubbers, composites, interlaced steel plates and trellised steel. The latter two have been the most successful in deviating from a straight line, but research continues, particularly into possibilities for variable speed lanes of up to 5m/s. However, there could be a danger if bunching were to occur at the exit point.

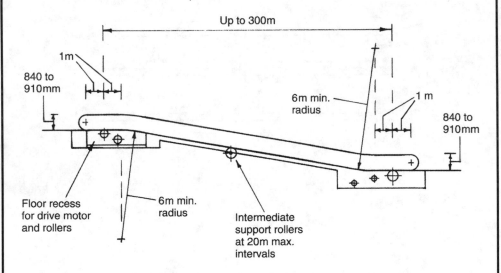

Capacity 6500 to 10800 persons per hour

Typical inclined travelator

Ref. BS 5656-2: Escalators and moving walks.

BS EN 115-1 and 2: Safety of escalators and moving walks.

Stair lifts have been used in hospitals, homes for the elderly and convalescent homes for some time. In more recent years, manufacturers have recognised the domestic need and have produced simple applications which run on a standard steel joist bracketed to the adjacent wall. Development of Part M to the Building Regulations, 'Access to and use of buildings', provides that staircases in all future dwellings are designed with the facility to accommodate and support a stair lift or a wheelchair lift. This will allow people to enjoy the home of their choice, without being forced to seek alternative accommodation.

Standard 230 volt single-phase AC domestic electrical supply is adequate to power a stair lift at a speed of about 0.15 m/s. A 24 volt DC transformed low-voltage supply is used for push button controls. Features include overspeed brake, safety belt, optional swivel seat, folding seat and arm rests, and a manual lowering device. The angle of support rail inclination is usually within the range of 22°–50° within a maximum travel distance of about 20 m.

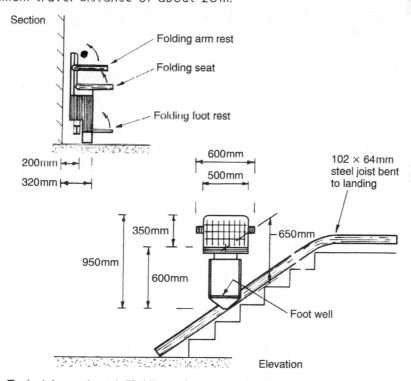

Typical domestic stair lift (dimensions approximate)

Ref. BS EN 81-40: Safety rules for the construction and installation of lifts. Special lifts for the transport of persons and goods. Stairlifts and inclined lifting platforms intended for persons with impaired mobility.

Ropeless Lifts

Standard lift roping has limitations on the height a car can travel, usually no more than 500 m (about 150 storeys). For the tallest buildings, passengers are required to transfer from one lift system to another, increasing transportation time. More than one lift shaft is cost and space inefficient as well as requiring accommodation for motorised drive equipment.

With ropeless lifts, heavy steel ropes are replaced by a system based on linear magnetic drive technology. This is capable of moving a lift car vertically within mega-high-rise buildings and along horizontal corridors. Where building shape has width and depth as well as height, a combination of shafts and floor level corridors provides for independently programmed cars to circumnavigate others. This type of drive technology suspends the car from electro-magnetic guide rails and permanent magnets located behind the lift car, known as magnetic levitation or 'maglev'. The electro-magnetic force either pulls the car up or regulates downward progress, or it can project it laterally.

Car speed is slower than with roped systems and capacity less, but compensated by multiple cars each independently powered by linear induction motors.

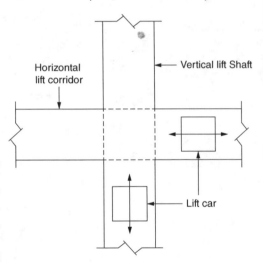

The overall building management system has a subsidiary safety control facility specifically to monitor the function of each lift car. Numerous sensors transmit data to a central control for computerised analysis during use. The computer models the systems and components, digitally replicating each car. Otherwise known as a digital twin that monitors real car movement, function and location. The control monitor can also determine any irregularity in operation and recognises if any malfunction is likely to occur.

The concept of electro-magnetic movement is an established technology, but application to transportation within buildings is relatively new and few examples exist. Case study is the OVG East Side Tower in Berlin.

13 FIRE PREVENTION AND CONTROL SERVICES

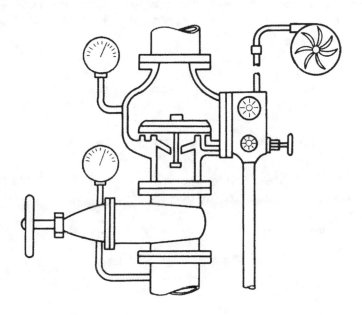

SPRINKLERS

DRENCHERS

HOSE REEL INSTALLATIONS

HYDRANTS

FOAM INSTALLATIONS

GAS EXTINGUISHERS

FIRE ALARMS

SMOKE, FIRE AND HEAT DETECTORS

FIRE DETECTION ELECTRICAL CIRCUITS

FIRE PREVENTION IN VENTILATING SYSTEMS

FIRE DAMPERS IN DUCTWORK

PRESSURISATION OF ESCAPE ROUTES

SMOKE EXTRACTION, VENTILATION AND CONTROL

PORTABLE FIRE EXTINGUISHERS

CARBON MONOXIDE DETECTORS

DOI: 10.1201/9781003434894-13

Sprinklers – The Principles

Water sprinklers provide an automatic spray dedicated to the area of fire outbreak. Sprinkler heads have temperature-sensitive elements that respond immediately to heat, discharging the contents of the water main to which they are attached. In addition to a rapid response which reduces and isolates fire damage, sprinklers use less water to control a fire than the firefighting service, therefore preventing further damage from excess water.

Sprinkler systems were initially credited to an American, Henry Parmalee, following his research during the late 1800s. The idea was developed further by another American, Frederick Grinnell, and the name 'Grinnell' is still associated with the glass-type fusible element sprinkler head.

Domestic pipework – solvent cement bonded, post-chlorinated polyvinyl chloride (CPVC).

Industrial and commercial pipework – threaded galvanised mild steel.

The simplest application is to attach and suspend sprinkler heads from a water main fixed at ceiling level. However, some means of regulation and control are needed, and this is shown in the domestic application indicated below.

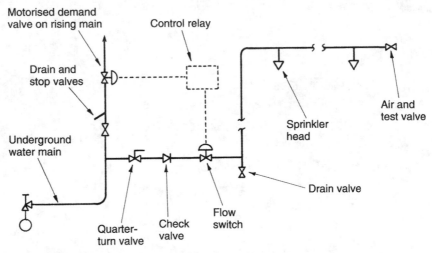

Note: When flow switch is activated, demand valve closes.

Typical domestic sprinkler installation

Refs. BS ENs 12259-1 to 5: Fixed firefighting systems. Components for sprinkler and water spray systems.

BS EN 12845: Fixed firefighting systems. Automatic sprinkler systems. Design, installation and maintenance.

Pipe materials – Copper tube – BS EN 1057
Post-chlorinated polyvinylchloride (CPVC).

System – mains supplied, wet.

Pipe sizes – 25mm min. i.d. incoming service to supply at least 60l/min. through any one sprinkler head, or 42l/min. through any two sprinkler heads operating simultaneously in the same room.

Sprinkler head spacing – area covered by one head, max. 12m^2.

Maximum distance between heads – 4m.
Maximum distance from wall to ceiling mounted head – 2m.
Minimum distance between heads in the same room – 2m (only one head per room is normal).

Operating pressure – Minimum 0.5 bar (50kPa).

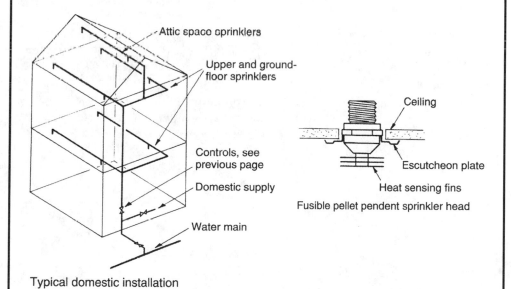

Attic space sprinklers

Upper and ground-floor sprinklers

Controls, see previous page

Domestic supply

Water main

Typical domestic installation

Ceiling

Escutcheon plate

Heat sensing fins

Fusible pellet pendent sprinkler head

Ref. BS 9251: Fire sprinkler systems for domestic and residential occupancies. Code of practice.

Types of Sprinkler Head

Quartzoid bulb – a glass tube is used to retain a water valve on its seating. The bulb or tube contains a coloured volatile fluid, which when heated to a specific temperature expands to shatter the glass and open the valve. Water flows on to a deflector, dispersing as a spray over the source of fire. Operating temperatures vary with a colour-coded liquid:

Orange – 57°C

Red – 68°C

Yellow – 79°C

Green – 93°C

Blue – 141°C

Mauve – 182°C

Black – 204 or 260°C

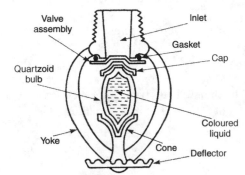

Quartzoid bulb-type head

Fusible strut – has two metal struts soldered together to retain a water valve in place. A range of solder melting temperatures are available to suit various applications. Under heat, the struts part to allow the valve to discharge water on the fire.

Duraspeed solder type – contains a heat collector which has a soldered cap attached. When heat melts the solder, the cap falls away to displace a strut, allowing the head to open. Produced in a range of operating temperatures.

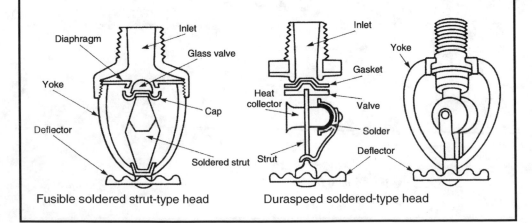

Fusible soldered strut-type head   Duraspeed soldered-type head

The specification of a sprinkler system will depend on the purpose intended for a building, its content, function, occupancy, size and disposition of rooms. Installations to commercial and industrial premises may be of the following type:

- Wet system – the simplest and most widely used application. The pipework is permanently charged with water. It is only suitable in premises where temperatures remain above zero, although small sections of exposed pipework could be protected by trace element heating. The maximum number of sprinklers on one control valve is 1000 (see page 655).
- Dry system – an air-charged system applied to unheated premises such as warehousing, where winter temperatures could drop below zero. The maximum number of sprinklers on one control valve is 250, but this may increase to 500 if the air controls include an accelerator (see page 656).
- Alternative wet and dry system – essentially a wet system, but due to the slightly slower response time as air precedes water on discharge, the pipework is charged with water for most of the year and only air charged in winter. The maximum number of sprinklers is the same as a dry system (see page 656).
- Tail-end system – used in a building with different internal functions, e.g. a mix of office accommodation with an unheated storage facility. The installation differs from an alternative wet and dry system, as most of the pipework is permanently charged with water. Only those pipes in parts of a building exposed to sub-zero temperatures are charged with air and these are designed as additions (tail ends) to a wet system. The wet and tail-end parts are separated by a compressed air control valve. As the system is essentially wet, the maximum number of sprinklers may be 1000. The maximum number after a tail-end air control valve is 100, with no more than 250 in total on tail-end air valves in one installation.
- Pre-action system – used where there is a possibility that sprinkler heads may be accidently damaged by tall equipment or plant, e.g. a fork-lift truck. To avoid unnecessary water damage, the system is dry. If a sprinkler head is damaged, compressed air discharges and an initial alarm is activated. Water will only be supplied to the damaged sprinkler if a ceiling-mounted heat detector senses a temperature rise. The sensor will open a motorised valve on the water supply and effect another alarm. Detectors have a lower temperature rating than the sprinkler; therefore, for a 68°C head, the detector will be set at about 60°C. Max. number of sprinklers is 1000.
- Recycling pre-action system – a variation of the pre-action system, designed as a damage-limiting installation. After sprinklers have subdued a fire, a heat detector responds to a lower temperature and disengages the water supply after a five-minute delay. If the fire restarts and temperature rises, the detector re-engages a motorised valve on the water supply. Maximum number of sprinklers is 1000.
- Cycling wet system — in principle similar to the recycling pre-action system except it is a normal wet system. It functions in conjunction with ceiling heat detectors which will disengage the water supply within a predetermined time of the temperature dropping. If the temperature rises, the water supply will be automatically turned on again.

Sprinkler Applications

In addition to the considerations for system selection given on the preceding page, the building insurer will probably have greatest influence in determining the final specification. Insurer's requirements are likely to be formulated from guidance in:

- The Building Regulations, Approved Document B: Fire safety.
- BS EN 12845: Fixed firefighting systems. Automatic sprinkler systems.
- The Loss Prevention Certification Board's Loss Prevention Standards.

The Loss Prevention Certification Board was formed in 1985 as an incorporation of the Fire Offices' Committee (FOC), the Fire Insurers' Research and Testing Organisation and the Fire Protection Association. Rules for design and installation originally produced by the FOC are now included in the British Standard.

Buildings are assessed by fire risk and categorised by fire load* as a hazard according to their purpose and content:

- Light hazard (LH) – low fire load and containing no single compartment exceeding 126m² floor area with fire resistance of at least 30min. Examples include educational premises, prisons and offices. Maximum protected area is 10000m² per control valve.
- Ordinary hazard (OH 1 to OH 4) – medium fire load category such as process or manufacturing premises.

  OH 1 – cement works, sheet metal processors, dairies, abattoirs, hospitals, hotels, offices, schools and restaurants.
  OH 2 – garages (car workshops), laboratories, bakeries, food processors, breweries, car parks and museums.
  OH 3 and 4 – industrial processors and warehouses with combustible stored products.
- High hazard – high fire load categories typical of warehouses containing combustible products in high racking systems. Fireworks factories and some chemical processes will also be included.

Note: Where specified, sprinklers should be installed in all parts of a building. Some exception may be permitted for toilets, washrooms and enclosed stairways.

*Fire load – an assessment of the combustible potential of materials contained within a building. Fire load is expressed as the heat potential per unit area, as a calorific value in joules/m².

Grade 1 (low) – 1150MJ/m², e.g. hotels, hospitals, schools, public libraries, offices, flats, restaurants, museums, sports centres and institutions.
Grade 2 (moderate) – 1150–2300MJ/m², e.g. retail premises, factories and workshops.
Grade 3 (high) – 2300–4600MJ/m², workshops, manufacturing processes and warehousing where combustible materials are deployed, e.g. timber and paper fabrication.

The wet system is used in heated buildings where there is no risk of the water in the pipework freezing. All pipework is permanently pressure charged with water and the sprinkler heads usually attach to the underside of the range pipes. Where water is mains supplied, it should be fed from both ends. If the main is under repair on one side, the stop valve and branch pipe can be closed and the sprinkler system supplied from the other branch pipe.

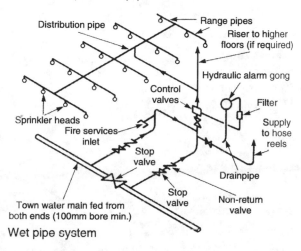

Wet pipe system

When a sprinkler head is fractured, water is immediately dispersed. Water will also flow through an annular groove in the alarm valve seating to a pipe connected to an alarm gong and turbine. A jet of water propels the turbine blades, causing the alarm gong to operate. Pipeline flow switches will alert the local fire service in addition to operating an internal alarm system. Except under supervised maintenance, the main stop valve is padlocked in the open position.

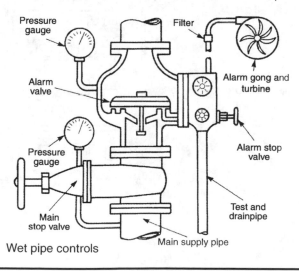

Wet pipe controls

Dry and Alternate Wet-and-Dry Sprinkler Installations

Dry or an alternate wet-and-dry sprinkler system may be used in buildings that are unheated.

Dry system – installation pipework above the differential valve is permanently charged with compressed air. When a fire fractures a sprinkler head, the compressed air escapes to allow the retained water to displace the differential valve and flow to the broken sprinkler.

Alternate wet-and-dry system – a wet system for most of the year, but during the winter months, it functions as a dry system.

The dry part of the system above the diaphragm or differential valve is charged with compressed air at about 200kPa. Any loss of pressure is automatically replenished by a small compressor, but this will not interfere with water flow if the system is activated. When a sprinkler is fractured, an automatic booster pump can be used to rapidly exhaust the air and improve water flow. Sprinkler heads are fitted above the range pipes which are slightly inclined to allow the system to be fully drained.

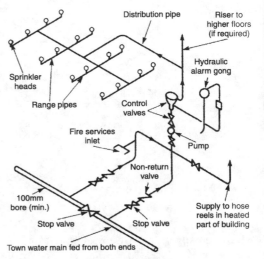

Dry pipe or alternate wet-and-dry pipe system

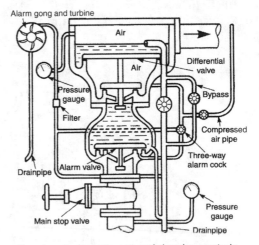

Dry pipe or alternate wet-and-dry pipe controls

Deluge system – used for specifically high fire hazards such as plastic foam manufacture, fireworks factories, aircraft hangars, etc., where there is a risk of intensive fire with a very fast rate of propagation. The pipework is in two parts: compressed air with quartzoid bulbs attached and a dry pipe with open-ended spray projectors. When a fire occurs, the quartzoid bulbs shatter and compressed air in the pipeline is released, allowing a diaphragm inside the deluge control valve to open and discharge water through the open pipe to the projectors.

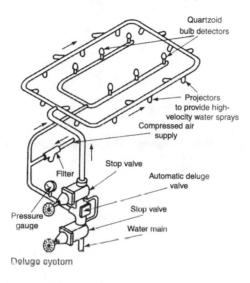

Deluge system

Multiple control system – a heat-sensitive sealed valve controls the flow of water to a small group of open sprayers attached to a dry pipe. When a fire occurs, the valve quartzoid bulb shatters, allowing the previously retained water to displace the valve stem and flow to the sprayers. An alternative to a heat-sensitive valve is a motorised valve activated by a smoke or fire detector.

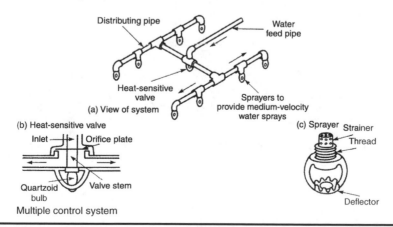

Multiple control system

Water Supplies for Sprinkler Systems (1)

Possible water sources –

- Local water authority's main:
  * Direct supply.
  * Indirect to a large break/suction tank with pumped delivery/outflow.
  * Indirect to a small break/suction tank with sufficient inflow to compensate for a pumped delivery/outflow.
  * Indirect to an elevated storage tank at sufficient height to provide a gravity delivery/outflow.
  * Indirect to a break/suction tank with delivery from an air pressurised vessel and pump.
- Elevated private reservoir.
- Recycled rainwater recovery system.
- Water course, e.g. river or canal.

The minimum design criteria for all applications are a sufficient source of water for the associated sprinkler system to operate at full demand for at least one hour.

Other design considerations –

Specific situations vary between building purpose, size and construction, and also, the potential that the content and processes therein have as a fire hazard. As previously indicated (page 654), exact requirements will be determined by insurers in liaison with the LPCB. The following table is provided to show typical data as guidance only:

Hazard group	Min. pressure at head (bar)	Design flow (l/m)	Design max. heads	Max. design flow (l/m)
Domestic	0.5	60	2	84
Residential	0.5	60	4	168
Light hazard	0.7	94	4	190
OH 1	1.0	375	6	540
OH 2	1.4	725	12	1000
OH 3	1.7	1100	18	1350
OH 3S	2.0	1800	30	2100

High hazard   See comment on page 654. This category has several sub-divisions to which the LPCB Standards should be applied.

Note: OH refers to Ordinary Hazard. With OH 3S, the S indicates Special. May also be classed as OH 4.

Direct supply – limited to domestic and other relatively small-scale installations. For factories, commercial premises and office buildings, the delivery requirements are likely to exceed that acceptable to the water supply authorities.

Elevated private reservoir – a reservoir of water on ground raised above the building height. Capacity requirements vary depending on the hazard category. This will range between 500 m^3 (500000 litres) and 1875 m^3 (1875000 l).

Gravity tank – usually located on a tower at a level raised above the building served. Design capacities range from 55 m^3 (55000 l) for light hazard applications up to 875 m^3 (875000 l) for high hazard. Capacities as little as 9 m^3 (9000 l) are shown in some design guides for extra light hazard situations. If a gravity tank is the only system water source, its supply from the main must be capable of refilling it in six hours.

Air pressurised vessel – can function with an indirect mains water supply from a storage break/suction tank, or from an elevated gravity tank. The capacity of a break/suction tank should be the same as for a gravity tank, but may be reduced where inflow from the mains is sufficient to ensure that pumped delivery from the tank is unimpeded. Pressure vessel water capacity is between 7 m^3 (7000 l) and 23 m^3 (23000 l) where it is the only supply source as applied to hazard classes up to and including OH 1. The remaining OH categories may be serviced from a pressure vessel, where that vessel is a supplement to other sources, i.e. a duplicate supply source is required and pressure vessel capacity can be reduced accordingly, up to 15 m^3 (15000 l).

Recycled rainwater reservoir or available water course, e.g. river or canal – an alternative that may be considered if availability of water is consistent and constant throughout the year. These sources may be subject to natural water levels, possibly the effects of drought and not necessarily reliable. Therefore, unlikely as a sole source but could be a useful supplement to others.

Refs: The LPCB have various related Standards. These should be consulted for information on design and installation procedures.

CIBSE Guide E: Fire safety engineering.

See also page 650.

Water Supplies for Sprinkler Systems (3)

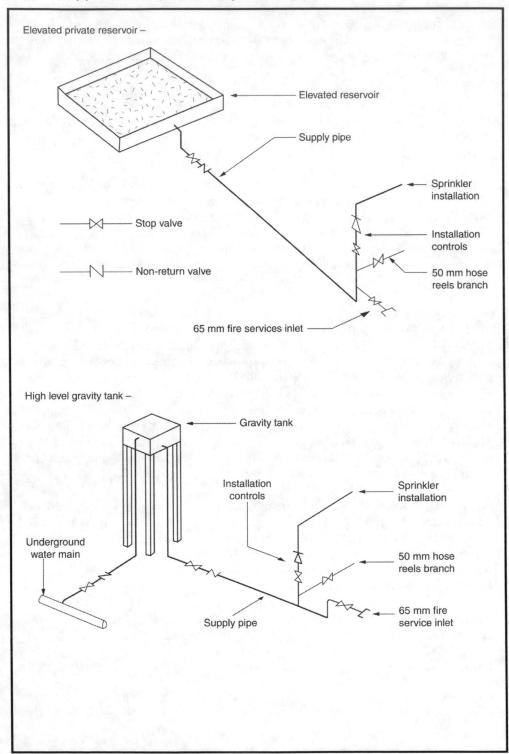

Elevated private reservoir –

Elevated reservoir

Supply pipe

Sprinkler installation

Stop valve

Installation controls

Non-return valve

50 mm hose reels branch

65 mm fire services inlet

High level gravity tank –

Gravity tank

Installation controls

Sprinkler installation

Underground water main

50 mm hose reels branch

Supply pipe

65 mm fire service inlet

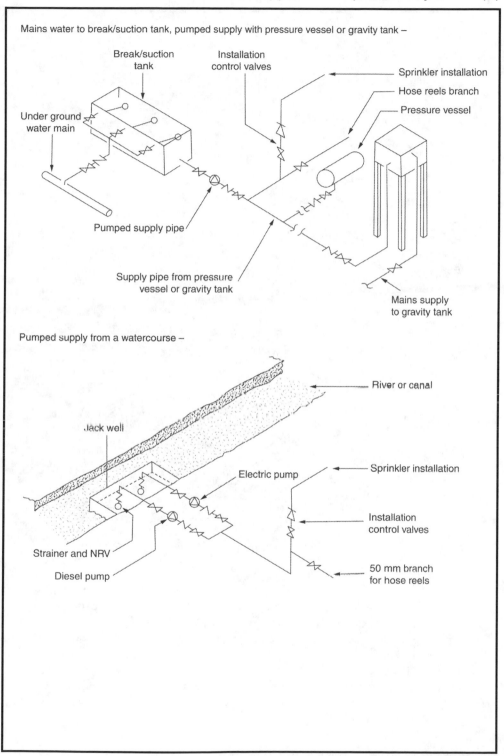

Mains water to break/suction tank, pumped supply with pressure vessel or gravity tank –

Break/suction tank

Installation control valves

Sprinkler installation

Hose reels branch

Pressure vessel

Under ground water main

Pumped supply pipe

Supply pipe from pressure vessel or gravity tank

Mains supply to gravity tank

Pumped supply from a watercourse –

River or canal

Jack well

Electric pump

Sprinkler installation

Installation control valves

Strainer and NRV

Diesel pump

50 mm branch for hose reels

Sprinkler and Hose Reel Installation

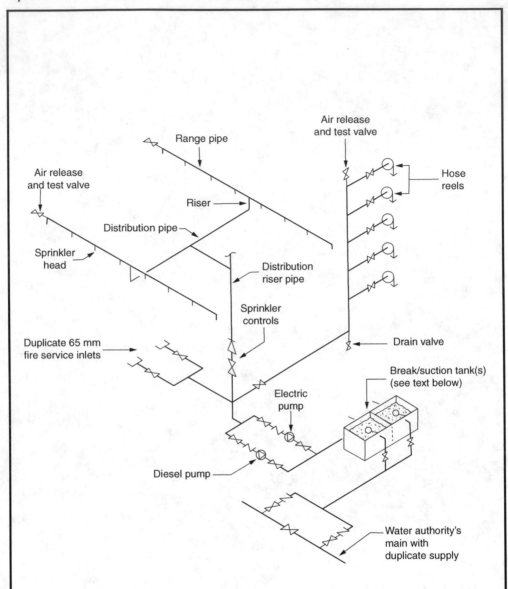

Break/suction tank – as a single unit, these should not exceed 1300 m³. Most individual systems are unlikely to require this amount of water, but if necessary tanks can be duplicated or sub-divided to a combined maximum of 2600 m³. Tanks are usually galvanised steel and may be EPDM or vitreous enamel lined for higher specification. GRP tanks are also acceptable. A tight fitting access cover is required for maintenance and inspection. Location may justify a frost thermostat activated immersion heater.

The arrangement of pipework will depend on the building shape and layout, the position of the riser pipe and the number of sprinkler heads required. To provide a reasonably balanced distribution, it is preferable to have a centre feed pipe. In practice, this is not always possible and end feed arrangements are used. The maximum spacing of sprinkler heads (s) on range pipes depends on the fire hazard classification of the building.

Hazard category	Max. spacing (s) of sprinkler heads (m)	Max. floor area covered by one sprinkler head (m²)
Light	4·6	21
Ordinary	4·0 (standard)	12
	4·6 (staggered)*	12
High	3·7	9

*See next page

For side wall-mounted sprinklers, the maximum floor area coverage by one sprinkler head is $17m^2$ for light hazard and $9m^2$ for ordinary hazard.

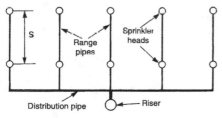

Two-end side with centre feed pipe

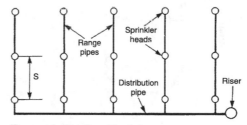

Three-end side with end feed pipe

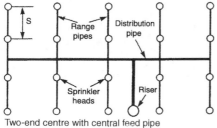

Two-end centre with central feed pipe

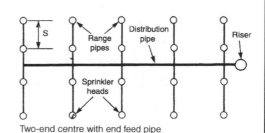

Two-end centre with end feed pipe

Further Pipework Distribution and Spacing Calculations

Staggered arrangement of sprinkler heads on an ordinary hazard installation:

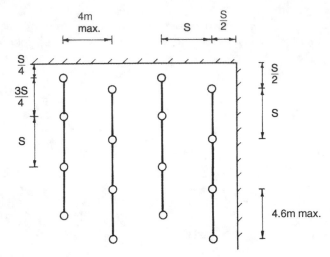

Calculating the number of sprinkler heads: e.g. an ordinary fire hazard category for a factory having a floor area 20m × 10m.

$$20 \times 10 = 200m^2$$

Ordinary hazard requires a maximum served floor area of 12m² per sprinkler head.

Therefore, 200 ÷ 12 = 16·67, i.e. at least 17 sprinkler heads.

For practical purposes, 18 could be installed as shown:

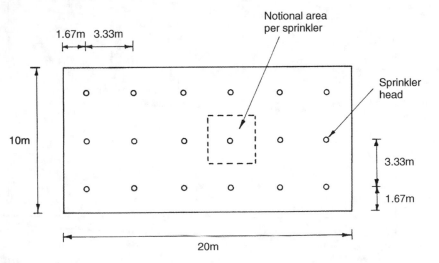

The maximum area served by each sprinkler head = 3·33m × 3·33m = 11·1m².

This is satisfactory, being less than 12m².

664

Sprinkler pipe installations downstream of the alarm and control valves should be sized by hydraulic calculation, with regard to system pressure and friction losses (see Part 2).

Tabulated data for pipe sizing are available in BS EN 12845 and CIBSE Guide E: Fire engineering. It is also possible to determine pipe diameters from the Hazen–Williams friction loss formula:

$$p = \frac{6 \cdot 05 \times 10^5 \times L \times Q^{1 \cdot 85}}{C^{1 \cdot 85} \times d^{4 \cdot 87}}$$

where p = pressure loss in pipe (bar)

$\quad$ L = equivalent length of pipework plus bends and fittings, i.e. effective pipe length (m)

$\quad$ Q = flow rate through the pipe (minimum 60 litres/minute)

$\quad$ C = constant for pipe material (see table)

$\quad$ d = pipe internal diameter (mm)

Pipe material	Constant (C)
Cast iron	100
Steel	120
Stainless steel	140
Copper	140
CPVC	150

Maximum water velocity through valves is 6 m/s. Through any other part of the system, 10 m/s.

By determining an acceptable pressure loss as a design prerequisite, the Hazen–Williams formula can be rearranged with the pipe diameter as the subject:

$$d = \sqrt[4 \cdot 87]{\frac{6 \cdot 05 \times 10^5 \times L \times Q^{1 \cdot 85}}{C^{1 \cdot 85} \times p}}$$

E.g. Calculate the diameter of 30 m effective length steel pipe, where the acceptable pressure loss is 0·02 bar with a water flow rate of 60 litres/minute.

$$d = \sqrt[4 \cdot 87]{\frac{6 \cdot 05 \times 10^5 \times 30 \times 60^{1 \cdot 85}}{120^{1 \cdot 85} \times 0 \cdot 02}}$$

$$d = \sqrt[4 \cdot 87]{\frac{353554 \cdot 56 \times 10^5}{140 \cdot 45}} = 53 \cdot 09 \text{ mm (i.d.)}$$

50 mm nominal inside diameter is just too small; therefore, a 65 mm nominal inside diameter steel pipe would be selected.

Drenchers

A drencher fire control system provides a discharge of water over roofs, walls and windows to prevent fire from spreading either from or to adjacent buildings. Automatic drenchers are similar in operating principle to individual quartzoid bulb sprinkler heads. A manually operated stop valve can also be used with dry pipes and open spray nozzles. This stop valve must be located in a prominent position with unimpeded access. Installation pipework should fall to a drain valve positioned at the lowest point above the stop valve. The number of drencher nozzles per pipe is similar to the arrangements for conventional sprinkler installations as indicated in BS EN 12845. For guidance, two drenchers can normally be supplied by a 25mm i.d. pipe. A 50mm i.d. pipe can supply ten drenchers, a 75mm i.d. pipe 36 drenchers and a 150mm i.d. pipe over 100 drenchers. An example of application is in theatres, where the drenchers may be fitted above the proscenium arch at the stage side to protect the safety curtain.

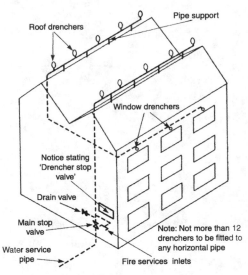

Typical drencher installation

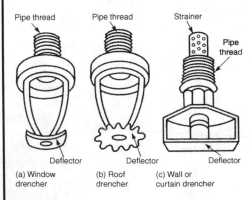

(a) Window drencher

(b) Roof drencher

(c) Wall or curtain drencher

Types of drencher

A hose reel is capable of delivering considerably more water than several portable extinguishers, especially with continuous replenishment from a mains or pumped water supply. Hose reels are a first-aid provision for building occupants to tackle an outbreak of fire. Fixed or swinging hose reels are located in wall recesses where occupants are least likely to be endangered by fire. In most instances, this will be a staircase landing. One hosepipe installation is normally required to service up to 800 m² of floor area.

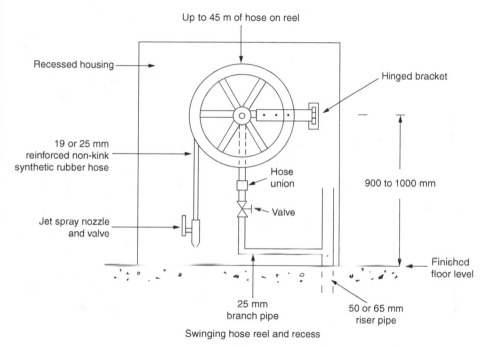

Swinging hose reel and recess

A minimum specification for the hose most distant from the source of water is to deliver at least 24 litres/minute (0.4 l/s) measured 6 metres from the hose nozzle, when the two most remote reels are operating simultaneously. A minimum pressure of 200 kPa is required at the highest reel. Where this pressure is unavailable from a direct mains water supply, an intermediate break or suction tank may be used as a reservoir from which booster pumps can supply the upper levels of a building. See illustrations on the next page.

Ref. BS EN 671: Fixed firefighting systems. Hose systems. Hose reels with semi-rigid hose.

Hose Reel Installations

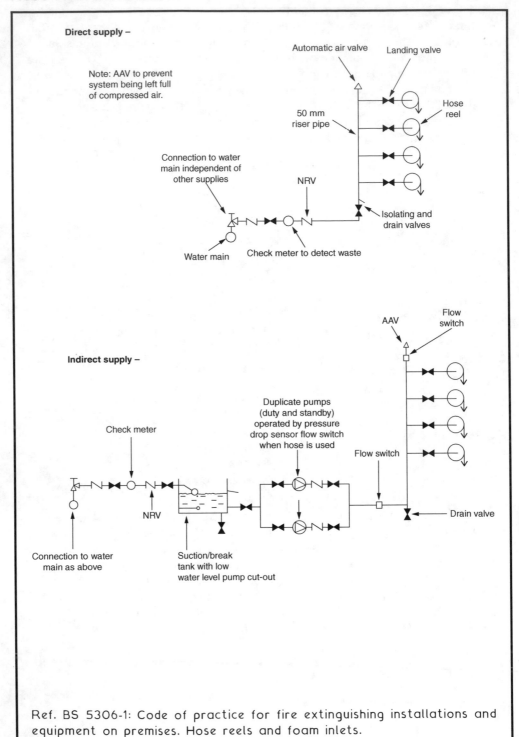

Direct supply –

Note: AAV to prevent system being left full of compressed air.

Automatic air valve

Landing valve

Hose reel

50 mm riser pipe

Connection to water main independent of other supplies

NRV

Isolating and drain valves

Water main

Check meter to detect waste

Indirect supply –

AAV

Flow switch

Check meter

Duplicate pumps (duty and standby) operated by pressure drop sensor flow switch when hose is used

Flow switch

NRV

Drain valve

Connection to water main as above

Suction/break tank with low water level pump cut-out

Ref. BS 5306-1: Code of practice for fire extinguishing installations and equipment on premises. Hose reels and foam inlets.

A dry riser is in effect an empty vertical pipe which becomes a fire-fighter's hose extension to supply hydrants at each floor level. Risers should be disposed so that no part of the floor is more than 60m from a landing valve. This distance is measured along a route suitable for a firefighting hose line, to include any dimension up or down a stairway. Buildings with floors up to 45m above fire service vehicle access level require one 65mm landing valve on each floor from a 100mm i.d. riser. Buildings between 45m and 60m with one or two landing valves per floor require a 150mm i.d. riser. For buildings above 60m, a wet riser must be installed. Two 65mm i.d. inlet hose couplings are required for a 100mm riser and four 65mm i.d. inlets are required for a 150mm riser. The riser must be electrically bonded to earth.

Note: A dry riser is installed either in unheated buildings or where the water main will not provide sufficient pressure at the highest landing valve. A hard standing for the fire service vehicle is required at the base of the riser. One landing valve is required for every 900m^2 of floor area.

Typical arrangement of a dry riser

(b) Front view of
Fire service inlets

(b) Front view of
Fire service inlet box

Note: Door fitted with
spring lock which opens
when the glass is broken

Details of dry riser inlet

Wet Riser

A wet riser is suitable in any building where hydrant installations are specified. It is essential in buildings where floor levels are higher than that served by a dry riser, i.e. greater than 60m above fire service vehicle access level. A wet riser is constantly charged with water at a minimum running pressure of 400kPa with up to three most remote landing valves operating simultaneously. A flow rate of 25l/s is also required. The maximum pressure with one outlet open is 500kPa to protect firefighting hoses from rupturing. Orifice plates may be fitted to the lower landing valves to restrict pressure. Alternatively, a pressure-relief valve may be incorporated in the outlet of the landing valve. The discharge from this is conveyed in a 100mm i.d. drainpipe.

To maintain water at the required pressure and delivery rate, it is usually necessary to install pumping equipment. Direct pumping from the main is unacceptable. A suction or break tank with a minimum water volume of 45m^3 is used with duplicate power source service pumps. One 65mm landing valve should be provided for every 900m^2 floor area.

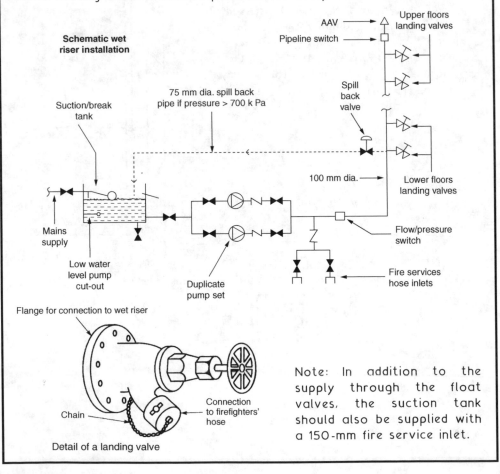

Schematic wet riser installation

AAV — Upper floors landing valves

Pipeline switch

75 mm dia. spill back pipe if pressure > 700 k Pa

Spill back valve

Suction/break tank

100 mm dia. — Lower floors landing valves

Mains supply

Flow/pressure switch

Low water level pump cut-out

Duplicate pump set

Fire services hose inlets

Flange for connection to wet riser

Connection to firefighters' hose

Chain

Detail of a landing valve

Note: In addition to the supply through the float valves, the suction tank should also be supplied with a 150-mm fire service inlet.

Combined Hose Reel and Hydrant System

Dual system supplied from a break/suction tank replenished from the water main. An elevated (roof top plant room) service tank is pumped supplied from the break tank. This provides a combined gravity/pumped delivery of water to hose reels and hydrants. A back-up dry riser inlet is also available for the fire services use.

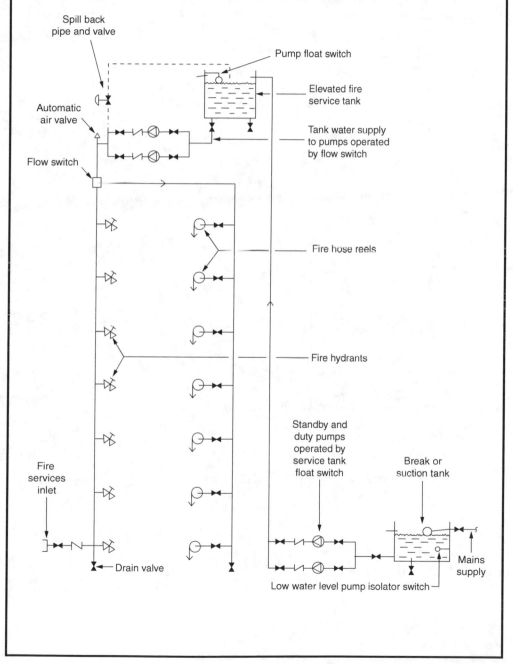

Fixed Foam Installations

A pump-operated mechanical foam installation consists of a foam concentrate tank located outside the area to be protected. The tank has a water supply pipe inlet and foam pipe outlet. A venturi is fitted in the pipeline to draw the foam out of the tank. When the water pump is switched on, the venturi effect causes a reduction in pressure at the foam pipe connection, resulting in a mixture of foam concentrate and water discharging through the outlet pipe.

A pre-mixed foam installation consists of a storage tank containing foam solution. When a fire occurs in the protected area, a fusible link is broken to release a weight which falls to open a valve on the carbon dioxide cylinder. Foam solution is forced out of the tank at a pressure of about 1000kPa to discharge over the protected equipment, e.g. an oil tank.

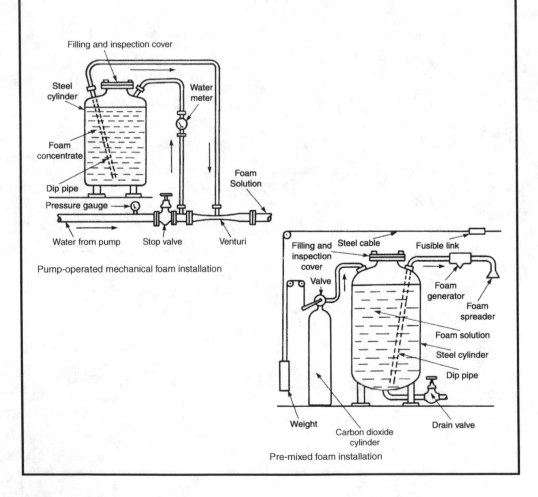

Pump-operated mechanical foam installation

Pre-mixed foam installation

A foam installation is used for application from remote points on to flammable liquid fire risks. This type of installation is often used with oil-fired boilers and oil storage tanks. A foam access box is built into the wall at an easily accessible place for firefighters to attach hoses from their foam-generating and mixing equipment. The box is usually located about 600 mm above adjacent ground and should be clear of any openings through which heat, smoke or flames can pass. The glass-fronted box can be broken and the lock released from inside. Two 65 mm diameter inlets may be used. A 65 or 75 mm i.d. galvanised steel pipe is normally used for the distribution. A maximum pipework length of 18 m is recommended, and this must slope slightly towards the spreaders. Vertical drop pipes are acceptable but vertically inclined pipes must not be used. Spreader terminals are positioned about 1 m above oil burners and about 150 mm above oil-spill level of stored fuel.

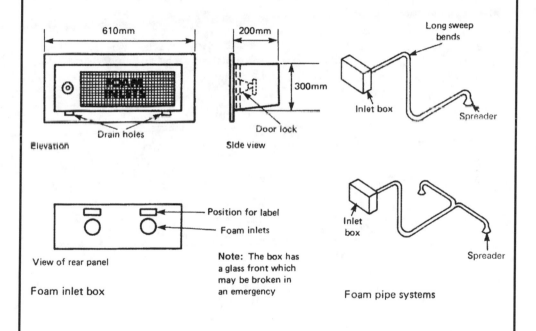

610mm

Drain holes

Elevation

200mm

300mm

Door lock

Side view

Long sweep bends

Inlet box

Spreader

Position for label

Foam inlets

View of rear panel

Foam inlet box

Note: The box has a glass front which may be broken in an emergency

Inlet box

Spreader

Foam pipe systems

Ref. BS EN 13565-2: Fixed firefighting systems. Foam systems. Design, construction and maintenance.

Gas Extinguishing Systems – Halon and Halon Substitutes

The majority of gas extinguishing systems have been either halon 1301 or carbon dioxide (see next page). Halons are electrically non-conductive and in this respect safe to use where personnel remain in an area of gas discharge. They are also more effective than carbon dioxide, being five times the density of air, while carbon dioxide is only one-and-a-half times. Unfortunately halon or bromochlorodifluoromethane (BCF) gases are a hazard to the environment, by contributing significantly to the depleting effect of the ozone layer. In 1987, a meeting of major countries at a Montreal convention agreed to phase out the use of these gases by 2002. Therefore, except for systems installed in less cooperative countries, new installations will contain halon substitutes. These include inergen and argonite, both mixtures of nitrogen and argon, the former containing a small amount of carbon dioxide.

In principle, the systems are suitable where there is a high density of equipment, e.g. tape libraries and computer suites where an alternative wet system would be considered too damaging. Gas is stored in spherical steel containers which can be secured in a ceiling or floor void or against a wall. When activated by smoke or heat, detectors immediately open valves on the extinguishers to totally flood the protected area with a colourless and odourless gas.

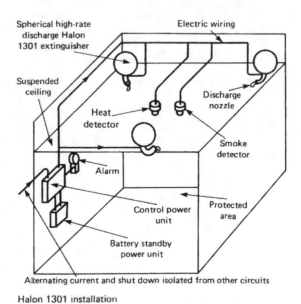

Halon 1301 installation

Ref. BS 5306-5.1: Code of practice for fire extinguishing installations and equipment on premises. Halon systems. Specification for halon 1301 total flooding systems.

Carbon dioxide is an alternative to halon as a dry gas extinguisher. It has been used as an extinguishing agent for a considerable time, particularly in portable extinguishers. As the gas is dry and non-conductive, it is ideal for containing fires from electrical equipment, in addition to textiles, machinery, petroleum and oil fires. Carbon dioxide is heavier than air and can flow around obstacles to effectively reduce the oxygen content of air from its normal 21% to about 15%. This considerably reduces an important component of the combustion process (see page 691). Integrated high- and low-pressure gas systems may be used, with the former operating at up to 5800kPa. Systems can be either electrical, pneumatic or mechanical with a manual override facility. Carbon dioxide is potentially hazardous to personnel; therefore, it is essential that the system is automatically locked off when the protected area is occupied. In these circumstances, it can be switched to manual control. Airtightness of a protected room is essential for the success of this system as total flooding relies on gas containment by peripheral means.

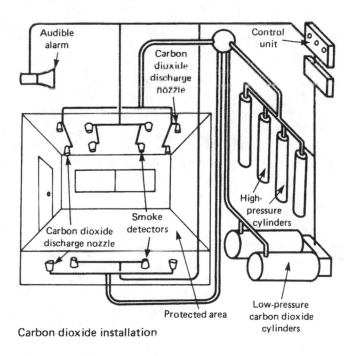

Carbon dioxide installation

Ref. BS 5306-4: Fire extinguishing installations and equipment on premises. Specification for carbon dioxide systems.

Fire Detection

In the UK, the Fire Service attend over half a million fires per year. These fires result in over 800 deaths and many more injuries. About one-tenth of all fires occur in homes and account for some 500 deaths and thousands of injuries. An early warning device to detect smoke and fire could significantly reduce the number of human casualties.

Since 1992 the Smoke Detectors Act requires all new homes to have a smoke detection facility. Detectors are available in two basic types. Each can be powered by a simple battery cell or by mains electricity. The latter will normally have battery back-up if the mains supply fails.

• Ionisation – an inexpensive device, sensitive to tiny smoke particles and fast-burning fires such as a flaming chip pan (page 679).

• Light scattering or optical – more expensive but more sensitive in slow-burning and smouldering fire produced by burning fabrics or upholstery and overheating PVC wiring (page 679).

• Combined – a unit containing both ionisation and optical detection.

Number and location – the more the better, as fires can start anywhere. Ideally detectors should be provided in every room except a bathroom, as dampness and steam can create a false affect. Likewise, for a kitchen, unless of sufficient volume to be unaffected by cooking appliances and washing-up facilities. Use in a garage can also be deceptive as exhaust fumes are likely to trigger the detector.

Minimum protection – one detector for every floor level positioned in a central hallway and/or landing. Building Regulation requirements for dwellings are summarised on pages 677–678. For other building purposes, brief mention only is given on page 677, as different situations have varying requirements. Therefore, the Approved Document should be consulted for specific applications.

Refs. Building Regulations, Approved Document B, Fire safety, Volume 1: Dwellings, and Volume 2: Buildings other than dwellings.
BS EN 54: Fire detection and alarm systems.

Fire detection and alarm systems may contain

- system control unit
- primary (mains) electrical supply
- secondary (battery or capacitor standby) power supply; an emergency generator could also be used
- alarm activation devices – manual or automatic
- alarm indication devices – audible and/or visual
- remote indication on a building monitoring system
- control relay via a building management system to effect fire extinguishers and ventilation smoke control actuators.

System control unit – an alarm panel which monitors the state of all parts (zones) of the installation. It identifies the point of origin of an alarm, displays this on the panel and communicates this to remote control locations.

Zones:

- Max. 2000m^2 floor area in one storey.
- No detachment of compartment areas within one floor area zone.
- Max. 30m search distance into a zone.
- Single occupancy of a zone where several separate business functions occur in one building.

Requirements for dwellings

Automatic fire detection and alarm systems min. Grade D2, Category LD3 are to be provided to the recommendations of BS 5839-6: Fire detection and alarm systems for buildings. Code of practice for the design, installation, commissioning and maintenance of fire detection and fire alarm systems in domestic premises. A smoke alarm system should comply with BS EN 14604: Smoke alarm devices. These should have primary mains and secondary power supplies.

Point detectors – individual heat or smoke detection units which respond to an irregular situation in the immediate vicinity.

Line detectors – a continuous type of detection comprising a pair of conducting cables separated by low-temperature melting insulation to permit a short-circuit alarm when the cables contact. Suitable in tunnels and service shafts.

Provision in large houses (> 1 storey):

Floor area	Storeys (exc. basement)	System
>200m²/storey	≥ 3	BS 5839-6, Grade A category LD2
>200m²/storey	≤ 2	BS 5839-6, Grade A category LD3

Note: prefixes used in the BS categories indicates that L is a specific application to protection of life, whereas P indicates that for property.

Application:
- Optical-type (photo-electric) detectors in circulation spaces, i.e. hallways, corridors and landings.
- Ionisation-type detectors in living and dining areas.

Preferred location of detectors:
- Over 300mm from light fittings.
- Min. one per storey.
- Loft conversions, with alarm linked to operate others and be operated by others in the dwelling.
- Circulation spaces between bedrooms.
- Circulation spaces <7.5m from doors to habitable rooms.
- Kitchens (with regard to heat/smoke-producing appliances).
- Living rooms.

Requirements for buildings other than dwellings:

This is less easy to define due to the variation in building types and patterns of occupancy. BS 5839-1 requirements may suit some buildings, but could cause panic in others, e.g. shopping centres, where people may be unfamiliar with the layout. In these situations, trained staff may be the preferred system of building evacuation. At building design stage, consultation between the local building control authority, the fire authority and the building's insurer is paramount, as alterations post-construction are always extremely expensive.

Refs. Building Regulations, Approved Document B: Fire safety. Section B1: Fire detection and alarm systems.

BS 5839-1: Fire detection and alarm systems for buildings. Code of practice for design, installation, commissioning and maintenance of systems in non-domestic premises.

Ionisation smoke detector – positive and negative charged plate electrodes attract opposingly charged ions. An ion is an atom or a group of atoms which have lost or gained one or more electrons, to carry a predominantly positive or negative charge. The movement of ions between the plates reduces the resistance of air, such that a small electric current is produced. If smoke enters the unit, particles attach to the ions slowing their movement. This reduction in current flow actuates an electronic relay circuit to operate an alarm.

Light scattering or optical smoke detector – a light beam projects onto a light trap into which it is absorbed. When smoke enters the detector, some of the light beam is deflected upward onto a photo-electric cell. This light energises the cell to produce an electric current which activates the alarm relay.

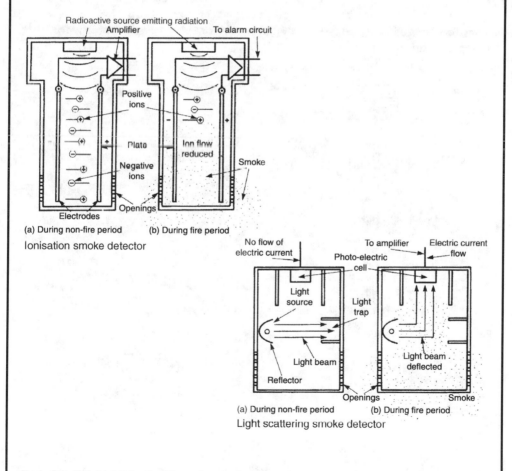

Ref. BS EN 14604: Smoke alarm devices.

Heat Detectors

Heat detectors are used where smoking is permitted and in other situations where a smoke detector could be inadvertently actuated by process work in the building, e.g. a factory. Detectors are designed to identify a fire in its more advanced stage, so their response time is longer than smoke detectors.

Fusible type – has an alloy sensor with a thin walled casing fitted with heat-collecting fins at its lower end. An electrical conductor passes through the centre. The casing has a fusible alloy lining and this functions as a second conductor. Heat melts the lining at a predetermined temperature, causing it to contact the central conductor and complete an alarm relay electrical circuit.

Bimetallic coil type – heat passes through the cover to the bimetal coils. Initially the lower coil receives greater heat than the upper coil. The lower coil responds by making contact with the upper coil to complete an electrical alarm circuit.

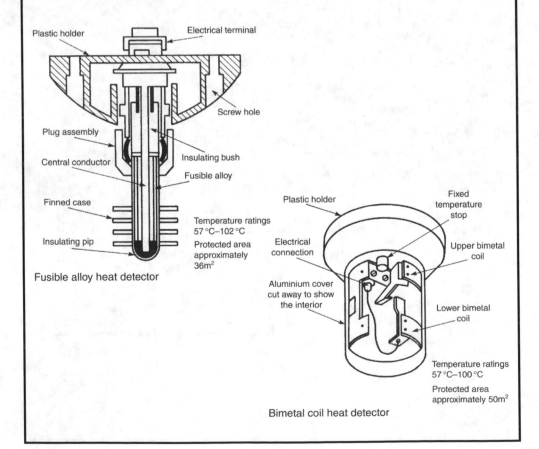

Fusible alloy heat detector

Bimetal coil heat detector

Light Obscuring and Laser Beam Detectors

Light obscuring – a beam of light is projected across the protected area close to the ceiling. The light falls onto a photo-electric cell which produces a small electrical current for amplification and application to an alarm circuit. Smoke rising from a fire passes through the light beam to obscure and interrupt the amount of light falling on the photo-electric cell. The flow of electric current from the cell reduces sufficiently to activate an alarm relay.

A variation is the light-scatter type. In normal use, the light is widely dispersed and no light reaches the photo-electric cell receptor. In the presence of smoke, particulates deflect light onto the receptor to energise the cell.

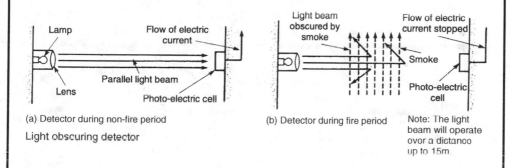

(a) Detector during non-fire period

Light obscuring detector

(b) Detector during fire period

Note: The light beam will operate over a distance up to 15m

Laser beam – a band of light which can be visible or infra-red projected onto a photo-electric cell. It does not fan out or diffuse as it travels through an uninterrupted atmosphere. The beam can operate effectively at distances up to 100m. If a fire occurs, smoke and heat rises and the pulsating beam is deflected away from the cell or reduced in intensity. As the cell is de-energised, this effects on alarm relay.

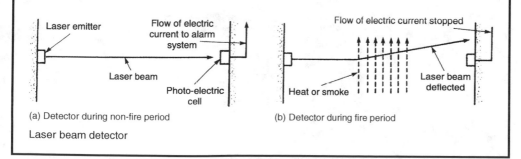

(a) Detector during non-fire period

Laser beam detector

(b) Detector during fire period

Radiation Fire Detectors

In addition to producing hot gases, fire also releases radiant energy in the form of visible light, infra-red and ultraviolet radiation. Radiant energy travels in waves from the fire.

Infra-red detector – detectors have a selective filter and lens to allow only infra-red radiation to fall on a photo-electric cell. Flames have a distinctive flicker, normally in the range of 4 to 15Hz. The filter is used to exclude signals outside this range. The amplifier is used to increase the current from the photo-electric cell. To reduce false alarms, a timing device operates the alarm a few seconds after the outbreak of fire.

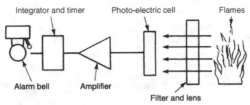

Components of an infra-red detector

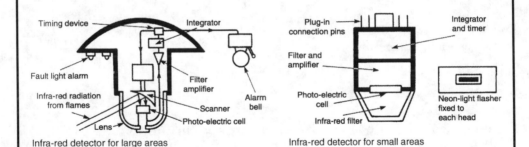

Infra-red detector for large areas

Infra-red detector for small areas

Ultraviolet detector – these detectors have a gas-filled bulb which reacts with ultraviolet radiation. When the bulb receives radiant energy, the gas is ionised to produce an electric current. When this current exceeds the set point of the amplifier, the alarm circuit closes to operate the alarm system.

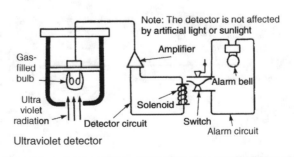

Ultraviolet detector

Fire alarm electrical circuits may be of the 'open' or 'closed' types. In addition, or as an alternative to automatic smoke- or fire-sensing switches, manual break-glass alarm switches can be wall mounted at about 1.5m above floor level in lobbies, corridors and other common access locations. No person should have to travel more than 30m to use an alarm. In large managed buildings, a sub-circuit will connect to the facilities manager's office or in more sophisticated situations the alarm can relay through telecommunications cables to a central controller and the fire service.

Open circuit – call points or detectors are connected to open switches, which prevent current from flowing through the circuit when it is on standby. Closing a switch on the detector circuit actuates a solenoid (electromagnet) to complete the alarm circuit. As there is no current flow while on standby, there is no electrical power consumption. The disadvantage of this system is that if part of the detector circuit is inadvertently damaged, some of the switches will not operate.

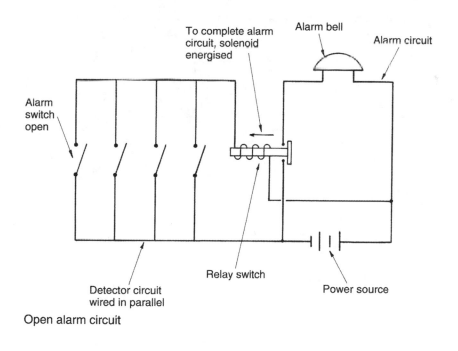

Open alarm circuit

Fire Detection Electrical Circuits – 2

Electrical power to 'open' or 'closed' fire alarm circuits should be separate from any other electrical installation. To isolate it completely from any interruption to mains supply, it is usually transformed to 24–60 volts DC and provided with a battery back-up system in the event of the fire damaging the mains source of power.

Closed circuit – call points or detectors may be regarded as closed switches allowing current to flow in the detector circuit. This permanent current flow energises a solenoid switch which retains a break in the alarm circuit. When a detector circuit switch is operated (i.e. opened), the solenoid is de-energised, allowing a spring mechanism to connect it across the alarm circuit terminals and effect the alarm.

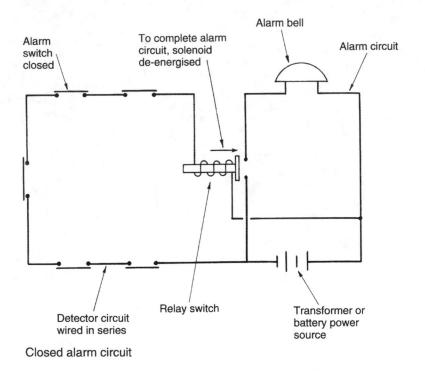

Closed alarm circuit

Ref. BS EN 54: Fire detection and fire alarm systems.

Ventilation of services enclosures is required to dilute flammable, toxic or corrosive gases. This can be taken to include smoke and hot gases that will occur as a result of fire, particularly where the void contains combustible PVC cable sheathing and uPVC pipes. To provide a safe level of ventilation and to prevent overheating in a restricted enclosure, permanent natural ventilation should be at least 0.05m² and 1/150 of the cross-sectional area for enclosure areas of less than 7.5m² and greater than 7.5m², respectively.

Openings and access panels into services enclosures should be minimal. The enclosure itself should be gas tight, and there must be no access from a stairway. Where access panels or doors are provided they should be rated at not less than half the fire resistance of the structure and have an integrity rating of at least 30 minutes (see BS 476-22). Fire doors should be fitted with self-closers.

Where ventilation ducts pass from one compartment to another or into a services enclosure, the void made in the fire-resisting construction must be made good with a suitable fire-stopping material. Automatic fire dampers are also required in this situation to prevent fire from spreading between compartments.

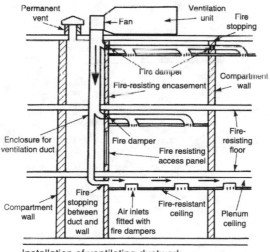

Installation of ventilating ductwork

Refs. BS 8313: Code of practice for accommodation of building services in ducts.

BS 9999: Fire safety in the design, management and use of buildings. Code of practice.

Building Regulations, Approved Document B3 (Section 9 Vol. 1 and Section 10 Vol. 2): Protection of openings and fire stopping.

Fire Dampers in Ventilation Ductwork

Fire dampers are required in ventilation and air-conditioning systems to prevent smoke and fire from spreading through the ductwork to other parts of the building. Dampers should be positioned to maintain continuity of compartmentation by structural division. They can operate automatically by fusible link melting at a predetermined temperature of about 70°C, to release a steel shutter. An electromagnet may also be used to retain the shutter in the open position. The electromagnet is deactivated to release the shutter by a relay circuit from a fire or smoke detector. The latter is preferable, as a considerable amount of smoke damage can occur before sufficient heat penetrates the ductwork to activate a heat detector or a fusible link.

An intumescent-coated honeycomb damper is an alternative. In the presence of heat, the coating expands to about a hundred times its original volume to form sufficient mass to impair the movement of fire through the duct. This type of damper has limited fire resistance and is only likely to be specified in low-velocity systems.

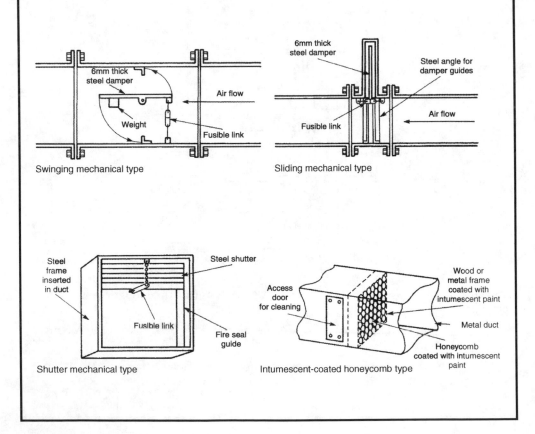

Swinging mechanical type

Sliding mechanical type

Shutter mechanical type

Intumescent-coated honeycomb type

In multi-storey buildings, stairways and lobbies may be air pressurised to clear smoke and provide an unimpeded escape route. The air pressurisation is usually between 25 and 50Pa depending on the building height and degree of exposure. This pressure is insignificant for movement of personnel. A number of pressurisation methods may be used:

- Pressurisation plant is disengaged, but it is automatically switched on by a smoke or fire detector.
- Pressurisation plant runs fully during hours of occupancy as part of the building ventilation system.
- Pressurisation plant runs continuously at a reduced capacity and output during the hours of building occupancy, but fire detection automatically brings it up to full output.

It is important to provide openings so that smoke is displaced from the escape routes to the outside air. This can be through purpose-made grilles or window vents. Pressurisation will help to limit entry of rain and draughts at external openings.

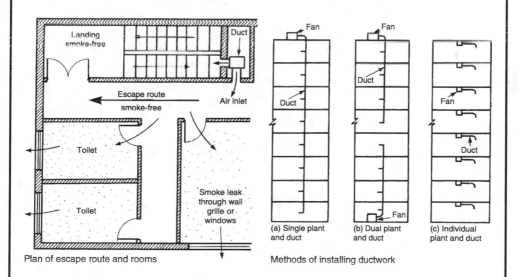

Plan of escape route and rooms

Methods of installing ductwork

(a) Single plant and duct

(b) Dual plant and duct

(c) Individual plant and duct

Ref. BS EN 12101-6: Smoke and heat control systems. Specification for pressure differential systems. Kits.

Smoke Extraction and Ventilation

Automatic fire ventilation is designed to remove heat, smoke and toxic gases from single-storey buildings. In large factories and shopping malls, the additional volume of air entering the building by fire venting is insignificant relative to the benefits of creating clear visibility. Parts of the roof can be divided into sections by using fireproof screens which may be permanent or may fall in response to smoke detection. Fire vents are fitted at the highest part of each roof section as is practical. Heat and smoke rise within the roof section above the fire outbreak. At a predetermined temperature, usually 70°C, a fusible link breaks and opens the ventilator above the fire. Heat and smoke escape to reduce the amount of smoke logging within the building. This will aid people in their escape and assist the fire service to see and promptly tackle the source of fire. The heat removed prevents risk of an explosion, flash-over and distortion to the structural steel frame.

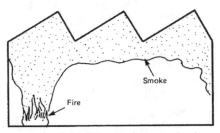

Fire in unvented building showing unrestricted spread of smoke

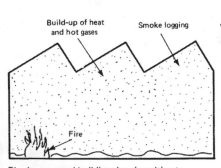

Fire in unvented building showing ultimate smoke logging

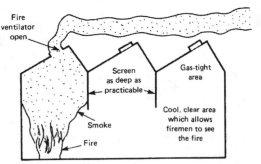

Fire in vented building showing restricted spread of smoke. The fire ventilator may also be used for normal ventilation.

Automatic smoke and fire ventilator:

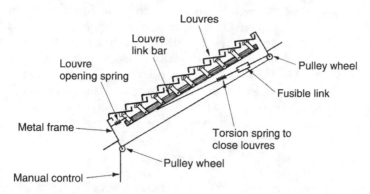

Number and area of ventilators – estimates are based on providing a smoke-free layer for about 3m above floor level.

E.g.

Floor to centre of vent height (m)	Ventilation factor (m)
4·5	0·61
7·5	0·37
10·5	0·27
13·5	0·23

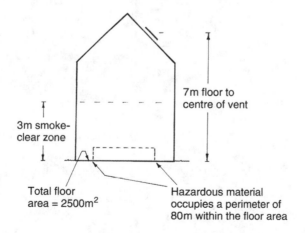

By interpolation, ventilation factor for 7m approximates to 0·41m.

Ventilator area can be taken as the perimeter occupied by hazardous material, multiplied by the ventilation factor, i.e. 80m × 0·41m. This approximates to 33m² or (33/2500 × 100/1) = 1·3% of the floor area.

Smoke Control in Shopping Malls

Most enclosed shopping centres have a mall with a parade of shops. The mall is the general circulation area and the obvious escape route from a fire. In these situations, a fire can generate a rapid spread of smoke and hot gases. It is therefore essential that some form of smoke control is adopted. If the central area has a normal (68°C) sprinkler system, the water may cool the smoke and hot gases to reduce their buoyancy and create an unwanted fogging effect at floor level. Therefore, consideration should be given to reducing the number of sprinkler heads and specifying a higher operating temperature. Smoke can be controlled by:

• Providing smoke reservoirs into which the smoke is retained before being extracted by mechanical or natural means.

• Allowing replacement cool air to enter the central area through low-level vents to displace the smoke flowing out at higher level.

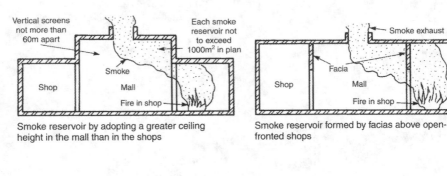

Smoke reservoir by adopting a greater ceiling height in the mall than in the shops

Smoke reservoir formed by facias above open-fronted shops

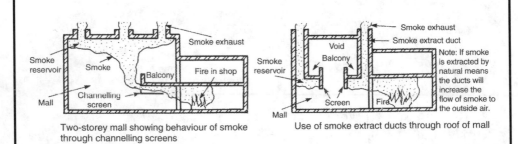

Two-storey mall showing behaviour of smoke through channelling screens

Use of smoke extract ducts through roof of mall

A portable fire extinguisher must contain the type of fire-extinguishing agent suitable for the fire it is required to extinguish. It must also be clearly identifiable by colour coding for its intended purpose.

Fires can be classified as:

Class A – organic solids, e.g. wood, paper, cloth.
Class B – flammable liquids, e.g. petrol, oil, paint.
Class C – flammable gases, e.g. methane, propane, acetylene.
Class D – flammable metals, e.g. zinc, aluminium, uranium.
Electrical – not specifically classed because it can apply to any of the other classifications.
Class F – cooking oil and fat.

Extinguishing agent	Extinguisher colour	Application/Class
Water	Red	A
Foam	Red with cream band	A and B
Carbon dioxide	Red with black band	B and electrical
Dry chemicals/powder	Red with blue band	A, B, C and electrical
Wet chemicals	Red with yellow band	A and F
Special powders	Red with blue band	D

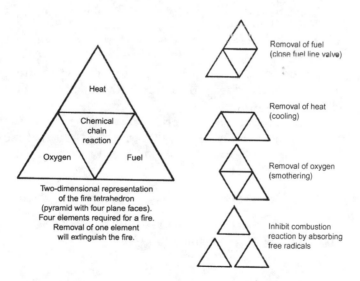

Heat
Chemical chain reaction
Oxygen
Fuel

Two-dimensional representation
of the fire tetrahedron
(pyramid with four plane faces).
Four elements required for a fire.
Removal of one element
will extinguish the fire.

Removal of fuel
(close fuel line valve)

Removal of heat
(cooling)

Removal of oxygen
(smothering)

Inhibit combustion
reaction by absorbing
free radicals

Refs. BS EN 3-8: Portable fire extinguishers.

BS 5306-10: Fire extinguishing installations and equipment on premises. Colour coding to indicate the extinguishers medium contained in portable fire extinguishers. Code of practice.

Extinguisher rating – performance rating and capability can be identified by a letter relative to Class types A to D and F, and a number. The higher the number, the larger the fire that the extinguisher is capable of controlling as determined under British Standard test conditions, e.g. 13A and 55B.

Some extinguishers have two or three letter ratings to indicate the range of capability. Class F fire extinguishers are rated relative to tests based on 5, 15, 25 and 75 litre quantities of sunflower oil. The oil is heated to a state of self- or automatic ignition and allowed to burn for two minutes and then extinguished. To qualify, no reignition is to occur within 10 minutes.

Ref. BS EN 3-7: Portable fire extinguishers. Characteristics, performance requirements and test methods.

Siting of extinguishers –

• In a conspicuous location that is easily accessible.
• Not in cupboards or behind doors.
• Not above cookers or other heat emitters, or in any place of excessive heat or cold.
• Hung on wall brackets within easy reach, not placed on floor.
• Carrying handle 1 metre above floor for heavier extinguishers (liquid based) and 1.5 metres for others.
• Along escape routes near to a door leading to a place of safety.
• Positioned in a wall recess so as not to obstruct general movement.
• A maximum distance of 30 metres from the site of a possible fire.
• Repeated location on each storey.

Ref. BS 5306-8: Fire extinguishing installations and equipment on premises. Selection and positioning of portable fire extinguishers. Code of practice.

Maintenance – after use, even if only partially, extinguishers must be serviced, i.e. recharged in accordance with the manufacturer's directives. Extinguishers should be labelled to record the last service check (usually annually) and a log-book endorsed for retention by the building facilities manager.

Sand and water buckets are no longer acceptable as a first-aid fire treatment facility. Purpose-provided extinguishers are now commonplace in public and commercial buildings. Under the obligations of the Health and Safety at Work, etc., Act, employees are required to undertake a briefing on the use and selection of fire extinguishers. Water in pressurised cylinders may be used for carbonaceous fires and these are commonly deployed in offices, schools, hotels, etc. The portable soda–acid extinguisher has a small glass container of sulphuric acid. This is released into the water cylinder when a knob is struck. The acid mixes with the water which contains carbonate of soda to create a chemical reaction producing carbon dioxide gas. The gas pressurises the cylinder to displace water from the nozzle. The inversion type of extinguisher operates on the same chemical principle.

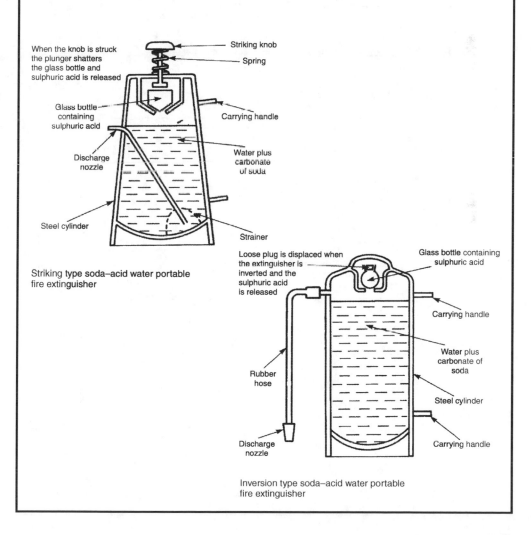

Striking type soda–acid water portable fire extinguisher

Inversion type soda–acid water portable fire extinguisher

Although water is a very good cooling agent, it is inappropriate for some types of fire. It is immiscible with oils and is a conductor of electricity. Therefore, the alternative approach of breaking the fire tetrahedron by depleting the oxygen supply can be achieved by smothering a fire with foam. Foam is suitable for gas or liquid fires.

Chemical foam type of extinguisher – foam is formed by chemical reaction between sodium bicarbonate and aluminium sulphate dissolved in water in the presence of a foaming agent. When the extinguisher is inverted, the chemicals are mixed to create foam under pressure which is forced out of the nozzle.

Carbon dioxide extinguisher – carbon dioxide is pressurised as a liquid inside a cylinder. Striking a knob at the top of the cylinder pierces a disc to release the carbon dioxide which converts to a gas as it depressurises through the extinguisher nozzle.

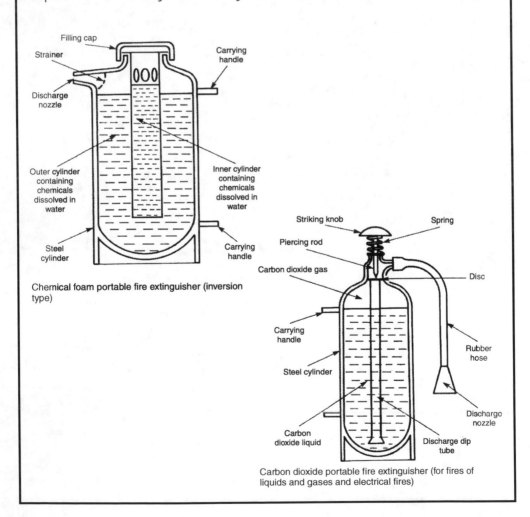

Chemical foam portable fire extinguisher (inversion type)

Carbon dioxide portable fire extinguisher (for fires of liquids and gases and electrical fires)

Carbon monoxide (CO) gas is colourless, invisible, tasteless and odourless. Where allowed to accumulate, it cannot be detected by human perception or senses. With sufficient exposure, it can be deadly, hence its common reference as the 'silent killer'. It is the primary cause of death by accidental poisoning in the UK, with estimates in excess of 20 persons per year and some 200 others seriously injured. About half of these incidents are attributed to faulty fuel-burning appliances, either incorrectly serviced or improperly installed. It is not easy to determine the total numbers of people affected, as the symptoms and characteristics can be similar to other medical disorders.

Symptoms – limited exposure to carbon monoxide poisoning is often unrecognised. The symptoms can be superficially very similar to that of influenza and food poisoning, leading to wrong diagnosis in the absence of blood tests.

- Slight exposure – headache, nausea, vomiting, fatigue and aching limbs.
- Greater exposure – throbbing headache, drowsiness, confusion and increased heart rate.
- High level of exposure – unconsciousness, collapse, convulsions, cardio-respiratory failure, deep coma and ultimately death.

Note. Exposure, whether in small ongoing doses or occasional concentrated amounts, can result in permanent disability due to neurological damage and functional loss of brain cells.

Effect on the human body – the body's ability to transport oxygen to vital organs is impaired when exposed to carbon monoxide. Carbon monoxide bonds with the haemoglobin in blood to gradually replace oxygen. This prevents the uptake of oxygen into the blood and the body begins to suffocate.

Most at risk –
- Those at home for long periods, i.e. the house-bound.
- Elderly and infirm, particularly those with heart/respiratory problems.
- Pregnant women, children and pets.

Carbon Monoxide Detectors – 2

Appliances – all those fuelled from fossil resources, including wood, coal, charcoal, oil, gas (inc. LPG) and paraffin. Carbon monoxide is a product of incomplete combustion. To function efficiently, heat-producing appliances must have adequate oxygen supplied through purpose-made air vents to achieve complete combustion of fuel. The products of combustion should be exhausted safely through a correctly sized, undamaged and unobstructed flue system. The position of flue outlets and the location of outside appliances is important as carbon monoxide can permeate the structure.

Modern houses are extremely well sealed which may be advantageous in preventing the ingress of flue gases. However, unlike older houses, there is less natural air leakage through the structure to aid fuel combustion and to dilute escaping gases. Whatever, there is no safe option other than regular servicing and maintenance for all heat-producing appliances. CO detectors are an essential safety installation for all dwellings and other buildings containing combustion appliances.

Registered social landlords have a duty of care for their tenants' safety. This includes provision for protection against CO poisoning. E.g. registered student accommodation.

Types of detector/alarm – mains or battery powered. Audible, also available with a visual facility for people with hearing difficulties. Size and appearance resembles a domestic smoke alarm, but the sensor inside the unit differs, being any of the following:

- An electrochemical type of fuel cell that is energised in the presence of CO.

- Biomimetic – a synthetic haemoglobin that darkens in the presence of CO. The colour change activates a light cell.

- Semiconductor – an electric circuit of thin tin oxide wires on a ceramic insulator. The presence of CO reduces the electrical resistance, allowing greater current flow to activate the alarm.

The positioning and number of carbon monoxide detectors depends on the layout of rooms. Several individual battery powered detectors/ alarms is acceptable, but it is preferable to have a system or network of hard-wired mains powered interlinked detectors.

Location –

- In any room containing a fuel-burning appliance.
- Bedrooms, positioned at pillow height.
- Remote rooms, 1.5 to 2.0m above floor level.
- Room adjacent to a dedicated boiler room.
- In bed-sits, close to sleeping area and away from cooking appliance.
- Not in bathrooms or shower rooms.

Positioning –

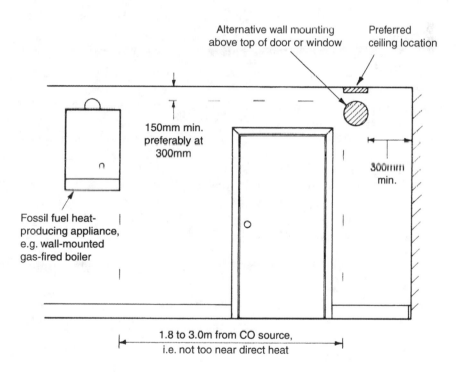

Ref. BS EN 50291-1: Gas detectors. Electrical apparatus for the detection of carbon monoxide in domestic premises. Test methods and performance requirements.

Carbon Monoxide – Detection Provision

The Building Regulations, Approved Document J: Combustion appliances and fuel storage systems include requirements for carbon monoxide alarms to be installed in buildings containing fossil fuel burning appliances, e.g. a hot water and/or central heating boiler. Specifically, where combustion appliances are a fixed installation, a carbon monoxide detection device should be provided to give warning if carbon monoxide gas is released into the building interior. A defective gas burner, broken or blocked flue, could be at fault.

Where a new or replacement fossil fuel burning appliance is installed in a dwelling, a means of carbon monoxide detection is required in the room where the appliance is located.

Applications:

AD J, Section J2: Solid fuel appliances with output rated up to 50 kW.

AD J, Section J3: Gas fired appliances with input rated up to 70 kW (net). Excluded are gas appliances used for cooking only.

AD J, Section J4: Oil fired appliances with a rated output up 45 kW.

Provision of carbon monoxide detection units relative to the appliance rating parameters ensures installation in all new-build and refurbished/renovated domestic dwellings subject to Building Regulation control.

Further considerations:

Details of CO detection unit types and applications are provided in BS EN 50292: Electrical apparatus for the detection of carbon monoxide in domestic premises, caravans and boats. Guide on the selection, installation, use and maintenance.

BS EN 50292 specifies two types of CO detection unit, selection depending on situation:

- Type A – visual (illuminated) and audible. Can also be linked to actuate extract ventilation and other ancillary devices, possibly for example, agency monitoring.
- Type B – most commonly installed in dwellings, visual and audible only.

    Both types mains electricity powered with battery back-up.

14 SECURITY INSTALLATIONS

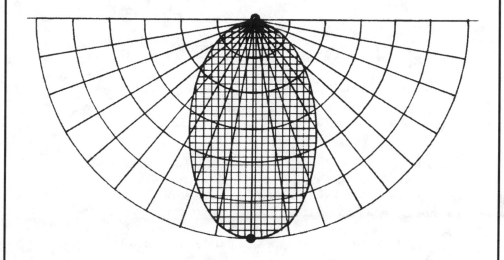

PHYSICAL SECURITY

INTRUDER ALARMS

MICRO-SWITCH AND MAGNETIC REED

RADIO SENSOR, PRESSURE MAT AND TAUT WIRING

ACOUSTIC, VIBRATION AND INERTIA DETECTORS

ULTRASONIC AND MICROWAVE DETECTORS

ACTIVE INFRA-RED DETECTOR

PASSIVE INFRA-RED DETECTOR

LIGHTNING PROTECTION SYSTEMS

SURGE PROTECTION

DOI: 10.1201/9781003434894-14

Physical Security

The first line of defence against intrusion includes physical measures that resist unwanted entry to the grounds and structure of a building. The second line of defence is an alarm system.

Some basic physical measures:

- Spring or night latch – the typical front door latch that is in place when the door is closed. Very vulnerable due to a simple two-lever spring mechanism that can be moved by other means than a key. Opened from the inside without a key, therefore no problem for an intruder seeking an easy escape.

- Deadlock – cannot function by just closing the door. Requires a key to lock it on leaving and when locked cannot be opened from the inside without a key. Espagnolette variation provides at least two more deadlocking positions throughout the height of the door or window frame, thus making it more difficult to force and smash the lock retainers/staples. Usually produced with a five-lever mechanism that is difficult to pick. No matter how sophisticated the ironmongery/brassware, weak and slender doors or window frames are always vulnerable to force.

- Telephone – with overhead supplies, these can easily be cut by would-be intruders. Most supplies are now under the ground in service ducts and therefore protected from abuse.

- Glazing – double panes are primarily to improve thermal and sound insulation. They also provide an intruder with more of a problem and a hazard than single panes. Double-glazed sealed units are difficult to break due to the air cushion between the panes.

- Window casement fasteners and stays – can be fitted with locking devices, but these can have a negative effect if they cannot be opened for an emergency escape. Sufficient provision must be made for access through these escape routes in an emergency and they should not be permanently sealed. See Building Regulations Approved Documents B1 and B2, Section 2: Means of escape.

Further reading: Part 5, 11th edition of the *Building Construction Handbook* (Chudley/Greeno). Specific references being Window Security and Door Security. Also Approved Document Q, Security — Dwellings.

Simple means of protection – intruders often target buildings at night, under the cover of darkness. They dislike being illuminated at work; therefore a sufficient deterrent may be achieved by leaving external lights on a timed control. A more selective variation uses an infra-red radiation movement detector (see pages 709–710) activated during hours of darkness. Another selective variation that could be used 24 hours a day incorporates a sound sensor or listening device that activates a light or radio transmitter in response to noise.

No amount of physical barriers, whether they be tall hedging, brick walls, fencing, door locks, etc., will prevent the most determined of intruders. The second line of defence, the intruder alarm, is regarded as the most effective deterrent, often just by its presence in the form of an alarm box mounted high on the face and rear of a building. These systems can be stand-alone (i.e. just used to raise a highly audible sound at the property), or they may be monitored systems, sometimes known as signalling systems. Monitored systems incorporate remote supervision through the telecommunications cables linked into the alarm circuit to a security company and possibly the local police.

Alarm installations – hard-wired (cable) systems to BS EN 50131 are the most reliable and conform to the NPCC alarm policy. An alternative is a wire-free or radio frequency installation that has appeal to the DIY market. These plug-in kits are relatively simple to install and have little disruptive effect to the finishes and structure. Wire-free systems satisfying BS EN 50131 are accepted by the NPCC alarm policy. Lesser classifications may well respond satisfactorily to an intruder but may also be triggered by radio frequencies transmitted by other equipment.

Refs. BS EN 50131-1: Alarm systems. Intrusion and hold-up systems. System requirements.

NPCC: National Police Chiefs Council.

Intruder Detection Systems – 2

Police response to alarm activations depends on the type of alarm system installed. There are a high number of false alarms as a result of user misuse and/or equipment error. Therefore, the NPCC have adopted a Unified Intruder Alarm Policy for responding to alarm signals. This policy designates alarm systems into two separate categories or types, each having a defined response:

- Type A – remote signalling or monitored alarm systems. Maintained and used in accordance with the recommendations of BS EN 50131. These systems are registered with an alarm-monitoring agent and with the police. When activated the monitoring agent will liaise with the police after telephoning the building owner or a nominated key holder (usually a neighbour for domestic systems) to establish whether the alarm activation is genuine or accidental. A password is determined and if the monitoring agent considers that the response is inappropriate or non-existent, they will confirm this with the police. Police response will depend on priority of other commitments at the time and a consideration of the number of recent false activations at the premises.

- Type B – stand-alone or audible-only unmonitored systems. These will only attract a police response if there is supplementary information in the form of an witness report that an offence is in progress. This category can also include some systems known in the industry as hybrids. These are monitored to a certain extent by including an automatic dialling facility to a security company.

Independent alarm inspection and certification authorities –

National Security Inspectorate (NSI).

Security Systems and Alarm Inspection Board (SSAIB).

Refs. BS 4737-4.3: Intruder alarm systems in buildings. Codes of practice. Code of practice for exterior alarm systems.

PD 6662: Scheme for the application of European Standards for intruder and hold-up alarm systems.

Some intruder alarm detection devices are considered between pages 703–710.

Video surveillance installations can provide recorded evidence of intrusion. The principles of application considered in BS 8418: Design, installation, commissioning and maintenance of detection activated video surveillance systems.

Intruder alarms have developed from a very limited specialist element of electrical installation work in high-security buildings to the much wider market of schools, shops, offices, housing, etc. This is largely a result of the economics of sophisticated technology surpassing the efficiency of manual security. It is also a response to the increase in burglaries at a domestic level. Alarm components are an alarm bell or siren activated through a programmer from switches or activators. Power is from mains electricity with a battery back-up. Extended links can also be established with the local police, a security company and the facility manager's central control by telecommunication connection.

Selection of switches to effect the alarm will depend on the building purpose, the extent of security specified, the building location and the construction features. Popular applications include

- Micro-switch
- Magnetic reed
- Radio sensor
- Pressure mat
- Taut wiring
- Window strip
- Acoustic detector
- Vibration, impact or inertia detector.

The alternative, which may also be integrated with switch systems, is space protection. This category of detectors includes:

- Ultrasonic
- Microwave
- Active infra-red
- Passive infra-red.

Circuit wiring may be 'open' or 'closed' as shown in principle for fire alarms – see pages 683 and 684. The disadvantage of an open circuit is that if an intruder knows the whereabouts of cables, the detector circuit can be cut to render the system inoperative. Cutting a closed circuit will effect the alarm.

The following references provide detailed specifications:

BS EN 50131-1: Alarm systems. Intrusion and hold-up systems. System requirements.

BS 9263: Intruder and hold-up alarm systems. Commissioning, maintenance and remote support.

Micro-Switch and Magnetic Reed

Micro-switch – a small component which is easily located in door or window openings. It is the same concept and application as the automatic light switch used in a vehicle door recess, but it activates an alarm siren. A spring-loaded plunger functions in a similar manner to a bell push button in making or breaking an electrical alarm detector circuit. The disadvantage is the constant movement and associated wear, exposure to damage and possible interference.

Magnetic reed – can be used in the same situations as a micro-switch but it has the advantage of no moving parts. It is also less exposed to damage or tampering. There are, however, two parts to install. One is a plastic case with two overlapping metal strips of dissimilar polarity, fitted into a small recess in the door or window frame. The other is a magnetic plate attached opposingly to the door or window. When the magnet is close to the overlapping strips, a magnetic field creates electrical continuity between them to maintain circuit integrity. Opening the door or window demagnetises the metal strips, breaking the continuity of the closed detector circuit.

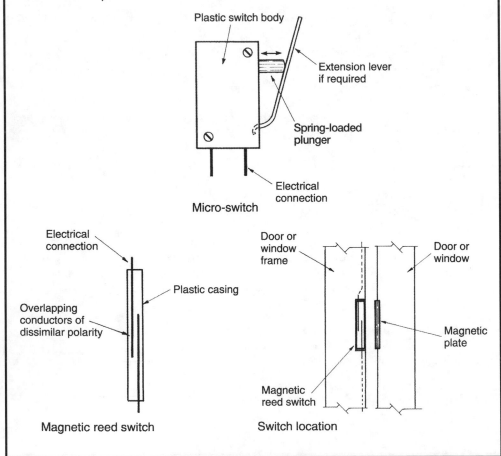

Plastic switch body

Extension lever if required

Spring-loaded plunger

Electrical connection

Micro-switch

Electrical connection

Plastic casing

Overlapping conductors of dissimilar polarity

Magnetic reed switch

Door or window frame

Door or window

Magnetic plate

Magnetic reed switch

Switch location

Radio sensor – these are surface mounted to windows and doors. They transmit a radio signal from an integral battery power source. This signal is picked up by a central control unit or receiver, which activates the alarm circuit. As these sensors are 'free wired' they can be moved, which is ideal for temporary premises or in buildings undergoing changes. A pocket or portable radio panic button transmitter is an option. The range without an aerial is about 60m; therefore, they can be used in outbuildings to a hard-wired system from a main building.

Pressure mat – these are a 'sandwich' with metal foil outer layers as part of a detector circuit. The inner core is a soft perforated foam. Pressure on the outer upper layer connects to the lower layer through the perforations in the core to complete the circuit and activate the alarm. Location is near entrances and under windows, normally below a carpet where a small area of underlay can be removed. Sensitivity varies for different applications, such as premises where household pets occupy the building.

Taut wiring – also available as a window strip. A continuous plastic-coated copper wire is embedded in floors, walls or ceilings, or possibly applied around safes and other secure compartments. As a window strip, silvered wire can be embedded between two bonded laminates of glass. Alternatively, a continuous self-adhesive lead or aluminium tape can be applied directly to the surface. In principle, it is similar to a car rear-heated window. When the wire or tape is broken, the closed circuit is interrupted which activates the alarm circuit.

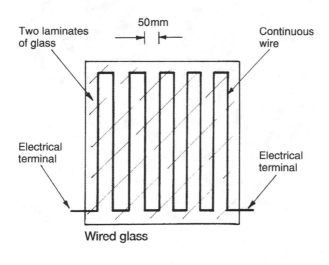

Acoustic, Vibration and Inertia Detectors

Acoustic – also known as sonic detectors. They are used mainly for protection against intruders in commercial and industrial premises. A sound receiver comprises a microphone, amplifier and an output relay. Also included is a filter circuit which can be tuned to respond to specific sound frequencies such as that produced by breaking glass.

Vibration – a slender leaf of steel is suspended between two electrical contacts. Hammering or structural impact produces vibration in the pendulum, sufficient for the contacts to meet and complete a detector circuit. Adjustment allows for a variety of applications, e.g. where a road or railway is adjacent and intermittent vibration would occur.

Inertia – these respond to more sensitive movements than vibrations, so would be unsuitable near roads, railways, etc. They are ideal to detect the levering or bending of structural components such as window sashes and bars. A pivotal device is part of a closed circuit, where displacement of its weight breaks the circuit continuity.

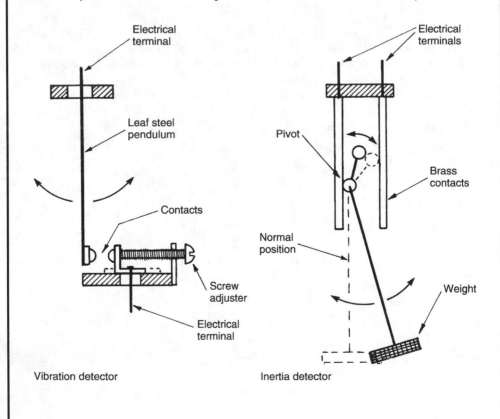

Vibration detector

Inertia detector

Ultrasonic – the equipment is simply a sound emitter and a receiver containing a microphone and sound processor. The sounds are at a very high frequency of between 20 and 40kHz (normal hearing limit is about 15kHz). Direct and indirect (reflected) sound distribution from the emitter to the receiver adopts a pattern which can be plotted as a polar curve. If an intruder encroaches the curve, the sound frequency will be disturbed. The receiver then absorbs the original frequency, the frequency reflected off the intruder and a mixture of the two. The latter is known as the 'beat note', and it is this irregularity which effects the detector circuit. Greatest detection potential is in the depth of the lobe; therefore, this should be projected towards an entry point or a window.

Microwave – operates on the same principle as ultrasonic detection, except that extremely high radio waves are emitted at a standard 10.7GHz. Emitter and receiver occupy the same unit which is mounted at high level to extend waves over the volume of a room, warehouse, office or similar internal area. An intruder penetrating the microwaves disturbs the frequency which effects the detector circuit. Unlike ultrasonic detectors, microwave detectors are not disturbed by air currents, draughts and ultrasonic sounds from electrical equipment such as computers. They are therefore less prone to false alarms.

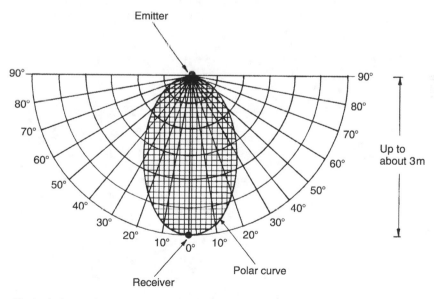

Typical ultrasonic detector response zone

Ref. BS EN 50131-2-3: Alarm systems. Intrusion and hold-up systems. Requirements for microwave detectors.

Active Infra-Red Detector

Otherwise known as an optical system, it uses a light beam from the infra-red part of the electromagnetic spectrum. This is imperceptible to the human eye. The system is based on a transmitter and receiver. The transmitter projects an invisible light beam at distances up to 300m onto a photo-electric cell receiver. An intruder crossing the beam will prevent the light from activating the cell. The loss of energy source for the cell effects an alarm relay. Even though the beam has extensive range, this system is not suitable for external use.

Atmospheric changes such as fog or birds flying through the beam can affect the transmission. Mirrors may be used to reflect the beam across a room or around corners, but each reflection will reduce the beam effectiveness by about 25%. Infra-red beams will penetrate glass partitions and windows, each pane of glass reducing the beam effectiveness by about 16%. The smarter intruder may be able to fool the system by shining a portable light source at the receiver. This can be overcome by pulsing the transmission, usually at about 200 pulses per second.

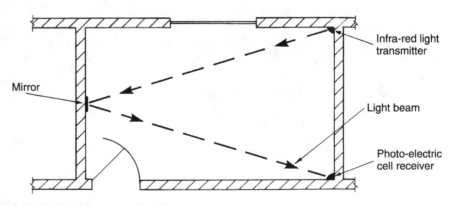

Infra-red light beam application

These detectors use highly sensitive ceramic infra-red receivers to recognise radiation from a moving body. Wall-mounted detector units focus the radiation through a lens which contains curved facets to concentrate the radiation onto two sensors. Image variation between the sensors generates a small electrical differential to effect an alarm relay. These systems have enjoyed widespread application, not least in the domestic market. Units of lower sensitivity can be used where pets occupy a home. A battery back-up energy source covers for periods of mains power isolation. PIR detectors can be used with other devices in the same system, e.g. radio pocket panic buttons, pressure mats, magnetic reeds, etc. PIR beam patterns vary in form and range to suit a variety of applications, both externally and internally.

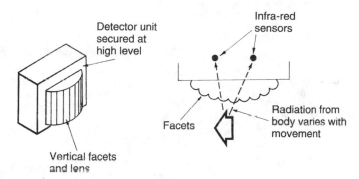

PIR detector unit, typically 75 × 50mm

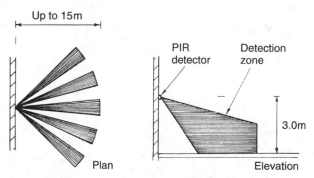

Typical pattern displacement for wall-mounted detector

Ref. BS EN 50131-2-2: Alarm systems. Intrusion and hold-up systems. Requirements for passive infrared detectors.

PIR Detector Displacements

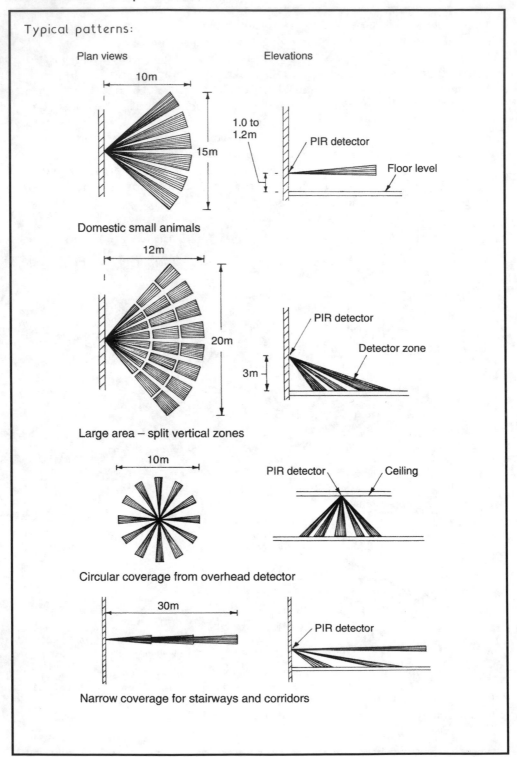

Typical patterns:

Plan views   Elevations

10m

15m

1.0 to 1.2m

PIR detector

Floor level

Domestic small animals

12m

20m

PIR detector

Detector zone

3m

Large area – split vertical zones

10m

PIR detector   Ceiling

Circular coverage from overhead detector

30m

PIR detector

Narrow coverage for stairways and corridors

Lightning occurs as a result of electrostatic discharge between clouds or between a cloud and the ground. The potential is up to 100MV with the current peaking at about 200kA. The average current is about 20kA. The number of days that thunderstorms occur in the UK varies between five and 20 per year, depending on location. Consequently, some degree of protection to buildings and their occupants is necessary.

As the risk of lightning striking a particular building is low, not all buildings are protected. Houses have least priority and are rarely protected, but other purpose groups will be assessed by their owners and their insurers. This will be on the basis of height, contents, function, type of construction (extent of metalwork, e.g. lead roofing), likelihood of thunderstorms in locality, extent of isolation and the general topography. Even where a lightning protection system is provided, it is unlikely to prevent some lightning damage to the building and its contents.

Function of a lightning protection system – to attract a lightning discharge which might otherwise damage exposed and vulnerable parts of a building. To provide a path of low impedance to an earth safety terminal.

Zone of protection – the volume or space around a conductor which is protected against a lightning strike. It can be measured at 45° to the horizontal, descending from the apex of the conductor. For buildings less than 20m in height the zone around a vertical conductor is conical. For buildings exceeding 20m, the zone can be determined graphically by applying a 60-m radius sphere to the side of a building. The volume contained between the sphere and building indicates the zone. See next page for illustrations.

Lightning Protection Systems – 2

Zones of protection:

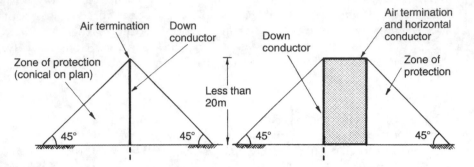

Protection zones for buildings < 20m height

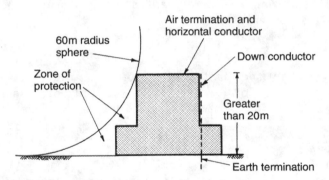

Protection zones for buildings > 20m height

Air terminations – these are provided to intercept a lightning strike. No part of a roof should exceed 5m from part of a termination conductor, unless it is a lower level projection which falls within the zone of protection. Metallic components such as aerials, spires, cooling towers, etc. should be connected to a terminal. Apart from specific apexes such as spires, air terminations are horizontal conductors running along the ridge of a pitched roof or around the periphery of a flat roof. If the roof is of sufficient size, a 20m × 10m grid or lattice of parallel terminations should be provided.

Down conductors – these provide a low impedance route from the air terminations to the earth terminal. They should be direct, i.e. vertical without bends and re-entrant loops. Spacing for buildings up to 20 m in height is 1 per 20 m of periphery starting at the corners and at equal distance apart. Buildings in excess of 20 m height require 1 per 10 m, at corners and equally spaced. All structural steelwork and metal pipes should be bonded to the down conductor to participate in the lightning discharge to earth.

Fixing centres for all conductors:

Horizontal and vertical – 1 m max.

Horizontal and vertical over 20 m long – 750 mm max.

25 m long – 500 mm max.

Minimum dimensions of conductors: 20 mm × 4 mm (80 mm^2) or

10 mm diameter (80 mm^2).

Conductor materials – aluminium, copper and alloys, phosphor-bronze, galvanized steel or stainless steel.

Earth termination – this is required to give the lightning discharge current a low-resistance path to earth. The maximum test resistance is 10 ohms for a single terminal and, where several terminals are used, the combined resistance should not exceed 10 ohms. Depth of terminal in the ground will depend on subsoil type. Vertical earthing rods of 10 or 12 mm diameter hard-drawn copper are preferred, but stronger phosphor-bronze or even copper-coated steel can be used if the ground is difficult to penetrate. Alternatively, a continuous horizontal strip electrode may be placed around the building at a depth of about 1 metre. Another possibility is to use the reinforcement in the building's foundation. To succeed, there must be continuity between the structural metalwork and the steel reinforcement in the concrete piled foundation.

Ref. BS EN 62305-1 to 4: Protection against lightning.

Lightning – Surge Protection

Purpose – the prime purpose of a lightning conductor is to prevent physical damage to a building and to persons within and close to it by diverting the impact of a lightning strike to ground/earth. The electrostatic discharge from a strike is capable of producing high-voltage spikes or surges, effecting a sudden energy release that can damage a building's electricity supply installation. A surge may be caused by a direct strike on a building or indirectly through overhead electrical supply or telephone lines. An increase in current surge for as little as a few micro-seconds is sufficient to generate a transient over-voltage capable of damaging consumer unit and distribution board installations. Over-heating of components and cables will be a fire risk that building insurers are aware of. In addition to insurance requirements, for safety and security of occupants and integrity of the structure, buildings considered at risk of lightning damage will have a surge protection device or surge arrester fitted to the electrical intake in parallel to the supply. Surge protectors are also fitted to exposed overhead power transmission lines and to substations.

Function – surge protectors limit the effect of a lightning charge by diverting excess electrical energy into a dedicated ground/earth rod – see page 553. The arrester unit's path to earth is through a metal oxide varistor that provides a selective path for an electrical surge. The response is instant, i.e. in nanoseconds, giving no time for an electrical surge to damage an installation. In normal use, the metal oxide varistor's high resistance allows current to flow unimpeded. If a lightning strike and/or an electrical surge occurs, the reference voltage of the surge arrestor will be exceeded causing the unit to react from high resistance to a low resistance. The excess electrical energy is conveyed into the ground, by-passing the normal installation leaving it undamaged.

Alternative use – surge arrestors are not solely limited to possible lightning strikes. They are also applied as a back-up safety measure to protect equipment against voltage spikes that may be caused by changes in operating conditions. Possibly, where numerous items of plant or machinery are used intermittently and there is an unusual response, i.e. all items engaged at the same time.

Ref. BS 7671 – IET Wiring Regulations. Should be provided where there is a risk of overvoltage leading to injury to personnel or interruption to commercial, industrial or public services.

15 ACCOMMODATION FOR BUILDING SERVICES

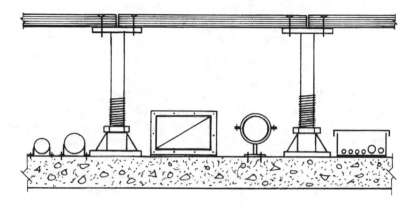

UNDERGROUND UTILITIES

DUCTS FOR ENGINEERING SERVICES

NOTCHING AND HOLING JOISTS

FLOOR AND SKIRTING DUCTS

MEDIUM AND LARGE VERTICAL DUCTS

MEDIUM AND LARGE HORIZONTAL DUCTS

SUBWAYS OR WALKWAYS

PENETRATION OF FIRE STRUCTURE BY PIPES

RAISED ACCESS FLOORS

SUSPENDED AND FALSE CEILINGS

FLUE PIPE LOCATION

DOI: 10.1201/9781003434894-15

Underground Utilities

Separate trenches for individual services are cost ineffective. A single excavation with sufficient space to accommodate several pipes and cables is less disruptive, less exposed to damage and uses plant and labour more effectively. Contact with the local public utilities suppliers to coordinate their facilities as a preliminary element of the work programme enables installations to be undertaken and backfilled without later disruption. Supplies can then be used during construction.

Separation between different services is necessary for maintenance accessibility. Plastic-piped water supplies should be laid at least 350mm from gas pipes, to prevent possible contamination from gas leakage. In addition, plastic-piped water services must not be located in the vicinity of hydrocarbon (oil, petrol and creosote) storage, as leakage of these can cause deterioration of the pipe.

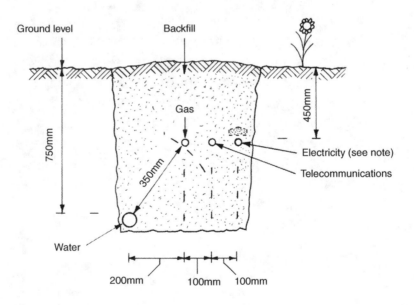

Note: High-voltage electricity cables to be protected, usually with clayware or concrete capping stones.
Common services/utilities trench, minimum dimensions given.

716

Ducts for Engineering Services – Below Ground

Before installing ducts for the entry of services into a building, it is essential to ascertain the location of pipes and cables provided by the public utilities companies. Thereafter, the shortest, most practicable and most economic route can be planned. For flexible pipes and cables, a purpose-made plastic pipe duct and bend is used to protect the service at its point of entering a building. Ducts are sealed at the ends with a plastic filler and sealant; otherwise subsoil, ground water and possibly vermin will access the duct. If the duct is blocked, its effectiveness to absorb differential settlement between the building and incoming service will be impaired. The details below show provision for water supply. For gas and electricity intakes, see pages 489 and 550, respectively.

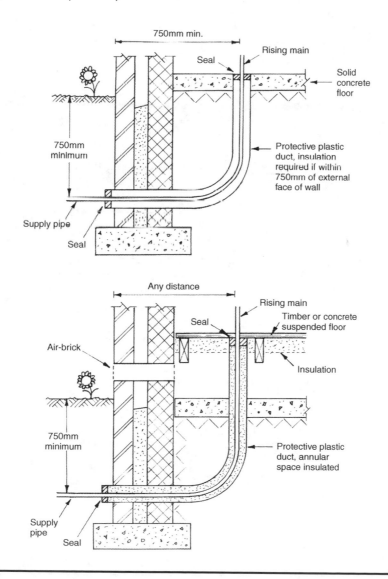

Ducts for Engineering Services – Above Ground

Water, gas and electrical services can be easily accommodated horizontally below timber and other wood product boarding by securing pipes and cables to the side of support joists. With solid flooring, a skirting or floor duct may be used. These are purpose made by the site joiner or purchased as standard manufactured items to be secured to the room periphery or set in concrete floors. Vertical services may be housed in a surface duct or a recess chased into the wall. A chase is only acceptable if its depth does not affect the structural strength of the wall. Also, depending on location, the potential to reduce the wall's thermal and sound insulation properties must be considered. Water installations and fittings must not be embedded in a wall or floor. They can be placed in ducts that are reasonably accessible by the removal of superficial finishes such as carpeting and tiling. Gas pipework should also be in purpose-made ducts, at least 25mm from any other service and 50mm from an electrical supply.

Timber suspended floor

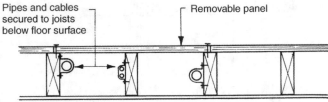

Solid concrete floor

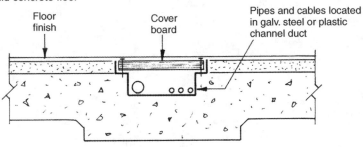

Purpose-made skirting duct

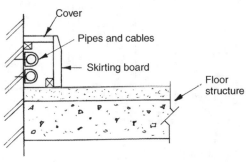

Services installations may be concealed within the structure by an access board or panel. The structure and its components should not be damaged to accommodate services but some nominal holing and notching will be unavoidable and is acceptable. Wherever possible, pipes and cables should run parallel and be secured to the sides of joists. Where services are at right-angles to joists the optimum location is through a hole in the joist centre or neutral axis. This is where compressive and tensile stresses are minimal. Holing is convenient for cables and flexible pipes, but notching the tops of joists is the only practical means for accommodating rigid pipes. Notching will reduce the strength of joists; therefore, where services are apparent, the structural designer should be informed and the joists oversized accordingly.

Restrictions and guidance – the principal areas to avoid notching and holing of joists are mid-span (maximum bending) and close to supports (maximum shear).

- Notches not greater than 0·125 × joist depth.
- Notches located between 0·07 and 0·25 times the span, from support.
- Hole diameter, maximum of 0·25 × joist depth.
- Holes a minimum of 3 × diameter apart.
- Holes located between 0·25 and 0·40 times the span, from support.

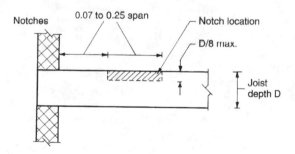

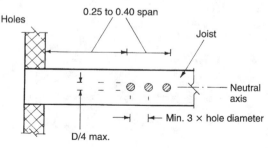

Recommended location of notches and holes in joists

Notching and Holing Fabricated Timber Joists

Manufactured timber joists/beams are frequently used in house construction as an economical alternative to standard timber sections. Notches should be avoided, but holing is acceptable as shown by the following guidance —

Laminated veneer beam —

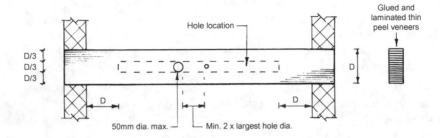

Parallel strand beam —

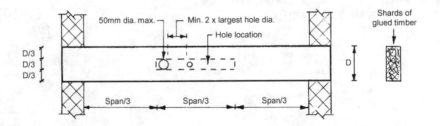

Engineered I beam —

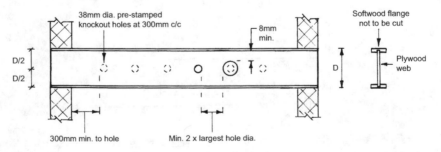

Max. hole diameter depends on joist depth – see manufacturer's data.
Square holes to have rounded corners.

A grid distribution of floor ducting is appropriate in open plan offices and shops where there is an absence of internal walls for power and telecommunications sockets. It is also useful in offices designed with demountable partitioning where room layout is subject to changes. Sockets are surface mounted in the floor with a hinged cover plate to protect them when not in use. The disruption to the structure is minimal as the ducts can be set in the screed, eliminating the need for long lengths of trailing cables to remote workstations. For partitioned rooms, a branching duct layout may be preferred. The branches can terminate at sockets near to the wall or extend into wall sockets. Where power supplies run parallel with telecommunications cables in shared ducts, the services must be segregated and clearly defined. For some buildings, proprietary metal, plastic or laminated plywood skirting ducts may be used. These usually have socket outlets at fixed intervals.

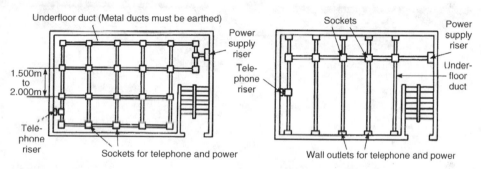

Grid layout floor duct

Branching layout floor duct

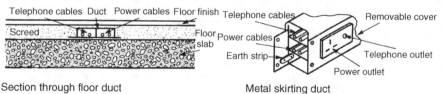

Section through floor duct

Metal skirting duct

See also page 718.

721

Medium and Large Vertical Ducts

The purpose of a service duct is to conceal the services without restricting access for inspection, repair and alterations. A duct also helps to reduce noise and protect the services from damage. When designing a service duct, the transmission of noise, possible build-up of heat in the enclosure and accessibility to the services must be considered. The number of ducts required will depend on the variation in services, the need for segregation and location of equipment served. Vertical ducts usually extend the full height of a building which is an important factor when considering the potential for spread of fire. The duct must be constructed as a protected shaft and form a complete barrier to fire between the different compartments it passes. This will require construction of at least 60 minutes' fire resistance with access doors at least half the structural fire resistance.

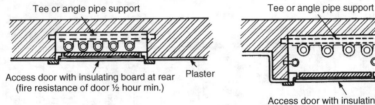

Tee or angle pipe support

Access door with insulating board at rear (fire resistance of door ½ hour min.)   Plaster

Recessed for medium-sized pipes and cables

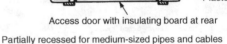

Tee or angle pipe support

Access door with insulating board at rear   Plaster

Partially recessed for medium-sized pipes and cables

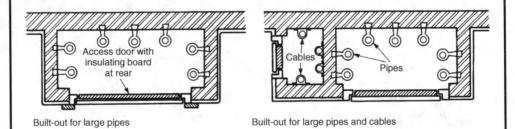

Access door with insulating board at rear

Built-out for large pipes

Cables   Pipes

Built-out for large pipes and cables

Refs. BS 8313: Code of practice for accommodation of building services in ducts.

Building Regulations, Approved Document B3: Internal fire spread (structure).

Floor trenches are usually fitted with continuous covers. Crawl-ways generally have access covers of minimum 600mm dimension, provided at convenient intervals. A crawl-way should be wide enough to allow a clear working space of at least 700mm and have a minimum headroom of at least 1m. Continuous trench covers may be of timber, stone, reinforced concrete, metal or a metal tray filled to match the floor finish. The covers should be light enough to be raised by one person, or, at most, two. Sockets for lifting handles should be incorporated in the covers. In external situations, the cover slabs (usually of stone or concrete) can be bedded and joined together with a weak cement mortar. If timber or similar covers are used to match a floor finish, they should be fixed with brass cups and countersunk brass screws. A trench has an internal depth of less than 1m. In internal situations where ducts cross the line of fire compartment walls, a fire barrier must be provided within the void and the services must be suitably fire stopped (see pages 476 and 725).

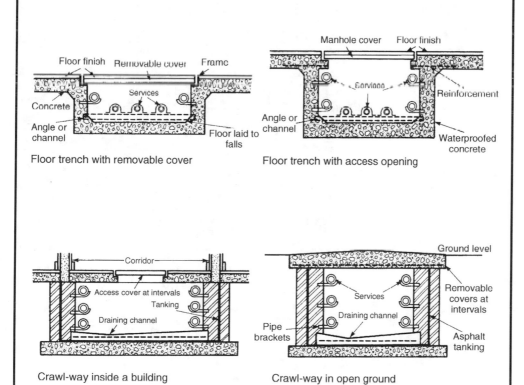

Floor trench with removable cover

Floor trench with access opening

Crawl-way inside a building

Crawl-way in open ground

Subways or Walkways

Access to a subway will normally be from a plant room, control room or a basement. Additional access from the surface should also be provided at convenient junctions and direction changes. See page 341 for provision of wall step irons. The design and construction of these ducts should adequately withstand the imposed loads and pressures that will occur under extreme working conditions. They should be watertight and where used internally have adequate resistance to fire. Ducts housing boiler or control room services must be provided with a self-closing fire door at the entry. Ventilation to atmosphere is essential, and a shallow drainage channel should convey ground water leakage and pipe drainage residue to a pumped sump or a gully connection to a drain.

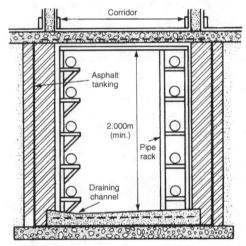

Subway inside a building

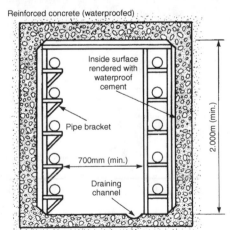

Subway in open ground

Note: Lighting may be provided operated at 110V.

The effect of fire spreading through the voids associated with internal pipework penetrating fire-resistant walls and floors can be considered in four areas:

1. Addition of fuel to the total fire load.
2. Production of toxic gases and smoke.
3. Risk of fire spread along the pipework.
4. Reduction in fire resistance of the building elements penetrated.

Guidance in Approved Document B3 to the Building Regulations is mostly applied to sanitation pipework penetrating the structure, but could affect other services, particularly in large buildings. Acceptable sleeving and sealing methods for uPVC discharge pipes are shown on page 476. Non-combustible pipe materials up to 160mm nominal i.d. (excluding lead, aluminium, aluminium alloys, uPVC and fibre cement) may have the structural opening around the pipe fire stopped with cement mortar, gypsum plaster or other acceptable non-combustible material. Where the pipe material is one of those listed in parentheses, and it penetrates a wall separating dwellings or a compartment wall or floor between flats, the discharge stack is limited to 160mm nominal i.d. and branch pipes limited to 110mm nominal i.d., provided the system they are part of is enclosed as shown.

* Any other materials (e.g. polypropylene) have a maximum nominal i.d. of 40mm.

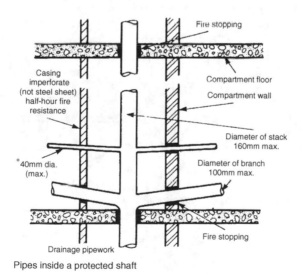

Pipes inside a protected shaft

Ref. Building Regulations, Approved Document B3: Internal fire spread (structure).

Raised Access Floors

Raised flooring provides discrete housing for the huge volumes of data and telecommunications cabling, electrical power cables, pipes, ventilation ducts and other services associated with modern buildings. Proprietary raised floors use standard 600mm square interchangeable decking panels, suspended from each corner on adjustable pedestals. These are produced in a variety of heights to suit individual applications, but most range between 100mm and 600mm. Panels are generally produced from wood particle board and have a galvanised steel casing or overwrap to enhance strength and provide fire resistance. Applied finishes vary to suit application, e.g. carpet, wood veneer, vinyl, etc. Pedestals are screw-threaded steel or polypropylene legs, connected to a panel support plate and a base plate. The void between structural floor and raised panels will require fire stopping at specific intervals to retain the integrity of compartmentation.

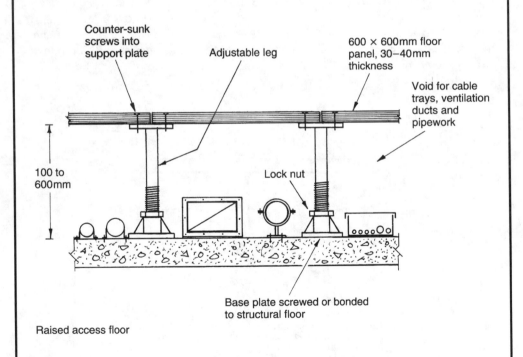

Counter-sunk screws into support plate

Adjustable leg

600 × 600mm floor panel, 30–40mm thickness

Void for cable trays, ventilation ducts and pipework

100 to 600mm

Lock nut

Base plate screwed or bonded to structural floor

Raised access floor

Ref. BS EN 12825: Raised access floors.

Building Regulations, Approved Document B: Fire safety, Vol. 2, Part B3, Section 9: Cavities.

A suspended ceiling contributes to the fire resistance of a structural floor. The extent of contribution can be determined by reference to Appendix A in Approved Document B of the Building Regulations. An additional purpose for a suspended ceiling is to accommodate and conceal building services, which is primarily the function of a false ceiling.

False ceiling systems may be constructed *in situ* from timber or metal framing. A grid or lattice support system is produced to accommodate loose-fit ceiling tiles of plasterboard, particle board or composites. Proprietary systems have also become established. These are a specialised product, usually provided by the manufacturer on a design and installation basis. Most comprise a simple metal framing with interconnecting panel trays. As with raised flooring, the possibility of fire spreading through the void must be prevented. Fire stopping is necessary at appropriate intervals as determined in Approved Document B3 to the Building Regulations.

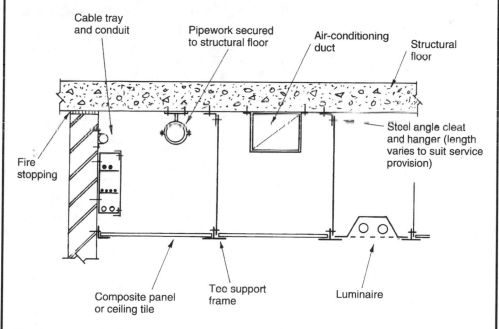

Simple suspended ceiling

Refs. BS EN 13964: Suspended ceilings. Requirements and test methods. Building Regulations, Approved Document B: Fire safety, Vol. 2, Part B3, Section 9: Cavities.

Accommodation of Flue Pipes

Chimney – structure enclosing a flue. Can be single-walled metal duct/pipe most suited to gas-burning appliances, or of twin-walled concentric metal with insulation between inner and outer liners for gas or oil fuels where flue temperature does not exceed 250°C. See pages 225 and 226 for solid fuels.

Flue – a void within a chimney that permits conveyance of fuel combustion products to the outside air.

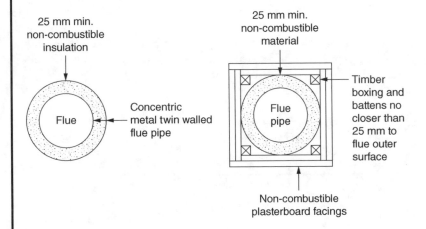

Protective measures – factory-made metal chimneys should be guarded if they could be a burn hazard to persons and/or adjacent combustible material, e.g. wood joists. Combustible material to be at least 25 mm distance from a hot flue, specific situations defined in BS EN 1856-1 and 2: Chimneys. Typically, where a flue passes through a floor or roof within the same dwelling. Where a flue penetrates a fire compartment or dwelling separating wall or floor, e.g. flats/apartments, Approved Document B: Fire Safety should be consulted to ascertain extent of fire resistant/fire stopping required. The 25 mm minimum is measured from the outer surface of a single-walled flue or the outer surface of the inner wall liner with a twin-walled concentric flue. Separation may be achieved using a non-combustible insulating sleeve.

See also page 514.

16 ALTERNATIVE AND RENEWABLE ENERGY

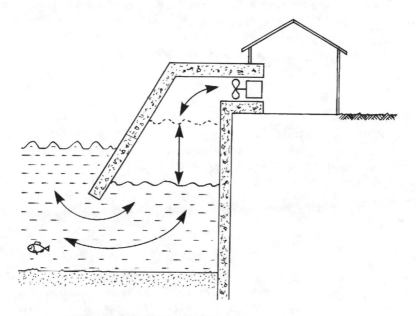

ENERGY PRODUCTION
ALTERNATIVE ENERGY
RENEWABLE ENERGY
ANAEROBIC DIGESTION
BIOGAS
WIND POWER
WIND POWER AND FUEL CELLS
WATER POWER
GEOTHERMAL POWER
MICRO-COMBINED HEAT AND POWER
SOLAR POWER
PHOTOVOLTAIC SYSTEMS
BIOMASS OR BIOFUEL
UNDERGROUND COAL GASIFICATION
CARBON CAPTURE AND STORAGE
HYDRAULIC FRACTURING
ANAEROBIC WASTE PROCESSING

DOI: 10.1201/9781003434894-16

Energy Production

Use of fuel as an energy resource is only cost-effective and efficient if the building in which it is used is constructed to a high standard of insulation to resist heat or cool energy losses through the fabric. It is also dependent on high-efficiency systems and equipment, e.g. mechanical ventilation with heat recovery, condensing boiler. A legislative strategy for energy efficiency in new buildings and alterations to existing buildings is established by Building Regulations, Part L – Conservation of fuel and power, Volumes 1 and 2.

Practical measures to ensure efficient use of fuel include double and triple glazing, airtightness and well-insulated external walls, floor and roof space. There are many construction techniques and procedures that can be used to apply these energy loss reductions through the external envelope. Numerous examples and applications are considered in some detail in the companion volume to this book, the *Building Construction Handbook*.

Where buildings are designed and constructed to limit fuel energy losses, consideration can also be given to on-site energy production as a viable alternative to reliance on conventional fossil fuels (coal, gas and oil). Alternatives are many and varied. On-site generation is dependent on geographical location, local climate, local utilities rates and availability of alternative fuels, e.g. biomass. Systems that qualify for government financial incentives such as 'feed-in tariff' are also an important factor.

Alternative energy – generally regarded as any type of usable energy that does not harm the environment, does not cause a decline in natural resouces and can be used as a replacement for fossil fuels, i.e. replaces fuels that have undesirable consequences when burnt.

Renewable energy – a natural, constantly replenished alternative energy resource including solar, wind, tidal, geothermal, hydro, biomass, biofuel and hydrogen.

Power-stations that burn conventional fossil fuels such as coal and oil, and to a lesser extent natural gas, are major contributors to global warming, production of greenhouse gases (including CO_2) and acid rain. Note: Acid rain occurs when the gaseous products of combustion from power-stations and large industrial plant combine with rainfall to produce airborne acids. These can travel hundreds of miles before having a devastating effect on forests, lakes and other natural environments. Current efforts to limit the amount of combustion gases in the atmosphere include

- CHP (cogeneration) and district heating systems (pages 171–173).
- Condensing boilers (page 102).
- Higher standards of thermal insulation of buildings (page 197 and Building Regulations, Approved Document L – Conservation of fuel and power).
- Energy management systems (pages 190–192).
- Recycling of waste products for renewable energy.

Renewable energy is effectively free fuel, but few of these installations exist in the UK. Other European states, particularly the Netherlands, Germany and Scandinavian countries, have waste segregation plant and selective burners as standard equipment at many power-stations. City domestic rubbish and farmers' soiled straw can be successfully blended with conventional fuels to power electricity generators and provide hot water for distribution in district heating mains. Small-scale waste-fired units from 60kW up to 8000kW are standard installations in many continental domestic and commercial premises, but are something of a rarity in this country.

Renewable and other alternative 'green' energy sources are also becoming viable. These include

- Wind power.
- Wind power and hydrogen powered fuel cells.
- Wave power.
- Geothermal power.
- Solar power.
- Biomass or biofuels.
- Anaerobic digestion and biogas.

Relative to 1990, the impact of power generation from 'green' sources in the UK has reduced CO_2 emissions by more than 30%. In 2009 the government introduced a Low Carbon Transition Plan committed to reducing carbon emissions by 80% by 2050. This has since been revised to bring all greenhouse gas emissions to a net zero target by 2050. See also pages 135 and 136.

Refs. The Energy Act.
The Climate Change Act.

Renewable Energy – Feed-in Tariff

Feed-in tariff – A UK government initiative that provides a financial incentive for users of low-carbon heating installations. In principle, households and communities can claim payment for producing electricity even if they consume it on their own premises. A payment can also be claimed for electricity exported to the market. This latter concept is known as Smart Export Guarantee scheme.

Eligible technologies –

Wind turbines up to 5MW.
Solar photovoltaic (PV) panels up to 5MW.
Hydro-power up to 5MW.
Anaerobic digestion up to 5MW.
Micro-combined heat and power (CHP) up to 50kW.

The following calculation is a theoretical example based on a solar panel system with a pay-back tariff of 4p/kWh. This figure is variable. See notes below.

Annual cost of electricity based on 3500kWh @ 30p/kWh = £1050.
Typical cost of a PV solar power installation = £8160.
Typical annual power produced by the PV system is 2000kWh.
With the power used being free (30p/kWh) plus the pay-back of 4p/kWh, a saving of 34p/kWh is achieved.
2000kWh × 34p = £680 annual income.
Pay-back time on the capital cost = £8160 ÷ £680 = 12 years.
Thereafter, the system is theoretically in profit. In reality, it is subject to energy price changes and SEG licensees terms of agreement, including contract duration.

Note 1: A figure for inflation is not included with this calculation, but pay-back tariffs may be updated in line with the Retail Price Index.

Note 2: The UK government regulator Ofgem (Office of gas and electricity markets) publish tables of tariff levels for different renewable energy installations (www.ofgem.gov.uk).

Note 3: Electricity cost at 30p/kWh is a typical figure. Some variation will be found between different suppliers.

Note 4: Pay-back tariffs are reviewed periodically by government; therefore, figures given here are for calculation guidance only.

Anaerobic digestion – an established technology that has been used for centuries as a process for producing methane gas.

Process – biomass products such as food waste, energy crops, crop residue and manure are compounded and stored in sealed containers. Here, in the absence of oxygen, naturally occurring micro-organisms digest the biomass and release methane gas that can be used as a fuel. After processing, there remains a residual solid waste. This by-product is rich in nutrients and can be used as a fertiliser. Timber biomass products cannot be processed in this way because the micro-organisms cannot break down the presence of lignin resin.

Biogas – composed mainly of methane (CH_4, approx. 60%) and carbon dioxide (CO_2, approx. 40%) with minor traces of other gases.

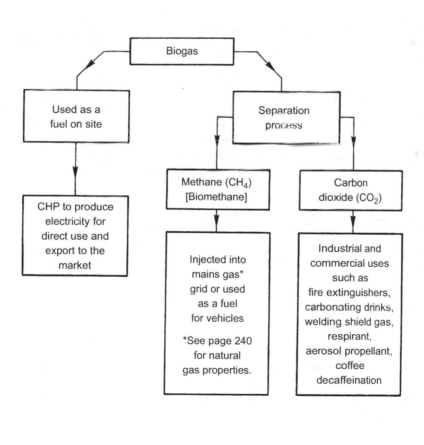

Anaerobic Digestion and Biogas Processing

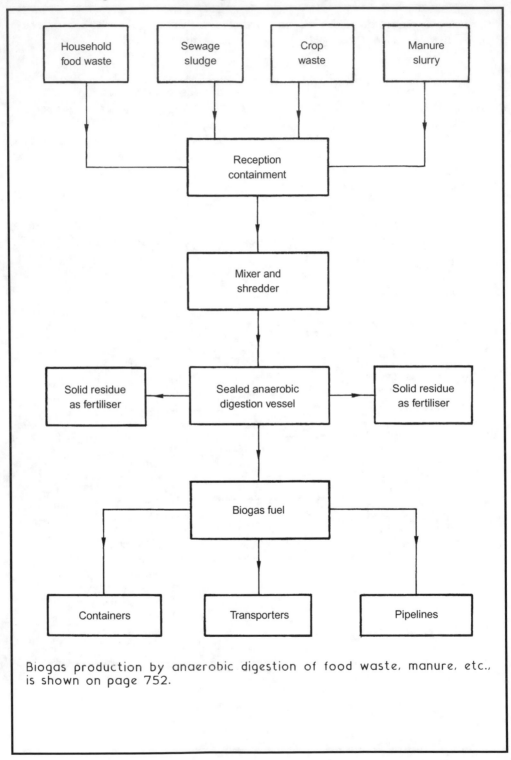

Biogas production by anaerobic digestion of food waste, manure, etc., is shown on page 752.

The development of wind power as an alternative energy source is well advanced. However, it is dependent on the fickle nature of the weather and can only be regarded as a supplementary energy source unless the surplus power produced is stored – see page 738.

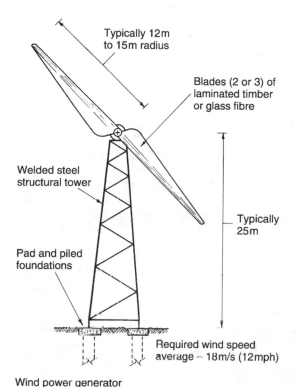

Typically 12m to 15m radius

Blades (2 or 3) of laminated timber or glass fibre

Welded steel structural tower

Typically 25m

Pad and piled foundations

Required wind speed average – 18m/s (12mph)

Wind power generator

The principle is simple enough. Wind drives a propeller, which rotates a shaft through a gearbox to drive an electricity generator. The generator produces direct current, similar in concept to a much smaller bicycle dynamo. Designs include two- and three-blade variants, elevated to between 25 and 45 metres from ground level to central axis. Blades are usually made from laminated timber or glass fibre and manufactured to tip diameters of between 6 and 60 metres (25 to 30m is typical). Electricity output is difficult to define, but claims are made of 300kW in a 25mph wind from one generator. This is enough electricity for about 250 houses. A wind farm of say 20 generators in an exposed location could produce 20GW of electricity an hour averaged over a year.

Environmental issues – no release of carbon, sulphur or nitrogen oxides, methane and other atmospheric pollutants. Conservation of finite fossil fuels. Aesthetically undesirable and noisy.

Costs – produces electricity for a minimal amount. Foundation costs are very high to anchor the units against lateral wind forces and dynamic forces during rotation. The capital cost of generators and their installation costs must be calculated against the long-term savings and environmental benefits. The purchase costs of wind turbines commence at about £1200 per kW of output, with a life expectancy of about 30 years. The smallest of units may take about a week to install.

Savings – estimates vary from speculative projections to realistic comparisons. A small generator such as that used at Wansbeck General Hospital, Northumberland can produce up to 450kW daily. On a greater scale, it is anticipated that by the year 2025, up to 20% of the UK's electrical energy requirements could be wind generated.

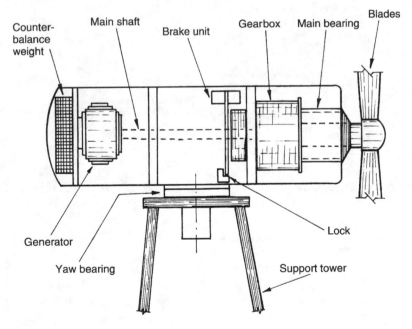

Main components of a wind turbine

Small wind powered electricity generators – wind powered micro-generators are used as a free energy source to charge batteries in yachts and navigation buoys. Larger variations with an output up to 15kW produce between 1000 and 50000kWh annually. These can be applied to individual buildings including houses, thereby benefiting from the UK government's feed-in tariff incentive (see page 732).

Advantages – application to off-grid remote areas offers the most benefits. Combined with solar panels, considerable fuel savings can be achieved compared with fuel oil powered generators. Individual small scale wind turbines can also be used in parallel with a grid supply, as shown on the next page. Lead-acid storage batteries receive a regulated DC charge. This is converted to 230 volts AC through an inverter transformer (see page 549).

Disadvantages –

- Efficiency – dependent on the fickle nature of the weather. Can be subject to sporadic output from wind turbulence when other buildings are in close proximity.
- Structural damage – movement vibrations will transfer into the support structure. Variable stresses can effect damage to an existing building's features and components. The original building design would not include measures to withstand the imposed loading from an aerodynamic device.
- Appearance – turbines need to be mounted as high as possible for maximum effectiveness. Therefore, not always perceived as an attractive feature to a dwelling house.
- Noise – aerodynamic sound from the rotating blades combines with mechanical noise from the generator gearing. Vibrations of operation may also produce sound.
- Flicker – caused by the sun shining through the rotating blades. This can be a nuisance as it effects a flickering shadow.
- Cost – capital outlay has to be balanced against pay back in terms of free electricity over time and potential gain from the feed-in tariff scheme.
- Wild life – unlikely to be affected, but in some situations, turbine installation could be prohibited where bird and bat habitats exist.
- Planning regulations – permission is nearly always required. It will depend on the size of unit, where mounted and potential to affect adjacent premises and people.

Wind Power and Fuel Cells

Wind is limited as a source of electrical power because of the unreliable nature of the weather. To use the potential of the wind effectively, it is necessary to store the energy generated when the wind blows and release it in response to demand.

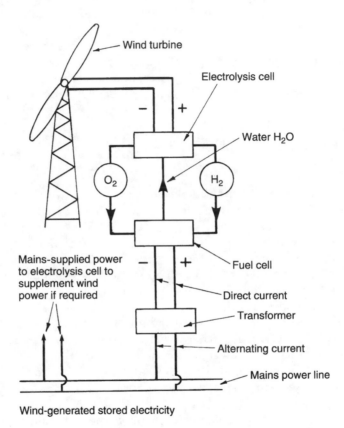

Wind-generated stored electricity

Instead of using the wind-generated electricity directly, it is used to electrolytically decompose water. This means separation of the hydrogen and the oxygen in water into different storage vessels. The stored hydrogen and oxygen are supplied to a fuel cell or battery in regulated amounts to produce a direct current. As the two gases combine they give water, which is returned to the electrolysis cell for reprocessing. Direct current is transformed to alternating current for compatibility with electricity distribution power lines.

The energy potential in differing water levels has been exploited for centuries through watermills and subsequently hydroelectric power. Another application is to build tidal barrages across major estuaries such as the Severn or Mersey. As the tide rises, the water would be impounded, to be released back as the tide recedes, using the head- or water-level differential as a power source. This has been used to good effect since the 1960s at La Rance near St Malo in France.

Another application uses a series of floats moored in the sea to generate an electrical potential as each float moves with the waves. Attempts have also been made to use the floats to rotate a crankshaft. There are limitations with this, not least the obstruction it creates in the sea.

Power potential from waves can also be harnessed by using their movement to compress air in shoreline chambers. Air pressure built up by the wave oscillations is used to propel an air turbine/electricity generator.

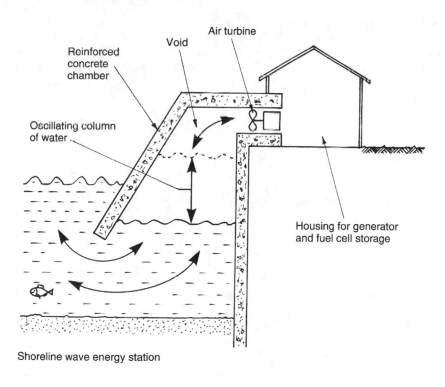

Shoreline wave energy station

Water Power – Tidal Lagoon

The power potential of water trapped within a lagoon is achieved by a difference in water level between open sea and that within a separating barrier or breakwater.

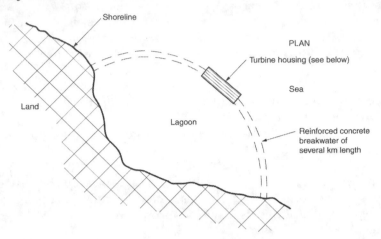

Housed mid-distance along the breakwater is sited a bank of propeller driven turbines. These rotate in response to the head differential between varying tidal sea levels and the level of water retained in the lagoon. Within the lagoon, whether water is relatively higher or lower than adjacent sea level, gravitational flow will propel the turbines to generate electricity.

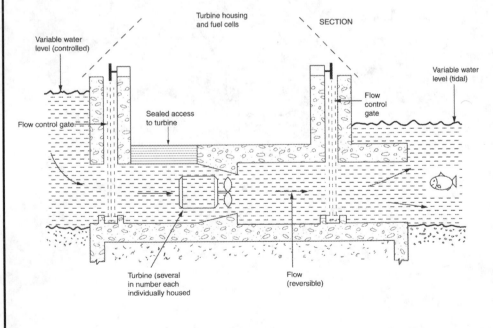

This is otherwise known as 'hot-dry-rock' technology, a name which gives some indication of the energy source. Heat energy is produced by boring two or more holes into the granite fissures found at depths up to 4·5 miles (7·2km) below the earth's surface. Cold water pumped down one borehole and into the fissures converts into hot water or steam which is extracted from the other borehole(s). The hot water can then be used directly for heating or it can be reprocessed into steam to drive turbines and electricity generators on the surface.

Enormous quantities of heat are believed to exist in underground rock formations throughout the world. New Zealand and Iceland are well known for having hot volcanic springs and established use of naturally occurring hot water from geysers. In the UK, there are a few isolated examples of spas, but the greatest potential lies below the impermeable granite substrata in the south-west corner of England. This concentrates in Cornwall and ranges up to Dartmoor and the Scilly Isles. Geological surveys suggest that the heat energy potential here is twice that elsewhere in the UK. Since the 1970s, the centre of research has been at Rosemanowes Quarry, near Falmouth. Indications from this and other lesser sites in the locality are that there may be enough geothermal energy in the West Country to provide up to 20% of the UK's electricity needs. Exploration by boreholes into aquifers in other parts of the country has met with some success. In Marchwood, Southampton, water at over 70°C has been found at depths of less than 2km. However, this resource was found to be limited and not cost-effective for long-term energy needs (see next page).

Exploitation of hot water from naturally occurring springs is not new. All over the world there are examples of spas which are known to have been enjoyed since Roman times. More recently in the early 1900s, a natural source of steam was used to generate electricity in Italy. Now it is very much a political and economic decision as to whether it is cost-effective to expend millions of pounds exploiting this possibly limited source of heat energy.

Geothermal Power – Installation

Location – during the 1970s and early 1980s, site boreholes were sunk at the Marchwood power-station site on Southampton Water and in Southampton centre.

Results – the second borehole near the city shopping centre provided greatest potential, with a water temperature of 76°C at 1800 metres.

Initial outcome – the Department of Energy considered the resource of limited economic value to make a significant contribution nationally.

Later outcome – Southampton City Council took the initiative to form a partnership with Utilicom, a French-owned energy management company, to develop a local district heating scheme. Utilicom's parent companies, IDEX and STREC, had considerable experience in operating geothermal district heating systems around Paris. In 1986, Utilicom and Southampton City Council formed the Southampton Geothermal Heating Company (SGHC).

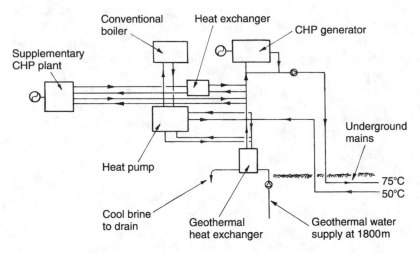

Schematic combined geothermal, CHP and conventional boiler district heating system

Energy use – the geothermal resource provides about 20%, with fuel oil and natural gas approximately 10% and 70%, respectively. A chilled water facility is also provided by the heat pump.

Clients – mainly corporations and commercial premises, although some housing estates and apartment blocks are included.

Commendation – received The Queen's Award for Enterprise: Sustainable Development 2001.

Micro-Combined Heat and Power (CHP)

Micro-CHP or microgeneration – an electricity generator at the point of use, independent of a mains grid supply of electricity, combined with a water heater. A micro-CHP unit is a gas-fired engine used to produce electricity with the heat energy used for domestic hot water requirements. Comparisons can be made with the power unit in a motor vehicle, except that the fuel energy conversion objective differs and any CHP function is secondary and uneconomical.

Function – fuelled by mains gas, although other fuel options are being developed. Comprises in one unit a condensing boiler to heat water and a Stirling engine to produce electricity. The Stirling engine is old technology, its invention dating back some 200 years. Motion occurs in response to combustion of gas to heat water, as explained:

- Helium in a sealed compartment is warmed as the gas burner heats water.
- Expansion of helium pushes down a magnetic piston.
- Cool water in the boiler absorbs the heat, allowing the helium to contract and the piston to rise.
- Heated water circulates through a heat exchanger to be replaced by cooler water in the return circuit.
- The magnetic piston moves up and down at 50 cycles/sec. between a generator coil, producing electricity by electromagnetic induction.
- For every 6kW of thermal energy produced, about 1kW of electricity is generated.
- Units are approximately 90% efficient and use 35% less primary energy, as the waste heat is used effectively and there are no transmission losses.

Operating principle –

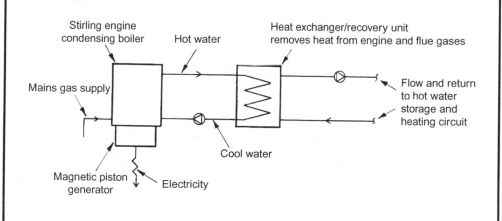

743

The potential of solar energy as an alternative fuel is underrated in the UK. It is generally perceived as dependent solely on hot, sunny weather to be effective. In fact it can be successfully used on cloudy days, as it is both the direct and diffused solar irradiation which is effective. The average amount of solar irradiation falling on a south-facing inclined roof is shown to vary between about 900 and 1300 kWh/m² per year depending on the location in the UK.

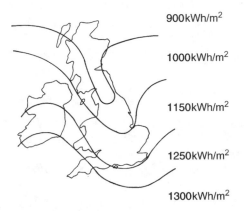

900kWh/m²

1000kWh/m²

1150kWh/m²

1250kWh/m²

Note: 1kWh = 3.6 MJ.

1300kWh/m²

Solar irradiation averaged over a year
for a 30° pitched roof facing south

The reluctance to accept solar panels in this country is understandable. The capital outlay is quite high and even though it is possible to achieve up to 40% of the average household's hot water requirements from solar energy, the pay-back period may be in excess of 10 years. It could also be argued that the panels are visually unattractive. The typical installation is shown on page 120. It has a flat plate 'black radiator' solar panel to absorb solar energy in water, which is transferred for storage in an insulated cylinder. From here it supplements hot water from a conventional boiler source. This application is also suitable for heating swimming pools.

An improvement uses collectors inside clear glass vacuum cylinders. These 'evacuated tube collectors' are capable of absorbing more heat at low levels of light. Other types of solar panel which can be used to power batteries or fuel cells include the photovoltaic system. This uses expensive crystalline silicon as a power generator. A less expensive alternative is amorphous silicon. Although less efficient, it is still capable of providing a trickle feed to batteries.

The flat plate 'black radiator' solar panel referred to on the previous page is not limited to roof-top applications. Any reasonably large, flat black surface can be effective. For example, asphalted road surfaces are very effective solar energy collectors. With piped water circuits installed close to the road surface, summer heat transfer to the sub-surface coils can be pumped through heat exchangers in adjacent buildings to provide hot water in storage. In addition, if the geology permits, the hot water generated at the surface can be pumped deep into the ground through heat exchangers located in an aquifer, thereby creating a heat store for winter use.

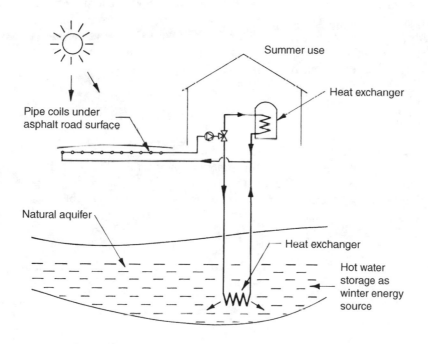

Photovoltaic (PV) cells use light as a source of energy. A small-scale application is to handheld calculators with an integral PV window as the power source instead of a conventional dry cell battery. On the larger scale and as a viable means for producing electrical energy in buildings, PV cells are arranged into a large array of panels that can be located on the roof slope. With sufficient output, surplus electricity can be stored for use during periods of limited or no light and may also be traded with the grid supply.

Principle – requires only daylight not direct sunlight to generate electricity. Output varies with the intensity of light. A PV cell processes natural light into electrical energy through the intermediary of a semiconductor. Suitable semiconductors include amorphous silicon, gallium arsenide, copper indium diselenide and cadmium telluride. Crystalline silicon is generally regarded as the most cost-effective. Light received by the cell produces an electric field over its layers to generate a direct current of about 12 volts.

Cell function – a PV cell comprises two thin layers, one with a positive charge and the other a negative charge. Light hitting the cell energises electrons that move towards the layer faces to produce an electrical imbalance between the layers, as shown in the diagram below.

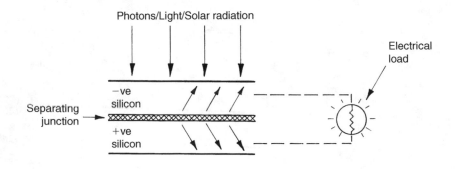

Potential output in the UK for a typical south-facing roof-top panel of 10–15m² can be about 750kWh of electrical energy. This is approximately one-quarter of the annual requirements for a typical three- to four-bedroom family house.

Systems may be grid connected or independent:

• Grid connected – at times when only a limited amount of electrical energy is required, for example, during a factory closure for maintenance, surplus energy from a PV installation can be used to supplement and be traded with the general supply from the national grid. Conversely, the grid can supplement the limitations of a PV system, particularly at night when there is no natural light source to activate the cells.

• Independent – suitable for use with isolated buildings detached from the grid. Rechargeable solar batteries will be required for storing electrical energy for use when the PV system is inactive, i.e. at night. This can be particularly beneficial in agricultural and farm buildings. Smaller applications include traffic information boards, advertising and car-park displays, navigation buoys and the many situations applicable to developing parts of the world that are without a conventionally generated mains supply.

Principle of PV installation –

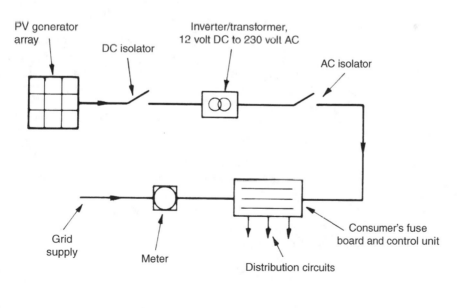

Biomass or Biofuel

Biomass is current terminology for the combustion of traditional fuels such as wood, straw and cow dung. The difference is that today we have the facility to process and clean the waste products. Gas scrubbers and electrostatic precipitators can be installed in the flues to minimise atmospheric pollution. Intensive farming methods produce large quantities of potentially harmful residues, including straw and chicken droppings. The latter combines with wood shavings and straw from the coops. Instead of burning these as waste, they can be reprocessed. A pioneer scheme at Eye in Suffolk burns the waste in a 10MW steam turbine electricity generator and sells the ash as an environmentally friendly fertiliser. This has the additional benefits of:

- Eliminating the traditional unregulated burning of farm waste which contaminates the atmosphere with carbon dioxide.

- Destroying the harmful nitrates which could otherwise be released into the soil.

- Destroying the potential for methane generation from decomposition. When this is released into the atmosphere, it is far more active than carbon dioxide as a greenhouse gas.

Farm wastes can also be used to produce methane gas for commercial uses. The waste is processed in a controlled environment in large tanks called digesters. The gas is siphoned off and used for fuel, while the remains are bagged for fertiliser.

The potential for forest farming wood as a fuel for power generation is also gaining interest. Trees naturally clean the atmosphere by absorbing carbon dioxide. However, when they die, they rot, releasing as much carbon dioxide as is absorbed during growth and a significant amount of methane. By controlled burning, the carbon dioxide is emitted, but the gains are destruction of the methane and an economical, sustainable fuel supply.

For applications, see pages 220-222.

UCG is not a new concept as shown by references to William Siemen's research from the mid-nineteenth century. The earliest recorded experimental work is that undertaken by the Scot, William Ramsay, during the early years of the twentieth century. Since then, development has been limited to periodic investigations at various sites throughout the world. Concentrated efforts have been cost restrained and at times curtailed when new finds of natural gas and oil resources have reduced the importance. However, that situation cannot be sustained, especially with the trend for increasing energy demands for industrial and commercial needs and from population expansion. By 2030, it is estimated that the world will need 50% more energy than that required in 2010. Therefore, the urgency for alternative fuel resources indicates that UCG is a viable development.

Principle of UCG – to convert unworked coal into a combustible gas that is processed to release CO_2 and to create a source of clean energy. Coal waste/ash remains underground.

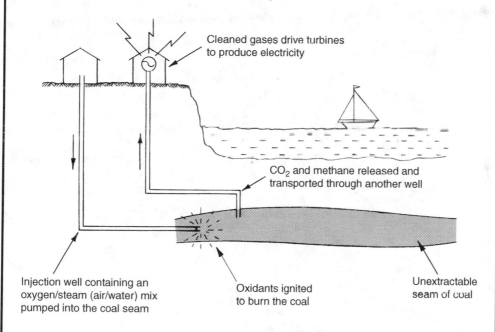

Cleaned gases drive turbines to produce electricity

CO_2 and methane released and transported through another well

Injection well containing an oxygen/steam (air/water) mix pumped into the coal seam

Oxidants ignited to burn the coal

Unextractable seam of coal

The most important part of the process is to remove the CO_2 from the generated gases. This is known as carbon capture and storage. The technology exists and is being developed.

From the perspective of the UK, estimates of coal resources vary but most agree that no more than 25% has ever been extracted. The remainder is not cost viable to remove but could hold considerable potential for UCG.

Carbon Capture and Storage (CCS)

Combustion of fossil fuels releases excessive amounts of carbon dioxide into the atmosphere. From power stations, the potential damage to the atmosphere is unacceptable. Considerable research has been undertaken to develop methods that control the CO_2 by cleaning and containment.

CCS commences by blasting the fuel with oxygen and steam at a temperature of 700°C. This is undertaken in a 'gasifier' to produce carbon monoxide and water, which is then converted to carbon dioxide and hydrogen. The CO_2 is cooled until it becomes liquid. Thereafter it is pumped into underground storage voids that remain from previous extraction of natural gas and oil below the sea. Inland depleted coal-mines have also been considered as suitable storage spaces.

The CO_2 would be stored indefinitely. A controversial issue, as there is no certainty that the gas would not seep out, even from hundreds of metres below the surface.

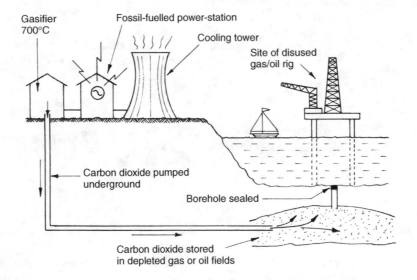

A variation of CCS processes the combustion products by 'scrubbing'. This extracts and separates the CO_2 from the other flue gases before they are released into the atmosphere. Thereafter the CO_2 is transported and stored as previously described.

Induced hydraulic fracturing – otherwise known as hydrofracking or just simply, fracking. Not alternative energy, but an alternative means for extraction of oil and gas contained in rock formations hundreds of metres below the surface. It originated from experimental work in the late 1940s, but diminishing accessible fuel resources have promoted much greater use since the 1990s.

Process – a deep well bored into rock formations that contain hydrocarbon fuel resources such as coal and shale gas. At selected depths which can be some 2000 metres, the bore continues horizontally. Pressurised fluid is injected into the horizontal bore to crack the rock formation. The fluid contains sand or ceramic granular proppants that remain to hold open the cracks. Pressurisation ceases and a mix of fluid and gases flow into the well for extraction.

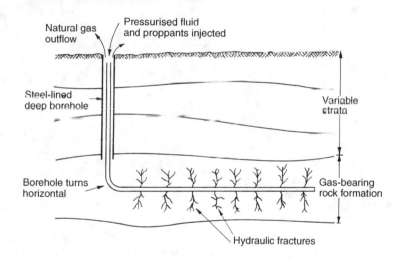

Environmental concerns –
• Possibility of contaminating ground water.
• Surface migration of gases, oils and fracturing chemicals, leading to contamination of the atmosphere, surface water and soil.
• Earth tremors (minor recordings occurred near Blackpool in 2011).

Anaerobic Digester – Biogas Processing

Biogas – approximately 60–70% methane (CH_4). Methane is the primary constituent of natural gas. Biogas also contains small quantities of carbon dioxide (CO_2) and hydrogen sulphide (H_2S) plus a minor presence of trace gases and water vapour.

Low-value constituents can be removed by filtration, the remaining gas being similar in composition with natural gas – see page 242. The compatibility of biogas makes it viable as a supplement to mains gas supplies. It can also be used as a vehicle fuel and for other energy products such as electricity generation.

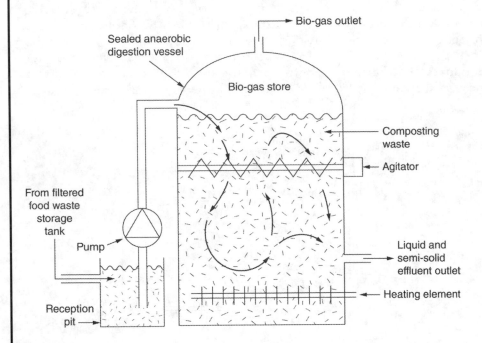

Biogas recovery units can be singular as shown in the illustrative principle above. These can be used where food waste, composted material and manure are only available in relatively small volumes. Variants covering large areas, sometimes known as big gas farms, can maximise the cost-effectiveness by economy of scale, providing a huge contribution to renewable energy. Once the capital cost of installation is expended, the financial returns are potentially high relative to maintenance and running costs.

17 APPENDICES

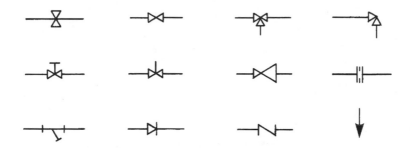

Appendix 1 – Glossary of Common Abbreviations (1)

BBA – British Board of Agrément. The function of the BBA is to assess, test and establish the quality of new products and innovations not represented by existing British (BSI) or European (CEN) Standards.

BRE – Building Research Establishment. Critically examines products and materials applicable to construction and issues certificates of conformity. Publishes research digests, good practice guides and information papers.

BS – British Standard. Publications issued by the British Standards Institution as support documents and recommendations for minimum practice and product manufacturing standards. Materials and components which comply are kitemarked:

BS EN – A British Standard which is harmonised with the European Standards body, CEN.

C€ – Communauté Européenne (European Community). This is a product mark which indicates presumption of conformity with the minimum legal requirements of the Construction Product Regulations 2011. Compliance is manufacture to a British Standard, a harmonised European Standard or a European Technical Approval (ETA).

CEN – Comité Européen de Normalisation. European standardisation body recognised by the European Commission (EC) for harmonising standards of product manufacturers in support of the CPR. Membership of CEN is composed of the standardisation bodies of the participating members of the European Union (EU) and the European Free Trade Association (EFTA). The standardisation body representing the UK is the BSI.

CIRIA – Construction Industry Research and Information Association. An independent research organisation which addresses all key aspects of construction business practice. Its operating principles are on a 'not-for-profit' basis for the benefit of industry and public good.

754

CPR – Construction Products Regulation. An EU regulator of harmonised conditions for marketing construction products. It clarifies CE marking as a basis for construction products that comply with current design factors, not least health, safety and environmental protection. CPR is particularly applicable to manufacturers, distributors and importers.

EC – European Commission. The executive organisation of the European Union (EU).

EEA – European Economic Area. Includes the member states of the European Union and three of the four states of the European Free Trade Association (EFTA): Iceland, Norway and Liechtenstein (Switzerland excluded).

EOTA – European Organisation for Technical Approvals. Operates over the same area as CEN, complementing the work of this body by producing guidelines for new and innovative products.

ETA – European Technical Approval. A technical assessment of products which indicates suitability and fitness for use for the CPR. Authorised bodies working with ETA include the BBA and WIMLAS Ltd (now part of BRE Certification). These bodies also produce technical specifications against which product compliance can be measured for approval.

EU – European Union. A unification of states. Before 2004 comprising 15 countries: Austria, Belgium, Denmark, Finland, France, Germany, Greece, Ireland, Italy, Luxemburg, Netherlands, Portugal, Spain, Sweden and the UK. Thereafter, a further 13: Bulgaria, Croatia, Cyprus, Czech Republic, Estonia, Hungary, Latvia, Lithuania, Malta, Poland, Romania, Slovakia and Slovenia. In 2016, the UK elected to leave.

ISO – International Organization for Standardization. This authority issues standards which are appropriate throughout the world. Products are identified with a number following the prefix ISO. Some of these may be adopted by the CPR, e.g. BS EN ISO 5667: Water quality and BS EN ISO 10960: Rubber and plastic hoses.

UKAS – United Kingdom Accreditation Service. An independent certification body that may be used by manufacturers to test and assess the suitability of their material products. UKAS issue certificates to show that materials conform to the criteria required of a recognised document, appropriate for the intended product use and application.

WRC – Water Research Council. A specialist testing agency with its own established brand of approval.

Appendix 2 – Abbreviations for Pipework

Design and installation drawings would be cluttered with writing, if there were not a simple format for representing pipes, fittings and accessories with abbreviations or symbols. The British and European Standards authorities have produced several documents recommending specific notations and symbols for all types of pipework installations, e.g. BS 1553-1 and 2 (see Appendix 5). Nevertheless, many offices prefer to use their own established procedures and variations from the standards. The following indicate some of the established alphabetical representations that may be used on engineering services drawings.

Service	Abbreviation
Boiler feed water	BFW
Brine	B
Chilled water	CHW
Cold water main	MWS
Cold water down service	CWS
Cold water drinking	DWS
Cold water flushing	FWS
Cold water pressurised	PWS
Treated water	TW
Waste water	WW
Condensate	C
Compressed air	CA
Cooling water	CLW
Fire service extinguisher	FE
Fire service hydrant	FH
Fuel	F
Liquefied petroleum gas	LPG
Nitrous oxide gas	N_2O
Oxygen	O_2
Refrigerant gas	R_0
Low-pressure hot water heating	LPHW
Medium-pressure hot water heating	MPHW
High-pressure hot water heating	HPHW
Hot water (domestic)	HWS
Steam	S
Vacuum	V

Appendix 3 – Abbreviations for Pipework Components

Component	Abbreviation
Cold feed	CF
Cold water storage cistern	CWSC
Drain or draw off	DO
Expansion vessel	ExVl
Feed and expansion	F & E
Hot water storage cylinder	HWSC
Open vent	OV
Strainer	S
Tundish	T
Warning pipe/overflow	WP
Control valves –	
Air release	ARV
Air	AV
Anti-vacuum	AVV
Automatic air	AAV
Check	CV
Double-check	DCV
Drain tap	DT
Expansion	EV
Float	FV
Gule	GV
Lockshield	LSV
Non-return	NRV
Plug cock	PC
Pressure-reducing valve	PRV
Servicing	SgV
Stopcock or valve	SC or SV
Temperature and pressure relief	TPRV
Thermostatic radiator	TRV
Wheel valve	WV
Pipework positions and direction –	
Flow	F
Return	R
From above	FA
To above	TA
From below	FB
To below	TB
High level	HL
Low level	LL

Appendix 4 – Abbreviations Used for Drainage Systems

Component and service	Abbreviation
Drains –	
Foul water	FWD
Surface water	SWD
Sewers –	
Foul water	FWS
Surface water	SWS
Effluents –	
Foul water	FW
Radioactive water	RAW
Rainwater	RW
Surface water	SW
Means of access, etc. –	
Access cover	A/C
Backdrop	BD
Cleaning or rodding eye	CE or RE
Fresh air inlet	FAI
Half-round channel	HRC
Invert	INV
Manhole	MH
Rainwater head	RWH
Rainwater shoe	RWS
Shallow access chamber	SAC
Gullies –	
Access	AG
Back inlet	BIG
Grease trap	GT
Road	RG
Sealed	SG
Yard	YG

Appendix 5 – Abbreviations Used for Sanitation Systems

Component and service	Abbreviation
Pipes –	
Discharge pipe	DP
Rainwater pipe	RWP
Soil and ventilating pipe or stack	SVP or SVS
Vent pipe or stack	VP or VS
Waste pipe	WP
Sanitary fittings –	
Access cap	ac
Air admittance valve	aav
Bath	b
Bidet	bt
Drinking fountain	df
Flushing cistern	fc
Shower	sh
Sink	s
Urinal	u
Wash basin	wb
Water closet	wc
Materials –	
Acrylonitrile butadiene styrene	ABS
Cast iron	CI
Copper	Cu
Heavy duty polypropylene	HDPP
High-density polyethylene	HDPE
Medium-density polyethylene	MDPE
Modified unplasticised polyvinyl chloride	MUPVC
Polypropylene	PP
Unplasticised polyvinyl chloride	uPVC or PVCu

*   *   *   *   *   *   *   *   *   *   *   *

Further references for specific applications –

BS 1553: Specification for graphical symbols for general engineering.

      Part 1: Piping systems and plant.

      Part 2: Graphical symbols for power-generating plant.

BS 1635: Recommendations for graphic symbols and abbreviations for fire protection drawings.

BS EN ISO 6412-3: Technical product documentation. Simplified representation of pipelines.

BS EN 1861: Refrigerating systems and heat pumps. System flow diagrams and piping and instrument diagrams. Layout and symbols.

Appendix 6 – Graphical Symbols for Pipework

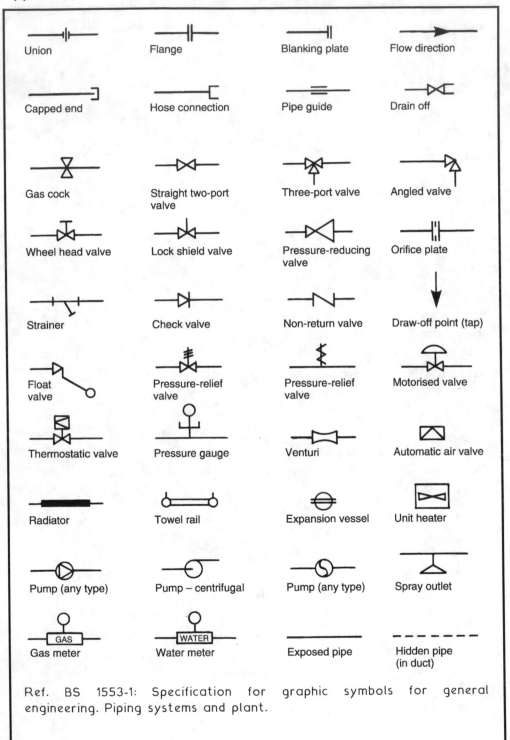

Union	Flange	Blanking plate	Flow direction
Capped end	Hose connection	Pipe guide	Drain off
Gas cock	Straight two-port valve	Three-port valve	Angled valve
Wheel head valve	Lock shield valve	Pressure-reducing valve	Orifice plate
Strainer	Check valve	Non-return valve	Draw-off point (tap)
Float valve	Pressure-relief valve	Pressure-relief valve	Motorised valve
Thermostatic valve	Pressure gauge	Venturi	Automatic air valve
Radiator	Towel rail	Expansion vessel	Unit heater
Pump (any type)	Pump – centrifugal	Pump (any type)	Spray outlet
Gas meter	Water meter	Exposed pipe	Hidden pipe (in duct)

Ref. BS 1553-1: Specification for graphic symbols for general engineering. Piping systems and plant.

Where a large quantity of piped services are deployed in boiler rooms, process plant service areas, etc., identification of specific services (e.g. compressed air, chilled water, etc.), can be very difficult and time consuming. The situation is not helped when installation drawings are lost or may not even have existed. In addition, modifications may have occurred since the original installation. This is made more difficult where a common pipe material such as galvanised steel is used for a variety of services.

The recommendations of BS 1710 have improved the situation considerably by providing a uniformly acceptable colour coding. This has also been endorsed by the Health & Safety (Safety Signs & Signals) Regulations which require visible markings on all pipework containing or transporting dangerous substances. Direction of flow arrows should also complement coloured markings. Colours can be applied by paint to BS 4800 schedules or with proprietory self-adhesive tape.

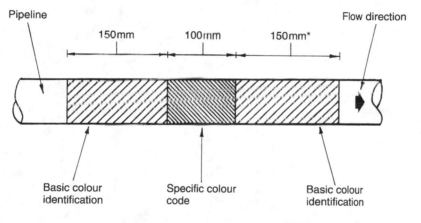

Pipeline | Flow direction

150mm | 100mm | 150mm*

Basic colour identification | Specific colour code | Basic colour identification

* Dimensions approximate

Refs. BS 1710: Specification for identification of pipelines and services.
BS 4800: Schedule of paint colours for building purposes.
Health & Safety (Safety Signs & Signals) Regulations 1996.

Appendix 7 – Identification of Pipework (2)

Contents	Basic i.d. colour	Specific colour	Basic i.d. colour
Water –			
Drinking	Green	Auxiliary blue	Green
Cooling (primary)	Green	White	Green
Boiler feed	Green	Crimson.White.Crimson	Green
Condensate	Green	Crimson.Emerald green. Crimson	Green
Chilled	Green	White.Emerald green. White	Green
Heating <100°C	Green	Blue.Crimson.Blue	Green
Heating >100°C	Green	Crimson.Blue.Crimson	Green
Cold down service	Green	White.Blue.White	Green
Hot water supply	Green	White.Crimson.White	Green
Hydraulic power	Green	Salmon pink	Green
Untreated	Green	Green	Green
Reclaimed	Green	Black	Green
Fire extinguishing	Green	Red	Green
Oils –			
Diesel fuel	Brown	White	Brown
Furnace fuel	Brown	Brown	Brown
Lubricating	Brown	Emerald green	Brown
Hydraulic power	Brown	Salmon pink	Brown
Transformer	Brown	Crimson	Brown
Refrigeration –			
Refrigerant 12	Yellow ochre	Blue	Yellow ochre
Refrigerant 22	Yellow ochre	Green	Yellow ochre
Refrigerant 502	Yellow ochre	Brown	Yellow ochre
Ammonia	Yellow ochre	Violet	Yellow ochre
Others	Yellow ochre	Emerald green	Yellow ochre
Other pipelines –			
Natural gas	Yellow ochre	Yellow	Yellow ochre
Compressed air	Light blue	Light blue	Light blue
Vacuum	Light blue	White	Light blue
Steam	Silver grey	Silver grey	Silver grey
Drainage	Black	Black	Black
Conduit/ducts	Orange	Orange	Orange
Acids/alkalis	Violet	Violet	Violet

Appendix 8 – Graphical Symbols for Electrical Installation Work

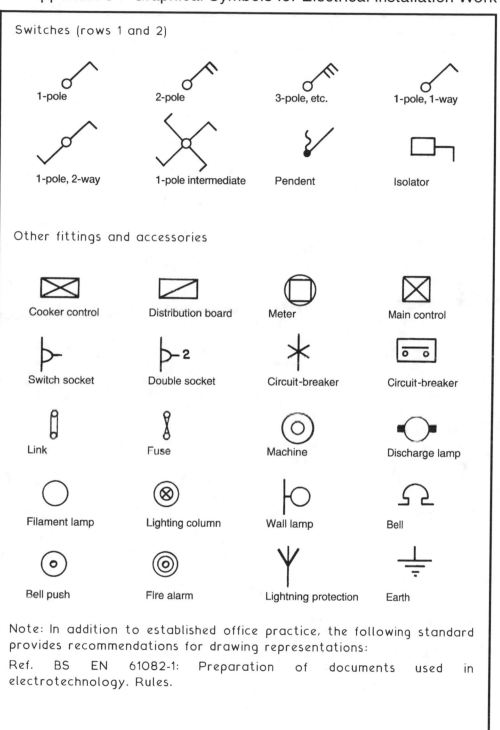

Switches (rows 1 and 2)

1-pole | 2-pole | 3-pole, etc. | 1-pole, 1-way

1-pole, 2-way | 1-pole intermediate | Pendent | Isolator

Other fittings and accessories

Cooker control | Distribution board | Meter | Main control

Switch socket | Double socket | Circuit-breaker | Circuit-breaker

Link | Fuse | Machine | Discharge lamp

Filament lamp | Lighting column | Wall lamp | Bell

Bell push | Fire alarm | Lightning protection | Earth

Note: In addition to established office practice, the following standard provides recommendations for drawing representations:

Ref. BS EN 61082-1: Preparation of documents used in electrotechnology. Rules.

Appendix 9 – Metric Units (1)

Metric measurements have been officially established in the UK since the Council of Ministers of the European Community met in 1971 to commit member countries to an International System of Units (SI). This has been endorsed by the International Organization for Standardization (ISO).

Basic or primary units:

Quantity	Unit	Symbol
Length	metre	m
Mass	kilogram	kg
Time	second	s
Electric current	ampere	A
Temperature	kelvin	K
Luminous intensity	candela	cd

Some commonly used supplementary and derived units:

Quantity	Unit	Symbol
Area	square metre	m^2
Volume	cubic metre	m^3
Velocity	metres per second	m/s
Acceleration	metres per second squared	m/s^2
Frequency	hertz (cycles per second)	Hz
Density	kilogram per cubic metre	kg/m^3
Force	newton	N
Moment of force	newton metre	N/m
Pressure	newton per square metre	N/m^2 (pascal – Pa)
Work, energy and heat	joule	J
Power, heat flow rate	watt	W (J/s)
Temperature – customary unit	degree Celsius	°C
Temperature – interval	degree kelvin	K

Note: degree Celsius and kelvin have the same temperature interval. Kelvin is absolute temperature with a zero factor equivalent to -273.15°C, i.e. 0°C = 273.15K.

Further derived units:

Quantity	Unit	Symbol
Density of heat flow	Watt per square metre	W/m^2
Thermal conductivity	Watt per metre degree	W/m K
Heat transfer ('U' value)	Watt per square metre degree	W/m^2 K
Heat capacity	Joule per degree	J/K
Specific heat capacity	Joule per kilogram degree	J/kg K
Entropy	Joule per degree	J/K
Specific entropy	Joule per kilogram degree	J/kg K
Specific energy	Joule per kilogram	J/kg

Derived units for electrical applications:

Quantity	Unit	Symbol
Electric charge	Coulomb	C (As)
Potential difference	Volt	V (W/A)
Electromotive force	Volt	V (W/A)
Electric field strength	Volt per metre	V/m
Electric resistance	Ohm	Ω (V/A)
Electric capacitance	Farad	F (As/V)
Magnetic flux	Weber	Wb (Vs)
Magnetic field strength	Ampere per metre	A/m
Inductance	Henry	H (Vs/A)
Luminous flux	lumen	lm
Luminance	candela per square metre	cd/m^2
Illuminance	lux (lumens per square metre)	lx (lm/m^2)

Appendix 9 – Metric Units (3)

Multiples and submultiples:

Factor	Unit	Name	Symbol
One billion (see note below)	10^{12}	Tera	T
One million million	10^{12}	Tera	T
One thousand million	10^{9}	Giga	G
One million	10^{6}	Mega	M
One thousand	10^{3}	kilo	k
One hundred	10^{2}	hecto	h
Ten	10^{1}	deca	da
One-tenth	10^{-1}	deci	d
One-hundreth	10^{-2}	centi	c
One-thousandth	10^{-3}	milli	m
One-millionth	10^{-6}	micro	µ
One-thousand millionth	10^{-9}	nano	n
One-million millionth	10^{-12}	pico	p
One-billionth	10^{-12}	pico	p
One-thousand billionth	10^{-15}	femto	f
One-trillionth	10^{-18}	atto	a

Common units for general use:

Quantity	Unit	Symbol
Time	minute	min
	hour	h
	day	d
Capacity	litre	l (1l = 1dm^3)
		(1000l = 1m^3)
Mass	tonne or kilogram	t (1t = 1000kg)
Area	hectare	ha (100m × 100m)
		(10000m^2)
Pressure	atmospheric	atm (1atm = 101·3kN/m^2)
Pressure	bar	b (1 bar = 100kN/m^2)

Note: Historically in British English, one billion equates to one million million (10^{12}). In American English, one billion has always equated to one thousand million (10^{9}). For consistency, one thousand million is now preferred in British English.

Appendix 10 – Water Pressure and Head – Comparison of Units

Head (metres)	Pressure kN/m² (kPa)	Pressure mbar or bar
1.00	9.81	98.7 mbar
1.02	10.00	100.00
2.00	19.61	196.13
2.04	20.00	200.00
3.00	29.42	294.19
3.06	30.00	300.00
4.00	39.23	392.26
4.08	40.00	400.00
5.00	49.03	490.33
5.10	50.00	500.00
6.00	58.84	588.39
6.12	60.00	600.00
7.00	68.65	686.46
7.14	70.00	700.00
8.00	78.45	784.53
8.16	80.00	800.00
9.00	88.26	882.59
9.18	90.00	900.00
10.00	98.07	980.66
10.20	100.00	1.00 bar
11.00	107.87	1.08
11.22	110.00	1.10
12.00	117.68	1.18
12.24	120.00	1.20
13.00	127.49	1.27
13.26	130.00	1.30
14.00	137.29	1.37
14.28	140.00	1.40
15.00	147.10	1.47
15.30	150.00	1.50
16.00	156.91	1.57
16.32	160.00	1.60
18.00	176.52	1.77
18.36	180.00	1.80
20.00	196.13	1.96
20.40	200.00	2.00
30.00	294.20	2.94
30.59	300.00	3.00
50.00	490.33	4.90
50.99	500.00	5.00
100.00	980.66	9.81
101.97	1.00 MN/m²	10.00
200.00	1.96	19.61
203.94	2.00	20.00

Appendix 11 – Conversion of Common Imperial Units to Metric (1)

Length
1 mile = 1·609km
1 yd = 0·914m
1 ft = 0·305m (305mm)

Area
1 sq. mile = 2·589km^2 or 258·9ha
1 acre = 4046·86m^2 or 0·404ha
1 yd^2 (square yard) = 0·836m^2
1 ft^2 (square foot) = 0·093m^2
1 in^2 (square inch) = 645·16mm^2

Volume
1 yd^3 (cubic yard) = 0·765m^3
1 ft^3 (cubic foot) = 0·028m^3
1 in^3 (cubic inch) = 16387mm^3 (16·387cm^3)

Capacity
1 gal = 4·546 l
1 qt = 1·137 l
1 pt = 0·568 l

Mass
1 ton = 1·016 tonne (1016kg)
1 cwt = 50·8kg
1 lb = 0·453kg
1 oz = 28·35g

Mass per unit area
1 lb/ft^2 = 4·882kg/m^2
1 lb/in^2 = 703kg/m^2

Mass flow rate
1 lb/s = 0·453kg/s

Volume flow rate
1 ft^3/s = 0·028m^3/s
1 gal/s = 4·546 l/s

Pressure
1 lb/in^2 = 6895N/m^2 (68·95mb)
1 in (water) = 249N/m^2 (2·49mb)
1 in (mercury) = 3386N/m^2 (33·86mb)

Appendix 11 – Conversion of Common Imperial Units to Metric (2)

Energy

1 therm = 105·5MJ
1 kWh = 3·6MJ
1 Btu (British thermal unit) = 1·055kJ

Energy flow

1 Btu/h = 0·293W (J/s) (see note below)

Thermal conductance

1 Btu/ft^{2}h °F = 5·678W/m^2 ('U' values)

Thermal conductivity

1 Btu ft/ft^{2}h °F = 1·730W/m K

Illumination

1 lm/ft^2 = 10·764lx (lm/m^2)
1 foot candle = 10·764lx

Luminance

1 cd/ft^2 = 10·764cd/m^2
1 cd/in^2 = 1550cd/m^2

Temperature

32°F = 0°C
212°F = 100°C

Temperature conversion

Fahrenheit to Celsius
(°F − 32) × 5/9

e.g. 61°F to °C
(61 − 32) × 5/9 = 16·1°C

Temperature conversion

Fahrenheit to kelvin
(°F + 459·67) × 5/9
e.g. 61°F to K
(61 + 459·67) × 5/9 = 289·26K,
i.e. 289·26 − 273·15 = 16·1°C

Note regarding energy flow:
Useful for converting boiler ratings in Btu/h to kW,
e.g. a boiler rated at 65000 Btu/h equates to:
65000 × 0·293= 19 045W, i.e. approx. 19kW.

Appendix 12 – Conversion of Common Metric Units to Imperial

Length

1 km = 0.621 mile
1 m = 1.093 yard
1 cm = 0.393 inch
1 mm = 0.039 inch

Area

1 ha = 2.471 acres
1 km^2 = 247.105 hectares
1 m^2 = 1.196 yd^2
1 cm^2 = 0.155 inch^2

Volume

1 m^3 = 1.308 yd^3
1 cm^3 = 0.061 in^3

Capacity

1 l = 0.220 gallon

Mass

1 tonne = 0.984 ton
1 kg = 2.204 pound (lb)
1 g = 0.035 ounce

Density

1 kg/m^3 = 0.062 lb/ft^3

Velocity

1 km/h = 0.621 mile/hr
1 m/s = 3.280 ft/s

Acceleration

1 m/s^2 = 3.280 ft/s^2

Force

1 N = 0.225 lb

Torque

1 Nm = 0.737 lb ft

Pressure

1 N/m^2 = 0.000145 lb/in^2
1 kN/m^2 = 20.885 lb/ft^2

Energy

1 J = 0.737 ft lb
1 kJ = 0.277 W h
1 kJ = 0.947 Btu
1 MJ = 0.0095 therm

INDEX

Index

Index

Index

Index

Index

Index

Index

Printed in the United States
by Baker & Taylor Publisher Services